INTEGRATED ENERGY SYSTEMS FOR MULTIGENERATION

INTEGRATED ENERGY SYSTEMS FOR MULTIGENERATION

IBRAHIM DINCER
Ontario Tech University

and

YUSUF BICER
Hamad Bin Khalifa University

ELSEVIER

Elsevier
Radarweg 29, PO Box 211, 1000 AE Amsterdam, Netherlands
The Boulevard, Langford Lane, Kidlington, Oxford OX5 1GB, United Kingdom
50 Hampshire Street, 5th Floor, Cambridge, MA 02139, United States

Notices

Knowledge and best practice in this field are constantly changing. As new research and experience broaden our understanding, changes in research methods, professional practices, or medical treatment may become necessary.

Practitioners and researchers must always rely on their own experience and knowledge in evaluating and using any information, methods, compounds, or experiments described herein. In using such information or methods they should be mindful of their own safety and the safety of others, including parties for whom they have a professional responsibility.

To the fullest extent of the law, neither the Publisher nor the authors, contributors, or editors, assume any liability for any injury and/or damage to persons or property as a matter of products liability, negligence or otherwise, or from any use or operation of any methods, products, instructions, or ideas contained in the material herein.

Library of Congress Cataloging-in-Publication Data
A catalog record for this book is available from the Library of Congress

British Library Cataloguing-in-Publication Data
A catalogue record for this book is available from the British Library

ISBN: 978-0-12-809943-8

For information on all Elsevier publications visit our website at https://www.elsevier.com/books-and-journals

Publisher: Joe Hayton
Acquisition Editor: Lisa Reading
Editorial Project Manager: Gabriela Capille
Production Project Manager: Sruthi Satheesh
Cover Designer: Mark Rogers

Typeset by SPi Global, India

Contents

Preface

We are in an era where integration has been essential in almost everything. Everything is now more integrated in every process, system, application and service in a smart way. That's why things are now more integrated. Integrating things more makes things smarter. It has been the main focus in this book to implement system integration concepts for multigenerational purposes. It is clearly evident that there will be no sustainable solutions without integrated systems for multigeneration. This critical theme is really the hearth of the book which will open a new domain for readers and take them beyond the conventional systems and applications.

This book on integrated energy systems will benefit the students, researchers, scientists and practicing engineers for better understanding the thermodynamic fundamentals, basic concepts about energy sources, energy systems and useful energy outputs, resource utilization criteria, performance assessment and evaluation, system flexibility, economic aspects, environmental dimensions and sustainable development. It also provides descriptions, models, analyses, assessments, evaluations, etc. to help better comprehend the topic and benefit from that to achieve better sustainability. Finally, its content makes it a distinguished textbook for senior undergraduate students, post graduate students, researchers, scientists and practicing engineers in the field. The chapters are prepared in a methodological manner to ease understanding of the concepts with examples and case studies. The book dwells on 3S (source-system-service) approach for energy domains.

This book consists of seven chapters starting from energy, environment and sustainability issues, fundamental thermodynamic principles and cycles, system integration and multigeneration aspects until conventional-, nuclear- and renewable-based integrated energy systems for multigeneration. The introductory chapter provides the interrelations among energy, environment and sustainable development. It is also important to understand the main motivation behind proposing integrated energy systems for achieving sustainable development. Chapter 2 offers the fundamental information about energy, entropy, and exergy concepts used in integrated energy systems by introducing basic thermodynamic principles and balance equations for common devices and cycles. Chapter 3 presents the main improvement options of energy systems by considering crucial system design criteria, integration and multigeneration aspects. In addition, it introduces a smart hierarchical thermodynamic analysis approach to facilitate the analysis. Chapter 4 focuses on integration of conventional systems for multigeneration, in which there are coal, natural gas as well as oil related case studies. In addition, it emphasizes the significance of transition from single generation to multigeneration systems. Chapter 5 presents the fundamentals of nuclear energy systems and then provides various options of system integration in nuclear energy for multigeneration applications. Chapter 6 classifies and introduces the fundamental renewable energy sources and presents multiple case studies on renewable-based integrated energy systems capable of producing several useful commodities. Chapter 7 closes the book by presenting further analysis for assessing sustainability of integrated energy systems through various techniques, namely, exergoeconomics, exergetic life cycle assessment, exergo-environmental, exergo-sustainability and exergo-enviro-sustainability assessment. In addition, it presents the future directions of integrated energy systems for multigeneration.

We hope this book provides fundamental information as well as presents new dimensions to integrated energy systems and helps the community implement better solutions for achieving sustainable development target.

Ibrahim Dincer
Yusuf Bicer

Acknowledgments

For the case studies and illustrative examples, we have extensively benefited from our previously published studies, especially carried out by the current and past graduate students of Prof. Dincer, namely, Canan Acar, Murat Emre Demir, Md Ali Tarique, Magd Nadeem Dinali, Maan Al-Zareer, Rami Salah El-Emam, Shahid Islam, Azzam Abu-Rayash, and Ron R. Roberts, as well as, Calin Zamfirescu, Shoukat Alim Khan, Usman Bin Shahid and Amro Mamoon Osman Mohamed.

In addition, Farrukh Khalid's help is gratefully acknowledged for materializing and preparing the several case studies, especially for Chapter 5.

Last but not least, we warmly thank our wives, Gulsen Dincer and Elif Derya Bicer, our children Meliha, Miray, Ibrahim Eren, Zeynep, Ibrahim Emir, and Erva Hatice, Zumra Ayse, our parents Fatma and Hasan Dincer and, Sema and Mehmet Bicer, in addition to siblings of Dr. Bicer, Betul and Zeynep. They have been a great source of support and motivation, and their patience and understanding throughout this book have been most appreciated.

CHAPTER

1

Energy, environment and sustainable development

1.1 Introduction

Sustainable development is considered a very diverse domain, covering the dimensions of energy, environment, economy, education and resources along with the social aspects. This can be simplified based on human needs for energy, water, environment, food and economy. Due to depletion or degradation of these fundamental aspects, human beings attempt to secure resources and substances, and are therefore concerned with the security of water, energy, the environment, and food. Sustainable and resilient communities require an integrated approach that accounts for all of these crucial factors. Energy plays an overarching role in water, food and environmental security for achieving sustainable development.

Water scarcity is recognized as one of the critical challenges worldwide. Hence, seawater desalination is an essential process that requires a significant amount of energy. Water security encompasses water treatment, water distribution, water quality and water required for cooling as well as many other subsystems. Although water is an abundant substance on Earth, availability of drinkable, potable, pure, clean water is a major problem, especially for the regions lacking in freshwater sources. Desert climatic regions are among the top locations suffering from limited freshwater availability and supply. Seawater is considered a significant source for desalination plants; however, these plants consume huge amounts of energy, mainly from fossil fuels. The sustainability level of thermal desalination plants is thus low. In order to address this issue, this book covers alternative desalination technologies integrated into renewable energy plants.

Food security encompasses food quality, food supply, fertilizers, food processing, food storage, irrigation and many other aspects. Those countries that are dependent on external food resources are in fact in urgent need of securing their food production and supply. Water security is one of the main concerns of food security. Therefore, these two aspects have a strong relationship and multiple interactions. Hot desert climates are short of water,

Integrated Energy Systems for Multigeneration
https://doi.org/10.1016/B978-0-12-809943-8.00001-7

causing irrigation difficulties. Therefore, humid air harvesting and on-site water production arise as alternative solutions for irrigation. Instead of relying on local production, these regions tend to import their food from external resources. As food and water are two of the fundamental requirements of mankind, their sustainable production and supply play a significant role in building resilient communities. Hence, this book covers several sustainable systems for food production and processing that are integrated into renewable energy systems.

Environmental security is concerned with preserving nature and reducing harmful emissions to acceptable levels. Energy production, conversion and consumption have a direct impact on the environment. Hence, there is a great need for environmentally friendly routes for the energy sector. Environmental security includes air quality, greenhouse gas emissions, the marine environment, land use, climate, vegetation, plant and animal health, as well as many other impacts related to the environment. The environment is a concern for all people and countries in the world. Increasing emissions and climate change impacts have led decision makers to implement cleaner and hence more sustainable energy systems.

The dimensions of energy are quite diverse, ranging from energy production to energy management, and cover multiple aspects related to energy fundamentals, materials, sources, generation, conversion, transmission, distribution, consumption, and management, which directly affect water, food and environmental security. Therefore, a secure supply of energy is usually accepted as a necessary but not sufficient factor for sustainable development of a society. Sustainable development requires not only a secure but also a sustainable supply of energy resources. Energy is considered the most vital part among the other sustainable development aspects, because in order to produce clean water as well as plant, harvest and process food, energy is the main requirement. Many of the environmental emissions are due to energy production and conversion, which means that emission reduction potential is a major factor in the development of sustainable energy systems. The significance of energy systems for sustainable communities is depicted in Fig. 1.1.

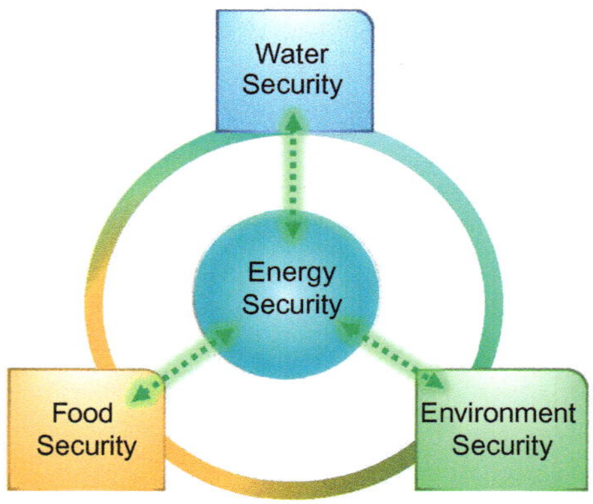

FIG. 1.1 Interactions among energy, environment, water and food for sustainable development.

Sustainable growth in a civilization needs energy resources that, in the long term, are on-demand and sustainably accessible at an acceptable price and that can be used for all obligatory duties without producing negative social impacts. The supply of fossil fuels such as natural gas, oil and coal as well as uranium is normally recognized to be limited; however, other energy sources such as solar, wind and hydropower are usually evaluated as renewable and sustainable over a comparatively extensive period.

Environmental issues are major factors in sustainable development due to several reasons, in particular because actions that continuously disrupt the environment are considered non-sustainable over periods of time. The growing effects of such actions on the environment lead to various health, ecological and other problems over time.

A part of the ecological impact of a civilization is related to the operation of its energy sources. In an ideal world, a civilization seeking sustainable development would exploit simple energy resources that do not have any ecological influence [1]. On the other hand, as all energy sources have some environmental issues, it makes more sense to reduce the associated emissions by improving the efficiency of the processes employed. In this way, some of the issues related to environmental impact can be addressed. Obviously, there is a strong relationship between energy efficiency and environmental impact. If the energy efficiency increases, eventually we consume less of the energy source for the same amount of products or services and reduce the associated pollutions. Improving energy efficiency reduces energy losses. Most productivity and efficiency enhancements offer direct environmental benefits in two ways. Initially, the energy input required by the energy system is reduced for the same output, resulting in a higher efficiency, and the released pollutants are reduced accordingly. Secondly, in view of the whole life cycle of energy resources and technologies, higher productivity results in reducing the impact on the environment over most life-cycle phases.

Various energy-related criteria considered vital in successful sustainable development in a society are listed here [1–3]:

- employing exergy analysis as a potential tool
- education and training on the environment and sustainable development
- policy development for sustainable energy implementation
- appropriate assessment, evaluation and monitoring tools
- strategic energy and exergy policies for better performance
- environmentally friendly policies, strategies and regulations
- incentives and promotions for environmentally benign practices
- exergy conservation practices, rather than energy conservation practices
- efficient energy utilization practices
- cleaner technologies for fossil fuels
- renewable energy technologies
- alternative and green fuels
- hydrogen and carbon free fuels, such as ammonia
- system integration for multigenerational applications
- system optimization
- research, innovation and commercialization programs
- academia-industry-government partnership programs
- roadmap developments for sustainable society
- artificial intelligence and machine learning tools

1.2 Energy classification

An energy resource is a form of energy existent in the world, which can be transformed into other forms of energy (e.g., mechanical, electrical, etc.). There are several types of energy resources available on Earth, namely fossil fuels, nuclear, renewables, wastes and others, as shown in Fig. 1.2.

First, we classify these energy resources and summarize their current status and potential in the world. This will help us to understand the importance of energy, environmental and sustainability relationships. Energy resource management is very important. Demand and supply of energy should be controlled and balanced. Market regulation is therefore necessary. Environmental emissions associated with any energy transformation process should be monitored. Actions to minimize pollution in the energy sector should also be taken. In this context, each jurisdiction should strive to achieve a balance between energy supply, demand and environmental effects, affecting the development of policies and strategies. Energy efficiency increase yields a decrease in energy need and then an extension of the existing reserves. In addition, utilization of energy sources in such a balanced manner leads to more stable energy consumption, which is helpful for sustainability and the environment.

Each energy-related action has a cost on efficiency, which is a key factor since cost on efficiency implies more savings or decreased expenditures for the identical amenities delivered or commodities generated. Nonetheless, cost savings and environmental contamination are related to each other, since creating real capital requires production activity, and any production activity leads to definite environmental influences. Consequently, energy security has turned into another significant aspect. Better energy security suggests advancement of energy policies and geopolitic policies, which ultimately guarantee proper access to energy resources and thus improved sustainability and environment.

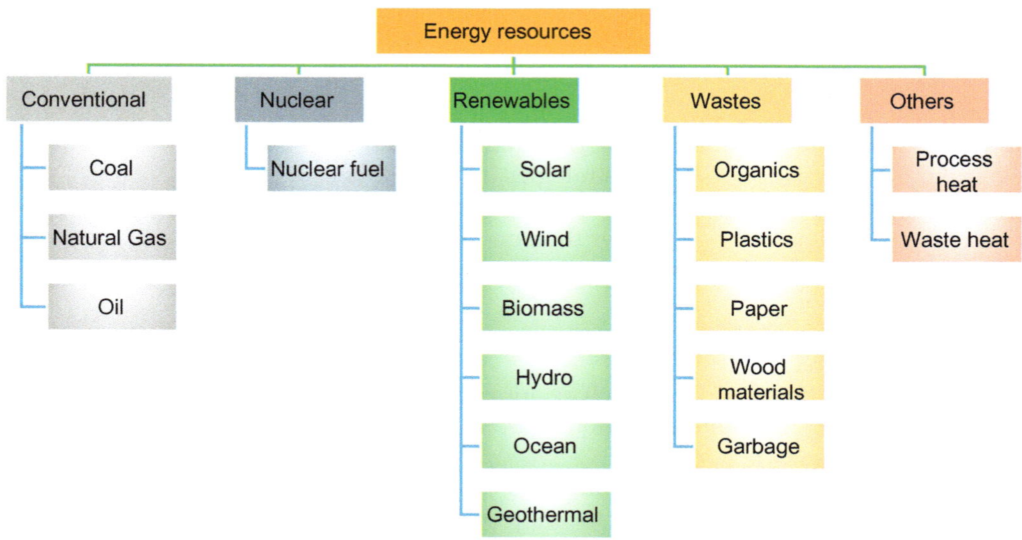

FIG. 1.2 Classification of energy resources for integrated energy systems.

In addition to renewable energy sources, other types of energy resources accessible to human beings are those resultant from fossil fuels, nuclear fuel and waste. Non-renewable energy sources are currently the most utilized source of energy globally with about 75% of the total electricity production as depicted in Fig. 1.3. Nuclear energy has about 6% of global energy production. The number of countries utilizing nuclear energy resources is limited. Nevertheless, a vast portion of global power generation is accomplished by fossil fuel-based power plants. Hydropower is also considered under the renewable energy category, representing about 16.6% of the worldwide power production. Among other renewables, bio-power is accountable for approximately 2% of global electricity production, followed by photovoltaics and other renewables. In fact, the highest quality form of energy is electricity, as illustrated in Fig. 1.4. The quality of energy is mostly associated with the conversion efficiencies. Hence, natural gas has one of the highest qualities among conventional sources. Conversion from most of the renewable resources (except hydropower) is in the low/medium efficiency region. One of the important advantages of solar photovoltaics is the direct conversion from resource to electricity. Among renewables, biomass has one of the highest qualities due to the combustion or gasification process. It can yield high temperature combustion gases. From a quantity point of view, geothermal power plants have larger scales than other renewables due to integration into conventional power cycles such as the Rankine cycle. However, solar and wind energy systems are more modular, which makes it easier to scale-up and scale-down.

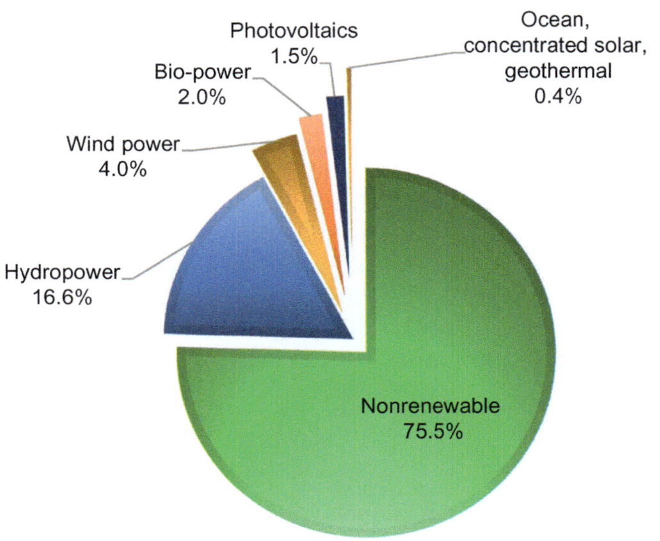

FIG. 1.3 Global electricity production shares. *(Data from Renewables 2017 Global Status Report REN21 Renewables Glob Status Rep 2017. http://www.ren21.net/gsr-2017/ (accessed 14 December 2018)).*

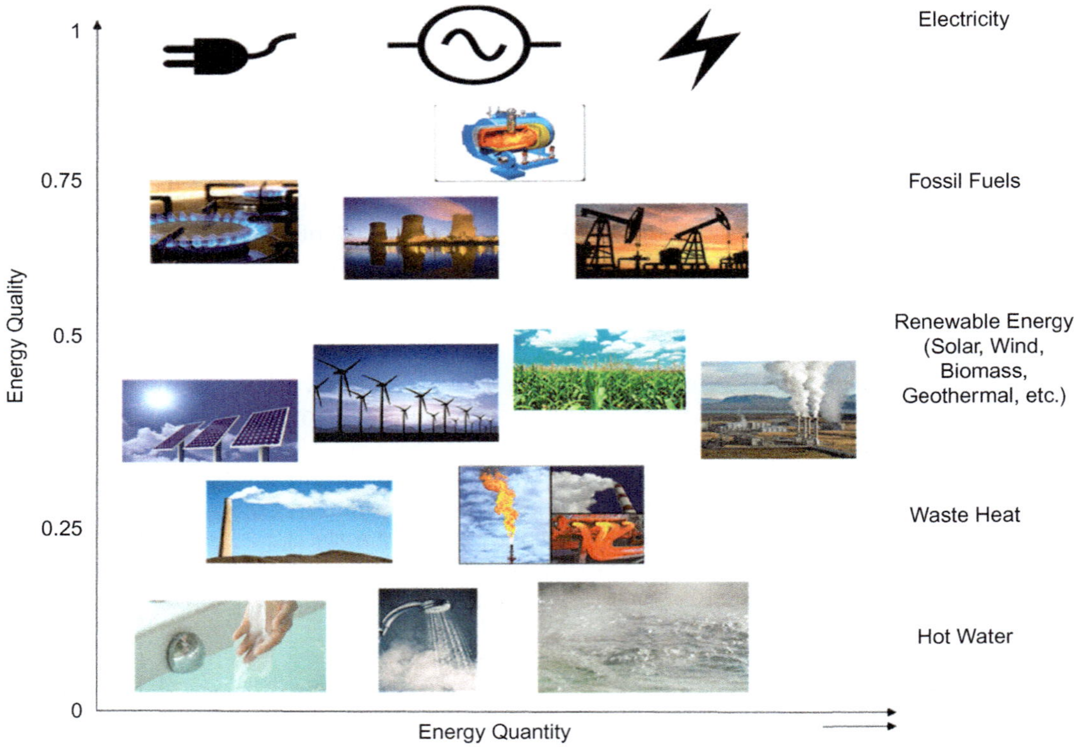

FIG. 1.4 Energy quality versus energy quantity for conventional and renewable sources.

1.2.1 Hydropower

Most of the energy sources are derived from solar energy, such as hydropower, wind energy, ocean currents and waves, biomass energy, and direct solar radiation. Almost 22% of the incoming solar radiation can be recovered as hydro-energy that is in fact a type of potential energy (convertible into kinetic energy) produced due to the height of water level [2]. Hydropower can be obtained from the natural water cycle, which is powered by solar irradiation. Solar energy causes water to vaporize from oceans and then forms into rainfall at higher heights leading to the structures of river basins and lakes. The projected worldwide electricity production potential through hydropower is the highest among other types of renewable energy sources, corresponding to about 30 PW [2]. This number is calculated based on a 22% hydropower breakdown, assuming 80% hydraulic energy conversion.

1.2.2 Wind power

Wind energy is another type of renewable source that originates from solar radiation. Wind is a mechanical movement of air mass that is caused by pressure gradient. The difference in pressure between two different locations results from differential heating produced due to sunlight exposure. Wind energy represents only 0.21% of the input solar energy from which it originates [2]. The available energy in wind is commonly extracted via wind turbines that have the slightly higher efficiency (compared to fossil fuel-based thermal power plants) of wind-to-power conversion. Nevertheless, due to wind fluctuations throughout the year, the capacity factor, meaning the yearly operational hours, is limited to about 30–40% [2]. Wind power has the potential to generate electricity corresponding to approximately 0.22 PW worldwide with an average wind turbine's efficiency and capacity factor.

1.2.3 Biomass

Biomass is mainly available from wood, plants, grass, straw, cane, manure, charcoal, domestic waste, waste paper, etc. In fact, they all represent a chemical type of energy. The heating value (HV) of biomass is commonly between 4 and 30 MJ/kg [4]. The global total energy potential of the biomass is quite low, approximately 0.02 PW, assuming that the energy conversion efficiency is about 50%. The energy available in biomass can be extracted as thermal energy mainly by combustion. A very small fraction of sunlight (approximately 0.02%) is used for photosynthesis by the plants on Earth. The process of photosynthesis is of low efficiency, in the range of 1–5%. Note that photosynthesis occurs in the visible light region of the solar spectrum. However, due to the high formation of sucrose, glucose, cellulose and other chemical compounds, photosynthesis is a vital supply of energy and food throughout the world. The processes of photosynthesis cause production of biomass. Furthermore, biofuels can be produced from biomass. Biofuels are suitable to be used in various applications such as transportation, power generation and heating. Some types of biomass can be converted into alcohols by aerobic fermentation. Many biomass sources are suitable and could be used to produce liquid fuel by transesterification, Fischer-Tropsch synthesis or other methods. Biomass, like wood, or other plants, is directly combustible. Synthetic fuels, such as biogas or liquid biofuels, can be burned for producing high-temperature thermal energy, which can drive a power cycle, e.g. Brayton cycle.

1.2.4 Solar energy

Solar energy has the second greatest capacity for electricity production among renewables, if it is assumed that conversion efficiency from solar to electricity is about 25%. In this case, the projected worldwide electricity production from solar energy can reach about 18 PW [2]. Solar energy has several forms, such as photonic, thermal and electrical. Photonic energy can be converted into more useful types of energy such as electricity and heat. Using the photovoltaic effect, solar energy can be converted into direct electricity. Solar radiation can be transformed into thermal energy by solar collectors. There are numerous types of solar collectors. Low-temperature solar collectors can be used for water heating, absorption cooling and other similar applications. In addition, high-temperature solar collectors utilize concentrated solar irradiation and they are mainly suitable for power generation as well as industrial heat requirements.

1.2.5 Ocean energy

There are other potential energy sources such as ocean energy, which can be harnessed in different forms such as ocean thermal, ocean waves, and ocean flows [4]. There are various types of technologies to utilize ocean energy. Ocean currents and waves can be converted into mechanical energy and then to electricity. Similarly, the temperature gradient of the ocean can be used in ocean thermal energy conversion (OTEC) plants to produce electricity by organic Rankine cycles. Several examples of ocean energy conversion systems are covered in Chapter 6.

1.2.6 Waste energy

Waste is an important energy source that is not normally used. Therefore, waste energy conversion systems are very important for sustainable communities. A number of flammable materials can be extracted from industrial and domestic areas containing plastic, paper, wood, garbage, etc. Waste stored in landfills produces landfill gas. It is actually a flammable gas, which can be used for combustion. The recovered plastic materials are also alternatively transformed into liquid fuel via the help of a catalyst. Other waste materials can be burned to produce high-temperature heat. This heat can drive several industrial processes, power cycles or material synthesis. In addition, wastes can be converted into chemical fuels (hydrogen, methanol, ammonia, etc.) to be used in fuel cells, engines, generators, etc.

1.2.7 Other sources of energy

Another form of available energy is waste heat. Many industrial processes produce waste heat that can be used as a source of energy for various purposes. Wastes such as manure or particular types of biomass materials could be transformed into biogas by anaerobic digestion. Other sources of heat from renewable energy sources are not anthropogenic but rather are naturally available, for instance direct sunlight, ocean thermal and geothermal sources. Waste heat can include the flare gas stream in oil and gas plants and exothermic reactions emitting their heat to the ambient environment, etc.

1.2.8 Conventional energy

The conventional power-producing techniques transform energy from fuels in a way that has severe implications for the environment. For instance, electricity production through fossil fuels is one of the most polluting techniques. Coal-fired power plants emit huge amounts of carbon dioxide, particulate substances and additional contaminants to the sky, ground and water. Thus, actions and strategies of management toward applying healthier and more ecological energy methodologies have become crucially significant, and a detailed plan of energy policies is vital.

The availability of fossil fuels has become an important concern due to depletion. Fig. 1.5 depicts a graphical comparison of fuel availability based on available reserves and current production/extraction ratio. Typically, coal, oil, and natural gas are evaluated as fossil fuels under conventional sources. Furthermore, some other kinds of alternative fossil

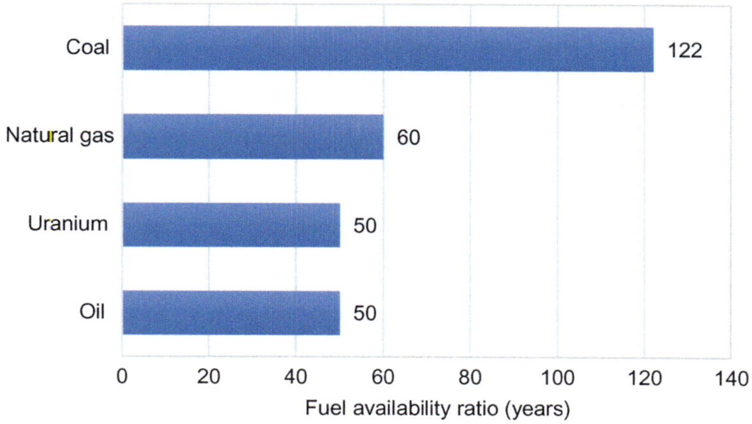

FIG. 1.5 Fuel availability ratio of various conventional fuels. *(Data from Dincer I, Zamfirescu C. Advanced Power Generation Systems. Oxford, United Kingdom, Elsevier; 2014).*

fuels have recently become available and are being explored such as oil sands, shale gas, coalbed gas, etc. The main reason for wide fossil fuel utilization is the flexibility of usage in differing applications. For instance, fossil fuels are the primary fuels used in numerous burning devices, like furnaces, gas turbines, internal combustion engines, etc.

Coal consists of mainly organic substances originating from fossilized plants with rooted mineral additions. The main chemical ingredient of coal is carbon, corresponding to about 70% of the total weight on average. (This can change based on the coal type: lignite, anthracite, bituminous, etc.). Coal is primarily burned or gasified. Hence, the heating value plays an important role in coal characteristics. The heating value of coals ranges from 20 to 35 MJ/kg based on the location and type. The lowest heating value is mostly observed in lignite type coals, whereas the highest heating value is mainly observed in anthracite type coals.

The fuel availability may be calculated using the proven reserves and the amounts needed as per the following equation:

$$\text{Fuel availability ratio} = \frac{\text{Reserves}}{\text{Requirements}} \tag{1.1}$$

The proven coal resources in the world are equal to the energy of about 23.1×10^6 PJ. This corresponds to about 122-year fuel availability based on the current state [2]. The fuel availability ratio is calculated based on the amount of resources, preferably expressed in units of energy, and the rate of production (or consumption) of fuel expressed in units of energy per year. Coal is primarily used in coal-based electricity generation stations, metallurgy and the cement industries.

Oil is a naturally existing hydrocarbon-based substance. It is mostly recovered as liquid. However, there are also non-conventional sources of oil in different forms such as bitumen, tar or oil sands. Alkanes, paraffin aromatics, naphthalenes and cycloalkanes are the principal elements of oil. World oil resources comprise 30% conventional oil, 25% extra heavy oil, 15% heavy oil and other oils such as bitumen, oil sands and shale oil making up the remaining 30% [2].

Petroleum derivatives are fundamentally consumed as fuel in transportation. The fuel availability ratio of oil is calculated as 50 years.

Natural gas is a naturally available substance containing primarily methane. It is a combustible material used in many natural gas-fired power plants worldwide. In addition, it is one of the main feedstocks for several industries, such as hydrogen production, fertilizer production (e.g. ammonia and urea), fuel for power generation in power plants, and source of energy for heating and cooking. Approximately 20% of the global energy generation results from natural gas burning [2, 5]. The fuel availability ratio of natural gas is about 60 years with the current proven reserves.

In power plants and engines, fossil fuels are eventually converted into electricity. The phases of this conversion involve a burning process and release of nitrogen oxides, greenhouse gases, sulfur dioxide and other pollutant/contaminants.

1.2.9 Nuclear energy

Another type of fuel, which is usually not classified under fossil sources, is nuclear fuel. Similar to coal, it is also a mined fuel. Atomic energy is a fissile uranium ^{235}U, which is naturally occurring in the form of U_3O_8 ore. Approximately 1 ton of U_3O_8 ore can be consumed for about 6 kg of fissile uranium equal to 144 TJ of electricity. However, the amount of electricity produced from 1 ton of coal is about 14,000 times less compared to the nuclear case [2]. This implies that nuclear fuel has a very high energy density compared to many other fuels. The fuel availability ratio of nuclear fuel is calculated as 50 years, as depicted in Fig. 1.5. It is noted that more non-conventional nuclear fuels are also available, mostly in the form of thorium. Nonetheless it is not currently used in nuclear reactors but can be used in the next-generation nuclear stations.

Fossil fuels as well as nuclear fuels are mined. The operation of each mining process brings additional environmental impacts. Mines have large processes and different devices that create difficulties and environmental issues. Many organic and inorganic emissions are released to the environment throughout the excavating process, because many kinds of motor vehicles are used during mining, and in the fuel processing and fuel delivery systems.

Nuclear fuels are usually transformed into electricity via conventional Rankine cycles. Though there are no direct emissions caused by nuclear fission, some environmental impacts still exist. In the power plant, the lake/ocean ecosystem is affected by the heat released from the plant's cooling system. The indirect and safety systems of nuclear power plants continue to produce combustion gas released to the ambient environment due to combustion. However, compared to fossil fuel-based power plants, the emissions are much less.

The details of conventional, renewable and nuclear energy types will be further discussed in the next chapters.

1.2.10 Energy consumption

Population growth leads to an upsurge in demand of several products including electricity, water and food.

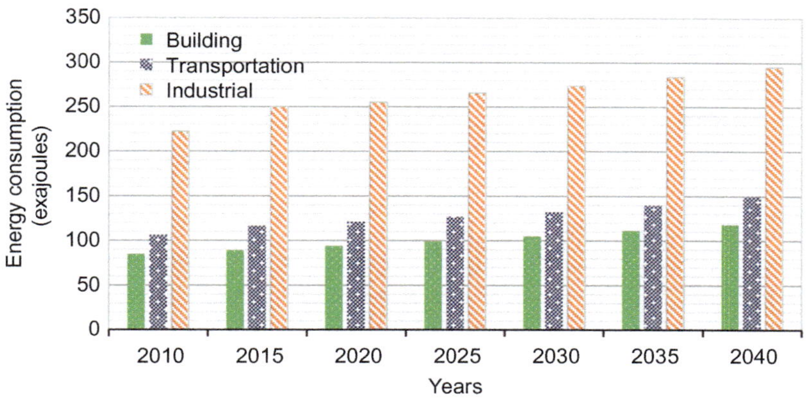

FIG. 1.6 Projection of world energy consumption by end-use sector. *(Data from International Energy Outlook 2018. US Government: 2018).*

The most widespread way to monitor the social needs of a product is to use energy. This is because energy is the source of the main human actions. The demand for energy depends on the density of population and the economic growth of the region.

The world's forecasted energy demands based on the major sectors are shown in Fig. 1.6. The greatest demand is in the industrial subdivision, which is mainly from non-OECD nations. While the industrial sector remains the world's major energy-consuming segment in the coming future as shown in Fig. 1.6, energy demand in all other sectors is rising more rapidly. World industrial sector energy use rises by 0.7%/year from 2015 to 2040, compared with an increase of 1.0%/year for transportation and 1.1%/year for buildings. Transportation is the second largest energy demand sector after industry, and the associated demand will continue to rise. In order to satisfy this demand increase, there will be new energy generation plants in the future. The forecast for future energy consumption data is presented in Fig. 1.7. Although coal remains constant in the long term, natural gas and petroleum still have increasing trends. The share of renewable energy will continue to rise in the coming decades.

An important source of energy for the transport sector is liquid fuel. Other types of liquid fuel uses are in the industrial sector with comparatively less in residential and commercial divisions. The trend of liquid fuel consumption per each sector is shown in Fig. 1.8. The transportation sector seems to be the maximum consumer of liquid fuels. The trends indicate that demand for liquid fuels in the transport and industrial sectors will increase in the future.

1.3 Energy and environment

All energy systems usually have an impact on the environment whether renewable or non-renewable. In recent years, energy has been associated with environmental problems ranging from global to local issues. Especially in developing or industrialized countries, where energy consumption levels are high and environmental management is not fully integrated into the infrastructure, environmental problems can be very serious. Currently,

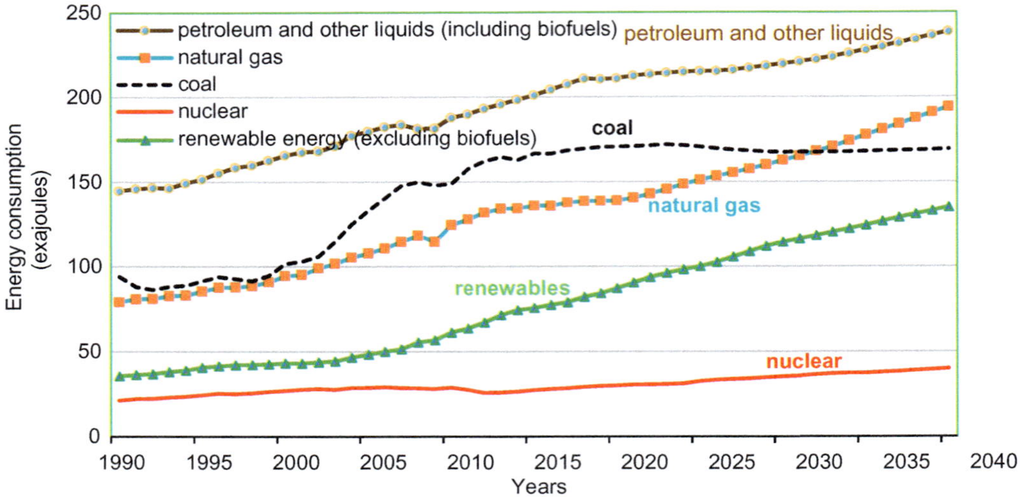

FIG. 1.7 Projection of world energy consumption by energy source. *(Data from International Energy Outlook 2018. US Government: 2018).*

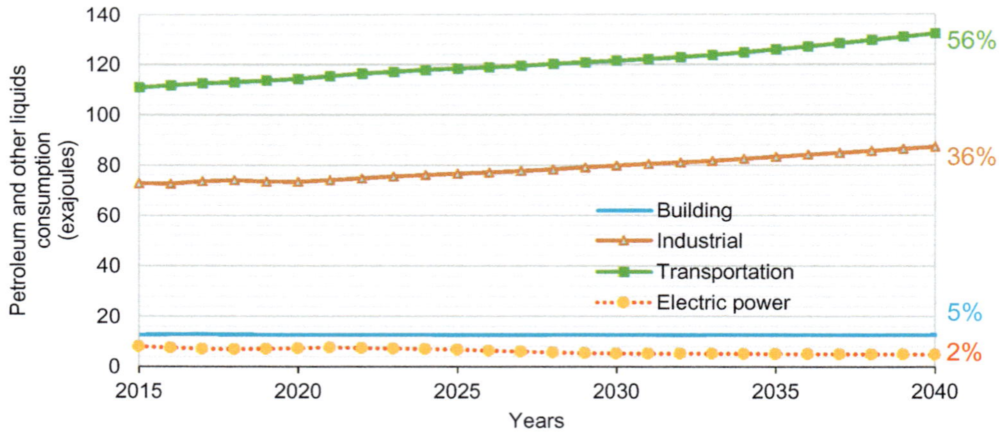

FIG. 1.8 Projection of refined petroleum and other liquids consumption by end-use sector. *(Data from International Energy Outlook 2018. US Government: 2018).*

industrialized countries are primarily responsible for air pollution, ozone depletion and carbon emissions.

The contaminants released by power generation systems cause many air pollution hazards. Extreme events can be caused by global warming or climate change effects, such as heavy rainfalls, marine level increases, etc. Acid vapors form acid rains, influencing the soil and water. In addition, aerosols such as volatile organic compounds (VOCs) and particulate matters (PM) are constantly released to the air and concentrate in its upper layers. Aerosols contribute to the Earth's *albedo*. Because of the availability in the atmosphere, aerosols reflect and scatter back a portion of the incident solar radiation into the extraterrestrial space. In fact, this reduces the amount of sunlight captured by the ground as well as solar collectors. This effect is called the *albedo effect*. In order to utilize the reflected irradiation from the ground, researchers have started to investigate and implement bi-facial solar modules. Although less sun radiation on the ground level can reduce the temperature, the balance between the greenhouse and albedo effects establishes the Earth's long-term temperature regime and regulates the Earth's climate [2]. This is a natural climate control mechanism occurring continuously. Nevertheless, starting from the Industrial Revolution, the anthropogenic impact on climate has been more obvious because of the increased emissions of greenhouse gases from many sectors mentioned earlier (energy, transportation, industry), hence causing global warming.

The releases of SO_2 and NO_x, which are usually caused by the supply of energy, have a direct environmental impact due to acidity. These gases could contribute to the complexity of chemical changes in the air, leading to acid rain. Road transport is also an important source of NO_x emissions. Therefore, selective catalytic converters are being used in vehicles for reducing NO_x into N_2 through NH_3. Other NO_x emissions are mainly caused by burning of fossil fuels in stationary applications. Countries that are generally engaged in energy-related activities contribute significantly to acid rains. A major problem with acid rain is that the effects often occur in a country other than the one in which the rain appears.

The chief effects of acid rainfall are as follows [2]:

- effect of sulfate aerosols on physical and optical characteristics of clouds
- toxic effects on plants due to excessive acid concentration
- harm to aquatic life and fish
- acidification of ground waters, lakes and streams
- effects on the forests and agricultural crops
- degradation of building materials as well as fabrics
- corrosion of exposed construction

Harmful wastes threaten health and environment, and mainly consist of metals. Commercial use of solid waste from construction and transportation products is limited to market size. In the case of pressure on atmospheric air quality, dangerous air pollutants will often be released in small quantities.

Lead is considered the main dangerous contaminant. Most of the world's lead contamination originates from the use of lead-based gasoline additives to raise octane ratings. Because of this issue, lead additives into gasoline products have been phased out. However, trace amounts of heavy metals can still be present in gasoline because of regulations. Lead contact can cause neurological damage. The number of risky and unsafe contaminants is huge.

Some toxic pollutants are cadmium, mercury, and polyclinic aromatic hydrocarbons (PAHs). Many energy issues derive from harmful air pollutants, such as hydrocarbons, being released during the extraction and processing of oil and gas. Emissions of hydrocarbons and dioxins due to burning of gasoline and diesel fuel, and small amounts of arsenic, mercury, beryllium and radionuclides, which are released due to coal and fuel oil, are among the major environmental issues. In addition, emissions of mercury, chlorinated dioxin and furan from disposing of household waste are also important to mention [2].

Acid-starting materials are primarily from fossil fuels, especially combustion of coal and oil, and non-metal ore being transmitted remotely through the atmosphere and stored in different ecosystems. The bulk of SO_2 emissions is generated by the power plants using fossil fuels, and the majority of NO_x emissions originates from transport. One of the main sources of acid rain is the processing of H_2S, which reacts with SO_2 after being exposed to air. SO_2 and NO_x dry deposits can form acid in soil or sea water. Direct depositions can take place from 1 to 2 km distance from the origin. The reaction of acid suppliers with water steam driven by sunlight stimulates the formation of high acid levels, such as sulfuric acid and nitric acid. Different types of acids fall to the ground as rain, fog, snow or other forms.

Water is also severely affected by the acidification. Acids as well as other types of contaminants produced by energy systems can impact the quality of waters including groundwater. Groundwater plays a major role in the supply of drinking and irrigation water for many countries. Substantial concerns exist about the quality and quantity of accessible water resources including groundwater. Therefore, intensive efforts are ongoing to reduce and regulate energy-related emissions. Among these emissions, we can count geothermal fluids containing toxic chemicals; acid drainage from mines; coal wastes; effluents containing hazardous chemicals from power plants and refineries; and thermal pollution from the discharges of cooling systems of power stations.

Starting from the Industrial Revolution, there has been a steep increase in the anthropogenic impact on climate because of the increased emissions of greenhouse gases by many different sectors such as energy, transportation, and industry. This is the leading cause of the global warming that humanity is facing today. The main forms of emissions due to energy systems are tabulated in Table 1.1 with their impacts and explanations.

Table 1.2 contains some parts of the environmental impact and the units employed to measure each category. The impact indicators are determined based on the respective quantitative measures.

Greenhouse gas emissions represent the main environmental pollution produced by power generation systems. Water vapor is one of the most important greenhouse gases. However, the water vapor concentration in the atmosphere fluctuates rapidly and water vapor absorbs radiation for a wide infrared spectrum. In addition to water vapor, the principal greenhouse gases (GHGs) emitted in the atmosphere are listed in Table 1.3.

The three gases shown in the table, carbon dioxide, methane gas and nitrogen dioxide, are part of the natural carbon and nitrogen cycles. They are part of anthropogenic release as well. As a result, they existed in the environment before the industrial era, which is considered to begin in 1750. The gases in 1750 were over zero. The other two types of artificial gases are shown in Table 1.3, which are examples of cooling gases (CFCs). Some examples of the CFCs are: 1,1,2-trichloro-1,2,2-trifluoroethane (CFC113), chlorodifluoromethane (HCFC22), CFC141b, CFC 142b, 1, 1, 1-trichloroethane (CH3CCl3), tetrachloride carbonone (CCl4), 1,1,1,2-tetrafluoroethane (HCFC134a). Other gases are also the result of human activities, but they have been released in smaller amounts, and therefore their concentrations are low.

Radiative forcing is described as the net change in radiation balance at the tropopause, formed by a definite cause. By convention, when the radiative forcing is positive it causes a rise in terrestrial temperature, and when negative, it is the other way around. The units of amount of radiative forcing are the same as the units of radiation energy rate per square

TABLE 1.1 Common atmospheric emissions released by various energy systems.

Pollutant type	Details
Greenhouse gases (GHG)	These are the gases that produce the greenhouse effect. The main greenhouse gases are H_2O, CO_2, CH_4 and N_2O. The greenhouse effect is the main cause of global warming.
Carbon monoxide (CO)	It is mainly caused due to incomplete combustion of fuels, and causes significant impacts on the health and life of animals and human beings upon inhalation. Strong release of CO in open areas can impact birds, while emissions in close areas (residences, garages) can cause death of exposed people
Sulfur dioxide (SO_2)	It is a corrosive gas that harms human health and endangers the natural environment. It is mainly caused by the combustion of coal and fuel oil, smelting of nonferrous metal ores, oil refining, electricity generation, and pulp and paper manufacturing. It causes respiratory difficulties, damages green plants, and is a precursor of acid precipitation.
Nitrogen oxides (NO and NO_2) (NO_x)	They are mainly produced by combustion of fuels at high temperature in all combustion devices such as motor engines, furnaces etc since air is required for combustion. The exposure can cause respiratory problems in humans and animals and form acids in high-altitude atmospheres.
Volatile organic compounds (VOCs)	Volatile organic particles are mainly caused by hydrocarbon combustion in engines. They can cause harmful effects in the atmosphere and delay formation of stratospheric ozone.
Particulate matter (PM)	There are several substances categorized under particulate matter including fly ash, sea salt, dust, metals, liquid droplets, soot. They originates from a variety of natural and human-made sources and emitted by factories, power stations, vehicles, and are produced in the atmosphere by condensation or chemical transformation of emitted gases. The exposure can cause of health and environmental effects, including acid precipitation, damage to plant life and human structures, loss of visibility, toxic or mutagenic effects on people, and possibly non-accidental deaths.
Chlorofluorocarbons (CFC, HCFC, HFC)	These are primarily refrigerants or propellants. They can contribute to stratospheric ozone layer depletion as well as greenhouse effect.

Source: *Dincer I, Zamfirescu C. 1.3 Environmental dimensions of energy. In Dincer I, editor. Compr. Energy Syst., 1st ed., Cambridge, United States: Elsevier; 2018. p. 49; Dincer I. Refrigeration systems and applications. 3rd ed. Chichester, UK: John Wiley & Sons, Ltd; 2017. https://doi.org/10.1002/9781119230793.*

meter of Earth's surface, where the Earth's surface, by convention, is the area of the sphere having the average radius of the planet. The typical symbol for the radiative force is ΔF. Usual numbers of radiative forcing are in the range of 0 to 2 W/m^2 [2, 6]. Based on the concentration of the gas in the atmosphere, there are three types of radiative forcing: low-, moderate- and high-concentration. The low-concentration type concerns gases existent in the atmosphere in parts per billion (ppb) or parts per trillion (ppt) levels. In this case, the forcing is proportional to the concentration change, $\Delta F \sim \Delta C = C - C_0$. CFCs and the tropospheric ozone fall in this category.

Calculating the influence of a specific atmospheric gas on climate is a multidimensional problem. For instance, once the gas absorbs more in the infrared spectrum, its greenhouse effect is related to a positive radiative forcing. Nevertheless, another parameter is also

TABLE 1.2 Main environmental impact categories and their associated indicators for quantification of impacts.

Impact category	Description	Unit
Climate change	This is measured as the global warming potential (GWP) for which the unit is kg of CO_2 equivalent. It is estimated based on the releases of greenhouse gases and the potential for the expansion of global warming.	kg $CO_{2,eqv}$
Acidification	The effects of NH_3, NO_x and SO_2 in the rain and their associated impacts on soil and liquids. It corresponds to SO_2 equivalent.	kg $SO_{2,eqv}$
Ozone depletion	This is usually caused by chlorine photoreactions. It concerns the depletion of ozone layer. It is measured by CFC-11 (trichlorofluoromethane) equivalent.	kg $CFC11_{eqv}$
Toxicity	It is the level of toxicity created in air, freshwater, seawater, or terrestrial. It is quantified with respect to the toxicity of DCB (1,4-dichlorobenzene) and measured in kg DCB equivalent.	kg DCB_{eqv}
Photo-oxidant formation	It is based on photochemical ozone formation potential commonly expressed with respect to the reference matter: ethylene. The measurement unit is kg of C_2H_4 equivalent.	kg $C_2H_{4,eqv}$
Eutrophication	Eutrophication includes a high nutritional substances where, most important nutrients are nitrogen (N) and phosphorus (P). The measurement unit is PO_4 equivalents.	kg $PO_{4,eqv}$
Abiotic depletion	Abiotic substances include all non-living sources such as coal, oil, iron ore, etc. It is calculated in kg Sb (antimony) equivalent.	kg Sb_{eqv}

Source: *Dincer I, Zamfirescu C. 1.3 Environmental dimensions of energy. In: Dincer I, editor. Compr. Energy Syst., 1st ed., Cambridge, United States: Elsevier; 2018, p. 49.*

TABLE 1.3 Main greenhouse gases and their associated global warming potential.

Gas	Absorption spectrum (cm^{-1})	Atmospheric concentration	Atmospheric lifetime	GWP		
				20 years	100 years	500 years
CO_2	550–800	387 ppm	50–200 years	1	1	1
CH_4	950–1650	1750 ppb	12 years	72	25	7.6
N_2O	1200-1350	314 ppb	120 years	289	298	153
$CFCl_3$	800–900	251 ppt	50 years	6730	4750	1620
CF_2Cl_2	875–950	538 ppt	102 years	11,000	10,900	5200

Source: *Intergovernmental Panel on Climate Change (IPCC). Climate change 2007: Synthesis report. Valencia, Spain: 2007.*

significant, that is the atmospheric period. The lifetime of the gas in the atmosphere is important with respect to balancing control of radiative force as shown in Table 1.3.

The energy sector accounts for 26% of the atmospheric emissions, followed by 17% from industry [7, 8]. The increased risk of using fossil fuels leads to an increase in the greenhouse gas in the atmosphere, indicating the significance of environmental impacts and global warming. Fig. 1.9 shows the structure and distribution of carbon dioxide as a result of the

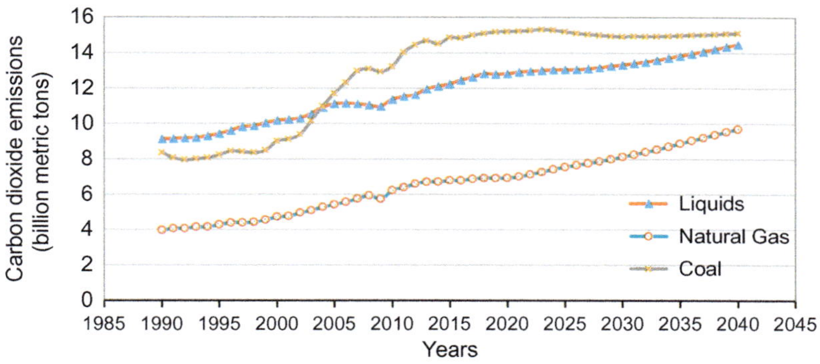

FIG. 1.9 Energy-related carbon dioxide emissions projected for the coming decades due to coal, petroleum liquids and natural gas. *(Data from International Energy Outlook 2018. US Government: 2018).*

consumption of three types of fossil fuels: oil, natural gas and coal. The emissions are evolving as clearly observed, with a dominant contribution from coal.

1.3.1 Environmental impact mitigation

The environmental issues related to the energy sector are desired to be reduced by utilizing alternative clean energy sources and fuels. One of the important factors in comparing various fuels is their potential to diminish greenhouse gases and other releases harmful to the environment and health. In this respect, the ratio of carbon to hydrogen is a good indicator because carbon will eventually contribute to global warming as well as other environmental impact categories. The carbon-hydrogen ratios of various fuels are shown in Fig. 1.10. Although biodiesel (B100) has a similar carbon content to petroleum diesel, this content is largely neutral in combustion depending on the feedstock [2]. Compared to other fuels used in combustion applications, ammonia has the highest hydrogen energy density (even higher than pressurized and liquefied hydrogen fuel) based on the current storage methods. It contains zero carbon atoms, and therefore has a global warming potential (GWP) of zero, and produces only nitrogen and water when combusted.

Reducing the carbon-hydrogen ratio is a target of the energy sector. Hence, Fig. 1.10 shows the pathway to the decarbonization of the world energy supply. Wood has the highest carbon-to-hydrogen ratio, which is about 10. The carbon-to-hydrogen ratio of coal can vary depending on the type of coal. For instance, high-grade anthracite coal can have a carbon-to-hydrogen ratio of about 2.66. Reducing this ratio will enable fewer emissions in the combustion processes. This graph reflects a trend in the direction of hydrogenization of the world energy supply, which is a progression toward global use of hydrogen as an energy carrier. This hydrogenization aspect will be further discussed in the next chapter.

The hydrogen economy promotes the wide implementation of sustainable energy resources such as hydroelectric power, wind power, solar, biomass, geothermal, ocean, and nuclear energy. Hydrogen can be generated from water, hydrocarbons or other hydrogen

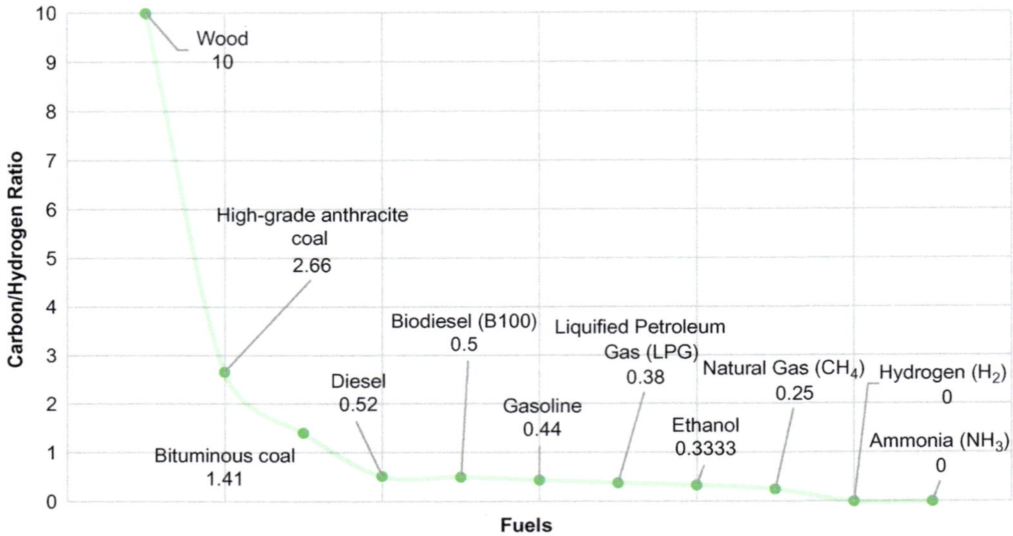

FIG. 1.10 Carbon to hydrogen ratio of various fuels. *(Data from Dincer I, Hogerwaard J, Zamfirescu C. Clean rail transportation options. Cham: Springer International Publishing; 2016. https://doi.org/10.1007/978-3-319-21726-0).*

containing materials. The use of hydrogen as an energy carrier and as a chemical intermediate is quite diverse, for example fertilizers, synthetic fuels, plastics, etc.

There are several routes to produce hydrogen from hydrogen containing materials, including thermal, electrical, and photonic based techniques. For example, heat with medium temperature can be used for thermochemical cycles of water separation. Several types of chemical and electrochemical reactions can occur in thermochemical cycles, which are a series of processes. Water, heat and, in some cases, electricity are the main inputs. Within the cycle, most substances are recycled, but hydrogen and oxygen are obtained as useful products when water is used as the feedstock.

Hydrogen can also be produced using photoelectrolysis, photoelectrochemical and photocatalysis methods. In these techniques, photons play an important role in supplying the required energy for water splitting.

In order to quantify the environmental benefits of reducing the carbon-to-hydrogen ratio, we can use several techniques and methods. The environmental impact assessment methods are among the suitable techniques for energy systems. They can be classified into four categories [1]:

- Environmental tools, including impact assessment and ecological footprints
- Thermodynamic tools, including performance indicators, energy, exergy and material flux analyses
- Sustainability tools, including life cycle assessment, sustainable process index, exergetic sustainability index, exergetic improvement potential, greenization factor, etc.
- Risk assessment tools

Some of these methods will be further discussed in the next chapters.

1.4 Energy policy and sustainability

The term "sustainable" can be interpreted as "persistent in time." An energy perspective for sustainability means sustainable development on three levels: energy generation, energy distribution and energy consumption [1]. Sustainability science has a diverse culture that includes the following concepts: physics, chemistry, biology, medicine, society and science and technology. One of the objectives of sustainable development science is to develop social, economic and environmental relationships and to preserve the wealth of humanity [1].

Sustainable development has been defined by the United Nations (UN) as "development that meets the needs of the present without compromising the ability of future generations to meet their own needs." [1]. Because of its diverse characteristics, ranging from the environment to society, it is not an easy task to assess sustainability based on quantifiable indicators.

In the past, efforts have been made to demonstrate social sustainability, using an "environmental sustainability index," "wellbeing index," "human development index" and others. Product-related assessment techniques have also been proposed based on product material flow analysis, life cycle assessment and thermodynamic analysis. Among those, the sustainability, environmental and energy assessment based on exergy analysis can broaden the scope of assessment by implying more effective resource use with reduced waste rejection into the environment.

There are several elements that contribute to achieving sustainable development. One of the substantial elements is the supply of energy resources in an entirely sustainable manner. In addition, efficiency improvement is considered as critical as sustainable energy supply.

Sustainability for any system or process means that the system or process is sustainable in the long-term. An engineering system can be considered sustainable if it does not damage the environment and does not reduce natural resources, but effectively operates over the long term without compromising the ability of future generations to meet their needs of natural resources and clean environment.

An increased level of sustainability of energy systems fulfills the following criteria [1]: (i) minimal or zero negative impact on the environment or social environment, (ii) preservation of natural resources, (iii) capability of meeting the energy needs of the current and future population, (iv) fairness and efficiency, (v) preservation of air, land and water, (vi) low or no net emissions of carbon or other greenhouse gases, (vii) securing of energy without burdening future generations. Other criteria can also be considered which take into account prosperity, local availability, cost-effectiveness, reliability, safety and environmental friendliness.

While determining sustainable energy policies and strategies, it is significant to comprehend the advantage of clean energy technologies. The kinds of energy technologies with zero or minimum environmental impact are generally called green energy. Considering the benefits of green energy, sustainability of the green energy supply is assumed to be a key element in the interactions between nature and society.

A significant increase in the number of policy tools and their effectiveness should be constantly sought to achieve sustainable solutions. It is important to have theoretical tools to quantify the development of green energy. These tools are desired to meet the following criteria [1]:

- Supporting detailed strategy and policies,
- Understanding basic concepts and concepts related to the use and maintenance of clean energy,
- Establishing links between the use and sustainable development of clean energy,
- Supporting strategic use and protection of clean energy sources,
- Ensuring energy security, implementation and development measures,
- Strengthening motivation for the implementation of clean energy strategies for better energy supply,
- Providing a solution for emissions reduction by looking at the potential clean energy strategies,
- Preparing a scientific platform to discuss potential clean energy strategies for the sector's use.

There is a need for policy development to reduce environmental pollution, and ensure energy security. Strategies are also required at a higher level because a strategy includes a plan or method toward achieving a specific target. Once a strategy is established, the policy, which represents the course of actions to implement the strategy, must be elaborated. Any policy will specify "policy instruments," which represent specific measures taken to implement a policy. Some examples of policy instruments are as follows [1]:

- Developing common exergy tools
- Increasing public awareness
- Imposing efficiency standards
- Setting public procurement policies
- Imposing appliance labelling norms
- Encouraging renewable energy sources
- Supporting research and development in demonstration projects
- Creating innovative programs
- Increasing engineering innovation and entrepreneurship
- Employing cutting-edge and smart technologies
- Implementing integrated systems
- Incorporating artificial intelligence and machine learning tools

In our context, the target of energy systems is to become sustainable. Some of the solutions can be: canceling subsidizations for fossil fuel-based energy production, rearranging the energy sector, supporting energy sector innovation, encouraging energy efficiency, supporting rural energy production, promoting decentralized energy generation, and refining access to efficient and clean fuel.

Some aspects affect clean energy policies and strategies, such as the following: [1]:

- Competition between sustainable energy systems can be disadvantageous as market changes may favor certain participants.
- Increasing energy efficiency will lead to lower investment cost compared to new systems.

- Large energy projects can shadow smaller renewable energy projects.
- The commercialization of clean energy technologies does not develop quickly.
- Critical public support is needed for significant sustainable development.
- It is important for the energy market to bring liberalization for opportunities to change.

Sustainable development is a confluence of resources sustainability, technological sustainability, energetic sustainability, environmental sustainability, economic sustainability and social sustainability, as shown in Fig. 1.11. Here, energetic sustainability includes several aspects, namely energy production and consumption, energy distribution, energy management, and energy supply. Technological sustainability includes technical feasibility, system capacity, supply-demand management, operation and maintenance, etc. Environmental sustainability includes industrial ecology, life cycle assessment, environmental planning, environmental control strategies, reduction of environmental pollutants and several other environment-related factors. Social sustainability includes satisfaction of social standards, awareness and education, political support, social acceptance and numerous other society related aspects. Economic sustainability includes financial analysis, cost-benefit analysis, affordable and cost-effective systems, exergoeconomic analysis and several other economy-related parameters. Resources sustainability includes sustainable use of available reserves (e.g. hydrocarbons, renewables, nuclear), minimization of resource depletion, effective mining and extraction, etc.

Clean and renewable energy resources and systems are vital elements of sustainable development for several causes such as: (i) having less environmental impact compared to other energy sources, (ii) diversity of clean energy resources, (iii) having non-depletion characteristics and being capable of supplying energy in a stable manner with reliable storage options, (iv) enabling decentralization and providing local solutions independent

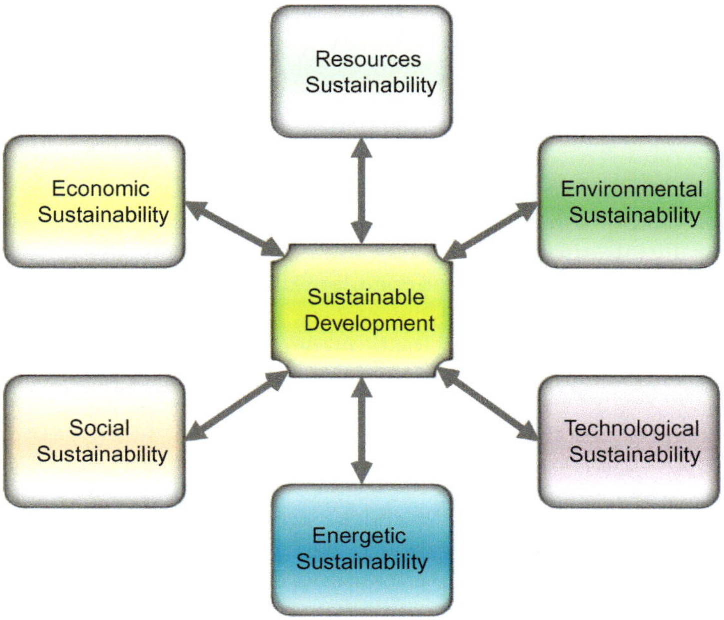

FIG. 1.11 The confluence of six main criteria for sustainable development.

from the grid, (v) providing economic benefits to small isolated populations, (vi) reducing the time required from initial design to operation, (vii) delivering superior flexibility in reacting to unpredictable increase and/or changes in energy demand. The leading element which will govern the specific role of clean energy technologies will possibly be energy demand.

Therefore, to make up for energy needs, it is possible to produce clean energy from renewable energy sources, such as hydropower, solar power, wind power, geothermal energy, wave energy, biomass, etc. Growth in the world's population requires efficient utilization of clean energy technologies. A clean energy system can [1]:

- offer a more environmentally friendly and sustainable future,
- improve energy security,
- facilitate the development of new and clean technologies,
- reduce pollution of air, water and soil as well as deforestation,
- reduce energy-related diseases and death,
- reduce or stop energy conflicts between countries.

Hence, clean energy and related systems are desired to contribute to global stability by decreasing the damaging impacts of fossil-based energy systems. For developing and publicizing the sustainable clean energy technologies, the subsequent significant clean energy strategies should be implemented [1]:

- industrial and technological support for the transition to clean energy technologies,
- control of the projection and analysis of clean energy sources,
- governmental and public support for clean energy economy,
- production, consumption, distribution, conversion, management and marketing of clean energy,
- research, development and application of sustainable clean energy technologies,
- support for the availability, productivity and reliability of clean energy and technologies, and
- design and fabrication of clean energy-based environmental and ecological applications.

While not all renewable energy resources are inherently clean, there is such a diversity of choices that a shift to renewables carried out in the context of sustainable development could provide a far cleaner system than would be feasible by tightening controls on conventional energy. Furthermore, being naturally site-specific, they favor the power system's decentralization and locally applicable solutions more or less independent of the national network. This enables citizens to perceive the positive and negative aspects of energy consumption. Consequently, a smaller scale of equipment often makes the time required from initial design to operation shorter, providing greater adaptability in responding to unpredictable growth and/or changes in energy demand.

The exploitation of renewable energy resources and technologies is a key component of sustainable development. There are three significant reasons for this as follows [1]:

- They have much less environmental impact compared to other sources of energy even though there is no energy source with zero environmental impact.
- Renewable energy resources cannot be depleted unlike fossil fuel and uranium resources. If used wisely in appropriate and efficient applications, they can provide reliable and

sustainable supply of energy. In contrast, fossil fuel and uranium resources are finite and can be diminished by extraction and consumption.

- They favor power system decentralization and locally applicable solutions more or less independent of the national grid network, thus enhancing the flexibility of the system and the economic power supply to small isolated settlements. That is why many different renewable energy technologies are potentially available for use in urban and remote areas.

Energy resources and their utilization are intimately linked to sustainable development. For societies to attain or try to attain sustainable development, much effort must be devoted not only to discover sustainable energy resources but also to increase the energy efficiencies of processes utilizing these resources. Under these circumstances, increasing the efficiency of energy-utilizing devices is important. Due to increased awareness of the benefits of efficiency improvements, many institutes and agencies have started working along these lines with some of the potential measures given in Fig. 1.12.

Many energy conservation and efficiency improvement programs have been and are being developed to reduce present levels of energy consumption. To implement these programs in a beneficial manner, an understanding is required of the patterns of "energy carrier" consumption, for example, the type of energy carrier used, factors that influence consumption, and types of end-uses.

Environmental concerns are an important factor in sustainable development. For a variety of reasons, activities that continually degrade the environment are not sustainable over time,

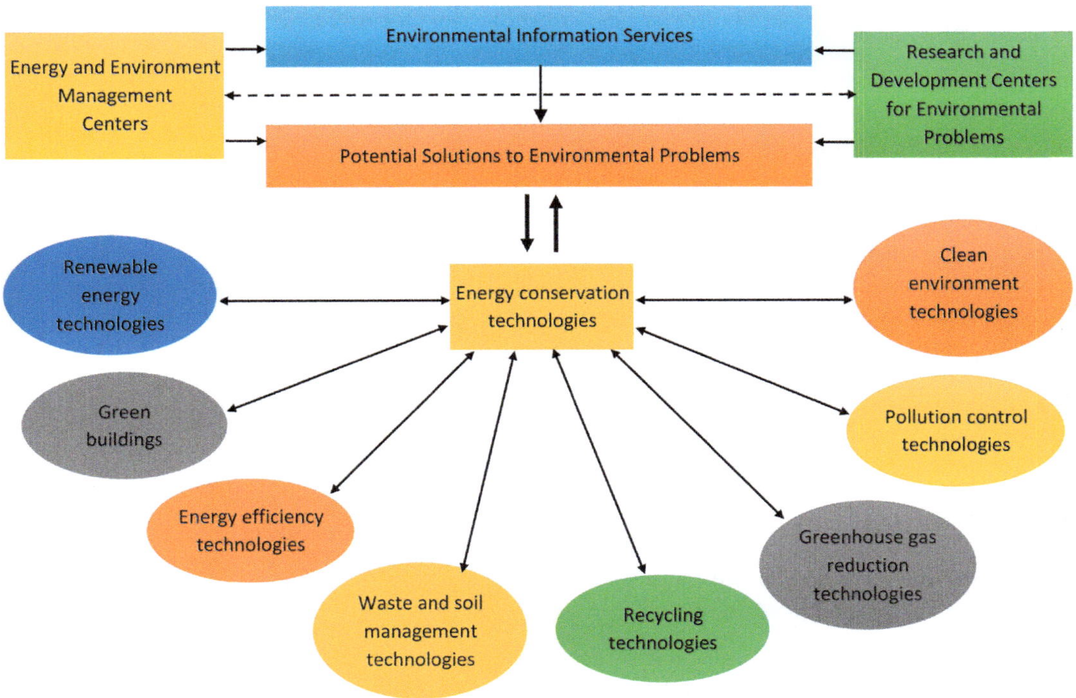

FIG. 1.12 The interrelations among possible environmental and energy conservation technologies.

e.g., the cumulative impact on the environment of such activities often leads to a variety of health, ecological and other problems over time. A large portion of the environmental impacts in a society is associated with the utilization of energy resources.

Ideally, a society seeking sustainable development utilizes only energy resources that cause no environmental impact (e.g., that release no emissions to the environment). However, since all energy resources lead to some kind of environmental impact, it is reasonable to suggest that some (not all) of the concerns regarding the limitations imposed on sustainable development by environmental emissions and their negative impacts can be in part overcome through increased energy efficiency. Clearly, as shown in Fig. 1.12, a strong relationship exists between energy efficiency and environmental impacts because, for the same services or products, less resource utilization and pollution is normally associated with increased energy efficiency.

1.5 Sustainability indicators

Table 1.4 contains a selection of significant indicators that can be used to elaborate aggregate sustainability indexes. In principle, the indexes can be elaborated according to the purpose of the analysis, assessment and comparative evaluation.

Sustainability indicators have some degree of subjectivity, because it is unlikely that exact definitions and understandings can be formulated of social changes, environmental protection, energy, resources, technology and economic development. The United Nations has established >130 sustainability indicators. Among them, 30% are social indicators, 17% are economic indicators, 41% are environmental indicators, and 12% are institutional indicators. Approximately 33% of sustainability indicators are drivers, 39% are state indicators and 18% response-kind indicators [1]. Some aspects of sustainability assessment tools can be stated as follows [1]:

- Sustainability assessment includes environmental assessment and system life habitats for the present and future generations.
- It is necessary to take into consideration the length of time horizon for the long-term ecosystem and consequences.
- Sustainability indicators should include direct relations with society goals.
- Sustainability assessment includes the following models: economic models, stress-response models, multiple models of capital, social models, environmental models, human and ecosystems models.
- Determination of the reference environment (or condition) is required to determine sustainability.
- Evaluation of costs associated with environmental degradation should be included.

Fig. 1.13 signifies a generic model of an energy system, which is beneficial for sustainability assessment. The energy system can be a power plant in which the chemical energy of a fuel is transformed into electricity. Other energy conversion techniques are also possible such as wind-to-power, solar-to-power, etc. In addition to energy transformation, an energy system can also comprise energy distribution and consumption.

TABLE 1.4 Groups and types of indicators affecting sustainability assessment.

Category	Driver indicators	State indicators	Responses indicators
Social	Unemployment rate Population growth rate Adult literacy rate Motor fuel consumption per capita Loss rate due to natural disasters	Poverty gap index Income inequality index School life expectancy House price per income Floor area per person	GDP on education Childhood immunization Infrastructure expenditure per capita Health expenditure per capita Hazardous chemicals in foods
Economic	Per capita GDP Investment share of GDP Annual energy consumption Natural resource consumption Capital goods imports	Proven mineral reserves Fossil fuel reserves Lifetime of energy reserves Share of renewable energy Manufacturing added value	Environmental protection expenditure Funding rate on sustainable development Amount of funds on sustainability Percentage of new funds on sustainability Funding grants on technology
Environmental	Water consumption per capita Generation of wastes ODP substances emission Emission of greenhouse gas, SO_2,NO_x Energy use in agriculture	Groundwater reserves Monthly rainfall index Desertification rate Pollutants concentration Acute poisoning	Waste-water treatment expenditures Natural resource management Air pollution mitigation expenditures Waste management expenditures Number of restricted chemicals
Institutional		Researchers number per capita Engineers number per capita Internet access per capita Telephone line access Other information channels	Policy on sustainable development Environment protection programs GDP share of R&D expenditures Number of R&D personnel per capita Ratification/implementation of agreements

Source: *United Nations (UN). Indicators of sustainable development: guidelines and methodology. 3rd ed. New York, NY: United Nations; 2007.*

Two categories of inputs are available for an energy system, namely the resources and the *process enablers* as shown in Fig. 1.13.

Process enablers (facilitators) are three types: human operators, machines and power systems covering all technical modules and the environment describing certain temperatures, pressures and compositions. For human operators, issues of sustainability are security, health and education. Energy and sustainability indicators should take into account cyclicality and environmental impact of a specific construction system.

There are three types of inputs [1]: building materials and production systems, oxidizing agents (if any), such as combustion air and energy. For each input, you can determine the appropriate indicators, quantities, and properties. For example, building materials are classified by type, specific structure, shape and size. The input form of energy is important such as work and heat.

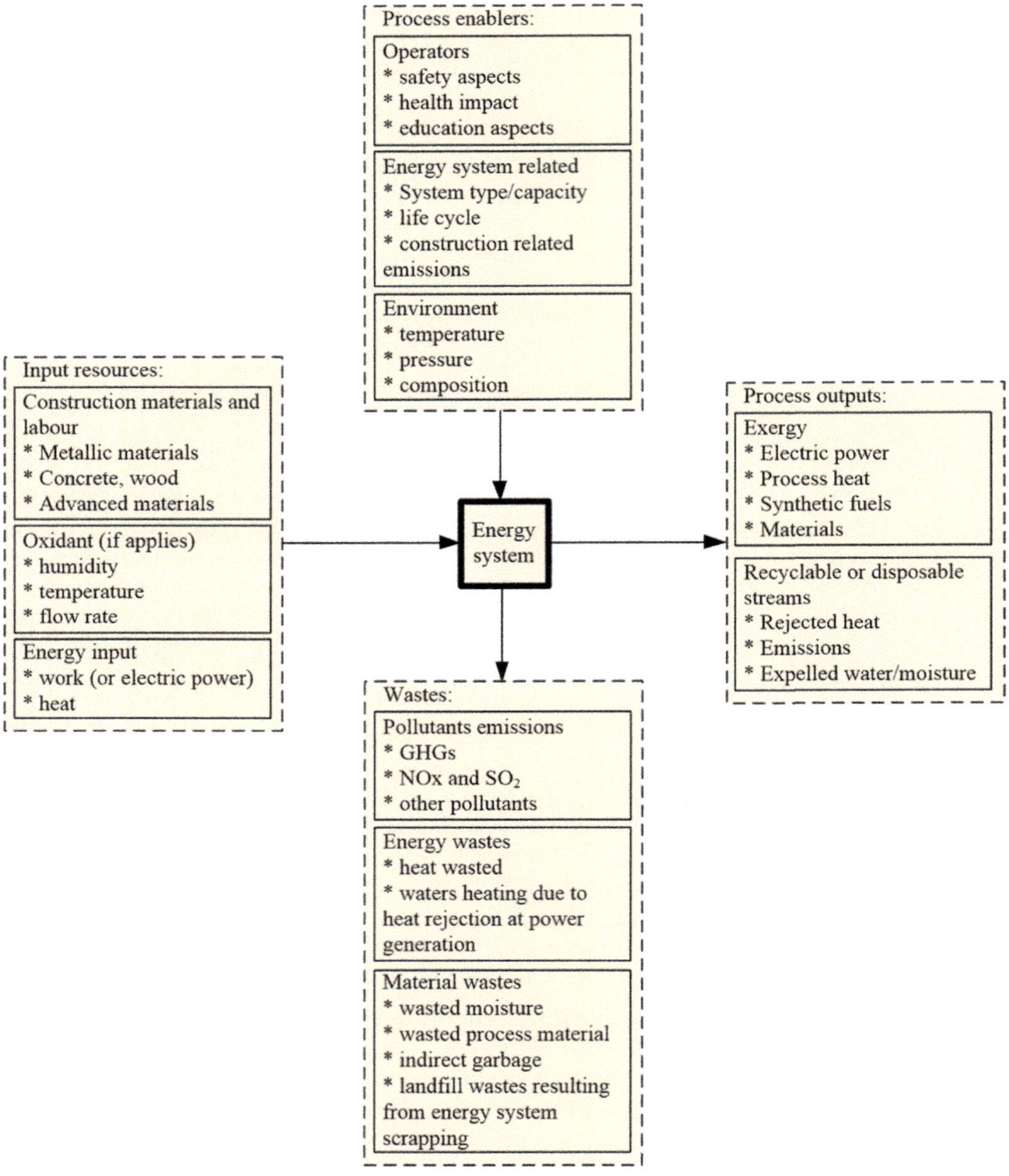

FIG. 1.13 Various models considered in sustainability assessment of energy systems based on material and energy balances. *(Modified from Linke B, Das J, Lam M, Ly C. Sustainability indicators for finishing operations based on process performance and part quality. Procedia CIRP 2014;14:564–9. https://doi.org/10.1016/J.PROCIR.2014.03.017).*

The outputs are basically two different types [1]: process results and process waste. The main output of the energy system is usually generated electricity. Three types of waste can be stated, which are environmental pollutants, waste energy and waste materials. Emissions are directly or indirectly associated with combustion processes related to the conversion of energy. In addition, emissions of exhaust gases at the end of their lifetime represent a type of pollutant. A portion of energy waste can be evaluated by the method of heat rejection. This heat is commonly rejected to ambient via the boundary of the system. The second form is through waste energy from hot water in the processes to the lakes, rivers and seas, mostly released from power plants. It is known that the cooling water used in the condensers of power stations affects the aquatic environment by altering the local temperature, which negatively impacts the life systems. Noise is another form of energy waste with environmental impact. The material wastes comprise similarly landfill material produced at the product scrapping phase.

The area of power production has made significant efforts to increase energy efficiency and recycle and reuse available resources. For example, waste and other solid waste can be used to create flammable materials for combustion. Resource recovery involves the life of important natural resources and is often natural and profitable for the environment. Energy efficiency and conservation of energy resources can eliminate deficiencies, reduce environmental impact and provide economic benefits. Efficient use of energy is important in developing countries, because the need for high capital investment can be foreseen. There is an enormous potential for energy efficiency and protection of lower energy consumption in the world, and therefore for the effect of energy consumption on the environment.

One of the most important factors is the overall cost of estimating and reducing CO_2 emissions. As for developing countries, cost and benefit arguments should take into account all recommendations for rapid economic growth. In order to achieve such a balance between economic development and emission reduction, internal policies are needed to improve energy efficiency and fuel switching for the engagement and implementation of international policies that enable easy access to advanced technologies and external resources.

It is expected that some countries would offer some tax reductions for businesses, especially those promoting renewable energy technologies, since these technologies are characterized by low or zero CO_2 levels. Several possible solutions include pre-cleaning of fossil fuels for burning, employing fluidized bed technology for coal, renewable energy usage, promoting hydrogen systems, incorporating energy storage technologies, applying energy efficiency solutions, promoting efficient public transport, etc.

1.6 Exergy and sustainability

Energy as well as exergy are key elements in the interactions between nature and society and they are considered key inputs for economic development and sustainable development. As illustrated in Fig. 1.14, there are several parameters to be decreased or increased to switch from non-sustainable to sustainable development. Energy use is very much governed by thermodynamic principles and, therefore, an understanding of thermodynamic aspects of energy

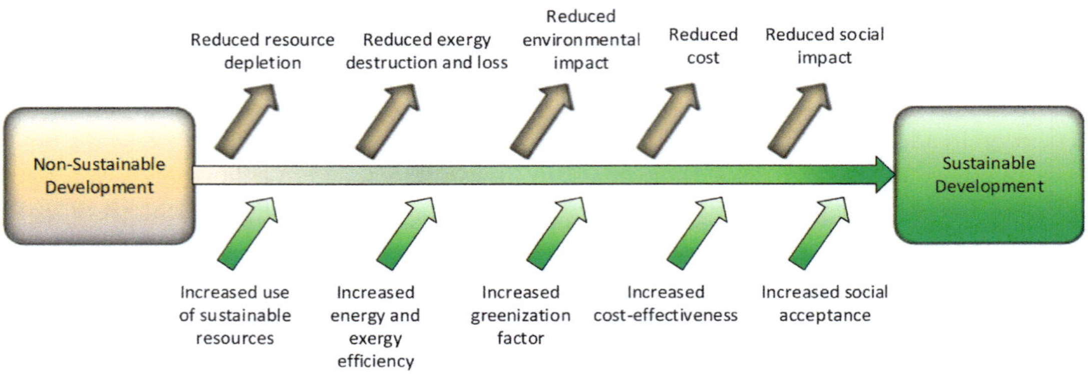

FIG. 1.14 Main steps toward sustainable development based on the fundamental pillars of sustainability.

can help us understand pathways to sustainable development [9]. The impact of energy resource utilization on the environment and the achievement of increased resource-utilization efficiency are best addressed by considering exergy. The exergy of an energy form or a substance is a measure of its usefulness or quality or potential to cause change and provide the basis for an effective measure of the potential of a substance or energy form to impact the environment. It is important to mention that in practice a comprehensive understanding of exergy, efficiency, environmental impact and sustainability of energy systems is required for the engineers or scientists working in the area of energy systems and the environment. During the past decade, the need to understand the linkages between exergy and energy and environmental impact has become increasingly significant [10, 11].

Exergy analysis suggests that it is the basis for assessing sustainability because exergy is associated with environmental conditions and measures the quality. Exergy analysis is very important in the evaluation of sustainability, because the system's performance as well as deficiencies are easily visible and quantifiable. It also enables possible ways to improve efficiencies, reduce destructions and emissions and design efficient energy systems.

Fig. 1.15 shows two major elements, namely use of exergy as a potential tool for analysis, design and performance improvement, and increased use of renewable (clean) energy systems, for increasing sustainable development. Behind any policy and decision, there must be a justification, which limits the decision spectrum. Besides numerous existing ideas, the status of technological development and the existing theoretical assessment and prediction tools play a key role in inducing policy making and high-level decisions.

Fig. 1.16 displays a model for sectorial exergy assessment. One of the largest segments is the power industry, which is responsible for the production and distribution of electricity. Energy sectoral assessment also includes energy generated for desalination plants, which is significant for countries located in desert areas such as Qatar, Saudi Arabia. The averaged world exergy efficiency of the electric utility subdivision is determined to be about 25% by Dincer and Zamfirescu [12].

The exergetic assessment of sectors has been extended to cover the environment and sustainable development by Rosen and Dincer [13]. It was shown that exergy is located at the confluence

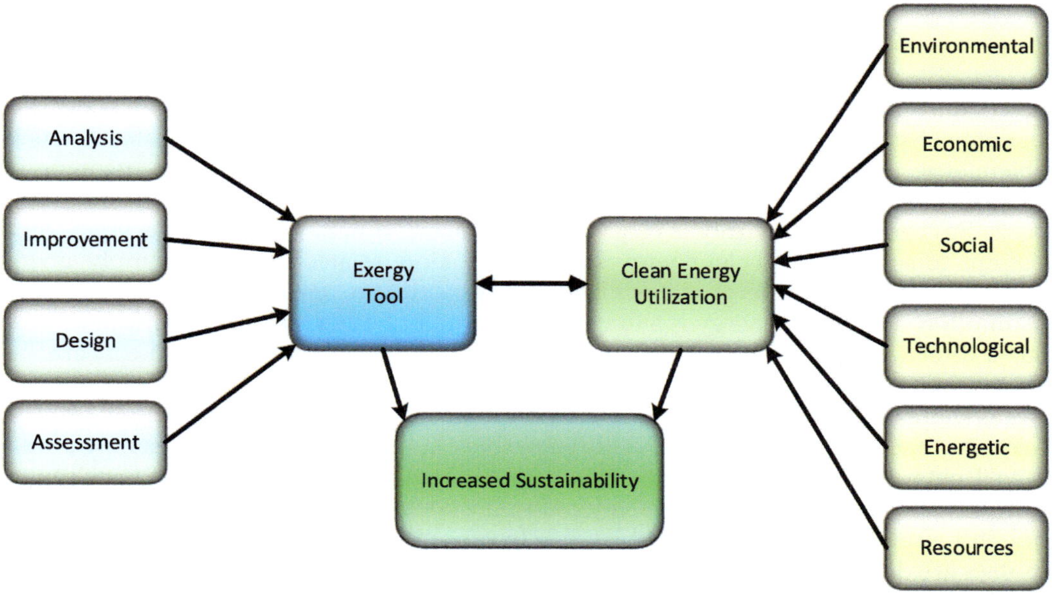

FIG. 1.15 The role of assessment tools and indicators in shaping the future sustainable society. *(Modified from Dincer I, Rosen MA. Exergy as a driver for achieving sustainability. Int J Green Energy 2004;1:1–19. https://doi.org/10.1081/GE-120027881).*

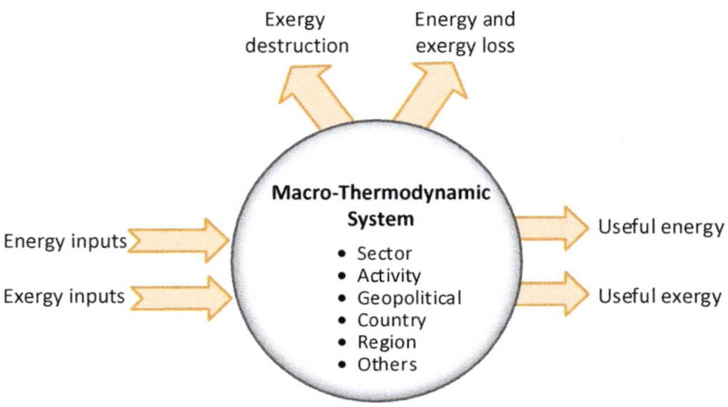

FIG. 1.16 Thermodynamic analysis at macro-scale based on energy and exergy concepts. *(Modified from Dincer I, Zamfirescu C. 1.4 Sustainability dimensions of energy. In: Dincer I, editor. Compr. Energy Syst. 1st ed., Cambridge, United States: Elsevier; 2018, p. 101).*

of energy, environment and sustainable development. The necessity to understand the relationship between energy and exergy, as well as the impact on the environment, is becoming increasingly significant, because lower exergetic destructions can yield reduced environmental impacts. In addition, since energy policy plays a highly important role in addressing issues of sustainability and a wide range of local, regional and global environmental problems, policymakers should also appreciate the concept of exergy and its connection to these problems. We will elaborate on the fundamentals and importance of energy and exergy in the next chapter.

1.7 Conclusions

Sustainable development includes several aspects such as energy, water, environment, food and economy. Securing each of these aspects is a significant challenge. Although water is an abundant substance on Earth, availability of drinkable, potable, pure, clean water is a major problem especially for the regions with few freshwater sources. Food security encompasses food quality, food supply, fertilizers, food processing, food storage, irrigations and many other aspects. Environmental security is concerned with preserving nature and reducing harmful emissions to acceptable levels. Energy security is mainly related to energy supply, energy generation, extraction, energy transmission, transportation, energy distribution, energy consumption. Therefore, energy is situated at the center of other sustainability aspects since it has a direct association with water, food and environment. Environmental concerns are an important factor in sustainable development. Improving efficiencies can reduce energy losses as well as environmental impacts. There are mainly three types of energy resources available on Earth: conventional, nuclear and renewables. Conventional fuels are mainly natural gas, coal, petroleum and its derivatives. The conventional energy production systems convert energy from fuels in a way that carries serious implications for the environment. Nuclear fuel is commonly uranium and thorium, which has fewer emissions than fossil fuels. Renewable resources are diverse ranging from solar to biomass. The science of sustainability is an interdisciplinary subject comprising the following main disciplines: physics, chemistry, biology, medicine, social and economic sciences and engineering. Therefore, sustainability assessment is a key tool for any energy system. This chapter dwelled on the importance of interrelations among environment, economy, society, energy, as well as exergy. The significance of these interrelations leads us to focus on sustainable energy systems with proper integration and management, to be discussed in the next chapters.

Nomenclature

ΔF radiative force
ΔC concentration change

Acronyms

CFC chlorofluorocarbons
EIA environmental impact assessment
GHG greenhouse gases
GWP global warming potential
ODP ozone depletion potential
OECD Organization for Economic Cooperation and Development
P production
PAHs polycyclic aromatic hydrocarbons
PJ picojoule
PM particulate matter
PW picowatt
R reserve
UN United Nations
VOC volatile organic compounds

References

[1] Dincer I, Zamfirescu C. 1.4 Sustainability dimensions of energy. In: Dincer I, editor. Compr. Energy Syst. 1st ed. Cambridge, United States: Elsevier; 2018. p. 101.
[2] Dincer I, Zamfirescu C. 1.3 Environmental dimensions of energy. In: Dincer I, editor. Compr. Energy Syst. 1st ed. Cambridge, United States: Elsevier; 2018. p. 49.
[3] Dincer I. Environmental and sustainability aspects of hydrogen and fuel cell systems. Int J Energy Res 2007;31:29–55. https://doi.org/10.1002/er.1226.
[4] Dincer I, Zamfirescu C. Advanced Power Generation Systems. Oxford, United Kingdom, Elsevier; 2014.
[5] Renewables 2017 Global Status Report REN21 Renewables Glob Status Rep 2017. http://www.ren21.net/gsr-2017/ (accessed 14 December 2018).
[6] Intergovernmental Panel on Climate Change (IPCC). Climate change 2007: Synthesis report. Valencia, Spain: 2007.
[7] International Energy Outlook 2018. US Government: 2018.
[8] International Energy Agency. Energy Statistics 2017;96. https://doi.org/10.1017/CBO9781107415324.004.
[9] Dincer I, Rosen MA. Thermodynamic aspects of renewables and sustainable development. Renew Sustain Energy Rev 2005;9:169–89. https://doi.org/10.1016/j.rser.2004.02.002.
[10] Dincer I, Rosen MA. A worldwide perspective on energy, environment and sustainable development. Int J Energy Res 1998;22:1305–21. https://doi.org/10.1002/(SICI)1099-114X(199812)22:15<1305::AID-ER417>3.0.CO;2-H.
[11] Dincer I, Rosen MA. Energy, environment and sustainable development. Appl Energy 1999;64:427–40. https://doi.org/10.1016/S0306-2619(99)00111-7.
[12] Dincer I, Zamfirescu C. Sustainable energy systems and applications. Springer Science & Business Media: New York, USA; 2011.
[13] Rosen MA, Dincer I. Exergy as the confluence of energy, environment and sustainable development. Exergy Int J 2001;1:3–13.

Further reading

[14] Dincer I. Refrigeration systems and applications. 3rd ed. John Wiley & Sons, Ltd: Chichester, UK; 2017. https://doi.org/10.1002/9781119230793.
[15] Dincer I, Hogerwaard J, Zamfirescu C. Clean rail transportation options. Cham: Springer International Publishing; 2016. https://doi.org/10.1007/978-3-319-21726-0.
[16] United Nations (UN). Indicators of sustainable development: guidelines and methodology. 3rd ed. New York, NY: United Nations; 2007.
[17] Linke B, Das J, Lam M, Ly C. Sustainability indicators for finishing operations based on process performance and part quality. Procedia CIRP 2014;14:564–9. https://doi.org/10.1016/J.PROCIR.2014.03.017.
[18] Dincer I, Rosen MA. Exergy as a driver for achieving sustainability. Int J Green Energy 2004;1:1–19. https://doi.org/10.1081/GE-120027881.

CHAPTER

2

Fundamentals of energy systems

2.1 Introduction

This chapter provides fundamental information about energy, entropy, and exergy concepts used in integrated energy systems. In order to analyze the integrated energy systems, basic thermodynamic principles are explained in detail. The main balance equations are given for common system components and several practical applications and implications of these thermodynamic systems are elaborated.

The roots of the term energy originate from the Greek expression *energia* stated by Aristotle for the initial meaning of activity. The source of this word *energia* could have been from the noun *ergon* meaning action, labor or work used by Aristotle [1]. Aristotle defines energia as the property of any perceived entity of "acting something out" for a goal. The contributions of Rumford, Carnot, Clausius, William Thompson and others were directed to the comprehensive formulation in 1851 of the dynamic theory of heat that was called thermodynamics later on. Thermodynamics is a subdivision of physics having four laws (0th, 1st, 2nd, and 3rd laws) with two of them being the main ones (1st and 2nd). The first law of thermodynamics is known as conservation of energy and the second law is non-conservation of exergy (available energy). The second law of thermodynamics summarizes that the processes are not reversible and the total entropy increases. Thermodynamics identifies that energy and matter can move in or out of the system and therefore a global balance or imbalance of streams can be expressed. There are three different forms of energy transfer through a system boundary, namely: (i) mass, (ii) heat, and (iii) work. The energy transfer through heat is associated with temperature and eventually entropy. Entropy is a physical terminology presented by Clausius when formulating the second law of thermodynamics in the form of an imbalance.

Many forms of energy are available that are convertible to each other. Thermodynamics plays a vital role in the analysis of processes, systems, and components in which energy transfers and energy conversions take place. In fact, the Industrial Revolution was driven by the finding of how to implement energy on a large scale and how to transform heat into work. The laws of thermodynamics permit the conversion of work completely into thermal energy, but thermal energy cannot be entirely converted into work.

Integrated Energy Systems for Multigeneration
https://doi.org/10.1016/B978-0-12-809943-8.00002-9

Thermodynamics is an important tool for detailed understanding, design and optimization of energy systems. It has been widely employed for many other engineered systems. Based on the main laws of thermodynamics, there are two assessment methods: first law analysis (energy) and second law analysis (exergy). In this book, we present both energy and exergy analyses approaches for performance investigation of integrated energy systems. In the forthcoming section, we provide the basics of thermodynamics starting from fundamental units to efficiency definitions.

2.2 Fundamentals of thermodynamics

The physical properties are of two kinds, extensive and intensive properties. An extensive physical property is the summation of individual quantities in entire elements of a specified system. Intensive properties are independent of the scope of the system. Mass, volume, electric charge, energy are categorized under extensive properties, whereas temperature, pressure, and specific volume are categorized under intensive properties. The definitions of fundamental physical properties, as well as their standard units, are given in Table 2.1.

Seven fundamental physical properties are defined as the amount of substance (molar), mass, time, luminous intensity (measured in candela), electric current, thermodynamic temperature and length by the International Standard of Units [2]. The fundamental properties cannot be derived from any other quantities applying constitutive relations. However, any non-fundamental physical property can be derived from the fundamental properties. Hence, they are called derived physical properties. The units of measure of any derived physical property can be calculated using the fundamental units and the constitutive relationships based on the dimensional analysis method. In this respect, the

TABLE 2.1 Definitions of the fundamental physical properties

Quantity	Symbol	Unit	Definition
Mass	m	kilogram	It is a quantitative measurement of inertia which is the resistance to acceleration
Length	l	meter	It is geometric distance between two points in space
Electric current	I	Ampere	It is electric charge flow rate
Luminous intensity	I_v	candela	It is (luminous) flux of a light source (per direction and solid angle) for a typical model of human eye sensitivity
Time	t	second	A measurable period of the progress of observable or non-observable events
Thermodynamic temperature	T	Kelvin	A measure of kinetic energy stored in a substance. It has a minimum of zero (Kelvin)
Amount of substance	n	mole	The quantity of explicitly identified units of a substance such as molecules, electrons, atoms, etc.

Source: *The International System of Units (SI). Standards. 8th ed., 2006.*

dimensional symbols for the fundamental quantity must be used to express the dimension for the derived quantities. The International System of Units (SI) provides the standard definitions, symbols, and units for fundamental properties. Among the fundamental properties, temperature can be considered as one of the most significant for thermodynamics since temperature denotes a grade of the average kinetic energy per degree of freedom of the molecular elements of a substance.

All substances show a molecular element with a degree of freedom and a relationship with the physical aggregate. In addition to temperature, pressure is also very significant in thermodynamics.

There are several fundamental physical constants (e.g., universal gas constant, Boltzmann constant, Faraday constant) that evolved throughout the historical developments. Some of the most important ones for thermodynamic analysis are listed in Table 2.2.

TABLE 2.2 Major constants and some standard parameters of substance for thermodynamic analysis

Constant/ Parameter	Value	Definition
Elementary charge	$e = 1.60218 \times 10^{-19}$ C	Electrical charge carried by a single proton
Stefan-Boltzmann constant	$\sigma = 5.670373 \times 10^{-8}$ W/ m^2 K^4	Constant in Stefan-Boltzmann law stating the proportionality between fourth power of temperature and black body's emissive power
Faraday's constant	$F = 96{,}485$ C/mol	Electric charge of one mole of electrons
Planck's constant	$h = 6.626 \times 10^{-37}$ kJ s	Specifies the degree of energy of a quanta which states the proportionality between frequency of a photon and its energy according to $E = h \bullet \nu$
Speed of light in vacuum	$c = 299{,}792{,}458$ m/s	Maximum speed at which matter and information can be transported in the known cosmos
Boltzmann constant	$k_B = 1.3806 \times 10^{-23}$ J/K $k_B = \mathcal{R}/N_A$	Signifies a measure of kinetic energy of one molecule of ideal gas
Avogadro number	$N_A = 6.023 \times 10^{26}$ molecules/kmol	Ratio of element entities of a bulk substance to the amount of substance. $N_A = N/n$
Standard atmospheric pressure	$P_0 = 101.325$ kPa	Pressure of the terrestrial atmosphere at sea level in standard conditions
Gravitational acceleration	$g = 9.80665$ m/s^2	Standard gravitational acceleration signifies the gravitational force (G) per unit of mass. $g = G/m$
Universal gas constant	$\mathcal{R} = 8.314$ J/mol K $\mathcal{R} = Pv/T$	Signifies a measure of kinetic energy of one mole of an ideal gas at molecular level

Source: *Dincer I, Zamfirescu C. 1.5 Thermodynamic aspects of energy. In: Dincer I, editor. Compr. Energy Syst., 1st ed., Cambridge, United States: Elsevier; 2018, p. 153–211. doi: 10.1016/B978-0-12-809597-3.00105-X.*

Energy is a scalar quantity that cannot be measured directly but can be recorded and assessed by indirect measurements. The absolute value of the energy of a system is relatively challenging to measure, while the energy change is moderately easier to calculate. In the forthcoming section, the forms of energy, as well as fundamental thermodynamic laws, are introduced.

2.2.1 Forms of energy

Energy is available in the universe in many forms, which are either internal or transient [3]. Energy in the universe can be converted from one form to another form. In the thermodynamic analysis, the forms of energy can be classified into two groups: macroscopic and microscopic.

2.2.1.1 *Macroscopic*

Macroscopic forms of energy contain an overall system energy with respect to a reference frame, for instance, kinetic and potential energies. The macroscopic energy of an object increases with rising velocity and elevation. Hence, the macroscopic energy of a system is associated with motion and the influence of external effects such as gravity, magnetism, electricity, and surface tension [4]. The energy available as a result of its motion is called kinetic energy. The total potential energy of a system is the summation of the gravitational, centrifugal, electrical, and magnetic potential energies. The energy available as a result of its elevation in a gravitational field is called gravitational potential energy. Since the changes of other subtypes of potential energy are mostly minor or neglected, it is commonly referred to as potential energy. Potential energy can be transformed into other forms of energy such as kinetic energy. Kinetic and potential energy quantities are dependent on the environment in which the system exists.

2.2.1.2 *Microscopic*

Microscopic forms of energy are associated with the molecular structure of a system and the degree of molecular activity. They are independent of outside reference conditions. The summation of the entire microscopic forms of energy of a system is its internal energy. The internal energy of a system is dependent on the inherent qualities or properties of the materials in the system, such as composition and physical form, in addition to the environmental parameters such as temperature, pressure, electric field, magnetic field, etc. There are many forms of internal energy such as mechanical, chemical, electrical, magnetic, surface, and thermal.

The nature of thermodynamics includes the concept of equilibrium states. It hypothesizes that the change in the value of thermodynamic properties, such as internal energy, between two equilibrium states of a system does not depend on the thermodynamic path the system takes [3]. The change is determined by the initial and final equilibrium states of the system. Therefore, the internal energy change of a system is defined by the parameters specifying the system in its final and initial states. The parameters include pressure, temperature, magnetic field, surface area, mass, and others [4].

2.2.2 Fundamental laws of thermodynamics

This section summarizes the four main laws of thermodynamics with a primary focus on the second law. Although all of these four laws are very important, there are two governing laws (like constitutional laws), namely the first law of thermodynamics and the second law of thermodynamics, for design, analysis, development, assessment, evaluation and improvement.

2.2.2.1 *The zeroth law of thermodynamics*

The zeroth law of thermodynamics is an expression of thermodynamic equilibrium described as follows: "if two thermodynamic systems are in thermal equilibrium with a third, they are also in thermodynamic equilibrium with each other." [3]. Most of the thermodynamic systems are in thermodynamic equilibrium, but there are also non-equilibrium thermodynamic systems. Thermodynamic equilibrium includes three sub-parameters, namely thermal, chemical and mechanical. Two thermodynamic systems are in thermal equilibrium when their temperatures are equal, meaning that there is no heat exchange process. Two thermodynamic systems are in mechanical equilibrium when there is no work exchange. Two thermodynamic systems are in chemical equilibrium when there is no change in their chemical compositions. In addition, insulated thermodynamic systems are in thermodynamic equilibrium when there is no mass, heat, work, chemical energy, etc. exchange between any parts within the system.

2.2.2.2 *The first law of thermodynamics*

The first law of thermodynamics (FLT) is the law of the conservation of energy, which expresses that energy cannot be created or destroyed, but it can change forms [3]. The first law of thermodynamics describes internal energy as a state function and delivers a formal statement of the conservation of energy. Nonetheless, it does not provide data about the direction of processes that can spontaneously happen. This means that the reversibility characteristics of thermodynamic processes are not elaborated. In thermodynamics, there are two different type of systems, namely the open and closed types. The type of system is defined based on the mass transfer through the selected system boundary. If there is a mass transfer through the system boundary, it is called an open system, whereas if there is no mass transfer through the system boundary, it is called a closed system. Although there is no mass transfer in closed systems, the energy can still be transferred via heat or work crossing the boundary.

If energy is denoted as E (in kJ) and change of energy of the system is denoted as ΔE_{sys}, then the first law of thermodynamics for a closed system experiencing any type of process can be expressed as shown in Fig. 2.1.

The energy balance for the first law of thermodynamics can be written either on an amount basis or on a rate basis as follows:

On amount basis:

$$E_{in} - E_{out} = \Delta E_{sys} \tag{2.1}$$

On rate basis:

$$\dot{E}_{in} - \dot{E}_{out} = dE_{sys}/dt \tag{2.2}$$

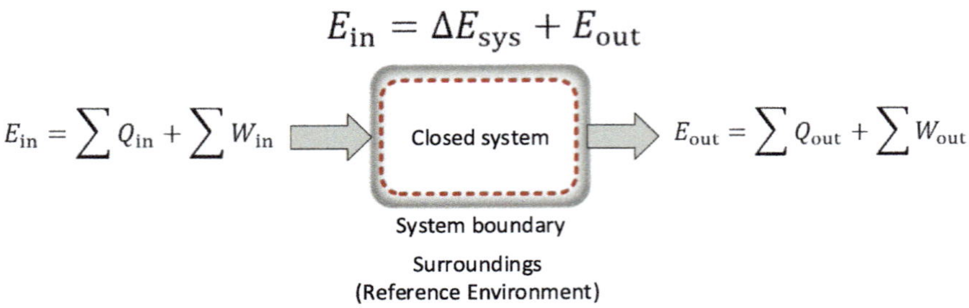

FIG. 2.1 The first law of thermodynamics written for a closed system.

For a closed system, the change in system energy can be associated with the specific energy and system mass as per the following equation:

$$\Delta E_{sys} = m\,(e_{out} - e_{in}) \tag{2.3}$$

where e is the specific total energy of a nonflow system, which can be written as

$$e_{nonflow} = u + \frac{1}{2}v^2 + gz \tag{2.4}$$

where u denotes the internal energy, $\frac{1}{2}v^2$ is the kinetic energy and gz is the potential energy of the system.

Internal energy consists of several microscopic energy types such as thermal, electrical, magnetic, vibrational, surface and chemical. The internal energy is proportional to the mean force that molecules bring to bear on the system boundary. Internal energy is an extensive property quantifying a state function. Based on thermodynamic principles, a state function is a property of a system depending only on existing state point values. When a variation takes place, the state function is not affected by the process in which the conversion is completed. Internal energy can be written using the specific internal energy:

$$u = \frac{U}{m} \tag{2.5}$$

where m is the system mass. The specific internal energy is an intensive property as can be concluded from the definition. Uppercase denotes the internal energy counting the mass and lower case denotes specific internal energy, which is per unit mass.

The change in internal energy of a closed system can be written as follows by using the specific internal energy:

$$\Delta U = m\,(u_2 - u_1) \tag{2.6}$$

In classical thermodynamics, based on the direction of energy flow, there is a sign rule for work and heat transfer. However, this can be eliminated by using an energy balance equation showing energy inputs are equal to energy outputs (accounting for energy generations and losses).

The input energy entering the closed system can be written as

$$E_{in} = \sum Q_{in} + \sum W_{in} \tag{2.7}$$

The output energy leaving the system is written as

$$E_{out} = \sum Q_{out} + \sum W_{out} \tag{2.8}$$

Consequently, the energy balance equation of the closed system is written as follows:

$$E_{in} = \Delta E_{sys} + E_{out} \tag{2.9}$$

$$\sum Q_{in} + \sum W_{in} = \Delta E_{sys} + \sum Q_{out} + \sum W_{out} \tag{2.10}$$

Contrary to closed systems, open systems have mass exchanges through the boundary as shown in Fig. 2.2.

In open systems, a flowing matter exists. Hence, the total energy of a flowing matter can be determined using the following expression:

$$e_{flow} = u + P\nu + \frac{1}{2}\upsilon^2 + gz \tag{2.11}$$

where ν is the specific volume, u is the internal energy, υ is the velocity and z is the elevation. For flowing systems, we commonly use an enthalpy term that is written as

$$h = u + P\nu \tag{2.12}$$

Hence the total energy of the flow can be written as

$$e_{flow} = h + \frac{1}{2}\upsilon^2 + gz \tag{2.13}$$

The energy balance equation for open systems needs to take into account the mass flow crossing the boundary, thence it can be formulated as follows:

$$\sum_{in} \dot{m} e_{flow} + \sum \dot{Q}_{in} + \sum \dot{W}_{in} = \sum_{out} \dot{m} e_{flow} + \sum \dot{Q}_{out} + \sum \dot{W}_{out} + \left[\frac{d(me)}{dt}\right]_{sys} \tag{2.14}$$

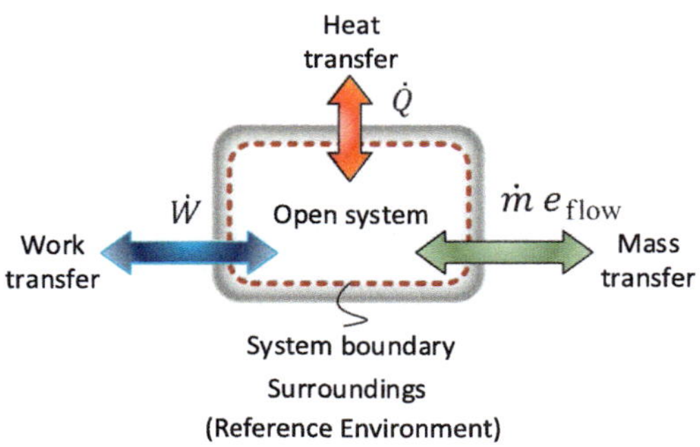

FIG. 2.2 A simple illustration of an open thermodynamic system showing three types of energy transfer mechanisms.

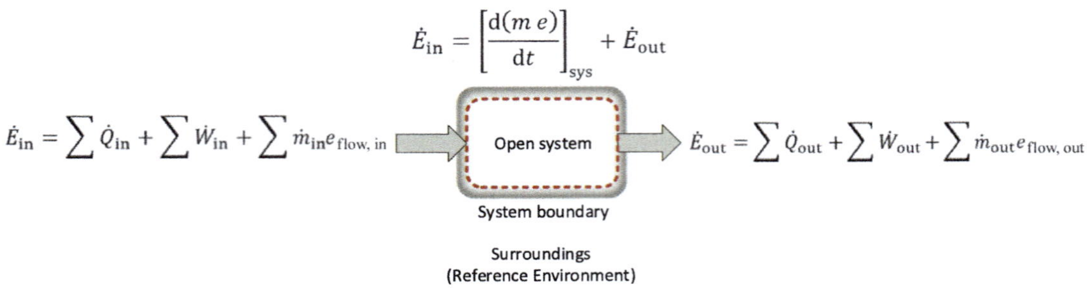

FIG. 2.3 The first law of thermodynamics written for an open system.

where $\frac{d(m\,e)}{dt}$ is the rate of change of total energy.

An illustration of an energy balance equation of an open system is shown in Fig. 2.3.

Since the mass flow rate, pressure, temperature, etc. do not change in time in steady flow operations, the energy balance equation for steady flow operations can be written in rate form as follows:

$$\dot{m}_1 e_1 + \sum \dot{Q}_{in} + \sum \dot{W}_{in} + \sum (\dot{m}\, e_{flow,in})_{in} = \dot{m}_2 e_2 + \sum \dot{Q}_{out} + \sum \dot{W}_{out} + \sum (\dot{m}\, e_{flow,out})_{out} \tag{2.15}$$

where e_{flow} is the total specific energy of the stream including the enthalpy, potential energy and kinetic energy. When there is no change in elevation and kinetic energy terms, they can be neglected in the total specific energy term. In this case, only enthalpy is considered as the flow energy.

In addition to steady flow systems, if the system is also steady state, meaning that there exists no net change over time in the open system, then the general energy balance equation for a steady-state steady-flow (SSSF) can be expressed as follows:

$$\dot{Q}_{in} + \dot{W}_{in} + \sum (\dot{m}\, e_{flow,in})_{in} = \dot{Q}_{out} + \dot{W}_{out} + \sum (\dot{m}\, e_{flow,out})_{out} \tag{2.16}$$

2.2.2.3 The second law of thermodynamics

The second law of thermodynamics (SLT) deals with the consideration of reversible and irreversible processes. A quasistatic process is a reversible process. The thermodynamic processes are considered to be reversible when both the thermodynamic system and its surroundings can be reverted to their initial states during a conversion process [1].

All microscopic changes over which the process changes must be reversible, such that the overall process is reversible. A process executed by a thermodynamic system is considered to be irreversible when it cannot be reverted to the initial state due to dissipation occurring (irreversibly) by friction, heat rejection, electrical, chemical, mechanical losses, etc.

In any actual process, dissipations and losses take place. Hence, reversibility is simply an idealization that is significant only in observing theoretical performances. There exist internal and external reversible processes based on the system boundary. They are briefly described as follows:

- Internally reversible process: In this process, there are no irreversibilities within the system boundary throughout the process.
- Externally reversible process: In this process, there are no related irreversibilities outside the system boundary throughout the process.
- Totally reversible process: In this process, there are no irreversibilities within the system boundary as well as surroundings.

The conventional expressions of the second law of thermodynamics explain that heat cannot be totally transformed into work even though the work can be totally transformed into heat. This is based on the Kelvin-Planck statement and Clausius statement, which are essentially the same [1, 3].

2.2.2.4 *The third law of thermodynamics*

The third law of thermodynamics is related to the entropy of a pure substance at absolute zero temperature, which is 0 K. It is based on the statement "the entropy of a pure crystalline substance at absolute zero temperature is zero." At this state, there is no uncertainty of state of the molecules. Determining the absolute entropy of a state is possible using this definition as an absolute reference point. It is important to note that the entropy of a non-pure crystalline at absolute zero temperature is not zero since there are more than one molecular configuration causing some uncertainty about the microscopic states. The entropy concept is discussed in detail in the next section.

2.2.3 Entropy

Entropy is associated with the degree of order and disorder, as well as reversibility and irreversibility. The degree of difference between irreversible and reversible processes is evaluated by the entropy concept in thermodynamics. If a system is isolated, the entropy of a system continuously upsurges because of irreversible processes and reaches the maximum probable value once the system reaches a state of thermodynamic equilibrium [4]. In the state of equilibrium, all irreversible processes finish. Once a system starts to transfer entropy with its surroundings, it moves off the equilibrium state and irreversible processes occur, causing entropy generation. Entropy transfer occurs either through heat or mass transfer. If there is no entropy accumulation within a system, then total entropy (system+surroundings) at the exit state is greater than the entropy at the inlet state due to the entropy generated by irreversible processes within the system. Entropy is an extensive property.

The terminology of disorder denotes the lack of information about exactly how much and what type of energy is related to each molecular or sub-molecular unit at any moment within a system. At the molecular and sub-molecular level, there is also ordered energy associated with the attractive and repulsive forces between units having fixed mean relative positions. Normally, a portion of this energy is available in work form at the macroscopic level under distinct conditions. Temperature is the property showing whether a system in equilibrium conditions will have a reduction or increase in its disordered energy when it is placed in another system that is in equilibrium. When the systems have different temperature levels, disordered energy will transfer from the higher temperature system to the lower temperature system.

Heat transfer into a system increases its disordered energy, while heat transfer from a system decreases its disordered energy. Reversible heat transfer is described using both the amount of energy transferred to or from the system and the temperature of heat transfer.

Reversible work is defined by the amount of energy transferred to or from the system independent of the temperature of the system.

Hence, the second law of thermodynamics expresses that the spontaneous or natural direction of energy transfer or conversion is toward increasing entropy. Low-entropy energy resources are anticipated as more useful. Energy forms of work and kinetic and gravitational potential energy are the more useful ones compared to heat. In fact, this implies that the second law of thermodynamics takes into account the quality of energy, unlike the first law of thermodynamics.

Although there are various formulations of the second law of thermodynamics, two are best-known, as follows [4]:

- Clausius statement: It is impossible for heat to move by itself from a low-temperature medium to a high-temperature medium. This means that heat transfer can only happen spontaneously in the direction of temperature drop.
- Kelvin-Planck statement: It is impossible for a system to accept a given amount of heat from a high-temperature medium and to deliver an equal amount of work output, meaning that a heat engine cannot have a thermal efficiency of 100%. On the other hand, a system converting work into an equivalent heat is possible.

The Clausius statement expresses that it is impossible to build a system operating in a cycle that transfers heat from a low-temperature medium to a high-temperature medium. The Clausius inequality delivers a mathematical expression of the second law of thermodynamics:

$$\oint \frac{dQ}{T} \leq 0 \tag{2.17}$$

where the circular integral specifies that the process is cyclic.

When the inequality becomes zero then the process is reversible, meaning that it is ideal. From this inequality, entropy generation can be defined. Hence, entropy generation is expressed as

$$S_{gen} = \oint \frac{dQ}{T} \tag{2.18}$$

Created entropy of a system throughout a process is the summation of entropy change of the thermodynamic system ΔS_{system} and the entropy change of the surroundings $\Delta S_{surroundings}$. Any actual process should yield positive total entropy generation. In some circumstances, the entropy change of a system may be negative but entropy change of the surroundings may be positive, yielding a positive total entropy generation. If entropy generation is zero, the process is called an ideal, reversible process. If entropy generation is positive, the process is a real, irreversible process. If entropy generation is negative, it is an impossible process, implying that such a process does not exist. Consequently, the entropy generation can be expressed as

$$S_{gen} = \Delta S_{system} + \Delta S_{surroundings} = \geq 0 \tag{2.19}$$

When there is a reversible process in which $S_{gen} = 0$, the entropy change of the system is the reverse of the entropy change of the surroundings:

$$\Delta S_{reversible} = -\Delta S_{surroundings} = \left(\frac{Q}{T}\right)_{reversible} \tag{2.20}$$

For an irreversible process the succeeding inequality is satisfied, as entropy generation is always positive.

$$\Delta S_{system} > \Delta S_{surroundings} = \left(\frac{Q}{T}\right)_{surroundings} \tag{2.21}$$

Though the change in entropy of the system and its surroundings may separately increase, decrease or remain constant, the total entropy change ($\Delta S_{system} + \Delta S_{surroundings}$) should be greater than zero for any process.

The entropy change of a process occurring between state point 1 and state point 2 can be defined as

$$S_{1-2} = \int_1^2 \frac{dQ}{T} \tag{2.22}$$

Hence by differentiation, one can obtain

$$dQ = T\,dS \tag{2.23}$$

The concept of random motion can be considered as the concept of order and disorder, and brings a relation between order and disorder and probability. Transfer of energy within a system can cause changes in the state points. Naturally, the direction of the change in the state points of a system is from a lower probability state to a higher probability state. As disordered states are more probable than ordered states, the natural direction of change of the state of a system is from order to disorder implying entropy generation. The entropy of a system at a specific state depends on its probability. Entropy measures the molecular random motion of the thermodynamic system and is correlated with the thermodynamic probability (p) of possible microscopic states, which is specified by the Boltzmann equation as follows:

$$S = k_B \ln p \tag{2.24}$$

where k_B is the Boltzmann constant.

The values of specific entropy are usually listed in thermodynamic tables similar to specific enthalpy and specific volume values. In some of the systems, the entropy change of a system may decrease, for instance charging a storage battery and freezing ice cubes [4]. The level of order in these systems increases and the level of disorder, entropy, decreases. However, the total degree of disorder of both the system and its surroundings always increases. Because, in order to charge a battery, there is a requirement of a certain amount of external energy having a certain quality (e.g., electricity) to achieve the chemical processes inside the battery. In the case of the battery, the input energy can be in the form of electricity. During charging the battery, a portion of electrical energy can be lost in the wires/connectors causing the production of high-entropy heat [4].

2.2.3.1 *Thermodynamic equilibrium*

There are several possibilities concerning the spontaneity and equilibrium state of a chemically reacting system, as listed in Table 2.3. A reaction may progress spontaneously in the direction of the consumption of the reactants and formation of products. A reaction may also progress toward non-spontaneous and thus does not proceed in the forward direction. The reaction can also be at equilibrium, where the backward and forward process have an identical rate and there will be no net change.

The criteria of spontaneous, nonspontaneous and equilibrium can be characterized depending on reaction enthalpy (ΔH) and entropy (ΔS). Using these parameters, the Gibbs free energy change of the reaction can be calculated as follows:

$$\Delta G = \Delta H - T\Delta S \tag{2.25}$$

Hence, the total entropy change of the reaction (system and surroundings) at constant temperature and pressure can be defined as follows:

$$\Delta S_{\text{total}} = -\Delta G/\text{T} \tag{2.26}$$

In general, both the enthalpy and entropy of reactants and products increase with the temperature. Therefore, the temperature change of reaction enthalpy and entropy is very slow. Thus, the temperature is the crucial factor affecting the reaction equilibrium for most of the reactions. If a system is at equilibrium, then the Gibbs free energy of the system is zero.

The equilibrium constant can be expressed using Gibbs free energy change of the reaction in standard conditions (ΔG^0) as follows:

$$K_{\text{equilibrium}} = \exp\left(-\frac{\Delta G^0}{RT}\right) \tag{2.27}$$

TABLE 2.3 Criteria for spontaneity and equilibrium of chemical reactions

Spontaneity	ΔG	ΔS_{tot}	ΔH	ΔS	T	ΔG	Spontaneity	ΔH	ΔS	T	ΔG	Spontaneity
Spontaneous (S)	<0	>0	<0	>0	any	<0	S	>0	<0	any	>0	NS
Nonspontaneous (NS)	>0	<0	<0	<0	low	<0	S	>0	<0	low	>0	NS
Equilibrium (E)	0	0	<0	<0	high	>0	NS	>0	>0	high	<0	S

Source: *Dincer I, Zamfirescu C. 1.5 Thermodynamic aspects of energy. In: Dincer I, editor. Compr. Energy Syst., 1st ed., Cambridge, United States: Elsevier; 2018, p. 153–211. doi: 10.1016/B978-0-12-809597-3.00105-X.*

The equilibrium constant is useful to define the mixture composition of the reaction at the equilibrium state [1]:

- When the equilibrium constant is $K_{\text{equilibrium}} > 10^3$, then the equilibrium mixture contains mostly reaction products.
- When the equilibrium constant is $K_{\text{equilibrium}} < 10^{-3}$, then the equilibrium mixture contains mostly reactants.
- When the equilibrium constant is $10^3 > K_{\text{equilibrium}} > 10^{-3}$, then the equilibrium mixture contains both reactants and products.

In relation to Le Châtelier's rule, if a reaction mixture at equilibrium is positioned upon a stress, the system reacts in the direction of relieving the stress.

For a reaction that diminishes the moles of gas:

- For an upsurge of pressure formed by reduction of volume, the direction of the reaction is toward reducing the moles of gas.
- An expansion of volume producing a reduction of pressure will shift the reaction toward an increase in the moles of gas.

A reaction can realize full transformation once all the reactants are spent and products formed in a stoichiometric amount. Most of the time, this is not the case for actual reactions due to a number of reasons. When a reaction has an equilibrium constant of approximately 1, there will not be a full conversion of the reactants into products since both reactants and products will be present in the reaction mixture. Furthermore, two or more reactions can take place concurrently, thus forming undesired products. The theoretical yield is the quantity of product that can be produced by a reaction based on the stoichiometry. The actual yield will be less than the theoretical yield. The percent yield is expressed as the ratio of actual yield to theoretical yield [1].

2.2.4 Exergy

The second law of thermodynamics presents an instrumental thermodynamic property that is called exergy. Exergy is defined as useful energy or available energy or work potential. The exergy of a system is described as the maximum work that can be done by the system at a specified reference environment. The reference environment is usually identified by reference temperature, pressure, and chemical composition. Hence, exergy is a thermodynamic property that also accounts for the reference environment.

Although most of the time the second law of thermodynamics is described using energy and entropy, Dincer [5, 6] defines thermodynamics as the science of energy and exergy due mainly to two reasons, namely: (i) energy and exergy have the same units, which allows comparison of quantities and qualities, and (ii) comparative assessment of energy and exergy is possible through energy and exergy efficiencies. However, there is no such concept as entropy efficiency. Hence, exergy and energy concepts need to be taken into account for comparison. The exergy efficiency is a true measure of system performance and quality and designates how much the actual system performance differs from the ideal system efficiency.

While a spontaneous process can progress only in a definite direction, the first law of thermodynamics does not provide evidence on the direction, as mentioned earlier. The first law is a measure of energy quantities involved and conserved regardless of the feasibility of the process. Therefore, there can be some processes satisfying the first law of thermodynamics but violating the second law of thermodynamics [4]. This signifies the importance of exergy analysis. For instance, the heat transfer from a low-temperature medium to a high-temperature medium without the required external energy input would not be possible.

Consequently, the second law of thermodynamics clarifies the difference in the quality of various forms of energy and enlightens why some processes can spontaneously occur while others cannot.

Some important features of exergy can be written as follows [4]:

- A system has zero exergy value when it is in complete equilibrium with its environment. In this complete equilibrium case, there is no difference in temperature, pressure, concentration, etc., hence there is no driving force for any process.
- The exergy of a system rises when it diverges from the environment. For example, a specified quantity of hot water has a higher exergy content during the winter than on a hot summer day. Similarly, cold water in a desert is more valuable than hot water. A chunk of ice carries little exergy in winter whereas it can have noteworthy exergy in summer.
- Exergy efficiency is a measure of ideality or reversibility.
- Exergy can commonly be deliberated as a valuable resource.
- Energy forms with high exergy content are usually more valuable and useful than energy forms having low exergy. For example, fossil fuels have high energy and exergy content whereas a waste heat close to ambient conditions has minor exergy content, even though the energy content might be high.

The exergy approach provides a distinctive understanding of the kinds, locations, and reasons of losses, thus serving to point out improvement potentials. Exergy analysis is an effective tool because of the following features [4]:

- It can help in achieving sustainable development by assessing the quality of the energy systems.
- It pinpoints the inefficiencies of the energy systems and helps design more efficient systems.
- It determines the locations, types, and true magnitudes of wastes and losses within an energy system.
- It combines the conservation of mass and conservation of energy principles.
- It addresses the impact of an energy system on the environment and other resource utilization which can help mitigate the impact.

2.2.4.1 *Reference environment*

The value of the exergy of a system or a flow depends on the state of both the system or flow and a reference environment. Reference environment is sometimes called ambient conditions, dead state, or reference conditions. Hence, a reference environment needs to be determined to perform exergy analysis. The environment is commonly modeled as a reference environment similar to the actual environment in which a system or flow exists. This potential of changing the reference environment to be consistent with the actual local environment is a significant advantage. A reference environment can vary depending on the location, season and several other conditions. For instance, an air conditioning system can function with varying performances at different operating conditions and in different climates [4]. By taking into account all the exergy streams of the system, it is possible to define the extent to which the system destroys exergy. The destroyed (or destructed) exergy is proportional to the generated entropy. Exergy is always destroyed in real processes, either partly or entirely based on the second law of thermodynamics. The destroyed exergy, or the generated entropy, is accountable for the less than ideal efficiencies of systems or processes. It is important to note that when there are chemical interactions within a process, reference chemical compositions need to be defined as well.

2.2.4.2 *Types of exergy*

There are several forms of exergy depending on the types of thermodynamic systems and equilibrium states. As previously written, there are two different type of energy systems: open and closed. Most of the open systems have mass, heat and work interactions and hence need to be accounted for in exergy balance equations. Though, in closed systems, there is no mass transfer crossing the boundary, so it is not considered in exergy balance equations. The exergy of a closed system can be termed non-flow exergy, whereas the exergy of an open system can be termed flow exergy.

2.2.4.2.1 Non-flow exergy

The exergy of a closed thermodynamic system, or the non-flow exergy, can be expressed as follows:

$$\mathrm{Ex}_{\mathrm{nonflow}} = \mathrm{Ex}_{\mathrm{physical,\,nonflow}} + \mathrm{Ex}_{\mathrm{kinetic}} + \mathrm{Ex}_{\mathrm{potential}} + \mathrm{Ex}_{\mathrm{chemical}} \tag{2.28}$$

where $\mathrm{Ex}_{\mathrm{potential}}$ is equal to potential energy, $\mathrm{Ex}_{\mathrm{kinetic}}$ is equal to kinetic energy, $\mathrm{Ex}_{\mathrm{chemical}}$ is equal to $\sum_i (\mu_{io} - \mu_{ioo})N_i$ and $\mathrm{Ex}_{\mathrm{physical,\,nonflow}}$ is equal to $(U - U_0) + P_0(V - V_0) - T_0(S - S_0)$ where the system has a temperature T, pressure P, chemical potential μ_i for species i, entropy S, energy E, volume V and number of moles N_i of species i. The system is within a reference environment in an equilibrium state with intensive properties T_o, P_o and μ_{ioo}. The quantity μ_{io} signifies the value of μ at the reference state.

The kinetic and potential energy, as well as exergy terms, can be defined as follows:

$$\mathrm{Ex}_{\mathrm{kinetic}} = \frac{1}{2} m \nu^2 \tag{2.29}$$

$$\mathrm{Ex}_{\mathrm{potential}} = m g (z - z_0) \tag{2.30}$$

where m is the system mass, and ν is its (macroscopic) velocity, z is the system elevation and z_0 is a reference elevation of the environment (e.g., ground level).

Physical non-flow exergy is the maximum work available from a system as it is brought to the environmental state in which it is at thermal and mechanical equilibrium with the environment, and chemical non-flow exergy is the maximum work available from a system as it is brought to the dead state in which it is in complete equilibrium with the environment.

If the system is not in chemical equilibrium with the reference environment, then the chemical exergy term should be considered. Chemical exergy characterizes the maximum work that can be removed during a process when the system composition changes to that of the environment [1]. There are two core components of chemical exergy: (i) exergy because of chemical reaction, and (ii) exergy because of concentration difference. If a chemical substance is allowed to interrelate with the environment, chemical reactions can occur, including unstable species. Ultimately, more stable species are made and further reaction is not possible. If a substance is not present in the atmosphere then the reference for zero chemical exergy is the most stable state of that substance in seawater. The standard chemical exergy values of common substances are tabulated in some of the related books, such as Kotas [7]. The standard chemical exergy of elements is beneficial for calculating the chemical exergy of chemical molecules if their Gibbs energy of formation is known. They are mostly tabulated in the units of kJ/mol. Hence, if it will be used on a mass basis, a conversion is required based on the molar mass of the element.

Therefore, the chemical exergy due to the difference in concentration of the gas component i having molar fraction y_i is written as follows:

$$\mathrm{Ex}_i^{\mathrm{ch}} = -\mathcal{R}\, T_0 \ln(y_i) \tag{2.31}$$

Rivero and Grafias [8] provided a common expression for the chemical exergy calculation of any chemical compound. For determining the chemical exergy of a material, it is essential to identify the standard specific Gibbs free energy of formation, Δg_f^0. Then, using Δg_f^0 and the standard exergy of the substances, the succeeding formula can be employed to define the chemical exergy of the substance:

$$\mathrm{ex}^{\mathrm{ch}} = \Delta g_\mathrm{f}^0 + \sum_{\mathrm{element}} \left(\varnothing \mathrm{ex}^{\mathrm{ch}}\right)_{\mathrm{element}} \tag{2.32}$$

where $\varnothing$ is the stoichiometric factor representing the number of moles of element per one mole of chemical compound.

A compromise can be made between the theoretical requirements of the reference environment and the actual performance of the natural environment. That is, a standard reference model can be adapted which is useful for exergy calculations. The reference environment is in stable equilibrium, acting as an infinite sink and source for heat and materials, experiencing only internally reversible processes with unaltered intensive states [1].

2.2.4.2.2 Flow exergy

The exergy value of a flowing stream of mass is the summation of non-flow exergy and the exergy associated with the flow work of the stream expressed as

$$\mathrm{Ex}_{\mathrm{flow}} = \mathrm{Ex}_{\mathrm{flow,physical}} + \mathrm{Ex}_{\mathrm{chemical}} + \mathrm{Ex}_{\mathrm{kinetic}} + \mathrm{Ex}_{\mathrm{potential}} \tag{2.33}$$

where physical, chemical, kinetic and potential parts are all accounted for.

The chemical and physical contents of exergy are expressed as

$$\mathrm{Ex}_{\mathrm{chemical}} = \sum_i (\mu_{\mathrm{io}} - \mu_{\mathrm{ioo}}) N_i \tag{2.34}$$

$$\mathrm{Ex}_{\mathrm{flow,physical}} = (H - H_0) - T_0(S - S_0) \tag{2.35}$$

When the kinetic and potential energy changes are neglected within an open system, the total exergy content becomes the summation of chemical and physical exergy terms:

$$\mathrm{Ex}_{\mathrm{flow}} = \mathrm{Ex}_{\mathrm{chemical}} + \mathrm{Ex}_{\mathrm{flow,physical}} \tag{2.36}$$

2.2.4.2.3 Thermal exergy

The flow of exergy associated with the heat transfer is called the thermal exergy and is denoted by Ex^{Q} as per the following equation, which depends on the temperature of heat transfer and reference state:

$$\mathrm{Ex}^{Q} = \int_{i}^{f} \left(1 - \frac{T_o}{T_s}\right) \Delta Q \tag{2.37}$$

where ΔQ is an incremental heat transfer, and the integral is from the initial state (i) to the final state (f). The thermal exergy is the minimum work necessary by the combined system of the control mass and the environment in taking the control mass from the dead state to the final state. When the temperature of the control mass is constant, the thermal exergy transfer related to a heat transfer is expressed as follows:

$$\mathrm{Ex}^{Q} = \left(1 - \frac{T_o}{T_s}\right) Q \tag{2.38}$$

Here, selection of the T_s value is important and depends on the direction of the heat transfer. Normally, the temperature T_s denotes the temperature of the location where heat transfer occurs, mostly the boundary temperature. In case the boundary temperature is not known, if there is heat penetration, the temperature T_s value can be taken as the source temperature from where heat is taken. If there is heat rejection, the temperature T_s value can be taken as sink temperature to where heat is being rejected. It is also possible to take the average temperature of the inlet and outlet streams for some cases.

2.2.4.2.4 Exergy of work

The exergy term associated with work (e.g., electrical, mechanical) can be denoted by Ex^{W} or simply by W since the energy content and exergy content of work is equal. The exergy transfer associated with work done by a system because of volume change is the net usable work due to the volume change, and can be signified by W_{net}.

2.2.4.2.5 Exergy destruction

The difference between the total exergy flows into and out of the system, minus the exergy accumulation in the system, yields the exergy destruction (Ex_d) for a process taking place in a system. The exergy destruction is expressed using reference temperature (T_0) and entropy generation (S_{gen}) as follows:

$$\mathrm{Ex}_{\mathrm{d}} = T_0 S_{\mathrm{gen}} \tag{2.39}$$

which points out that exergy destruction is proportional to entropy generation, and is known as the Gouy-Stodola relation.

2.3 Energy and exergy analyses

In this section, a thermodynamics analysis approach is presented including the first and second law analysis using four fundamental balance equations, namely mass balance equation, energy balance equation, entropy balance equation and exergy balance equation.

There is always a strong desire to make better devices, processes, and energy systems to achieve better sustainability. This can be achieved by implementing sustainable energy solutions that cover the following six key pillars [5]: (i) better design and analysis, (ii) better cost-effectiveness, (ii) better resources use, (iv) better efficiency, (v) better energy security, and (vi) better environment. Exergy analysis can play an important role in sustainable energy systems since exergy essentially considers environmental conditions in which a system works. As previously mentioned, for performing exergy analysis, a reference environment should be defined and it is presumed to be a thermodynamic system in the dead state. In addition, a reference environment is considered to be huge reservoir for which parameters do not vary. Exergy analysis enables calculation of the maximum reversible work, which is the work produced or expended within the system to attain the dead state at the conditions of the reference environment. By comparing the reversible work with actual work, it is possible to calculate the irreversibilities within the system or, in other words, the exergy destructions occurring within the system. Once an exergy analysis is conducted for an energy system such as a power plant, a refrigeration facility or a chemical processing plant, the thermodynamic losses can be quantified in terms of exergy destructions, which signify losses in the quality of energy or usefulness of energy. Subsequently, exergy analysis can help improve and optimize designs.

Herein, we provide the fundamental balance equations and introduce how to write balance equations in a conceptually correct manner.

2.3.1 Mass balance equation (MBE)

Based on the conservation of mass principle, the net mass transported to the system is identical to the net change in mass inside the system plus the net mass exiting an open thermodynamic system. The mass addition or removal to or from an open system brings energy, entropy and exergy transfer that is proportional to the flow rate of mass. The mass flow rate is determined as the quantity of mass passing through a cross-section of a flow stream per unit of time.

A sketch of a sample open system showing one inlet and two outlets is presented in Fig. 2.4. The total mass flow rate of all input streams is denoted as $\sum \dot{m}_{\text{in}}$ while the total mass flow rate of all output streams is denoted as $\sum \dot{m}_{\text{out}}$. In this case, the mass of the open system will vary with differential amount dm_{cv} for the non-steady state case. The mass balance equation (MBE) for a general non-steady state open system can be written as follows:

$$\sum \dot{m}_{\text{in}} = \sum \dot{m}_{\text{out}} + \frac{dm_{\text{cv}}}{dt} \tag{2.40}$$

For a steady flow system, there is no mass differential, hence it can be written as follows:

$$\sum \dot{m}_{\text{in}} = \sum \dot{m}_{\text{out}} \tag{2.41}$$

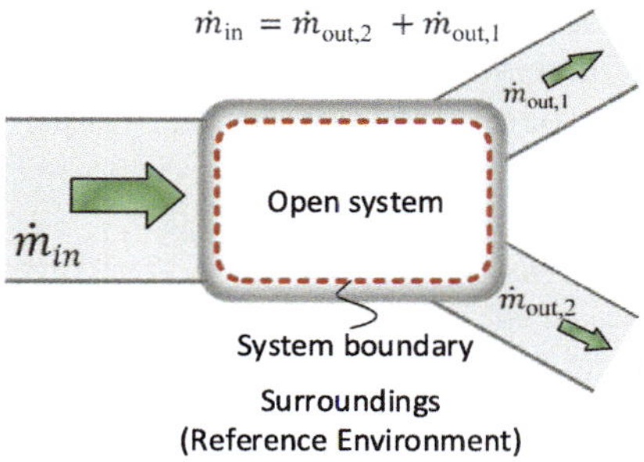

FIG. 2.4 A simple illustrative sketch for mass balance equation of a steady-state open system.

2.3.2 Energy balance equation (EBE)

The change of energy between states 1 and 2 for a general system can be written as

$$\Delta E_{system} = m\Delta e_{system} = m\left[\left(u_2 + \frac{1}{2}\nu_2^2 + gz_2\right) - \left(u_1 + \frac{1}{2}\nu_1^2 + gz_1\right)\right] \tag{2.42}$$

For a closed system, the energy balance equation is written using the total specific energy of a non-flowing thermodynamic system as follows:

$$\sum \dot{q}_{in} + \sum \dot{w}_{in} = \sum \dot{q}_{out} + \sum \dot{w}_{out} + \frac{d\dot{e}}{dt} \tag{2.43}$$

In amount formulation, the energy balance equation for a closed system that changes its state from 1 to 2 is:

$$e_1 + \sum q_{in} + \sum w_{in} = \sum q_{out} + \sum w_{out} + e_2 \tag{2.44}$$

The energy balance equation for open systems should also take into account the presence of flow work and boundary work and the rate of change of total energy $\left[\frac{d(m\,e)}{dt}\right]$ which can be written for non-steady state systems as follows:

$$\sum \dot{m}e_{flow,in} + \sum \dot{Q}_{in} + \sum \dot{W}_{in} = \sum \dot{m}e_{flow,out} + \sum \dot{Q}_{out} + \sum \dot{W}_{out} + \left[\frac{d(m\,e)}{dt}\right]_{system} \tag{2.45}$$

where e_{flow} is the total energy of flowing mass representing the summation of internal energy, flow energy, kinetic energy and potential energy as follows:

$$e_{flow} = u + P\,\nu + \frac{1}{2}\nu^2 + g\,z = h + \frac{1}{2}\nu^2 + g\,z \tag{2.46}$$

where $u + P\,\nu = h$.

In a steady-flow system, flows of mass and energy that cross the system boundary are steady. In addition, mass flow rate, pressure, temperature do not vary in time, therefore the combination of the following equations $d(m\,h) = (\dot{m}\,h)dt$, $dQ = \dot{Q}\,dt$, and $dW = \dot{W}\,dt$ between initial state 1 and a final state 2 of the open system is obvious. In a steady-flow regime, the energy balance equation (EBE) can be written in rate form:

$$\dot{m}_1 e_1 + \dot{Q}_{in} + \dot{W}_{in} + \sum (\dot{m}\,e_{flow,in})_{in} = \dot{m}_2 e_2 + \dot{Q}_{out} + \dot{W}_{out} + \sum (\dot{m}\,e_{flow,out})_{out} \tag{2.47}$$

In addition to steady-flow operation, if the system has a steady state, meaning that there is no net energy change associated with the open system, the system is called steady-state steady-flow (SSSF). In this case, the energy balance equation (neglecting the kinetic and potential energy changes) is written as follows:

$$\dot{Q}_{in} + \dot{W}_{in} + \sum(\dot{m}\,h)_{in} = \dot{Q}_{out} + \dot{W}_{out} + \sum(\dot{m}\,h)_{out} \tag{2.48}$$

Most energy systems commonly operate at steady-state or quasi steady-state except for periods of startup and shutdown. Since non-steady-state periods are minor compared to the time of steady operation, the most relevant balance equations for energy systems are written for steady-state steady-flow (SSSF) conditions.

2.3.3 Entropy balance equation (EnBE)

The second law of thermodynamics can also be articulated using the entropy balance equation (EnBE). For a thermodynamic system, entropy input in addition to generated entropy is equivalent to entropy output in addition to the change of entropy within the system. The entropy balance equation hypothesizes that the entropy change of a thermodynamic system is equivalent to entropy generation within the system in addition to the net entropy transmitted to the system through the boundary. Entropy can be transferred outside of the system boundary in mass and heat forms, but it cannot be transferred in work form. Entropy transfer of a system is illustrated in Fig. 2.5. Thence, the entropy balance equation is expressed as

$$\sum \dot{S}_{in} + \dot{S}_{gen} = \sum \dot{S}_{out} + \frac{dS_{system}}{dt} \tag{2.49}$$

The entropy transferred through the system boundary or during a process 1–2 is written as

$$S_{1-2} = \int_1^2 \frac{dQ}{T} \tag{2.50}$$

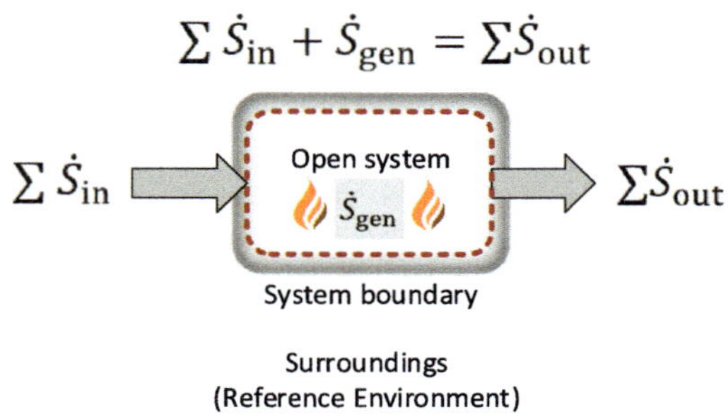

FIG. 2.5 A simple illustrative sketch for entropy balance equation for the steady-state steady-flow open system on rate basis.

The entropy balance equation of a closed system having heat transfer through the boundary can be written as

$$\sum\left(\int\frac{\mathrm{d}\dot{Q}}{T}\right)_{\mathrm{in}}+\dot{S}_{\mathrm{gen}}=\frac{\mathrm{d}S_{\mathrm{system}}}{\mathrm{d}t}+\sum\left(\int\frac{\mathrm{d}\dot{Q}}{T}\right)_{\mathrm{out}} \quad (2.51)$$

For a closed system, there is no mass transfer through the system boundary. Thus, there is only one way of entropy transfer, which is heat. If the closed system is adiabatic as well, then there is no entropy transfer because of mass or because of heat transfer. Therefore, $\dot{S}_{\mathrm{system}}=\dot{S}_{\mathrm{gen}}$.

The entropy balance equation in rate form for an open system can be written as follows:

$$\sum\left(\int\frac{\mathrm{d}\dot{Q}}{T}\right)_{\mathrm{in}}+\sum(\dot{m}s)_{\mathrm{in}}+\dot{S}_{\mathrm{gen}}=\frac{\mathrm{d}S_{\mathrm{CV}}}{\mathrm{d}t}+\sum\left(\int\frac{\mathrm{d}\dot{Q}}{T}\right)_{\mathrm{out}}+\sum(\dot{m}s)_{\mathrm{out}} \quad (2.52)$$

The entropy balance equation for a steady flow through a control volume must account for the fact that there are no temporal variations of parameters; thence the mass enclosed in the control volume and specific entropy of the control volume remains constant in time as

$$\sum\left(\int\frac{\mathrm{d}\dot{Q}}{T}\right)_{\mathrm{in}}+\sum(\dot{m}s)_{\mathrm{in}}+\dot{S}_{\mathrm{gen}}=\sum\left(\int\frac{\mathrm{d}\dot{Q}}{T}\right)_{\mathrm{out}}+\sum(\dot{m}s)_{\mathrm{out}} \quad (2.53)$$

Applying the mass balance equation, $\dot{m}_{\mathrm{out}}=\dot{m}_{\mathrm{in}}=\dot{m}$, the entropy balance equation simplifies to

$$\sum\left(\int\frac{\mathrm{d}\dot{Q}}{T}\right)_{\mathrm{in}}+m(s_{\mathrm{in}}-s_{\mathrm{out}})+\dot{S}_{\mathrm{gen}}=\sum\left(\int\frac{\mathrm{d}\dot{Q}}{T}\right)_{\mathrm{out}} \quad (2.54)$$

The entropy balance equation for a steady-state steady-flow open system is written as shown in Fig. 2.5.

As mentioned previously, there is no heat transfer crossing the boundary of the system if the process is adiabatic, so the entropy balance equation simplifies to

$$\dot{m}\,s_{\mathrm{in}}+\dot{S}_{\mathrm{gen}}=\dot{m}\,s_{\mathrm{out}} \quad (2.55)$$

The total entropy generation is the summation of entropy change of the system and of its surroundings.

2.3.4 Exergy balance equation (ExBE)

The exergy balance equation presents the term exergy destruction representing the maximum work potential that cannot be recovered for the useful operation because of irreversibilities. For a reversible system, there is no exergy destruction, as all work generated by the system can be made useful. The exergy destruction and entropy generation are related as per the following equation:

$$\text{Ex}_\text{d} = T_0\, S_\text{gen} \tag{2.56}$$

where T_0 is the reference temperature. Similar to the entropy case, when $\text{Ex}_\text{d} > 0$ then the process is irreversible, when $\text{Ex}_\text{d} = 0$ then the process is reversible and when $\text{Ex}_\text{d} < 0$ the process is impossible.

The total exergy input to a thermodynamic system should be balanced by the total exergy output from the system in addition to the change of exergy content of the system plus the exergy destruction. Exergy can be transferred to or from a system by three mechanisms: work, heat and mass. As opposed to entropy, exergy can also be transferred through work crossing the system boundary.

Based on the sample sketch shown in Fig. 2.6, the exergy balance equation can be expressed generally in rate form as

$$\sum\left[\dot{W} + \dot{m}\,\text{ex} + \left(1 - \frac{T_0}{T}\right)\dot{Q}\right]_\text{in} = \frac{\text{dEx}}{dt} + \sum\left[\dot{W} + \dot{m}\,\text{ex} + \left(1 - \frac{T_0}{T}\right)\dot{Q}\right]_\text{out} - P_0\frac{dV_\text{CV}}{dt} + \dot{\text{E}}\text{x}_\text{d} \tag{2.57}$$

where the total specific exergy is expressed as

$$\text{ex} = (h - h_0) + T_0(s - s_0) + \frac{1}{2}v^2 + g(z - z_0) + \text{ex}^\text{ch} \tag{2.58}$$

When there is no change in chemical conditions, kinetic energy and potential energy, these terms can be neglected in the balance equation.

As mentioned earlier, the exergy of work transfer (Ex^W) is identical to the energy in work $\text{Ex}^\text{W} = W$. Nevertheless, when the system imposes against a moving boundary, the exergy amount should be reduced associated with boundary work, so $\text{Ex}^\text{W} = W - P_0(V - V_0)$. The exergy amount due to mass transfer (Ex^m) is written $\text{Ex}^\text{m} = m\,\text{ex}$. The exergy associated with transfer of heat is written using the Carnot limit as per the following equation:

$$\text{Ex}^Q = \int_{\substack{\text{system}\\ \text{boundary}}} \left(1 - \frac{T_0}{T}\right) dQ \tag{2.59}$$

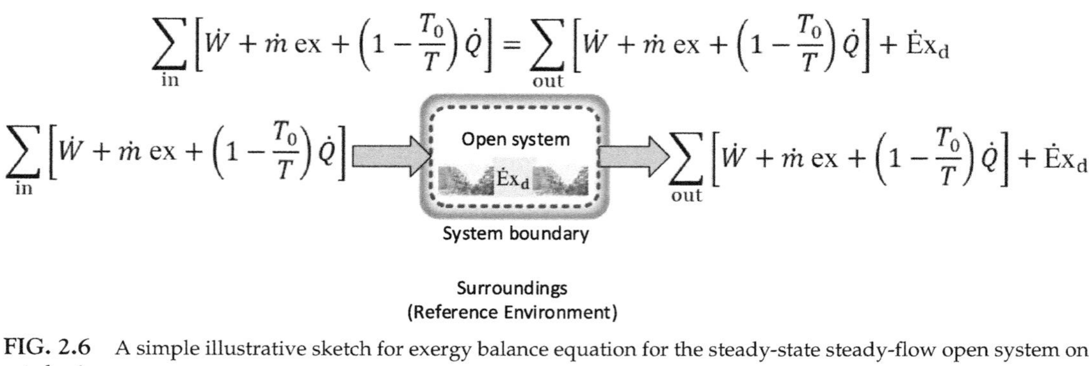

FIG. 2.6 A simple illustrative sketch for exergy balance equation for the steady-state steady-flow open system on rate basis.

Thus, the exergy balance equation for a thermodynamic system on a rate basis at steady-state steady-flow (SSSF) condition can be written as

$$\sum\left[\dot{W}+\dot{m}\,\mathrm{ex}+\left(1-\frac{T_0}{T}\right)\dot{Q}\right]_{\mathrm{in}}=\sum\left[\dot{W}+\dot{m}\,\mathrm{ex}+\left(1-\frac{T_0}{T}\right)\dot{Q}\right]_{\mathrm{out}}+\dot{\mathrm{E}}\mathrm{x}_{\mathrm{d}} \tag{2.60}$$

In order to find the reversible work of a system, the exergy destruction is set to zero. In addition, $\dot{W}=\sum\dot{W}_{\mathrm{out}}-\sum\dot{W}_{\mathrm{in}}$ and $\dot{Q}=\sum\dot{Q}_{\mathrm{in}}-\sum\dot{Q}_{\mathrm{out}}$. Therefore, the reversible work can be expressed as

$$\dot{W}_{\mathrm{rev}}=\dot{m}(\mathrm{ex}_1-\mathrm{ex}_2)+\sum\left(1-\frac{T_0}{T}\right)\dot{Q} \tag{2.61}$$

2.3.5 Efficiency definition

Efficiency assessment is critical for energy systems and it is broadly utilized for any systems and processes. It is a measure of performance and effectiveness of a system or component. The main approach to define efficiency is the ratio of useful output per required input. If the efficiency definition is based on the first law of thermodynamics, it is termed energy efficiency. If the efficiency definition is based on the second law of thermodynamics, then it is termed exergy efficiency.

Therefore, the energy efficiency of a component or system can be formulated as

$$\eta_{\mathrm{en}}=\frac{\sum\dot{E}_{\mathrm{useful}}}{\sum\dot{E}_{\mathrm{input}}} \tag{2.62}$$

If one wants to write the energy efficiency definition using the total energy losses, the equation can be written as follows:

$$\eta_{\mathrm{en}}=1-\frac{\sum\dot{E}_{\mathrm{loss}}}{\sum\dot{E}_{\mathrm{input}}} \tag{2.63}$$

The exergy efficiency is expressed as the proportion between total exergy amount in the useful output and the total exergy amount of the required input, so the general expression is written as follows:

$$\eta_{\mathrm{ex}}=\frac{\sum\dot{\mathrm{E}}\mathrm{x}_{\mathrm{useful}}}{\sum\dot{\mathrm{E}}\mathrm{x}_{\mathrm{input}}} \tag{2.64}$$

If one wants to write the exergy efficiency definition using the total exergy consumed, the equation can be written as follows:

$$\eta_{\mathrm{ex}}=1-\frac{\sum\dot{\mathrm{E}}\mathrm{x}_{\mathrm{consumption}}}{\sum\dot{\mathrm{E}}\mathrm{x}_{\mathrm{input}}} \tag{2.65}$$

where $\sum\dot{\mathrm{E}}\mathrm{x}_{\mathrm{consumption}}=\sum\dot{\mathrm{E}}\mathrm{x}_{\mathrm{destruction}}+\sum\dot{\mathrm{E}}\mathrm{x}_{\mathrm{loss}}$.

2.4 Common steady-flow devices

In this section, we briefly present the fundamental information about common steady flow devices and provide the approach of defining efficiencies utilized in thermodynamics analysis of integrated systems. Determining the energy and exergy efficiencies of components within an energy system is very significant in understanding the performances. Hence, in this section, we provide energy and exergy efficiencies of fundamental steady-flow components such as compressors, pumps, turbines, heat exchangers, nozzles, and diffusers.

2.4.1 Turbine

A turbine is a work producing device. Production of power is achieved by expanding a fluid. Mainly, there are two different types of turbines, steam and gas turbines. An ideal turbine has an isentropic process. Hence, the performance of a turbine having an isentropic process is expressed by isentropic efficiency. A high enthalpy flow enters the turbine; work is produced, and a lower enthalpy flow exits the turbine. The turbine efficiency quantifies various losses such as the isentropic losses, the heat losses from the turbine, and the friction losses. Exergy efficiency of a turbine is defined as the ratio of produced work and rate of exergy consumed.

A turbine is shown in Fig. 2.7 with input state 1 at temperature T_1 and pressure P_1 and an output state 2 at temperature T_2 and pressure P_2.

The mechanical work output from the turbine is the maximum when the fluid is expanded isentropically and adiabatically between the specified initial state and given exit pressure. The isentropic efficiency is defined as the ratio of actual power to the isentropic power or ideal power:

$$\eta_{\text{isentropic}} = \frac{\dot{w}_{\text{actual}}}{\dot{w}_{\text{isentropic}}} \cong \frac{h_1 - h_2}{h_1 - h_{2s}} \tag{2.66}$$

where h_{2s} is the enthalpy of the fluid at the turbine outlet if the process was isentropic. This enthalpy can be obtained by setting $s_1 = s_2$ and using outlet pressure. In this case, kinetic and potential energy changes are neglected.

The energy efficiency of an adiabatic turbine can be defined based on useful output and required input approach. For a turbine, the useful energy output is the mechanical (shaft)

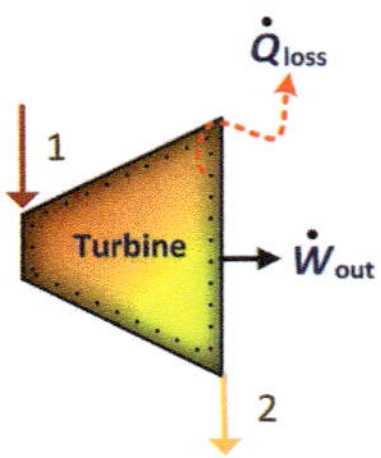

FIG. 2.7 Simple schematic diagram of a turbine with heat loss showing the component boundary.

work produced whereas the required input is the change in energy of the stream (gas or steam) at the inlet and outlet, that is the decrease in the energy of the stream (gas or steam) as it passes through the turbine. Hence, the energy efficiency of a turbine is written as follows:

$$\eta_{\text{en,Turbine}} = \frac{\dot{W}_{\text{out}}}{\dot{\text{E}}\text{n}_1 - \dot{\text{E}}\text{n}_2} = \frac{\dot{W}_{\text{out}}}{\dot{m}(h_1 - h_2)} \tag{2.67}$$

Similarly, the useful exergy output is the mechanical (shaft) work produced whereas the required input is the exergy change of stream (gas or steam) at the inlet and outlet, that is the decrease in the exergy of the stream (gas or steam) as it flows through the turbine. The exergy efficiency of a turbine can be expressed as

$$\eta_{\text{ex,Turbine}} = \frac{\dot{\text{E}}\text{x}_{\text{useful}}}{\dot{\text{E}}\text{x}_{\text{input}}} = \frac{\dot{W}_{\text{out}}}{\dot{\text{E}}\text{x}_1 - \dot{\text{E}}\text{x}_2} = \frac{\dot{W}_{\text{out}}}{\dot{W}_{\text{reversible}}} = \frac{\dot{m}(h_1 - h_2)}{\dot{m}(h_1 - h_2 - T_o(s_1 - s_2))} \tag{2.68}$$

It is important to note that isentropic efficiency and exergy efficiency are different definitions. Isentropic efficiency represents an ideal isentropic process between actual input state and an assumed hypothetical output state, whereas exergy efficiency represents an ideal reversible process between actual input state and an actual output state.

2.4.2 Compressor

A compressor is a component used for increasing the pressure of a fluid in gas form. It is a work consuming device, hence requires work input for the compression process. Similar to turbines, the performance of a compressor is typically denoted by isentropic efficiency.

A simple diagram of a compressor having input state 1 and output state 2 is shown in Fig. 2.8. The work input to the compressor would be lowest when the gas is compressed adiabatically and isentropically.

The isentropic efficiency can be defined as the ratio of the isentropic or ideal work to the actual work as written here:

$$\eta_{\text{isentropic,compressor}} \simeq \frac{\dot{w}_{\text{isentropic}}}{\dot{w}_{\text{actual}}} = \frac{(h_{2s} - h_1)}{(h_2 - h_1)} \tag{2.69}$$

where h_{2s} is the enthalpy of the fluid at the compressor exit when the process is isentropic. This enthalpy h_{2s} can be attained by equaling $s_2 = s_1$ and from output pressure. Kinetic

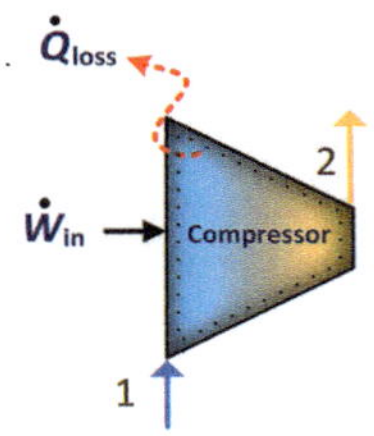

FIG. 2.8 A simple schematic diagram of a compressor with heat loss showing the component boundary.

and potential energy changes are neglected. If the gas is assumed as an ideal gas with constant specific heats, the isentropic power input can be calculated using

$$\dot{w}_{\text{isentropic}} = \frac{kR(T_2 - T_1)}{k-1} = \frac{kRT_1}{k-1}\left[\left(\frac{P_2}{P_1}\right)^{\frac{k-1}{k}} - 1\right] \tag{2.70}$$

where k is the specific heat ratio ($k = c_p/c_v$) for which the value is 1.4 for air at room temperature. Note that this equation is written independent of mass implying the unit of work is kJ/kg.

The gas is occasionally cooled as it is being compressed in a non-adiabatic compressor to decrease power input, since the power input is proportional to specific volume of the gas and cooling the gas reduces its specific volume. In this case, the isothermal efficiency is written as follows:

$$\eta_{\text{en, isothermal, compressor}} = \frac{\dot{w}_{\text{isothermal}}}{\dot{w}_{\text{actual}}} \tag{2.71}$$

The work input for the reversible and isothermal situation is given for an ideal gas with constant specific heats as

$$\dot{w}_{\text{isothermal}} = RT_1 ln\left(\frac{P_2}{P_1}\right) \tag{2.72}$$

where R is gas constant, T_1 is the input temperature of the gas, and P_1 and P_2 are the pressures at the input and output of the compressor, respectively. Sometimes, a reversible, polytropic process can also be employed as the ideal process for the actual compression. In this case, a polytropic efficiency is written as follows:

$$\eta_{\text{en, polytropic, compressor}} = \frac{\dot{w}_{\text{polytropic}}}{\dot{w}_{\text{actual}}} \tag{2.73}$$

Here the work input for the reversible, polytropic process can be written for an ideal gas with constant specific heats as follows:

$$\dot{w}_{\text{polytropic}} = \frac{nR(T_2 - T_1)}{n-1} = \frac{nRT_1}{n-1}\left(\left(\frac{P_2}{P_1}\right)^{\frac{n-1}{n}} - 1\right) \tag{2.74}$$

where n is the polytropic exponent and the value of n varies between 1 and specific heat ratio k.

Compressors are usually evaluated by the isentropic efficiency, which is expressed as the ratio of isentropic work and actual work. The useful output is the change in energy of the outlet and inlet stream. The required input is the work input to drive the shaft.

Similar to turbines, the energy and exergy efficiencies of a compressor can be calculated using useful output divided by input approach. In this case, the required work input is shaft work provided to the compressor, and useful output is the energy/exergy delivered to the working fluid of the compressor. Thus, the useful output is the rise in the energy/exergy values of fluid as it flows across the compressor. Hence the energy efficiency can be written as

$$\eta_{\text{en, compressor}} = \frac{\dot{\text{En}}_2 - \dot{\text{En}}_1}{\dot{W}_{\text{in}}} = \frac{\dot{m}(h_2 - h_1)}{\dot{W}_{\text{in}}} \tag{2.75}$$

Similarly, the exergy efficiency can be written as

$$\eta_{ex,compressor} = \frac{\dot{E}x_{useful}}{\dot{E}x_{input}} = \frac{\dot{W}_{reversible}}{\dot{W}_{in}} = \frac{\dot{E}x_2 - \dot{E}x_1}{\dot{W}_{in}} = \frac{\dot{m}(ex_2 - ex_1)}{\dot{W}_{in}} \tag{2.76}$$

2.4.3 Pump

A pump is commonly utilized to increase the pressure of a liquid. The main difference between pump and compressor is the form of the fluid flowing through. Pumps are designed for liquids and they are work consuming devices, hence a power input is required for pumping. The liquid can be considered as an incompressible fluid. Pumps also bear isentropic efficiencies. The work input for the isentropic situation can be calculated using specific volume and pressure at the inlet and outlet of the pump. A schematic diagram of a pump is shown in Fig. 2.9.

The liquid is incompressible and therefore the power required for pumping the liquid for a reversible process is $\dot{W}_{isentropic} = \dot{m}\,v(P_2 - P_1)$. Hydraulic turbines are devices that produce work from potential energy of a liquid.

When the changes in potential and kinetic energies of a liquid are negligible, the isentropic efficiency of a pump is defined as

$$\eta_{isentropic,pump} = \frac{\dot{w}_{isentropic}}{\dot{w}_{actual}} \cong \frac{v\,(P_2 - P_1)}{(h_2 - h_1)} \cong \frac{h_{2s} - h_1}{h_2 - h_1} \tag{2.77}$$

where v is the specific volume of the liquid, typically taken at the pump inlet, and h_{2s} is the enthalpy of the fluid at the pump outlet in case the process was isentropic.

When the actual enthalpy value at the pump outlet h_2 is required, it can be calculated using the isentropic efficiency of the pump (assumed or given) or from the measurement of actual work input.

The energy efficiency of a pump can be calculated using the work input and energy gained (enthalpy change) by the liquid due to the pressure increase:

$$\eta_{en,pump} = \frac{\dot{E}n_2 - \dot{E}n_1}{\dot{W}_{in}} = \frac{\dot{m}(h_2 - h_1)}{\dot{W}_{in}} \tag{2.78}$$

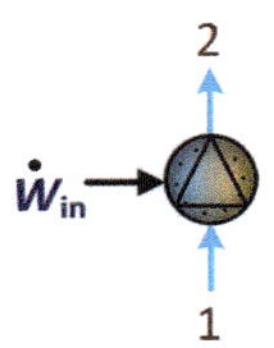

FIG. 2.9 A simple schematic diagram of a pump showing the component boundary.

The exergy efficiency of a pump is calculated as the reversible work divided by the actual work:

$$\eta_{ex,pump} = \frac{\dot{E}x_{useful}}{\dot{E}x_{input}} = \frac{\dot{E}x_2 - \dot{E}x_1}{\dot{W}_{in}} = \frac{\dot{m}\,(ex_2 - ex_1)}{\dot{W}_{in}} = \frac{\dot{W}_{reversible}}{\dot{W}_{in}} \tag{2.79}$$

2.4.4 Nozzle

A nozzle is used to accelerate a fluid as depicted in Fig. 2.10. It is typically an adiabatic component due to negligible heat transfer. Hence, similar to pumps, compressors and turbines, nozzles have isentropic efficiencies.

The isentropic efficiency of a nozzle can be defined as the ratio of the actual kinetic energy of the fluid at the nozzle outlet to the kinetic energy value at the outlet of an isentropic process for the same input state and output pressure as follows [9, 10]:

$$\eta_{isentropic,Nozzle} = \frac{KE_{exit,actual}}{KE_{exit,isentropic}} = \frac{V_2^2}{V_{2s}^2} \tag{2.80}$$

where V_{2s} is the velocity of the fluid at the nozzle outlet when the process is isentropic.

If the input velocity is insignificant, the isentropic efficiency of the nozzle can be written using enthalpies:

$$\eta_{isentropic,Nozzle} = \frac{h_1 - h_2}{h_1 - h_{2s}} \tag{2.81}$$

where h_{2s} is the enthalpy of the fluid at the nozzle outlet in case the process was isentropic.

A nozzle is a device for converting the enthalpy of a fluid into kinetic energy. In a nozzle, the energy/exergy recovered is the increase in the kinetic energy of the fluid. The energy input is the decrease in energy content of the fluid. Therefore, the energy efficiency of a nozzle can also be written as follows [9, 10]:

$$\eta_{en,Nozzle} = \frac{\frac{V_2^2}{2}}{h_1 - h_2} \tag{2.82}$$

In this expression, the enthalpy term at the exit can also be added in the numerator to represent the total energy at the outlet.

Similarly, exergy consumed is the exergy decrease of the fluid stream. Hence, the exergy efficiency of a nozzle can be written as follows [9, 10]:

$$\eta_{ex,Nozzle} = \frac{\dot{E}x_{useful}}{\dot{E}x_{input}} = \frac{\frac{V_2^2}{2} - \frac{V_1^2}{2}}{h_1 - h_2 - T_o(s_1 - s_2) + \frac{V_1^2}{2}} \cong \frac{\frac{V_2^2}{2}}{h_1 - h_2 - T_o(s_1 - s_2)} \tag{2.83}$$

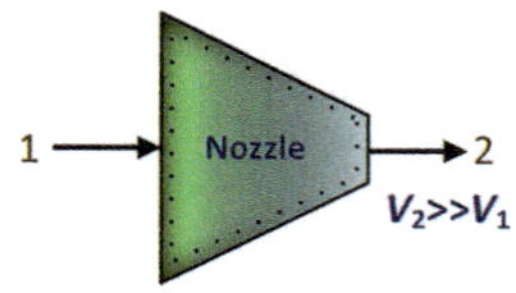

FIG. 2.10 A simple schematic diagram of a nozzle showing the component boundary.

Here, the inlet velocity is commonly very small compared to outlet velocity hence it can be neglected.

It is important to express that expended exergy corresponds to the exit kinetic energy in a reversible process. If the nozzle is reversible and adiabatic, the exergy efficiency becomes maximum. For adiabatic nozzles, isentropic and exergy efficiencies are the same. The unit conversion should be carefully performed since the kinetic energy commonly has different unit than kJ/kg.

2.4.5 Diffusers

Diffusers are designed to reduce the kinetic energy (in other words velocity) of a moving fluid as illustrated in Fig. 2.11. They are categorized under steady flow devices which raise the pressure of fluids. Commonly, diffusers do not have any heat interactions similar to nozzles since they are designed to be adiabatic.

The isentropic efficiency of a diffuser is written as the ratio of the final kinetic energy of fluid leaving the diffuser when the process is isentropic to the actual kinetic energy of the fluid leaving the diffuser [9, 10]:

$$\eta_{\text{isentropic, Diffuser}} = \frac{\text{KE}_{\text{exit, isentropic}}}{\text{KE}_{\text{exit, actual}}} = \frac{V_{2s}^2}{V_2^2} \tag{2.84}$$

where V_{2s} is the velocity of the fluid at the diffuser outlet in case the process was isentropic. In the isentropic case, we expect a lower velocity at the exit since the purpose of the diffuser is to reduce the velocity.

Diffusers are designed for converting the kinetic energy of the fluid into enthalpy increase of the fluid. Therefore, the energy and exergy recovered through the velocity decrease process is the rise in the fluid energy and exergy which has the enthalpy as the source.

Hence, the energy efficiency of a diffuser can be written as [9, 10]

$$\eta_{\text{en, Diffuser}} = \frac{h_2 - h_1}{\frac{V_1^2}{2}} \tag{2.85}$$

In this equation, the enthalpy term at the inlet can also be written in the denominator to represent the total energy at the inlet.

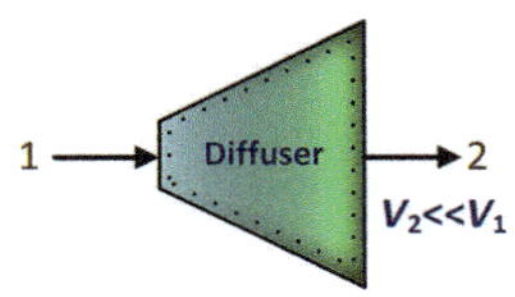

FIG. 2.11 A simple schematic diagram of a diffuser showing the component boundary.

Similarly, the exergy efficiency of a diffuser can be written as follows [9, 10]:

$$\eta_{ex.Diffuser} = \frac{\dot{E}x_{useful}}{\dot{E}x_{input}} = \frac{ex_2 - ex_1}{\frac{V_1^2}{2}} \tag{2.86}$$

It is important to mention that the exergy recovered does not contain the final kinetic energy, as the final kinetic energy is not the desired portion of exergy to be recovered out of the process. In addition, the unit conversion should be carefully performed since the kinetic energy commonly has different unit than kJ/kg.

2.4.6 Throttling/expansion valve

The purpose of a throttling or expansion valve is reducing the pressure of a fluid in a constant-enthalpy process. Typically, there is no heat or work interaction in this device as they are commonly adiabatic. Since the energy content of the mass entering and leaving the valve is the same, the energy efficiency seems to be 100%. However, it does not make much sense to express an efficiency based on the first law of thermodynamics. Nevertheless, the adiabatic expansion process in a throttling valve is a hugely irreversible process, and therefore an exergy efficiency can be determined. A schematic diagram of a throttling valve is illustrated in Fig. 2.12.

The input to the valve is the high pressure and temperature fluid, and exergy expended is the exergy delivered by the fluid, which is the reduction of the exergy of the fluid while it flows through the valve. In fact, there is no exergy recovered in any form, and therefore exergy recovered is zero. The expansion valve does not make use of the fluid exergy it expands. Hence exergy efficiency is [9, 10]

$$\eta_{ex,valve} = \frac{\dot{E}x_{useful}}{\dot{E}x_{input}} = \frac{0}{\dot{E}x_1 - \dot{E}x_2} = 0 \tag{2.87}$$

Alternatively, another exergy efficiency definition is also employed representing the output exergy divided by the input exergy [9, 10]:

$$\eta_{ex,valve} = \frac{\dot{E}x_{out}}{\dot{E}x_{in}} = \frac{\dot{E}x_2}{\dot{E}x_1} \tag{2.88}$$

In this case, exergy efficiency is not zero and has a value between 0 and 1.

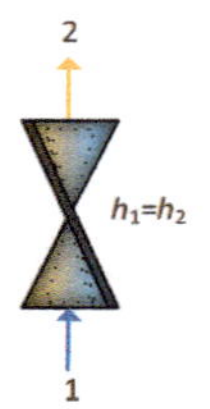

FIG. 2.12 A simple schematic diagram of an expansion/throttling valve showing the component boundary.

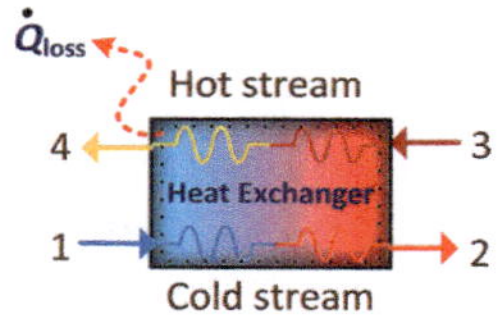

FIG. 2.13 A simple schematic diagram of a heat exchanger showing the component boundary.

2.4.7 Heat exchanger

In a heat exchanger, two streams of fluid exchange heat without mixing as depicted in Fig. 2.13. Heat exchangers have different designs such as parallel flow or counter flow. They can be well-insulated, meaning that the processes can become adiabatic. However, practically, they have heat losses through the component boundary.

This is a device that enables heat transfer between two fluids without mixing. It is known that heat exchangers are evaluated by their effectiveness parameter, which signifies the ratio between the actual amount of heat transfer and the maximum possible amount of heat transfer.

Defining energy and exergy efficiencies of heat exchangers is mostly based on the hot fluid stream as the source and the cold fluid stream as a potential energy/exergy recovery. Commonly, the hot fluid stream is considered to be the exergy source, and the exergy recovered is the exergy rise of the cold fluid stream. Then the energy and exergy efficiencies of a heat exchanger are expressed, respectively, as follows:

$$\eta_{en,HX} = \frac{\left(\dot{E}n_{out} - \dot{E}n_{in}\right)_{cold}}{\left(\dot{E}n_{in} - \dot{E}n_{out}\right)_{hot}} = \frac{\dot{m}_1(h_2 - h_1)}{\dot{m}_3(h_3 - h_4)} \tag{2.89}$$

$$\eta_{ex,HX} = \frac{\dot{E}x_{useful}}{\dot{E}x_{input}} = \frac{\left(\dot{E}x_{out} - \dot{E}x_{in}\right)_{cold}}{\left(\dot{E}x_{in} - \dot{E}x_{out}\right)_{hot}} = \frac{\dot{m}_1(ex_2 - ex_1)}{\dot{m}_3(ex_3 - ex_4)} \tag{2.90}$$

Note that due to the conservation of mass principle through the mass balance equation (MBE), $\dot{m}_3 = \dot{m}_4 = \dot{m}_{hot}$ and $\dot{m}_1 = \dot{m}_2 = \dot{m}_{cold}$.

In the exergy analysis of heat exchangers, all fluid streams experiencing a reduction in their exergy content are to be considered in the evaluation of the input exergy. Similarly, all fluid streams experiencing an increase in their exergy content are to be considered in the evaluation of the recovered/useful exergy.

There is another factor utilized in heat transfer analysis of heat exchangers, called heat exchanger effectiveness. The performance of heat exchangers is typically conveyed by their effectiveness defined as follows:

$$\eta_{effectiveness,HX} = \frac{\dot{Q}_{actual}}{\dot{Q}_{max}} = \frac{\dot{Q}_{actual}}{\left(\dot{m}c_p\right)_{min}\left(T_{hot,in} - T_{cold,in}\right)} = \frac{\left(\dot{m}c_p\Delta T\right)_{cold\ or\ hot}}{\left(\dot{m}c_p\right)_{min}\left(T_{hot,in} - T_{cold,in}\right)} \tag{2.91}$$

or using the enthalpies as follows:

$$\eta_{effectiveness,HX} = \frac{\dot{Q}_{actual}}{\dot{Q}_{max}} = \frac{\left(\dot{m}\Delta h\right)_{cold\ or\ hot}}{\left(\dot{m}\right)_{min}\left(h_{hot,in} - h_{cold,in}\right)} \tag{2.92}$$

where $(\dot{m}c_p)_{min}$ is the smaller of the heat capacity rate between hot and cold fluids, and $(\dot{m})_{min}$ is the smaller mass flow rate. The effectiveness of a heat exchanger becomes 100% if it is adiabatic and transferring the maximum amount of heat.

2.4.8 Mixing chamber

In a mixing chamber, multiple fluid streams mix to yield other fluid streams. Commonly, mixing chambers have two inlets and one outlet as depicted in Fig. 2.14. However, they can have several inputs and outputs.

Mixing chambers are commonly well insulated, hence they are typically assumed adiabatic. However, heat losses can also be considered in the analysis. In addition, there is no work interaction inside the mixing chambers unless there are additional impellers working with a shaft, hence mechanical energy. In that case, mechanical energy should also be considered as an input term. The kinetic and potential energies of the fluid streams are typically neglected.

There can be several efficiency definitions of mixing chambers due to the selection of useful output. Most of the time, two streams are mixed to obtain a desired temperature and/or pressure at the outlet. In this case, the energy and exergy efficiencies for the mixing chamber, shown in Fig. 2.14, can be written as follows:

$$\eta_{\text{en, Mixing Chamber}} = \frac{(\dot{\text{E}}\text{n}_3)}{(\dot{\text{E}}\text{n}_1 + \dot{\text{E}}\text{n}_2)} = \frac{\dot{m}_3 h_3}{\dot{m}_1 h_1 + \dot{m}_2 h_2} \tag{2.93}$$

$$\eta_{\text{ex, Mixing Chamber}} = \frac{(\dot{\text{E}}\text{x}_3)}{(\dot{\text{E}}\text{x}_1 + \dot{\text{E}}\text{x}_2)} = \frac{\dot{m}_3 \text{ex}_3}{\dot{m}_1 \text{ex}_1 + \dot{m}_2 \text{ex}_2} \tag{2.94}$$

If both arriving fluid streams are above the environment temperature, the exergy resource is the hot fluid, and the exergy input is the exergy reduction of the hot fluid. The exergy recovered is the exergy rise of the cold fluid. Taking state 1 as hot fluid inlet, state 2 as cold fluid inlet and state 3 as the state of the mixture, the energy and exergy efficiencies for the mixing chamber shown in Fig. 2.14 can also be respectively written as follows:

$$\eta_{\text{en, Mixing Chamber}} = \frac{(\dot{\text{E}}\text{n}_{out} - \dot{\text{E}}\text{n}_{in})_{cold}}{(\dot{\text{E}}\text{n}_{in} - \dot{\text{E}}\text{n}_{out})_{hot}} = \frac{(\dot{m}_3 h_3 - \dot{m}_2 h_2)}{(\dot{m}_1 h_1 - \dot{m}_3 h_3)} \tag{2.95}$$

$$\eta_{\text{ex, Mixing Chamber}} = \frac{\dot{\text{E}}\text{x}_{useful}}{\dot{\text{E}}\text{x}_{input}} = \frac{(\dot{\text{E}}\text{x}_{out} - \dot{\text{E}}\text{x}_{in})_{cold}}{(\dot{\text{E}}\text{x}_{in} - \dot{\text{E}}\text{x}_{out})_{hot}} = \frac{(\dot{\text{E}}\text{x}_3 - \dot{\text{E}}\text{x}_2)}{(\dot{\text{E}}\text{x}_1 - \dot{\text{E}}\text{x}_3)} = \frac{(\dot{m}_3 \text{ex}_3 - \dot{m}_2 \text{ex}_2)}{(\dot{m}_1 \text{ex}_1 - \dot{m}_3 \text{ex}_3)} \tag{2.96}$$

If the mixing chamber loses heat to a medium then this heat should be considered in the balance equations.

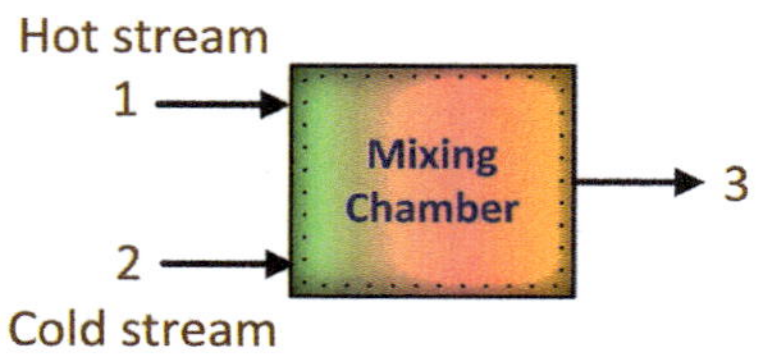

FIG. 2.14 A simple schematic diagram of a mixing chamber showing the component boundary.

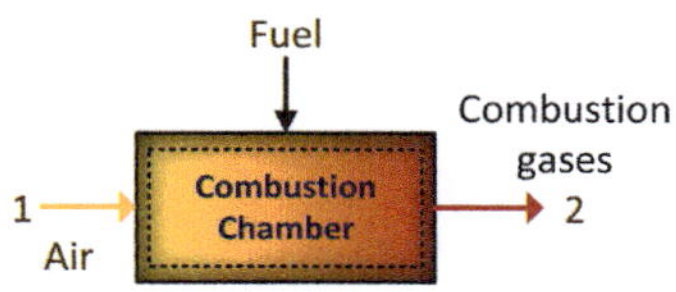

FIG. 2.15 A simple schematic diagram of a combustion chamber showing the component boundary.

2.4.9 Combustion chamber

Most of the conventional power cycles include a combustion process in which fuel and air (or oxygen) are supplied as illustrated in Fig. 2.15. The amount of heat produced during combustion is dependent on the type of fuel.

The efficiency of equipment involving the combustion of a fuel is based on the heating value of the fuel. Since most fuels include hydrogen, there is water/steam output at the end of the combustion process. Hence, the heating value of a fuel would vary, depending on whether the water in combustion products is in the liquid or vapor form. If the water at the exit is in vapor form, the heating value is called the lower heating value (LHV), while if the water in the combustion products is fully condensed, then it is called the higher heating value (HHV) meaning that the heat of vaporization of water is also recovered. For efficiency definitions of systems including the combustion process, it is necessary to clearly indicate whether LHV or HHV is used. For instance, lower heating values are commonly used for car engines since water exits as a vapor in the exhaust gases.

In a combustion chamber, the released heat is the useful output whereas the energy taken from the fuel is the required input. Hence, the energy efficiency of the combustion chamber can be expressed as

$$\eta_{\text{en,combustion}} = \frac{\dot{Q}_{\text{released}}}{\dot{m}_{\text{fuel}}\,\text{HV}_{\text{fuel}}} = \frac{\dot{m}_2 h_2}{\dot{m}_{\text{fuel}}\,\text{HV}_{\text{fuel}}} \tag{2.97}$$

where HV_{fuel} refers to the heating value of the fuel and can be taken either as lower heating value (LHV) or higher heating value (HHV) depending on the form of water at the exit of the combustion process. h_2 is the total energy content of combustion gases at T_2 and P_2 neglecting the kinetic and potential energy terms. Note that $\dot{m}_2 = \dot{m}_1 + \dot{m}_{\text{fuel}}$ due to the conservation of mass principle.

Similarly, the exergy efficiency of a combustion chamber can be written as

$$\eta_{\text{ex,combustion}} = \frac{\dot{\text{E}}\text{x}^{Q_{\text{released}}}}{\dot{m}_{\text{fuel}}\,\text{ex}_{\text{fuel}}} = \frac{\dot{Q}_{\text{released}}\left(1 - \frac{T}{T_{\text{cc}}}\right)}{\dot{m}_{\text{fuel}}\,\text{ex}_{\text{fuel}}} = \frac{\dot{m}_2 \text{ex}_2}{\dot{m}_{\text{fuel}}\,\text{ex}_{\text{fuel}}} \tag{2.98}$$

where T_{cc} is the combustion temperature or boundary temperature, which can be taken as the temperature of the outlet stream of the combustion chamber. ex_{fuel} is the exergy content of the fuel accounting for chemical as well as physical exergy. Depending on the chemical composition of fuel, chemical exergy of fuels varies. Chemical exergy content of most common fuels is tabulated in the literature [7, 11].

In fact, the air or oxygen is a fundamental requirement of combustion. Although the energy/exergy content can be minor compared to the heating value of fuel, it is also a required input. Thence, it can also be considered as an input term in the denominator of efficiency definitions as follows:

$$\eta_{en,combustion} = \frac{\dot{Q}_{released}}{\dot{m}_{fuel}\,HV_{fuel} + \dot{m}_1 h_1} = \frac{\dot{m}_2 h_2}{\dot{m}_{fuel}\,HV_{fuel} + \dot{m}_1 h_1} \tag{2.99}$$

$$\eta_{ex,combustion} = \frac{\dot{Q}_{released}\left(1 - \frac{T}{T_{cc}}\right)}{\dot{m}_{fuel}\,ex_{fuel} + \dot{m}_1 ex_1} = \frac{\dot{m}_2 ex_2}{\dot{m}_{fuel}\,ex_{fuel} + \dot{m}_1 ex_1} \tag{2.100}$$

Since the energy and exergy value of inlet air is relatively too low compared to the heating value and chemical exergy content value of fuel, the final results do not vary significantly.

2.4.10 Boiler

A boiler is one of the most common components used in conventional power cycles (e.g., Rankine cycle) to obtain high-temperature steam by transferring heat from burning fuel. A boiler can also be used to heat water similarly to a natural gas water heater or to heat industrial oil, or another fluid. Similar to the combustion chamber, the heat is taken from fuel burning. The useful output, in this case, is the energy gained by the working fluid (e.g., water). Typically, a phase change process occurs in the boiler to convert liquid into gas. A boiler can also behave as a heat exchanger when the heat is sourced from an external medium (e.g., waste heat, concentrated solar). This is commonly the case for organic Rankine cycle boilers. Otherwise, the combustion process occurs inside the boiler, where air is also supplied. There is no work interaction in a boiler. Sometimes, boilers are assumed to have no heat losses. However, the boiler can have associated heat losses crossing the boundary. A simple sketch of a boiler is depicted in Fig. 2.16.

The energy efficiency of a boiler can be expressed as

$$\eta_{en,boiler} = \frac{\dot{Q}_{useful}}{\dot{m}_{fuel}\,HV_{fuel}} = \frac{\dot{m}_{working\ fluid}(h_2 - h_1)}{\dot{m}_{fuel}\,HV_{fuel}} \tag{2.101}$$

where $\dot{Q}_{useful}$ is the rate of useful heat supplied to the working fluid, h_1 is the enthalpy of working fluid (commonly in saturated liquid or subcooled phase) at the input of the boiler

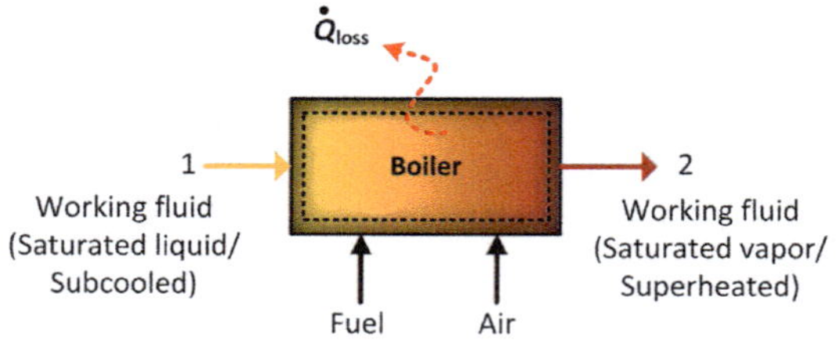

FIG. 2.16 A simple schematic diagram of a boiler showing the component boundary.

and h_2 is the enthalpy of working fluid (commonly in saturated vapor or superheated phase) at the exit of the boiler. Similar to a combustion chamber, air can also be considered as the input term in the efficiency definition. However, compared to the high energy content of fuel, it is considerably lower.

Similarly, the exergy efficiency of a boiler is written as

$$\eta_{ex,boiler} = \frac{\dot{E}x^{Q_{useful}}}{\dot{m}_{fuel}\,ex_{fuel}} = \frac{\dot{m}_{working\ fluid}(ex_2 - ex_1)}{\dot{m}_{fuel}\,ex_{fuel}} \tag{2.102}$$

2.4.11 Mechanical and electrical devices

The transfer of mechanical energy is typically accomplished by a rotating shaft, and therefore mechanical work is regularly referred to as shaft work. A pump or a fan takes shaft work, commonly from an electrical motor, and transfers it to the fluid as mechanical energy. A turbine does the reverse, transforming the mechanical energy of a fluid into shaft work. Assuming no irreversibilities such as friction, mechanical energy can be transformed completely from one mechanical form to another, and the mechanical efficiency of a device or process can be expressed as

$$\eta_{en,mechanical} = \frac{\dot{E}n_{mechanical,out}}{\dot{E}n_{mechanical,in}} = 1 - \frac{\dot{E}n_{mechanical,loss}}{\dot{E}n_{mechanical,in}} \tag{2.103}$$

The mechanical efficiency of a fan is expressed as follows [9, 10]:

$$\eta_{en,mechanical,\ fan} = \frac{\Delta\dot{E}n_{mechanical,\ fluid}}{\dot{W}_{shaft,in}} = \frac{\left(\frac{\dot{m}V_2^2}{2}\right)}{\dot{W}_{shaft,in}} \tag{2.104}$$

where $\dot{m}$ is the mass flow rate of air flowing through the casing of the fan, v_2 is the velocity of air at the outlet of the fan, and $\dot{W}_{shaft,in}$ is the input shaft power to the fan. Here, the inlet velocity is commonly very low, hence it is not written in the numerator.

The overall energy efficiency of the fan can be expressed as

$$\eta_{en,\ fan,overall} = \frac{\Delta\dot{E}n_{mechanical,\ fluid}}{\dot{W}_{electrical,in}} \tag{2.105}$$

where $\dot{W}_{electrical,in}$ is the electrical work received by the fan.

The overall exergy efficiency of a fan is written as

$$\eta_{ex,\ fan,overall} = \frac{\Delta\dot{E}x_{mechanical,\ fluid}}{\dot{W}_{electrical,in}} \tag{2.106}$$

It is noted that this efficiency is typically equal to the energy efficiency of a fan because exergy gained is equal to the kinetic energy and shaft work is equal to exergy of work.

If a mechanical efficiency is 95%, it means that 5% of the input mechanical energy is converted to non-useful forms of energy such as thermal energy as a result of frictional heating, causing a minor increase in the temperature of the fluid.

The quality of the conversion process between the mechanical work delivered or removed and the mechanical energy of the fluid is given by the pump efficiency and turbine efficiency, defined respectively as [9, 10]

$$\eta_{en,Pump} = \frac{\Delta \dot{E}n_{mechanical,\ fluid}}{\dot{W}_{shaft,in}} \tag{2.107}$$

$$\eta_{en,Turbine} = \frac{\dot{W}_{shaft,out}}{\Delta \dot{E}n_{mechanical,\ fluid}} \tag{2.108}$$

In these devices, electrical energy is normally transformed into rotating mechanical energy by electric motors to energize fans, compressors, etc. The efficiency of this transformation phase is evaluated by the motor efficiency, that is the ratio of the mechanical energy production of the motor to the electrical energy input. The difference between the electrical energy consumed and the mechanical energy delivered is dissipated as waste heat.

The mechanical efficiency should not be confused with the motor efficiency and the generator efficiency, which are defined as [9, 10]

$$\eta_{motor} = \frac{\dot{W}_{shaft,out}}{\dot{W}_{electrical,in}} \tag{2.109}$$

$$\eta_{generator} = \frac{\dot{W}_{electrical,out}}{\dot{W}_{shaft,in}} \tag{2.110}$$

It is noted that in this case, energy and exergy efficiencies of motor and generator devices are identical due to involvement of work only.

A pump is typically integrated to a motor, and a turbine is integrated to a generator as well. Thus, combined or overall efficiency of pump-motor and turbine-generator combinations can also be defined as [9, 10]

$$\eta_{Pump,overall} = \eta_{Pump}\,\eta_{motor} = \frac{\Delta \dot{E}n_{mechanical,\ fluid}}{\dot{W}_{electrical,in}} \tag{2.111}$$

$$\eta_{Turbine,overall} = \eta_{Turbine}\,\eta_{generator} = \frac{\dot{W}_{electrical,out}}{\Delta \dot{E}_{mechanical\ fluid}} \tag{2.112}$$

2.5 Energy and exergy analysis of common energy systems and components

In this section, we summarize the mass, energy, entropy and exergy balance equations for commonly used thermodynamics components as listed in Table 2.4. Furthermore, energy and exergy efficiencies of each component are also defined using the previously mentioned approach. In addition, the three most commonly employed thermodynamic cycles are analyzed, namely the Brayton cycle, Rankine cycle and refrigeration/heat pump cycle.

TABLE 2.4 Fundamental balance equations of some important thermodynamics components

Device	Equations
Heat exchanger (with heat loss) 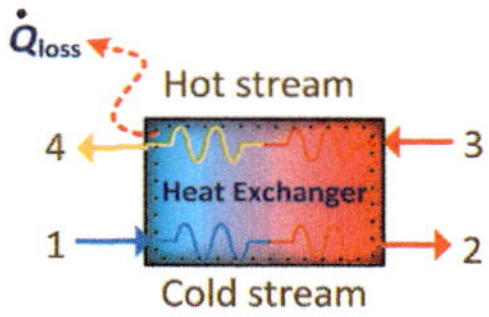 	Balance equations: MBE: $\dot m_3=\dot m_4=\dot m_{hot}$ *and* $\dot m_1=\dot m_2=\dot m_{cold}$ EBE: $\dot m_1 h_1+\dot m_3 h_3=\dot m_2 h_2+\dot m_4 h_4+\dot Q_{loss}$ EnBE: $\dot m_1 s_1+\dot m_3 s_3+\dot S_{gen}=\dot m_2 s_2+\dot m_4 s_4+\frac{\dot Q_{loss}}{T_s}$ ExBE: $\dot m_1 ex_1+\dot m_3 ex_3=$ $\dot m_2 ex_2+\dot m_4 ex_4+\dot Q_{loss}\left(1-\frac{T_o}{T_s}\right)+\dot{E}x_d$ Effectiveness: $\varepsilon=\frac{\dot Q_{cold}}{\dot Q_{max}}=\frac{(\dot m C_p \Delta T)_{cold}}{(\dot m C_p)_{min}(T_3-T_1)}$ Efficiencies: $\eta_{en}=\frac{\dot Q_{cold}}{\dot Q_{hot}}$ $\eta_{ex}=\frac{\dot m_{cold}(ex_2-ex_1)}{\dot m_{hot}(ex_3-ex_4)}$
Fan 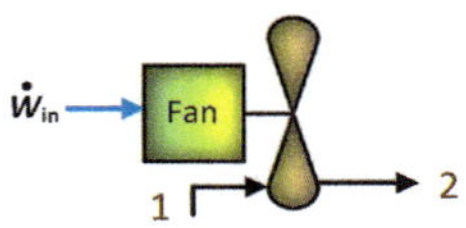 	Balance equations: MBE: $\dot m_1=\dot m_2=\dot m$ EBE: $\dot m_1\left(h_1+\frac{v_1^2}{2}\right)+\dot W_{in}=\dot m_2\left(h_2+\frac{v_2^2}{2}\right)$ EnBE: $\dot m_1 s_1+\dot S_{gen}=\dot m_2 s_2$ ExBE: $\dot m_1\left(ex_1+\frac{v_1^2}{2}\right)+\dot W_{in}=\dot m_2\left(ex_2+\frac{v_2^2}{2}\right)+\dot{E}x_d$ Efficiencies: $\eta_{en}=\eta_{ex}=\frac{\dot m_2 \frac{v_2^2}{2}}{\dot W_{in}}$ $\eta_{en}=\frac{\dot m_2\left(h_2+\frac{v_2^2}{2}\right)-\dot m_1\left(h_1+\frac{v_1^2}{2}\right)}{\dot W_{in}}$ $\eta_{ex}=\frac{\dot m_2\left(ex_2+\frac{v_2^2}{2}\right)-\dot m_1\left(ex_1+\frac{v_1^2}{2}\right)}{\dot W_{in}}$
Turbine (with heat loss) 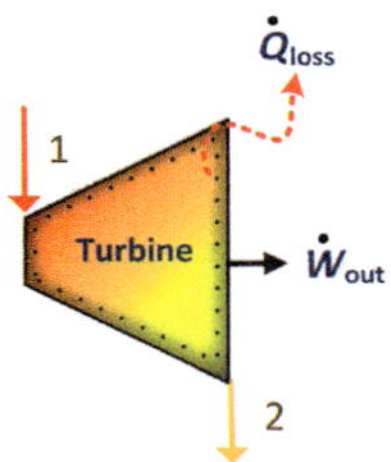 	Balance equations: MBE: $\dot m_1=\dot m_2=\dot m$ EBE: $\dot m_1 h_1=\dot W_{out}+\dot m_2 h_2+\dot Q_{loss}$ EnBE: $\dot m_1 s_1+\dot S_{gen}=\dot m_2 s_2+\frac{\dot Q_{loss}}{T_s}$ ExBE: $\dot m_1 ex_1=\dot W_{out}+\dot m_2 ex_2+\dot Q_{loss}\left(1-\frac{T_o}{T_s}\right)+\dot{E}x_d$ Efficiencies: $\eta_{isentropic}=\frac{\dot W_{actual}}{\dot W_{isentropic}}=\frac{\dot m(h_1-h_2)}{\dot m(h_1-h_{2s})}$ $\eta_{en}=\frac{\dot W_{out}}{\dot m(h_1-h_2)}$ $\eta_{ex}=\frac{\dot W_{out}}{\dot{E}x_{input}}=\frac{\dot W_{out}}{\dot m(ex_1-ex_2)}$

Continued

TABLE 2.4 Fundamental balance equations of some important thermodynamics components—Cont'd

<table>
<tr><th>Device</th><th>Equations</th></tr>
<tr><td>

Pump

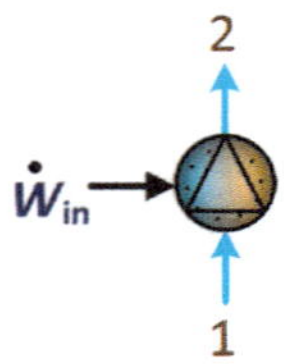

</td><td>

Balance equations:
MBE : $\dot{m}_1=\dot{m}_2=\dot{m}$
EBE : $\dot{m}_1h_1+\dot{W}_{in}=\dot{m}_2h_2$
EnBE : $\dot{m}_1s_1+\dot{S}_{gen}=\dot{m}_2s_2$
ExBE : $\dot{m}_1ex_1+\dot{W}_{in}=\dot{m}_2ex_2+\dot{E}x_d$
Efficiencies:
$\eta_{isentropic}=\frac{\dot{W}_{isentropic}}{\dot{W}_{actual}}=\frac{\dot{m}v(P_2-P_1)}{\dot{m}(h_2-h_1)}$
$\eta_{en}=\frac{\dot{m}(h_2-h_1)}{\dot{W}_{in}}$
$\eta_{ex}=\frac{\dot{m}(ex_2-ex_1)}{\dot{W}_{in}}$

</td></tr>
<tr><td>

Hydraulic turbine

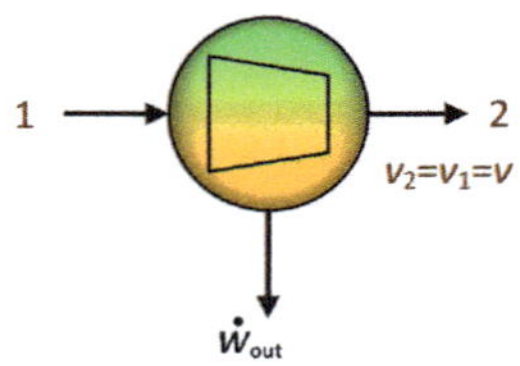

</td><td>

Balance equations:
MBE : $\dot{m}_1=\dot{m}_2=\dot{m}$
EBE : $\dot{m}_1h_1=\dot{W}_{out}+\dot{m}_2h_2$
EnBE : $\dot{m}_1s_1+\dot{S}_{gen}=\dot{m}_2s_2$
ExBE : $\dot{m}_1ex_1=\dot{W}_{out}+\dot{m}_2ex_2+\dot{E}x_d$
Efficiencies:
$\eta_{isentropic}=\frac{\dot{W}_{actual}}{\dot{W}_{isentropic}}=\frac{\dot{m}(h_1-h_2)}{\dot{m}v(P_1-P_2)}$
$\eta_{en}=\frac{\dot{W}_{out}}{\dot{m}(h_1-h_2)}$
$\eta_{ex}=\frac{\dot{W}_{out}}{\dot{E}x_{input}}=\frac{\dot{W}_{out}}{\dot{m}(ex_1-ex_2)}$

</td></tr>
<tr><td>

Expansion/Throttling valve

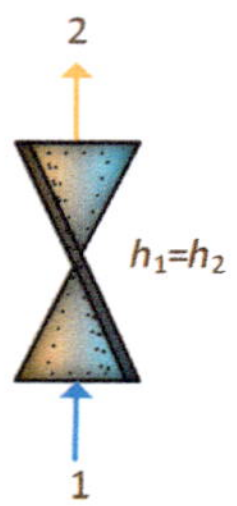

</td><td>

Balance equations:
MBE : $\dot{m}_1=\dot{m}_2=\dot{m}$
EBE : $\dot{m}_1h_1=\dot{m}_2h_2$
EnBE : $\dot{m}_1s_1+\dot{S}_{gen}=\dot{m}_2s_2$
ExBE : $\dot{m}_1ex_1=\dot{m}_2ex_2+\dot{E}x_d$
Efficiencies:
$\eta_{en}=\frac{\dot{m}_2h_2}{\dot{m}_1h_1}=100\%$ or 0% (no recovered energy)
$\eta_{ex,1}=\frac{\dot{m}_2ex_2}{\dot{m}_1ex_1}$
$\eta_{ex,2}=\frac{\dot{E}x_{recovered}}{\dot{m}_1ex_1}=\frac{0}{\dot{m}_1ex_1}=0\%$ (no recovered exergy)

</td></tr>
<tr><td>

Compressor (with heat loss or cooling)

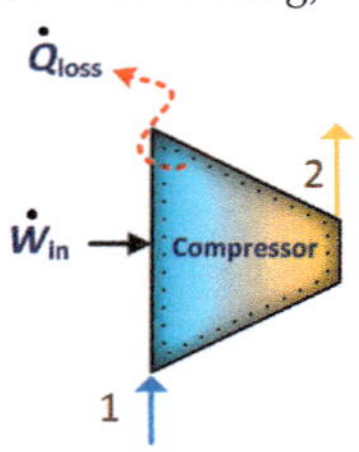

</td><td>

Balance equations:
MBE : $\dot{m}_1=\dot{m}_2=\dot{m}$
EBE : $\dot{m}_1h_1+\dot{W}_{in}=\dot{m}_2h_2+\dot{Q}_{loss}$
EnBE : $\dot{m}_1s_1+\dot{S}_{gen}=\dot{m}_2s_2+\frac{\dot{Q}_{loss}}{T_s}$
ExBE : $\dot{m}_1ex_1+\dot{W}_{in}=\dot{m}_2ex_2+\dot{Q}_{loss}\left(1-\frac{T_o}{T_s}\right)+\dot{E}x_d$
Efficiencies:
$\eta_{isentropic}=\frac{\dot{W}_{isentropic}}{\dot{W}_{actual}}=\frac{\dot{m}(h_{2s}-h_1)}{\dot{m}(h_2-h_1)}$
$\eta_{en}=\frac{\dot{m}(h_2-h_1)}{\dot{W}_{in}}$
$\eta_{ex}=\frac{\dot{m}(ex_2-ex_1)}{\dot{W}_{in}}$

</td></tr>
</table>

TABLE 2.4 Fundamental balance equations of some important thermodynamics components—Cont'd

Device	Equations
Diffuser and nozzle (adiabatic) 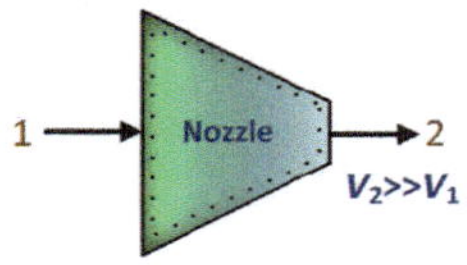 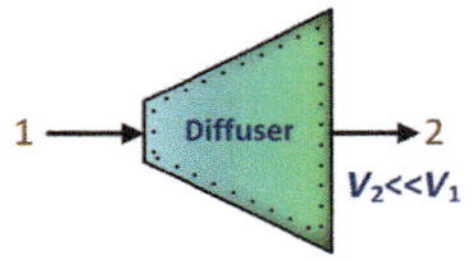	Balance equation: $\text{MBE}: \dot{m}_1 = \dot{m}_2 = \dot{m}$ $\text{EBE}: \dot{m}_1\left(h_1 + \frac{v_1^2}{2}\right) = \dot{m}_2\left(h_2 + \frac{v_2^2}{2}\right)$ $\text{EnBE}: \dot{m}_1 s_1 + \dot{S}_{gen} = \dot{m}_2 s_2$ $\text{ExBE}: \dot{m}_1\left(\text{ex}_1 + \frac{v_1^2}{2}\right) = \dot{m}_2\left(\text{ex}_2 + \frac{v_2^2}{2}\right) + \dot{\text{E}}\text{x}_\text{d}$ Efficiencies: $\eta_{\text{isentropic,nozzle 1}} = \frac{h_1 - h_2}{h_1 - h_{2s}}$ $\eta_{\text{isentropic,nozzle 2}} = \frac{V_2^2}{V_{2s}^2}$ $\eta_{\text{en,nozzle}} = \frac{\left(\frac{v_2^2}{2} - \frac{v_1^2}{2}\right)}{h_1 - h_2 + \frac{v_1^2}{2}}$ $\eta_{\text{ex,nozzle}} = \frac{\dot{\text{E}}\text{x}_\text{useful}}{\dot{\text{E}}\text{x}_\text{input}} = \frac{\left(\frac{v_2^2}{2} - \frac{v_1^2}{2}\right)}{\text{ex}_1 - \text{ex}_2 + \frac{v_1^2}{2}}$ $\eta_{\text{isentropic, diffuser 1}} = \frac{h_1 - h_{2s}}{h_1 - h_2}$ $\eta_{\text{isentropic, diffuser 2}} = \frac{V_{2s}^2}{V_2^2}$ $\eta_{\text{en, Diffuser}} = \frac{h_2 - h_1}{\frac{V_1^2}{2}}$ $\eta_{\text{ex, Diffuser}} = \frac{\dot{\text{E}}\text{x}_\text{useful}}{\dot{\text{E}}\text{x}_\text{input}} = \frac{\text{ex}_2 - \text{ex}_1}{\frac{V_1^2}{2}}$
Boiler (with heat loss, fuel and air represent total heat input) 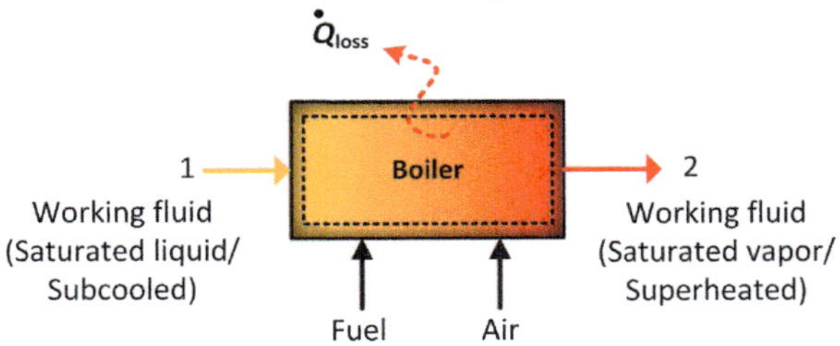	Balance equations: $\text{MBE}: \dot{m}_1 = \dot{m}_2$ $\text{EBE}: \dot{m}_1 h_1 + \dot{Q}_{in} = \dot{m}_2 h_2 + \dot{Q}_{loss}$ $\text{EnBE}: \dot{m}_1 s_1 + \frac{\dot{Q}_{in}}{T_{source}} + \dot{S}_{gen} = \dot{m}_2 s_2 + \frac{\dot{Q}_{loss}}{T_s}$ $\text{ExBE}: \dot{m}_1 \text{ex}_1 + \dot{Q}_{in}\left(1 - \frac{T_0}{T_{source}}\right) =$ $\dot{m}_2 \text{ex}_2 + \dot{Q}_{loss}\left(1 - \frac{T_0}{T_s}\right) + \dot{\text{E}}\text{x}_\text{d}$ Efficiencies: $\eta_{en} = \frac{\dot{m}_2 h_2 - \dot{m}_1 h_1}{\dot{Q}_{in}}$ $\eta_{ex} = \frac{\dot{m}_2 \text{ex}_2 - \dot{m}_1 \text{ex}_1}{\dot{Q}_{in}\left(1 - \frac{T_0}{T_{source}}\right)}$
Mixing chamber (with heat loss) 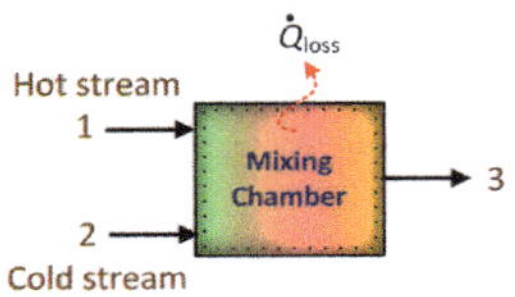	Balance equations: $\text{MBE}: \dot{m}_1 + \dot{m}_2 = \dot{m}_3$ $\text{EBE}: \dot{m}_1 h_1 + \dot{m}_2 h_2 = \dot{m}_3 h_3 + \dot{Q}_{loss}$ $\text{EnBE}: \dot{m}_1 s_1 + \dot{m}_2 s_2 + \dot{S}_{gen} = \dot{m}_3 s_3 + \frac{\dot{Q}_{loss}}{T_s}$ $\text{ExBE}: \dot{m}_1 \text{ex}_1 + \dot{m}_2 \text{ex}_2 = \dot{m}_3 \text{ex}_3 + \dot{Q}_{loss}\left(1 - \frac{T_0}{T_s}\right) + \dot{\text{E}}\text{x}_\text{d}$ Efficiencies: $\eta_{en} = \frac{\dot{m}_3 h_3}{\dot{m}_1 h_1 + \dot{m}_2 h_2}$ $\eta_{ex} = \frac{\dot{m}_3 \text{ex}_3}{\dot{m}_1 \text{ex}_1 + \dot{m}_2 \text{ex}_2}$

Continued

TABLE 2.4 Fundamental balance equations of some important thermodynamics components—Cont'd

Device	Equations
Combustion chamber (assuming no heat loss) 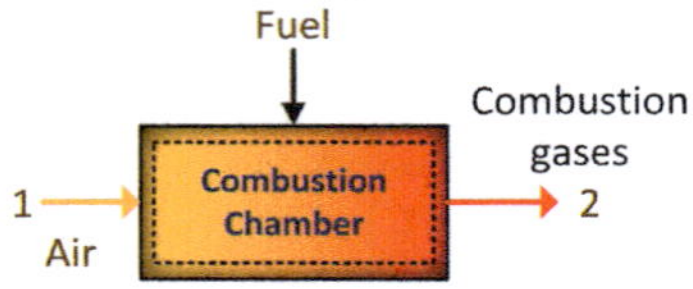 	Balance equations: MBE : $\dot{m}_1 + \dot{m}_{\text{fuel}} = \dot{m}_2$ EBE : $\dot{m}_1 h_1 + \dot{m}_{\text{fuel}}\,\text{HV}_{\text{fuel}} = \dot{m}_2 h_2$ EnBE : $\dot{m}_1 s_1 + \frac{\dot{m}_{\text{fuel}}\,\text{HV}_{\text{fuel}}}{T_{\text{source}}} + \dot{S}_{\text{gen}} = \dot{m}_2 s_2$ ExBE : $\dot{m}_1 \text{ex}_1 + \dot{m}_{\text{fuel}} \text{ex}_{\text{fuel}} = \dot{m}_2 \text{ex}_2 + \dot{E}\text{x}_d$ Efficiencies: $\eta_{\text{en},1} = \frac{\dot{m}_2 h_2 - \dot{m}_1 h_1}{\dot{m}_{\text{fuel}}\,\text{HV}_{\text{fuel}}}$ $\eta_{\text{en},2} = \frac{\dot{m}_2 h_2}{\dot{m}_{\text{fuel}}\,\text{HV}_{\text{fuel}} + \dot{m}_1 h_1}$ $\eta_{\text{en},3} = \frac{\dot{m}_2 h_2}{\dot{m}_{\text{fuel}}\,\text{HV}_{\text{fuel}}}$ $\eta_{\text{ex},1} = \frac{\dot{m}_2 \text{ex}_2 - \dot{m}_1 \text{ex}_1}{\dot{m}_{\text{fuel}} \text{ex}_{\text{fuel}}}$ $\eta_{\text{ex},2} = \frac{\dot{m}_2 \text{ex}_2}{\dot{m}_{\text{fuel}} \text{ex}_{\text{fuel}} + \dot{m}_1 \text{ex}_1}$ $\eta_{\text{ex},3} = \frac{\dot{m}_2 \text{ex}_2}{\dot{m}_{\text{fuel}} \text{ex}_{\text{fuel}}}$
Gasification chamber (assuming no heat loss) 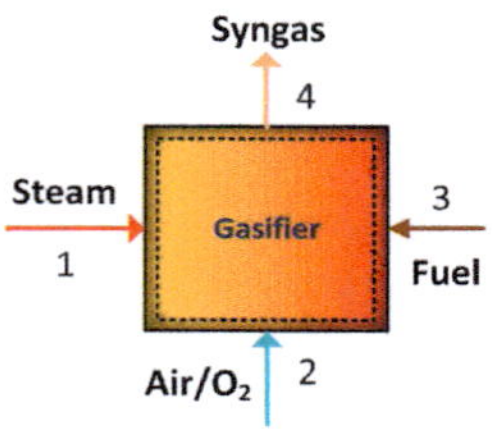 	Balance equations: MBE : $\dot{m}_1 + \dot{m}_2 + \dot{m}_3 = \dot{m}_4$ EBE : $\dot{m}_1 h_1 + \dot{m}_2 h_2 + \dot{m}_3\,\text{HV}_{\text{fuel}} = \dot{m}_4 h_4$ EnBE : $\dot{m}_1 s_1 + \dot{m}_2 s_2 + \dot{m}_3 s_3 + \dot{S}_{\text{gen}} = \dot{m}_4 s_4$ ExBE : $\dot{m}_1 \text{ex}_1 + \dot{m}_2 \text{ex}_2 + \dot{m}_3 \text{ex}_{\text{fuel}} = \dot{m}_4 \text{ex}_4 + \dot{E}\text{x}_d$ Efficiencies: $\eta_{\text{en}} = \frac{\dot{m}_4 \text{HV}_{\text{syngas}}}{\dot{m}_{\text{fuel}}\,\text{HV}_{\text{fuel}} + \dot{m}_1 h_1 + \dot{m}_2 h_2}$ $\eta_{\text{ex}} = \frac{\dot{m}_4 \text{ex}_{\text{syngas}}}{\dot{m}_{\text{fuel}} \text{ex}_{\text{fuel}} + \dot{m}_1 \text{ex}_1 + \dot{m}_2 \text{ex}_2}$
Pyrolysis chamber (assuming no heat loss and all products are useful) 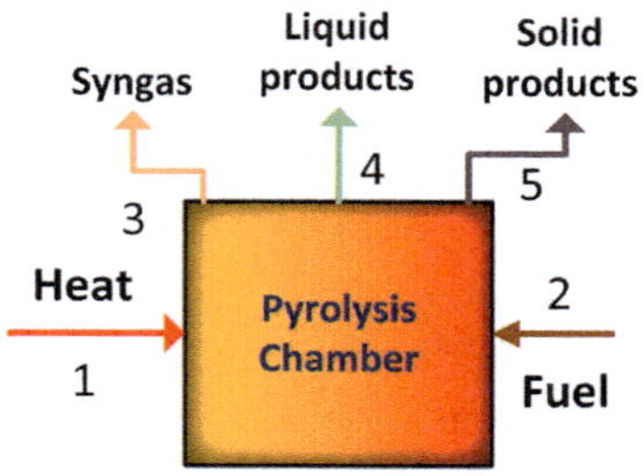 	Balance equations: MBE : $\dot{m}_{\text{fuel}} = \dot{m}_3 + \dot{m}_4 + \dot{m}_5$ EBE : $\dot{Q}_{\text{in}} + \dot{m}_2 h_2 = \dot{m}_3 h_3 + \dot{m}_4 h_4 + \dot{m}_5 h_5$ EnBE : $\dot{m}_2 s_2 + \frac{\dot{Q}_{\text{in}}}{T_{\text{source}}} + \dot{S}_{\text{gen}} = \dot{m}_3 s_3 + \dot{m}_4 s_4 + \dot{m}_5 s_5$ ExBE : $\dot{m}_2 \text{ex}_2 + \dot{Q}_{\text{in}}\left(1 - \frac{T_o}{T_{\text{source}}}\right) =$ $\dot{m}_3 \text{ex}_3 + \dot{m}_4 \text{ex}_4 + \dot{m}_5 \text{ex}_5 + \dot{E}\text{x}_d$ Efficiencies: $\eta_{\text{en}} = \frac{\dot{m}_3 h_3 + \dot{m}_4 h_4 + \dot{m}_5 h_5}{\dot{m}_2 h_2 + \dot{Q}_{\text{in}}}$ $\eta_{\text{ex}} = \frac{\dot{m}_3 \text{ex}_3 + \dot{m}_4 \text{ex}_4 + \dot{m}_5 \text{ex}_5}{\dot{m}_2 \text{ex}_2 + \dot{Q}_{\text{in}}\left(1 - \frac{T_o}{T_{\text{source}}}\right)}$

TABLE 2.4 Fundamental balance equations of some important thermodynamics components—Cont'd

Device	Equations
Separation device (taken as endothermic) (there can also be heat loss) 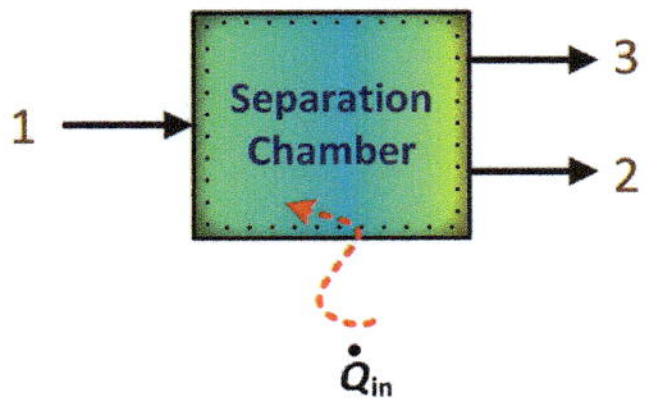	Balance equations: MBE: $\dot{m}_1 = \dot{m}_2 + \dot{m}_3$ EBE: $\dot{m}_1 h_1 + \dot{Q}_{in} = \dot{m}_2 h_2 + \dot{m}_3 h_3$ EnBE: $\dot{m}_1 s_1 + \frac{\dot{Q}_{in}}{T_{source}} + \dot{S}_{gen} = \dot{m}_2 s_2 + \dot{m}_3 s_3$ ExBE: $\dot{m}_1 ex_1 + \dot{Q}_{in}\left(1 - \frac{T_0}{T_{source}}\right) = \dot{m}_2 ex_2 + \dot{m}_3 ex_3 + \dot{E}x_d$ Efficiencies: (Both 3 and 2 are taken as useful outputs) $\eta_{en} = \frac{\dot{m}_2 h_2 + \dot{m}_3 h_3}{\dot{m}_1 h_1 + \dot{Q}_{in}}$ $\eta_{ex} = \frac{\dot{m}_2 ex_2 + \dot{m}_3 ex_3}{\dot{m}_1 ex_1 + \dot{Q}_{in}\left(1 - \frac{T_0}{T_{source}}\right)}$
Chemical reactor (All possible work and heat interactions are included) 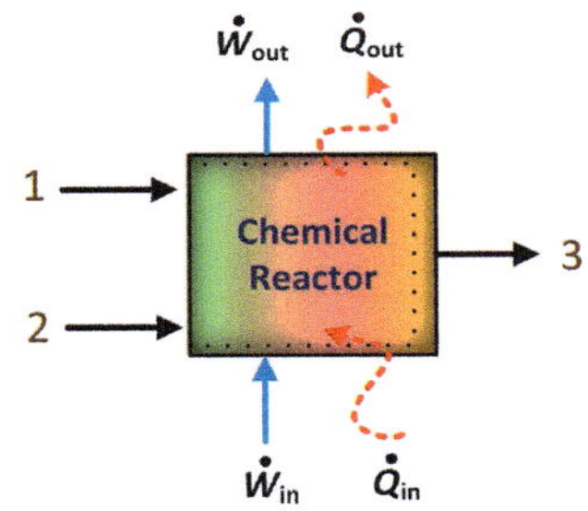	Balance equations: MBE: $\dot{m}_1 + \dot{m}_2 = \dot{m}_3$ EBE: $\dot{m}_1 h_1 + \dot{m}_2 h_2 + \dot{Q}_{in} + \dot{W}_{in} = \dot{m}_3 h_3 + \dot{Q}_{out} + \dot{W}_{out}$ EnBE: EBE: $\dot{m}_1 s_1 + \dot{m}_2 s_2 + \frac{\dot{Q}_{in}}{T_{source}} + \dot{S}_{gen} = \dot{m}_3 s_3 + \frac{\dot{Q}_{out}}{T_s}$ ExBE: $\dot{m}_1 ex_1 + \dot{m}_2 ex_2 + \dot{Q}_{in}\left(1 - \frac{T_0}{T_{source}}\right) + \dot{W}_{in} =$ $\dot{m}_3 ex_3 + \dot{Q}_{out}\left(1 - \frac{T_0}{T_s}\right) + \dot{W}_{out} + \dot{E}x_d$ Efficiencies: (Work, heat and state 3 are considered to be useful outputs) $\eta_{en} = \frac{\dot{m}_3 h_3 + \dot{Q}_{out} + \dot{W}_{out}}{\dot{m}_1 h_1 + \dot{m}_2 h_2 + \dot{Q}_{in} + \dot{W}_{in}}$ $\eta_{ex} = \frac{\dot{m}_3 ex_3 + \dot{Q}_{out}\left(1 - \frac{T_0}{T_s}\right) + \dot{W}_{out}}{\dot{m}_1 ex_1 + \dot{m}_2 ex_2 + \dot{Q}_{in}\left(1 - \frac{T_0}{T_{source}}\right) + \dot{W}_{in}}$

Footnote: Note that some processes can be adiabatic. The efficiencies are written for steady-state steady-flow (SSSF) conditions. There can be different efficiency definitions for components based on the concept and the purpose. If a chemical reaction occurs, chemical exergy contents of the species should also be taken into account.

2.5.1 Refrigerators and heat pumps

A refrigerator or a heat pump is a system/cycle that is used for transferring heat from a low-temperature reservoir to a high-temperature reservoir. These are mainly cyclic devices. It is important to note that these devices do not create heating or cooling, but rather they transfer thermal energy between two reservoirs. Therefore, we quantify the performances with the so-called Coefficient of Performance (COP). A simple layout of a refrigeration and heat pump system is depicted in Fig. 2.17.

The main difference between a refrigerator and heat pump cycle is the purpose of the system. If the cycle is used for refrigeration, then the evaporator is placed in a medium where the space is being cooled down. On the other hand, if the cycle is used for heating purposes, then the condenser is placed in an ambient location desired to be heated.

A working fluid, commonly a refrigerant, is used in these cycles due to their lower boiling temperatures. The working fluid arrives at the compressor as a vapor and is compressed to the condenser pressure. The high-temperature refrigerant cools in the condenser by rejecting

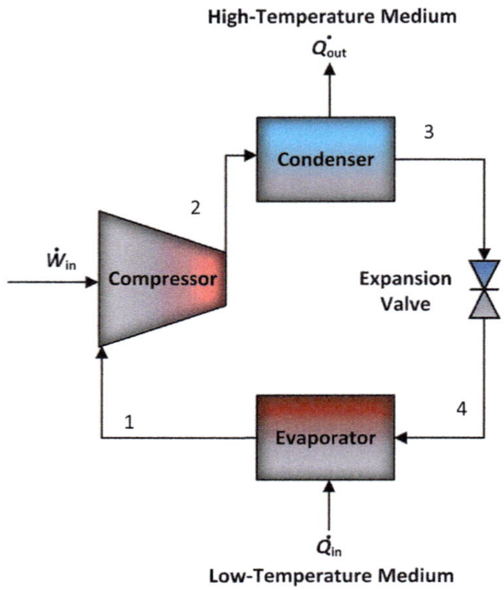

FIG. 2.17 A simple schematic diagram of a heat pump/refrigeration cycle.

heat to a high-temperature medium. The refrigerant arrives at the expansion valve as a liquid. It is expanded in an expansion valve and its pressure and temperature decrease. The refrigerant is a mixture of vapor and liquid at the inlet of the evaporator. It absorbs heat from a low-temperature medium as it flows in the evaporator. The cycle is completed once the refrigerant exits the evaporator as a vapor and arrives at the compressor. The energy and exergy analyses are performed using the fundamental mass, energy, entropy and exergy balance equations as listed in Table 2.5.

T_{space} is the temperature of the medium being cooled down (for the refrigeration cycle) and T_{sink} is the temperature of the medium being heated (for the heat pump cycle).

The performance of these cycles is measured via the coefficient of performance (COP) values rather than efficiencies, since values can be obtained that are greater than one. The coefficient of performance (COP) of a refrigeration cycle is expressed as the heat absorbed from the cooled space divided by the work input in the compressor:

$$COP_{en,\,Refrigeration} = \frac{\dot{Q}_{in}}{\dot{W}} = \frac{\dot{Q}_{in}}{\dot{Q}_{out} - \dot{Q}_{in}} \tag{2.113}$$

Similarly, exergetic COP of a refrigeration cycle can be written as

$$COP_{ex,\,Refrigeration} = \frac{\dot{E}x^{Q_{in}}}{\dot{W}} = \frac{\dot{E}x^{Q_{in}}}{\dot{E}x^{Q_{out}} - \dot{E}x^{Q_{in}}} \tag{2.114}$$

Since the purpose of a heat pump is transferring heat to a heated space to preserve it at a desired high temperature, the useful output is rejected heat from the condenser in this case. Hence, the energetic COP of a heat pump can be expressed as

$$COP_{en,\,Heatpump} = \frac{\dot{Q}_{out}}{\dot{W}} = \frac{\dot{Q}_{out}}{\dot{Q}_{out} - \dot{Q}_{in}} \tag{2.115}$$

TABLE 2.5 Fundamental balance equations for refrigeration and heat pump cycles

Adiabatic compressor	MBE: $\dot{m}_1 = \dot{m}_2$
	EBE: $\dot{m}_1 h_1 + \dot{W}_{in} = \dot{m}_2 h_2$
	EnBE: $\dot{m}_1 s_1 + \dot{S}_{gen} = \dot{m}_2 s_2$
	ExBE: $\dot{m}_1 ex_1 + \dot{W}_{in} = \dot{m}_2 ex_2 + \dot{Ex}_d$
Condenser	MBE: $\dot{m}_2 = \dot{m}_3$
	EBE: $\dot{m}_2 h_2 = \dot{m}_3 h_3 + \dot{Q}_{out}$
	EnBE: $\dot{m}_2 s_2 + \dot{S}_{gen} = \dot{m}_3 s_3 + \frac{\dot{Q}_{out}}{T_{sink}}$
	ExBE: $\dot{m}_2 ex_2 = \dot{m}_3 ex_3 + \dot{Q}_{out}\left(1 - \frac{T_o}{T_{sink}}\right) + \dot{Ex}_d$
Evaporator	MBE: $\dot{m}_4 = \dot{m}_1$
	EBE: $\dot{m}_4 h_4 + \dot{Q}_{in} = \dot{m}_1 h_1$
	EnBE: $\dot{m}_4 s_4 + \frac{\dot{Q}_{in}}{T_{space}} + \dot{S}_{gen} = \dot{m}_1 s_1$
	ExBE: $\dot{m}_4 ex_4 + \dot{Q}_{in}\left(1 - \frac{T_o}{T_{space}}\right) = \dot{m}_1 ex_1 + \dot{Ex}_d$
Expansion valve	MBE: $\dot{m}_3 = \dot{m}_4$
	EBE: $\dot{m}_3 h_3 = \dot{m}_4 h_4$
	EnBE: $\dot{m}_3 s_3 + \dot{S}_{gen} = \dot{m}_4 s_4$
	ExBE: $\dot{m}_3 ex_3 = \dot{m}_4 ex_4 + \dot{Ex}_d$

The exergetic COP of a heat pump can be expressed as follows:

$$COP_{ex,\,Refrigeration} = \frac{\dot{E}x^{Q_{out}}}{\dot{W}} = \frac{\dot{E}x^{Q_{out}}}{\dot{E}x^{Q_{out}} - \dot{E}x^{Q}} \tag{2.116}$$

It can be easily shown that for given values Q_{in} and Q_{out} the COPs of a refrigerator and a heat pump are related to each other by

$$COP_{en,Heatpump} = COP_{en,Refrigerator} + 1 \tag{2.117}$$

This shows that the COP of a heat pump is >1. The COP of a refrigerator can be less than 1 or >1.

2.5.2 Brayton cycles

A simple Brayton cycle of an open type consists of a compressor, a combustion chamber and gas turbine. In an actual combustion turbine, the working fluid is signified by pressurized combustion gases, which expand and produce power. The air delivered to the combustion chamber is typically about 4–50 times more than the fuel mass flow [1]. Henceforward, air is highly existent in combustion gases, thus, resembling the features of combustion gases with that of air is a common approach for the analysis. If the working fluid is modeled as an ideal gas (air) with specific heats corresponding to 298 K ($k = 1.4$), then the complete cycle is called the air-standard Brayton cycle. If all internal processes are reversible, then the cycle is called the ideal standard-air Brayton cycle. Thus, herein for simplification of the analysis, the

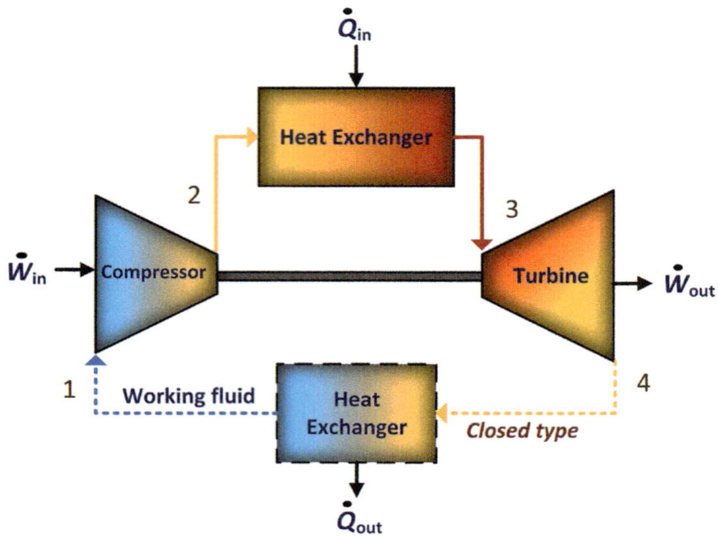

FIG. 2.18 A simple schematic diagram of open and closed type Brayton cycle based power plant.

working fluid is assumed to be standard air. In addition, the compressors, turbines and heat exchangers work with zero internal irreversibilities for this example.

The simple ideal Brayton cycle consists of four fundamental processes: isentropic compression, isobaric heat addition, isentropic expansion and isobaric heat rejection, as shown in Fig. 2.18. The processes taking place during addition and rejection of heat are isobaric rather than isothermal in the Brayton cycle, so this cycle is externally irreversible. The Brayton cycle can be employed both as an internal combustion engine system and as an external combustion system, based on the method in which the isobaric heat transfer takes place at the heat source side.

As a result of the direct link with the atmosphere, the entry pressure and gas ejecting pressure are identical to the atmospheric pressure, implying a constraint of turbine work. This limitation can be eliminated once the cycle is closed via a heat exchanger for a constant pressure heat rejection process. Therefore, the turbine can release in vacuum, meaning that more work is produced and an improved efficiency can be attained [1]. The turbine exit pressure could be varied indirectly by changing the temperature of the heat sink. On the other hand, the closed Brayton cycle typically operates with an inert working fluid such as dry air or helium. The heat source can be a combustion process, but it needs to be exchanged externally via a heat exchanger. The combustion gases do not mix with the working fluid in this case, and the Brayton cycle based system becomes an external combustion engine. It is important to note that the open and closed Brayton cycles are equivalent from the thermodynamic point of view. In the open cycle, the atmosphere plays the role of cooler, where the isobaric cooling process occurs, while in a closed cycle the same process occurs in a heat exchanger.

The energy and exergy balance equations can be written for each of the cycle processes. It is assumed that the surrounding atmosphere is at standard conditions defined by T_o, P_o which define the reference state for exergy calculations. The energy (EBE), entropy (EnBE) and exergy (ExBE) balance equations for each process are given in Table 2.6. Although the ideal cycle is internally reversible, meaning that the compression and expansion processes are

TABLE 2.6 The fundamental balance equations for the ideal standard air simple Brayton cycle

Process	MBE	EBE	EnBE	ExBE
Compression (1–2)	$\dot{m}_1 = \dot{m}_2$	$\dot{m}_1 h_1 + \dot{W}_c = \dot{m}_2 h_2$	$\dot{m}_1 s_1 + \dot{S}_{gen} = \dot{m}_2 s_2$ $\dot{S}_{gen,C} = 0$	$\dot{m}_1 ex_1 + \dot{W}_c = \dot{m}_2 ex_2 + \dot{E}x_{d,C}$ $\dot{E}x_{d,C} = 0$
Heat Addition (2–3)	$\dot{m}_2 = \dot{m}_3$	$\dot{m}_2 h_2 + \dot{Q}_{in} = \dot{m}_3 h_3$	$\dot{m}_2 s_2 + \frac{\dot{Q}_{in}}{T_3} + \dot{S}_{gen,CC} = \dot{m}_3 s_3$	$\dot{m}_2 ex_2 + \dot{Q}_{in}\left(1 - \frac{T_0}{T_3}\right) = \dot{m}_3 ex_3 + \dot{E}x_{d,CC}$
Expansion (3–4)	$\dot{m}_3 = \dot{m}_4$	$\dot{m}_3 h_3 = \dot{m}_4 h_4 + \dot{W}_T$	$\dot{m}_3 s_3 + \dot{S}_{gen,T} = \dot{m}_4 s_4$ $\dot{S}_{gen,T} = 0$	$\dot{m}_3 ex_3 = \dot{m}_4 ex_4 + \dot{W}_T + \dot{E}x_{d,T}$ $\dot{E}x_{d,t} = 0$
Heat Rejection (4–1)	$\dot{m}_4 = \dot{m}_1$	$\dot{m}_4 h_4 = \dot{m}_1 h_1 + \dot{Q}_{out}$	$\dot{m}_4 s_4 + \dot{S}_{gen,HX} = \dot{m}_1 s_1 + \frac{\dot{Q}_{out}}{T_1}$	$\dot{m}_4 ex_4 = \dot{m}_1 ex_1 + \dot{Q}_{out}\left(1 - \frac{T_0}{T_1}\right) + \dot{E}x_{d,HX}$
Overall	N/A	$\dot{W}_c + \dot{Q}_{in} = \dot{W}_t + \dot{Q}_{out}$	$\frac{\dot{Q}_{in}}{T_3} + \dot{S}_{gen} = \frac{\dot{Q}_{out}}{T_1}$	$\dot{W}_c + \dot{Q}_{in}\left(1 - \frac{T_0}{T_3}\right) = \dot{W}_t + \dot{Q}_{out}\left(1 - \frac{T_0}{T_1}\right) + \dot{E}x_d$

isentropic, the heat transfer processes at heat sink and heat source are presumed internally reversible but externally irreversible, meaning that the system boundary at the source is set to the highest temperature (T_3) within the cycle, that is after the combustion chamber or heat exchanger and the sink temperature is set to the lowest temperature within the cycle (T_1). Hence, there is entropy generation and associated exergy destruction within the cycle.

Note that since the compression and expansion processes are assumed to be adiabatic in this case, associated entropy generation and exergy destruction terms are zero.

The net work obtained from the cycle is signified by the difference of turbine work and compressor work $\dot{W}_{net} = \dot{W}_T - \dot{W}_c$. In addition, the net produced work is written based on the back work ratio (BWR). The BWR signifies the ratio between the work consumed by the compressor and the work produced by the turbine as follows:

$$BWR = \dot{W}_c / \dot{W}_t \tag{2.118}$$

The energy efficiency of the Brayton cycle for power production can be characterized by the ratio of net work production and the rate of heat input. Therefore, the energy and exergy efficiencies are expressed respectively as follows:

$$\eta_{en} = 1 - \frac{\dot{Q}_{out}}{\dot{Q}_{in}} = \frac{\dot{W}_{net}}{\dot{Q}_{in}} \tag{2.119}$$

$$\eta_{ex_1} = \frac{\dot{Q}_{out} - \dot{Q}_{in}}{\dot{Q}_{in}\left(1 - \frac{T_0}{T_3}\right)} = \frac{\dot{W}_{net}}{\dot{Q}_{in}\left(1 - \frac{T_0}{T_3}\right)} = \frac{\eta_{en}}{1 - \frac{T_0}{T_3}} > \eta_{en} \tag{2.120}$$

Here, the exergy efficiency (η_{ex1}) of the cycle is presented as the ratio of useful work output to the total exergy input at the heat source. In the exergetic heat equation, the temperature is commonly taken as the highest temperature in the cycle which is after the combustion chamber. It is in fact the source temperature, which can be taken as the combustion temperature for the Brayton cycle.

Alternatively, the exergy efficiency can be defined with respect to the exergy taken by the cycle, which is signified by the exergy input at source (Ex_{in}) and the exergy output at sink (Ex_{out}) as follows:

$$\eta_{\text{ex}_2} = \frac{\dot{W}_{\text{net}}}{\dot{\text{E}}\text{x}_{\text{in}} - \dot{\text{E}}\text{x}_{\text{out}}} = 1 - \frac{\dot{\text{E}}\text{x}_{\text{d}}}{\dot{Q}_{\text{in}}\left(1 - \frac{T_0}{T_3}\right) - \dot{Q}_{\text{out}}\left(1 - \frac{T_0}{T_1}\right)} \tag{2.121}$$

There is another way of writing the exergy efficiency of the cycle, namely as the ratio of its actual energy efficiency and the efficiency of a reversible heat engine operating between T_1 and T_3 heat medium, as follows:

$$\eta_{\text{ex}_3} = \frac{\eta_{\text{en}}}{1 - \frac{T_1}{T_3}} = \frac{1 - \frac{\dot{Q}_{\text{out}}}{\dot{Q}_{\text{in}}}}{1 - \frac{T_1}{T_3}} \tag{2.122}$$

It is noted that the preceding three exergy efficiency definitions are identical $\eta_{\text{ex1}} = \eta_{\text{ex2}} = \eta_{\text{ex3}}$ once $T_1 = T_0$. This characteristic is mostly applicable for the open type Brayton cycle. Hence it is usual to accept that the reference temperature is identical to the ambient air. For specific circumstances with closed Brayton cycles, the temperature T_1 can be different from T_0, thus $\eta_{\text{ex1}} \neq \eta_{\text{ex2}} \neq \eta_{\text{ex3}}$.

For the standard air Brayton cycle, the ideal gas equation of state is utilized to develop a broader methodical formulation for cycle efficiency. The ideal gas equation of state undertakes constant specific heats, meaning that it is independent of temperature. For standard air, the specific heats are estimated at 298.15 K. Using the general equation, $\Delta h = c_p \Delta T$ and the energy balance equation (EBE), the energy efficiency becomes

$$\eta_{\text{en}} = 1 - \frac{c_p(T_4 - T_1)}{c_p(T_3 - T_2)} = 1 - \left(\frac{T_1}{T_2}\right)\frac{\left(\frac{T_4}{T_1}\right) - 1}{\left(\frac{T_3}{T_2}\right) - 1} \tag{2.123}$$

Another common equation for ideal gas defines the relationship between pressure and temperature along an isentropic process 1–2 according to $T_1 P_2^{\gamma} = T_2 P_1^{\gamma}$, with

$$\gamma = (k-1)/k \tag{2.124}$$

and k the specific heat ratio. Since processes 1–2 and 3–4 are isentropic and change between the same pressures, $P_1 = P_4$ and $P_3 = P_2$, one has that

$$\frac{T_3}{T_4} = \left(\frac{P_3}{P_4}\right)^{\kappa} = \left(\frac{P_2}{P_1}\right)^{\kappa} = \frac{T_2}{T_1} \rightarrow \frac{T_4}{T_1} = \frac{T_3}{T_2} \tag{2.125}$$

Consequently, the energy efficiency of a standard air Brayton cycle can also be written as

$$\eta = 1 - r^{-\kappa} \tag{2.126}$$

where the pressure ratio is defined by $r = \frac{P_2}{P_1} > 1$.

2.5.3 Rankine cycles

The steam Rankine cycle is one of the highest utilized types of thermodynamic cycles for electricity production in the world [1]. The steam power stations run on Rankine cycles using steam as a working fluid. The simple ideal Rankine cycle is in fact the most basic vapor power cycle. This ideal thermodynamic cycle consists of four main components: an ideal pump that works isentropically and adiabatically, an ideal heat exchanger system for vapor generation, a turbine operating isentropically and adiabatically, and a condenser without pressure losses and infinite heat transfer surface. The vapor generator is commonly shown as a boiler. The main purpose is to obtain superheated steam. Hence, it may contain three different phases: preheating phase, boiling phase and vapor superheating phase. It works without pressure losses and without temperature difference between the heat source and the working fluid. Usually, steam is utilized as the working fluid for conventional steam power plants. If the working fluid is steam, the cycle is called the simple ideal steam Rankine cycle. The simple layout of a Rankine cycle is shown in Fig. 2.19.

The simple ideal Rankine cycle is not a totally reversible cycle since the heat addition process at the source part is not isothermal. Thus, there is external irreversibility because of heat transfer at the source part. As it is externally reversible, an exergy efficiency less than 1 should be observed.

The mass, energy, entropy and exergy balance equations for components of the cycle and the overall balance equations for the cycle are written in Table 2.7. In a simple ideal Rankine cycle, there are no exergy destructions in the condenser, pump and turbine because of isentropic and adiabatic processes. However, there is exergy destruction in the boiler of the cycle. The irreversibilities in the vapor generator are estimated according to the temperature of the heat source. However, if the processes are not assumed to be adiabatic and isentropic, it is expected to have higher exergy destructions.

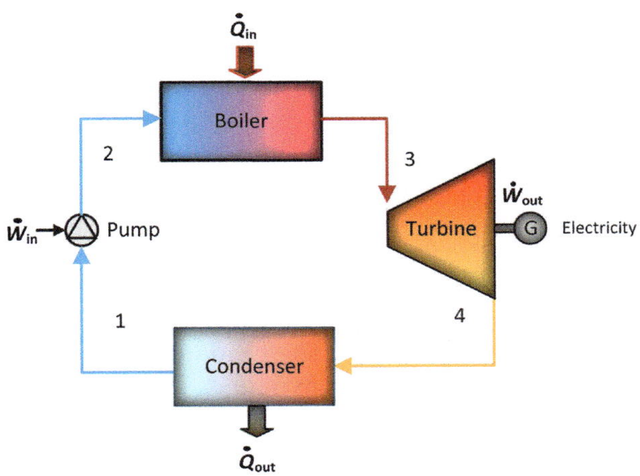

FIG. 2.19 A simple layout of an ideal steam Rankine power plant.

TABLE 2.7 Fundamental balance equations for simple ideal Rankine cycle

Component	Balance equations
Pump (isentropic process 1–2)	MBE: $\dot{m}_1 = \dot{m}_2$ EBE: $\dot{m}_1 h_1 + \dot{W}_{in} = \dot{m}_2 h_2$ EnBE:$\dot{m}_1 s_1 + \dot{S}_{gen} = \dot{m}_2 s_2$ ExBE:$\dot{m}_1 ex_1 + \dot{W}_{in} = \dot{m}_2\, ex_2 + \dot{Ex}_d$
Turbine (adiabatic process 3–4)	MBE: $\dot{m}_3 = \dot{m}_4$ EBE: $\dot{m}_3 h_3 = \dot{m}_4 h_4 + \dot{W}_{out}$ EnBE: $\dot{m}_3 s_3 + \dot{S}_{gen} = \dot{m}_4 s_4$ ExBE: $\dot{m}_3 ex_3 = \dot{m}_4 ex_4 + \dot{W}_{out} + \dot{Ex}_d$
Condenser (isobaric process 4–1)	MBE: $\dot{m}_4 = \dot{m}_1$ EBE: $\dot{m}_4 h_4 = \dot{m}_1 h_1 + \dot{Q}_{out}$ EnBE: $\dot{m}_4 s_4 + \dot{S}_{gen} = \dot{m}_1 s_1 + \frac{\dot{Q}_{out}}{T_s}$ ExBE: $\dot{m}_4\, ex_4 = \dot{m}_1 ex_1 + \dot{Q}_{out}\left(1 - \frac{T_0}{T_s}\right) + \dot{Ex}_d$
Boiler (isobaric process 2–3)	MBE: $\dot{m}_2 = \dot{m}_3$ EBE: $\dot{m}_2 h_2 + \dot{Q}_{in} = \dot{m}_3 h_3$ EnBE: $\dot{m}_2 s_2 + \frac{\dot{Q}_{in}}{T_{source}} + \dot{S}_{gen} = \dot{m}_3 s_3$ ExBE: $\dot{m}_2 ex_2 + \dot{Q}_{in}\left(1 - \frac{T_0}{T_{source}}\right) = \dot{m}_3 ex_3 + \dot{Ex}_d$

The energy efficiency of the cycle is written using net work output and the total heat input from the source as follows:

$$\eta_{en} = \frac{\dot{W}_{net}}{\dot{Q}_{in}} = \frac{\dot{W}_{out} - \dot{W}_{in}}{\dot{Q}_{in}} \tag{2.127}$$

The exergy efficiency implies the ratio between the net exergy output (net work production) and the actual exergy input. In the Rankine cycle, the exergy required is the input to the boiler, so the exergy efficiency can be expressed as follows:

$$\eta_{ex} = \frac{\dot{W}_{net}}{\dot{Ex}^{Q_{in}}} = \frac{\dot{W}_{out} - \dot{W}_{in}}{\dot{Ex}^{Q_{in}}} = \frac{\dot{W}_{net}}{\dot{Q}_{in}\left(1 - \frac{T_o}{T_{source}}\right)} \tag{2.128}$$

2.6 Conclusions

In this chapter, thermodynamic fundamentals are presented along with detailed balance equations of main components used in integrated energy systems. The fundamental laws of thermodynamics are first introduced. The Zeroth Law of Thermodynamics is a statement about thermodynamic equilibrium. The First Law of Thermodynamics (FLT) is the law of the conservation of energy. The Second Law of Thermodynamics (SLT) refers to the concept of reversible and irreversible processes and is summarized as non-conservation of exergy. The Third Law of Thermodynamics is related to the entropy of a pure substance at absolute zero temperature, that is 0 K. The degree of difference between irreversible and reversible processes is evaluated by the entropy concept in thermodynamics. Exergy is defined as useful energy or available energy or work potential. There are two different types of energy systems: open and closed. The open systems have mass, heat and work interactions whereas there is no mass transfer crossing the boundary in closed systems. Exergy analysis can have an important role for sustainable energy systems since exergy essentially considers environmental conditions in which a system works. The performance of a thermodynamic system can be evaluated using energy and exergy efficiencies. The general approach of defining efficiency is the ratio of the useful output and the required input. The fundamental balance equations for the most commonly used thermodynamic components are presented along with the efficiency definitions. These equations will be employed in the next chapters for specific system analyses.

The energy and exergy analysis of three key thermodynamics cycles, namely (i) Brayton cycle, (ii) Rankine cycle, (iii) refrigeration/heat pump cycle, are introduced. A refrigerator or a heat pump is a system/cycle that is used to transfer heat from a low-temperature reservoir to a high-temperature reservoir. A simple Brayton cycle is used to generate power through gas turbines where the source of heat is either combustion or constant pressure heat addition. Rankine cycles are the most commonly used thermodynamic cycle for power generation. The working fluid is typically steam, but other fluids can also be recycled, for which the cycle is called the organic Rankine cycle.

This chapter provides a basis for system integration and analysis to be discussed in the next chapters.

Nomenclature

μ_i chemical potential
C_p specific heat capacity at constant pressure (kJ/kg K)
C_v specific heat capacity at constant volume (kJ/kg K)
I_v luminous intensity (cd)
k_B Boltzmann constant (J/K)
N_a Avogadro number
E energy (J or kJ)
e specific total energy (kJ/kg)
Ex exergy (J or kJ)
g gravitational acceleration (m/s^2)
h enthalpy (kJ/kg)
h Planck's constant (J s)
I electric current (A)
K equilibrium constant
k specific heat ratio
L length (m)
m mass (kg)
n mole (mol)
p probability
P pressure (kPa)
Q heat (J or kJ)
R universal gas constant
S entropy (kJ/K)
T temperature (°C)
U internal energy (J or kJ)

V volume (m^3)
v specific volume (m^3/kg)
W work (J or kJ)
z elevation (m)
Δg Gibbs free energy (J or kJ)
K exponential parameter
ν velocity (m/s)
ex specific total exergy (kJ/kg)
c speed of light in vacuum (m/s)
T time (s)

Acronyms

BWR back work ratio
COP coefficient of performance
EBE energy balance equation
EnBE entropy balance equation
ExBE exergy balance equation
FLT first law of thermodynamics
HHV higher heating value
HV heating value
HX heat exchanger
KE kinetic energy
LHV lower heating value
MBE mass balance equation
NS nonspontaneous
PE potential energy
S spontaneous
SI International System of Units
SLT second law of thermodynamics
SSSF steady-state steady-flow

Greek letters

y_i molar fraction
$\varnothing$ stoichiometric factor
Δ change
η efficiency

Subscripts and superscripts

C compressor
cc combustion chamber
cv control volume
d destruction
En energy
Ex exergy
f final
gen generation
H high

HX heat exchanger
i initial
in input
L low
Max maximum
o reference state
out output
rev reversible
s source or sink
sys system
T turbine

References

[1] Dincer I, Zamfirescu C. 1.5 Thermodynamic aspects of energy. In: Dincer I, editor. Compr. Energy Syst. 1st ed. Cambridge, United States: Elsevier; 2018. p. 153–211. https://doi.org/10.1016/B978-0-12-809597-3.00105-X.

[2] The International System of Units (SI). Standards. 8th ed.

[3] Çengel YA, Boles MA. Thermodynamics: an engineering approach. 8th ed. US: McGraw-Hill Education; 2015.

[4] Dincer I, Rosen MA. Exergy: energy, environment and sustainable development. In: Exergy. 2nd ed. New York, NY: Elsevier; 2013. p. 51–73. https://doi.org/10.1016/B978-0-08-097089-9.00004-8.

[5] Dincer I. Exergization. Int J Energy Res 2016;40:1887–9. https://doi.org/10.1002/er.3606.

[6] Dincer I. 1.6 Exergy. In: Dincer I, editor. Compr. Energy Syst. 1st ed. Cambridge, United States: Elsevier; 2018. p. 212–64. https://doi.org/10.1016/B978-0-12-809597-3.00106-1.

[7] Kotas TJ. The exergy method of thermal plant analysis. 1st ed. Essex, Great Britain: Elsevier; 1985. https://doi.org/10.1016/C2013-0-00894-8.

[8] Garfias M. Standard chemical exergy of elements updated. Energy 2006;31:3310–26. https://doi.org/10.1016/J.ENERGY.2006.03.020.

[9] Kanoğlu M, Çengel YA, Dinçer İ. Efficiency evaluation of energy systems. New York, NY: Springer New York; 2012. https://doi.org/10.1007/978-1-4614-2242-6.

[10] Dincer I, Zamfirescu C. Chapter 1—Fundamentals of thermodynamics. In: Zamfirescu ID, editor. Adv. Power Gener. Syst. Boston: Elsevier; 2014. p. 1–53. https://doi.org/10.1016/B978-0-12-383860-5.00001-8.

[11] Szargut J, Morris DR, Steward FR. Exergy analysis of thermal, chemical, and metallurgical processes. 1st ed Hemisphere; 1988.

CHAPTER

3

System integration for multigeneration

3.1 Introduction

Energy is crucial to solving the most significant problems of this era and the future, such as climate change, health, environment, sustainable development, worldwide energy and food security, and environmental security. However, conventional energy systems do not satisfy the multidimensional and multidisciplinary needs of this century unless they are greenized or significantly improved.

Securing the energy supply is vital for the development and progress of all civilizations. Energy demand has increased exponentially since the Industrial Revolution because of worldwide population growth and the expanding life standards of humanity. These issues have directed scientists, researchers, engineers and many other people to focus on energetic and environmental factors for achieving a sustainable future.

Industrialization is greatly dependent on energy as well as raw material inputs. Transportation, processing and utilization of raw materials for various industries necessitate huge amounts of energy in fuel, heat, electricity or chemical energy form. The energy needs of residential and transportation segments are also two of the largest contributors to global energy demand [1]. The transportation sector is heavily dependent on fossil fuel resources, mostly gasoline, diesel, heavy fuel oil and kerosene. For residential requirements such as heating and cooling, the world is still dependent on conventional energy resources.

There are significant dimensions that need to be simultaneously considered while designing energy systems. The major ones are environmental sustainability, economic growth, and social acceptance without compromising on any characteristics of the energy 3S (source-system-service) chain. The additional criteria to be considered are convenience, cost, readiness, safety, health, climate, and environmental security [1].

For an effective transition from conventional to sustainable energy systems, the expectations are to design, develop, analyze, evaluate and employ sustainable solutions for enhancing the energetic, exergetic and environmental performances of the systems. There is a solid requirement to transform conventional methods into greener methods using the most promising techniques by accounting for multiple principles, for instance environmental impact,

Integrated Energy Systems for Multigeneration
https://doi.org/10.1016/B978-0-12-809943-8.00003-0

resource utilization, efficiency, and cost-effectiveness, which will help realize better sustainability throughout the world.

Humanity seeks sustainable conversion from energy and material resources to diverse types of energy and industrial products. The efficient transfer of energy using energy carriers is also a vital concern. Therefore, numerous forms of energy carriers like electricity, chemical fuels (hydrogen, ammonia, methanol, etc.), heat, etc. are being seriously considered. All of these energy carriers are crucial for transporting the energy produced in remote and rural locations to urban areas, where the energy demand is higher. To achieve sustainable energy systems, there is an urgent need to deliver novel, integrated, and cost-effective energy storage solutions using clean energy carriers.

The research and development activities are focused on achieving the following criteria for better sustainability [2]:

- Minimization of cost
- Reduction of environmentally harmful emissions
- Decrease of size and volume
- Improvement in reliability and durability
- Optimization of service, training, and maintenance plans
- Establishment of safety, standards and certification
- Enhancement of system efficiency

3.2 System design

Sustainable energy systems must possess a number of features simultaneously, such as efficiency, environment, resource use, cost, sustainability, integrability, commercial viability and others, as illustrated in Fig. 3.1. When designing sustainable energy systems, these aspects should all be carefully considered and taken into account. Careful design of renewable energy systems can address many of the sustainability criteria shown in Fig. 3.1.

Each of these dimensions is briefly explained here, based on Dincer and Acar [1]:

- *Energetically secure*: Securing diverse energy resources is vital for any country or society. If there is a high dependency on a single type of energy source, the energy supply is always vulnerable to interruptions. Sustainable energy systems need to make use of abundant, reliable, locally accessible, cost-effective and replenished sources in order to become safe, self-sufficient, effective and therefore secure. Only in this way can a secure energy supply to end users be broadly guaranteed.
- *Environmentally benign*: Any sustainable energy system must be clean and less polluting at every stage of energy production, conversion and utilization from source to final use. One of the advantages of sustainable energy systems is the ability of waste and loss recovery for both energy and materials. Having less waste and loss leads to more effective systems, lower emissions, and a better environment for sustainable development.

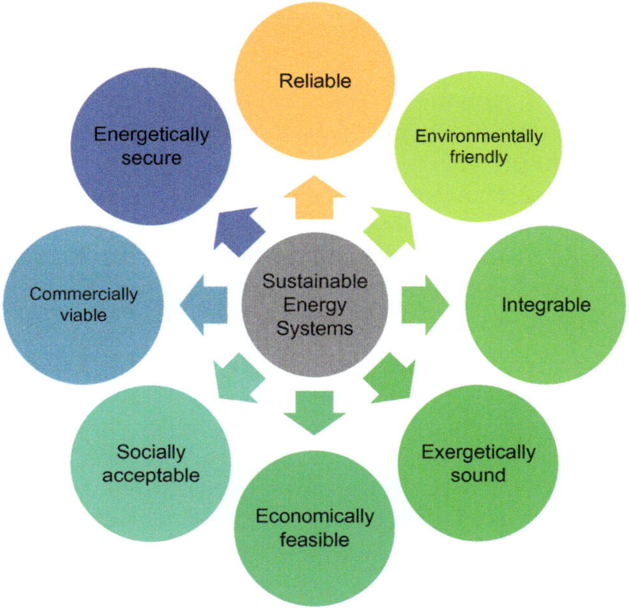

FIG. 3.1 Main features of sustainable energy systems.

- *Reliable*: In each energy-associated step, from extraction to final consumption, sustainable energy systems must be reliable, using simply available resources and reliable energy processing/transformation systems, and delivering reliable service for end use. Having reliable systems increases the probability of social acceptance.
- *Exergetically sound*: As a significant quality indicator, exergy is a vital element of sustainable energy systems. Exergetically sound systems tend to be highly sustainable because of the fact that higher exergy efficiencies bring lower environmental emissions and more reliable operations. The term exergetically sound implies maximum possible exergy efficiency and minimum exergy destructions. By considering exergetically sound systems, the quality of energy is also taken into account as well as the quantity.
- *Integrable*: System integration brings numerous benefits over non-integrated systems. Therefore, sustainable energy systems that are integrable can help achieve the integration of many systems for the production of multiple useful commodities. Integration enables a reduction in individual deficiencies and can make the overall system more effective. The integration process can be expressed as a definitive operation in which energy systems and/or sources are combined in a synergetic arrangement to attain cost effectiveness and better efficiency, environment and resources use.
- *Economically feasible*: Cost-effectiveness is one of the top decision criteria in energy systems. Sustainable energy systems need to be affordable and economically feasible through utilizing reliable, available, and abundant resources. Furthermore, especially in integrated energy systems, we can minimize losses and waste and maximize overall efficiencies and anticipated commodities to minimize the cost.

- *Commercially viable*: A sustainable energy system must utilize well developed and already existent or easily available sources/systems and must deliver the anticipated merchandise and services that are commercially viable. This can increase market competitiveness and make the system economically profitable. Although some novel systems may seem to bear high initial and capital costs, through continuous developments and enhancements, they may become commercially viable in future energy systems.
- *Socially acceptable*: In order to have a successful sustainable energy system, it needs to be recognized by the society to become a part of daily life. A sustainable energy system is expected to be socially acceptable to the local and global communities, as such systems can satisfy social needs and synchronize the possibilities. This is particularly true since sustainability is also concerned in the final end use stage of energy.

To achieve sustainable energy systems, current and future resources, technologies, knowledge, and policies need to be operated in collaboration with each other. Sustainable energy systems have numerous benefits, namely: better health and environment, better employment, better economies, better productivity, better social welfare, better infrastructure, and better energy security. A sustainable future requires a sustainable transition from current conventional energy systems to better ones having improved energy and exergy efficiencies at each stage of energy production, conversion and consumption and greater use of renewable and clean energy resources.

In this respect, the seven pillars of sustainable energy systems originally proposed by Dincer [3] can be summarized as shown in Fig. 3.2.

Production of a variety of anticipated outputs from energy sources is a favorable option to enhance the energy conversion efficiency from source to product. Multigeneration systems offer distinctive benefits, especially when combined with clean energy sources, such as renewables. Ultimately, multigeneration systems can reduce wastes and environmental impacts, while increasing system efficiencies and cost-effectiveness. Furthermore, in comparison to conventional single generation systems, multigeneration systems can deliver multiple outputs concurrently such as electricity, heating, cooling, chemical fuels, value-added products, energy storage, etc.

Designing sustainable energy systems using the seven pillars requires various design criteria originally proposed by Dincer [4] and illustrated in Fig. 3.3. They are categorized into eight different topics as explained in the forthcoming section: exergization [3], greenization [5], renewabilization, hydrogenization, integration, multigeneration, storagization and intelligization.

3.2.1 Exergization

Escalation of worldwide energetic and environmental matters has driven some researchers, scientists, engineers, and technologists to seek clean and sustainable energy solutions for execution. The solutions require exergization as a critical method of using exergy analysis as a potential technique in this discipline [3]. There have been studies on the utilization of exergy analysis in numerous disciplines and areas, extending from mechanical to chemical, from biology to medical and from economy to public relations. Hence, the process

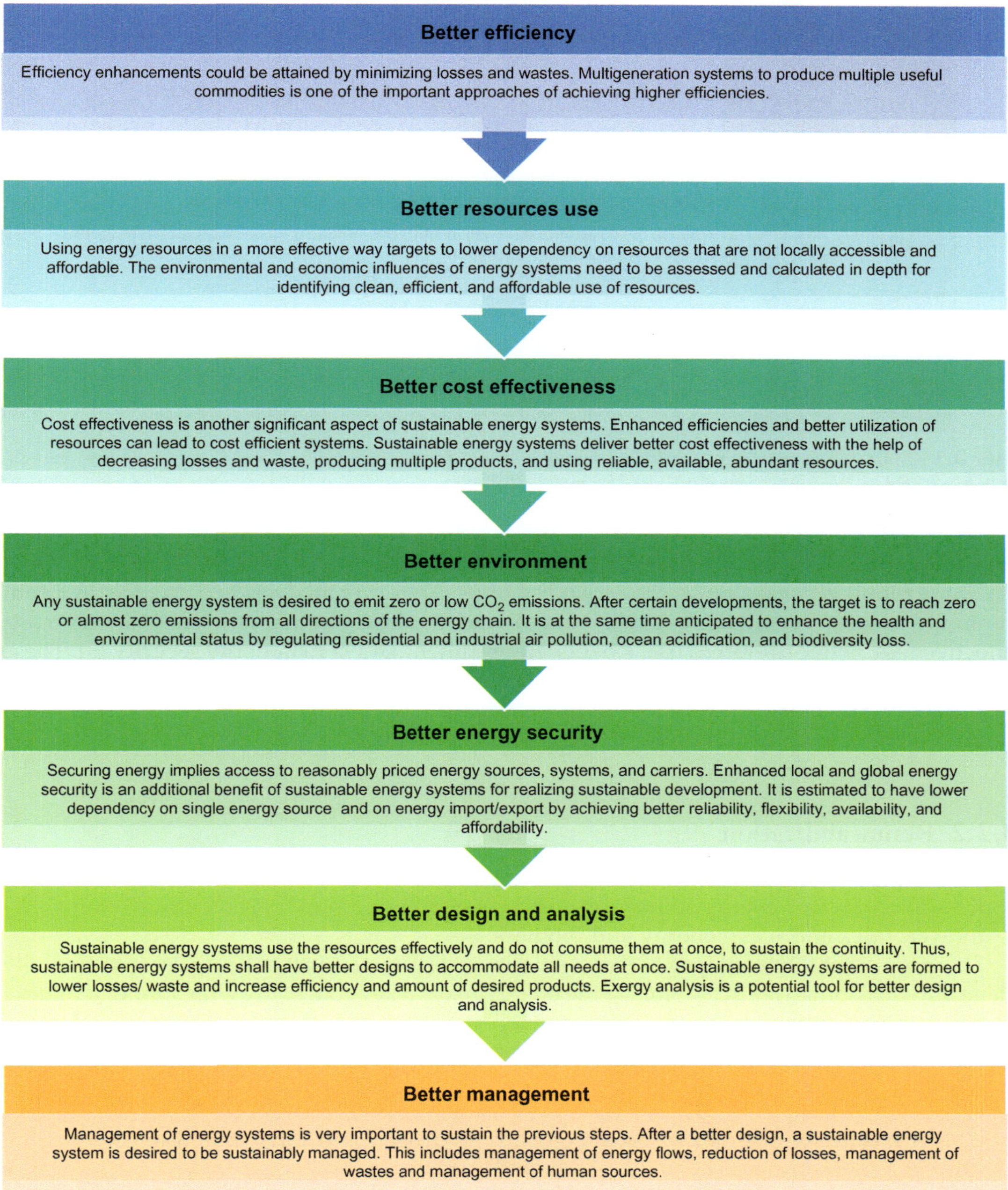

FIG. 3.2 Seven pillars of sustainable energy systems. *Modified after Dincer I. Exergization. Int J Energy Res 2016;40:1887–9. https://doi.org/10.1002/er.3606*

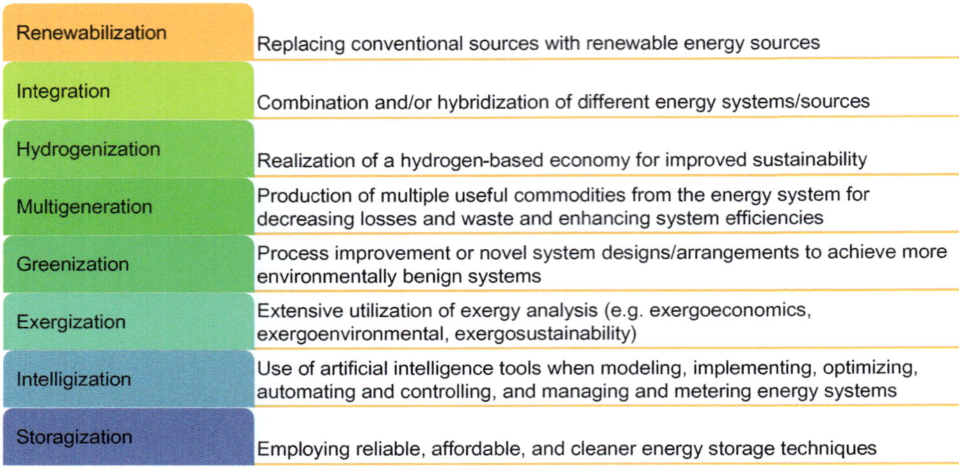

FIG. 3.3 Main design criteria of sustainable energy systems. *Adapted from Dincer I. Smart energy solutions. Int J Energy Res 2016;40:1741–2. https://doi.org/10.1002/er.3621*

of exergization proposes several routes in various dimensions, including efficiency, effectiveness, cost, environment, sustainability, size, etc.

Exergization is expressed as the capability of using exergy analysis and its tools in a conceptually correct manner for better design and analysis, better efficiency, better cost-effectiveness, better resources use, a better environment and better energy security [3]. These factors are important in realizing better sustainability, which decisively needs exergization as a crucial process. Due to its multicomponent structure, the process of exergization shall comprehensively be employed in each discipline by accounting for the interrelations.

3.2.2 Renewabilization

Renewabilization is described as the process of switching from conventional fossil fuels sources to renewable energy sources, e.g. solar energy, wind power, geothermal energy, hydropower, tides and wave energy and biomass.

There are several advantages of renewable energy systems, such as being environmentally clean and distributed sources. Storing the energy plays a key role in the wide implementation of renewables. Carbon-free energy carriers such as hydrogen and ammonia are crucial to improve the role of renewables in sustainable energy systems for realizing sustainable development. These energy carriers can possibly assist grid modernization and can be an energy storage media for renewable electricity, to be utilized as a fuel for numerous purposes or to deliver electricity at times of peak demand. Hydrogen and other renewable fuels such as ammonia can expedite the growth of renewable power generation, as they can store energy in chemical form and they can be produced by surplus electrical energy by electrochemical routes. The stored energy can either be utilized for other systems such as cogeneration, transportation fuel, heating, cooling, etc. or it can be transformed back into grid electricity, using fuel cells, generators or turbines. One of the best advantages of hydrogen and ammonia is the ability of being transported long distances through pipelines or trailers, hypothetically at higher efficiency and less expense than conventional long-distance electricity transmission. Therefore, carbon-free energy carriers can enable renewabilization via the transmission of renewable energy sources in remote sites to the locations of need very effectively. The difference between greenization and renewabilization is that greenization deals with the carbon neutrality accompanied by decreased negative environmental impact, and renewabilization deals with the resources being replaced at a higher (or equal) rate compared to the consumption rate [6].

3.2.3 Hydrogenization

Hydrogenization is described as the method in which hydrogen is intensively utilized as a clean energy carrier and fuel for systems and applications, to accomplish a hydrogen economy for better sustainability [6]. The role of hydrogen is in fact very significant in sustainable energy systems. As a storage medium, hydrogen can be compressed or liquefied or can be converted into other chemical fuels such as ammonia and methanol to serve as gaseous or liquid fuel. There are various methods for storing hydrogen, such as compressed gas in salt caverns, metal hydrides, and liquefied cylinders, etc. One of the most attractive hydrogen storage techniques is in the form of ammonia, which has three hydrogen atoms and requires lower storage pressures than hydrogen. Methanol is another alternative chemical to store hydrogen in tanks (liquid at atmospheric conditions) and it can be used as fuel for transportation as well. Converting hydrogen into useful commodities is mostly performed by fuel cells and hydrogen-powered internal combustion engines, which are expected to intensify hydrogenization in sustainable energy systems.

Large-scale hydrogen generation can help obtain competitive hydrogen prices. However, once the renewable energy systems become lowercost, hydrogen can also be produced on smaller scales on site using renewable resources. With sustainable hydrogen, the chemical industry can potentially decarbonize their chemical products and thus play a part in the greenhouse gas emission reduction. The roles of hydrogenization in sustainable energy systems are further explained in Fig. 3.4.

3.2.4 Integration

Integration in this context is defined as a method in which energy systems/sources are combined and/or hybridized to attain better efficiency, cost-effectiveness, resources use and environment [6]. Integrating individual subsystems can eliminate the deficiencies of each system. For example, the intermittent nature of solar energy can be overcome by integration with geothermal energy, which is considered a continuous energy supply. The integration concept is further elaborated in detail in Section 3 of this chapter, "Integration and multigeneration."

3.2.5 Multigeneration

Multigeneration refers to the method producing four or more useful commodities by using the same input or system, because there are already cogeneration as well as

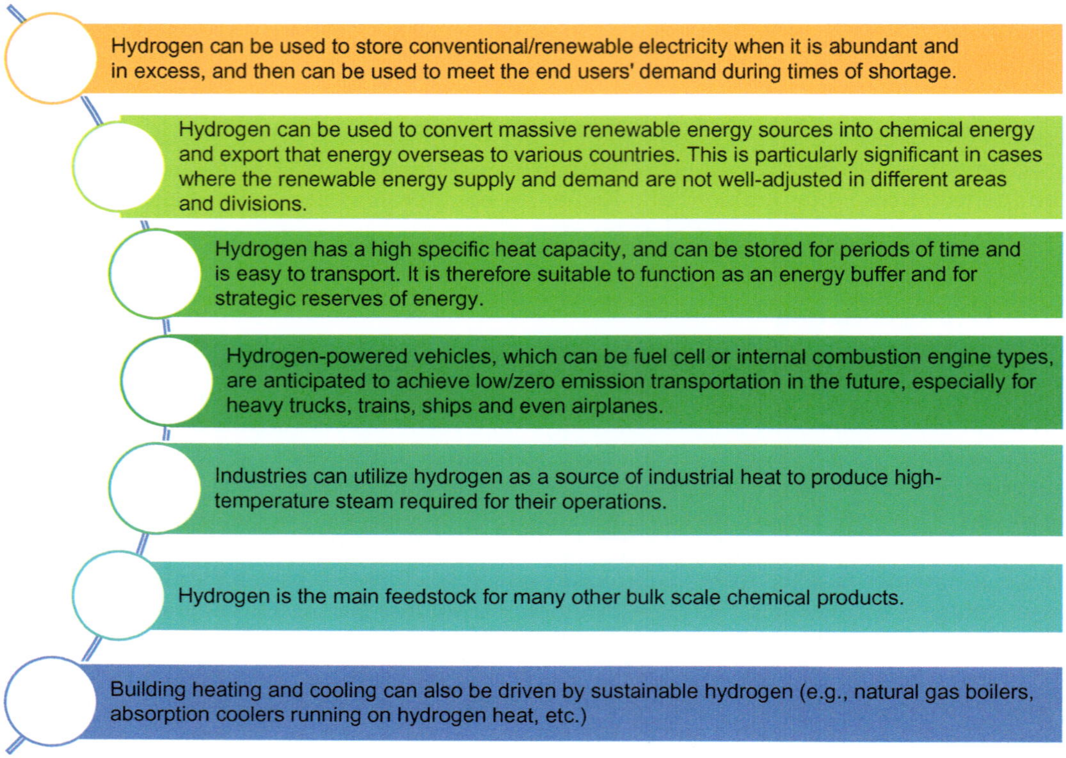

FIG. 3.4 Significance of hydrogenization in sustainable energy systems.

trigeneration plants having two and three outputs, respectively. The multigeneration approach offers various improvements: the losses are minimized, efficiencies are improved, and the negative environmental impacts of energy systems are substantially diminished.

Energy storage is one of the key points of multigeneration systems. Energy carriers such as hydrogen and ammonia are of great benefit to multigeneration systems. In order to take complete advantage of multigeneration, the points listed in Fig. 3.5 should be taken into account.

3.2.6 Storagization

Storagization can be defined as the method of employing energy storage routes to offset the discrepancy between demand and supply and to help the systems function in a continuous, more efficient, economic and environmentally benign manner [6].

Producing clean energy carriers using green energy sources can address the irregularity and discontinuity in supply. Thus, storagization is a crucial energy solution along the path of transitioning to sustainable systems, providing a way to meet the requirements of end

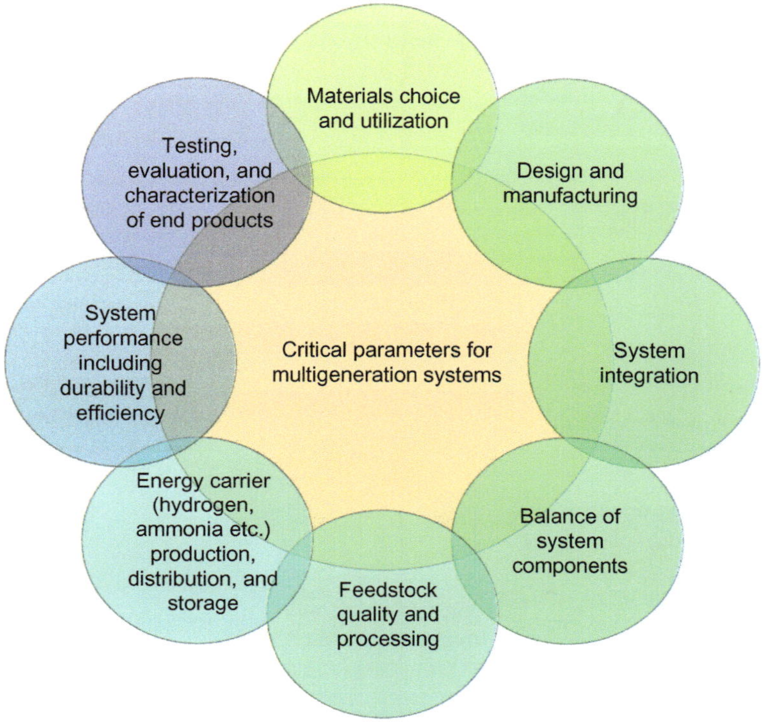

FIG. 3.5 Critical parameters to be considered in multigeneration systems.

users for power, heating, cooling, fresh water, drying, and transportation. The power-to-gas, gas-to-power, and the combined power-to-power energy chains offer substantial flexibility and control possibilities. This is realized by combining both electricity and energy carriers with reversible conversion.

In addition to the energy carriers classified under chemical storage, there are various energy storage techniques. They can be classified under 10 different categories, as shown in Fig. 3.6, and are explained in the following paragraphs.

3.2.6.1 *Pumped storage*

Pumped hydro storage is currently the highest capacity energy storage method employed in the world. In this technique, the excess electricity is used to pump water from a lower reservoir to a higher reservoir by converting electricity into potential energy as shown in Fig. 3.7. The water in the uphill reservoir can then be used to run water turbines (similar to hydropower) and produce electricity whenever required. Large-scale energy storage is possible using this storage option. This technique has proven itself economically viable, but the usage is limited by geographical conditions [7]. Transmission of energy is not possible since there is a fixed installation.

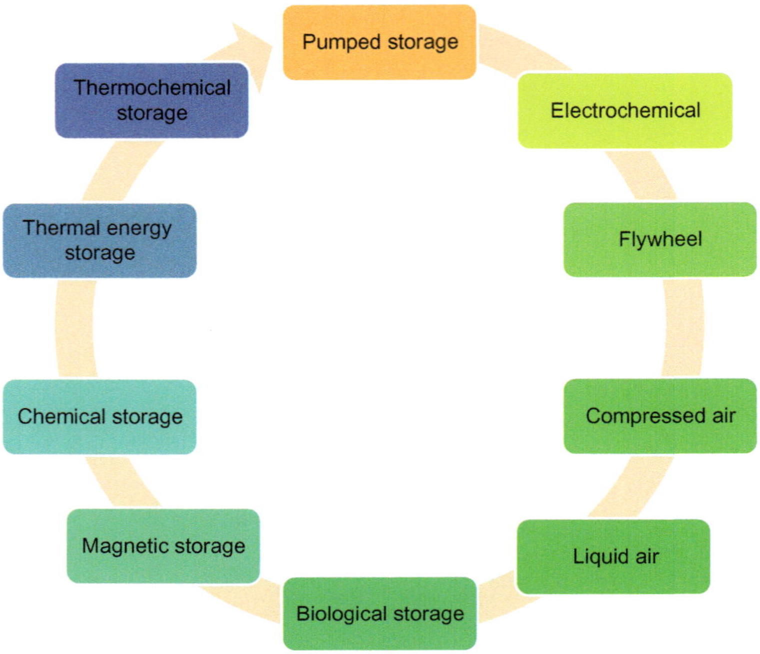

FIG. 3.6 Main energy storage techniques.

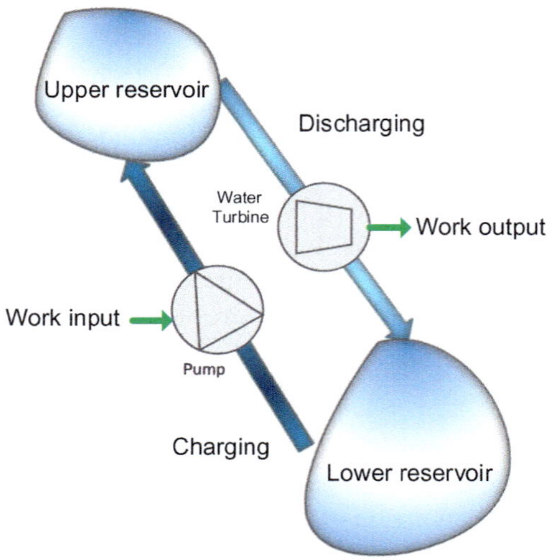

FIG. 3.7 Basic schematic illustration of a pumped hydro storage plant.

3.2.6.2 Electrochemical

Recent developments in battery technology have made batteries a possible energy storage alternative. Batteries chemically store energy and then provide electric current on demand. Three steps are involved: charging, storing and discharging; the operating principle is illustrated in Fig. 3.8. The efficiencies of each step are in fact quite high. This is essentially a steady form of storage, especially for battery electric vehicles, and can deliver the power needed for transportation. However, the main disadvantage is having a very low energy density (kJ/kg) compared to other materials and methods [7]. In addition, batteries are sensitive to low and high temperatures. Therefore, they require an effective thermal management. In hot arid climates, there is a need for external cooling, which consumes additional power. In cold climates, there is a need for external heating, which consumes additional energy.

Fuel cells are energy conversion devices capable of transforming the chemical energy of the fuel to electrical energy directly through an electrochemical reaction. The direct energy conversion path eliminates the need for intermediate thermodynamic processes and results in higher conversion efficiency. Fuel cells are considerably more efficient than conventional cycles (e.g., Rankine, Brayton) and generate electricity in one step, with no moving parts. In many cases, fuel cells operate at low temperatures. This contrasts with the combustion process employed in traditional power plants, where a fuel is burned at high temperature to create heat, which is converted to mechanical energy and then to electricity. However, some types of fuel cells can also operate at medium-high temperatures such as the solid oxide fuel cell (SOFC). They bring additional advantages such as feed fuel flexibility due to the high temperatures. There are various types of fuel cells, classified as shown in Fig. 3.9.

The reversible fuel cells are also categorized under electrochemical storage since they involve electrochemical conversion. The key difference between fuel cells and batteries is that a fuel cell is not consumed when it produces electricity and does not need recharging. Reversible fuel cells can also act as an electrolyzer, which can convert excess electricity into chemical

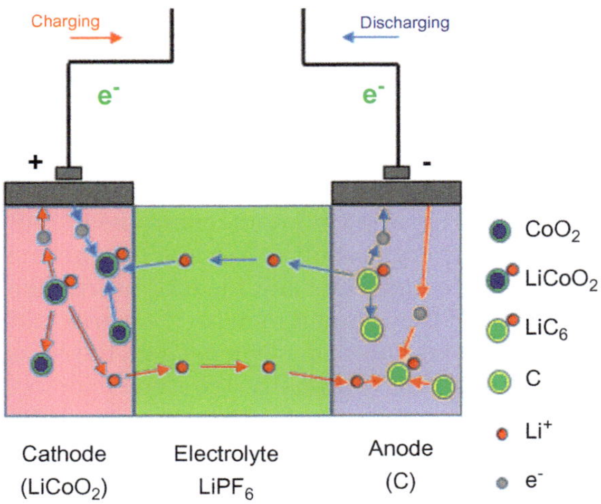

FIG. 3.8 Working principle of a sample battery storage unit. *Modified from Nikolaidis P, Poullikkas A. A comparative review of electrical energy storage systems for better sustainability. J Power Technol 2017;97*

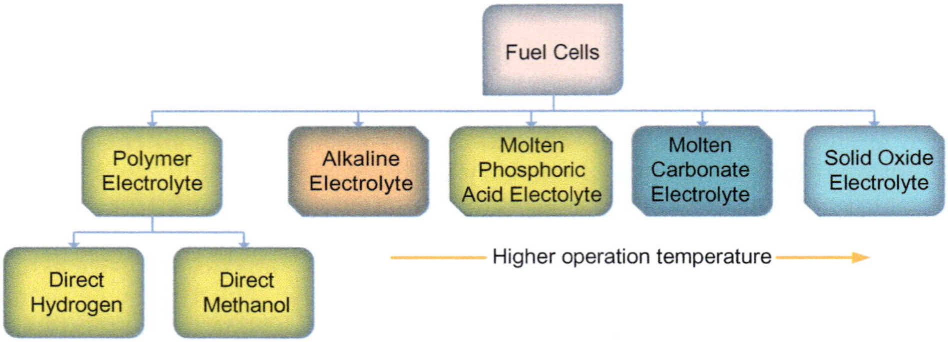

FIG. 3.9 Classification of fuel cells as electrochemical storage devices.

fuel (e.g., hydrogen). Later, fuel cells take the chemical energy stored in a fuel and convert it into useful electrical energy as shown in Fig. 3.10. They are considered better for large-scale energy storage compared to batteries owing to their high energy density and smaller size. They do not need external cooling or heating for operation unlike batteries (except start-up and shut-down). In addition, they can generate both thermal and electrical energy acting as a cogeneration system, which increases their overall efficiencies.

3.2.6.3 Flywheels

Flywheels are mostly used in transportation and for energy storage in road vehicles (e.g., busses). They can potentially be applied in road vehicles, particularly those with recurrent start and stop operations as in urban transportation [7]. The fundamental idea is that when decelerating, one does not transform the mechanical energy into heat by applying the brakes; rather, as shown in Fig. 3.11. As an alternative, the kinetic energy is stored by setting a wheel spinning. Then, the power surge needed for vehicle acceleration is delivered from the spinning wheel.

3.2.6.4 Compressed air

In this technique, compressed air is used to store energy. Excess electricity is used to compress the ambient air using compressors and store it, either in large tanks or underground caverns as depicted in Fig. 3.12. For a gas turbine to run appropriately, compressed, hot air must be supplied. Approximately two-thirds of the energy required to run a gas turbine is used by the compressor. It is proposed to decouple the compressor and the turbine, run the former during on-peak hours to produce compressed air, then store it in natural caverns, old oils or gas wells, or porous rock formations or even in tanks. Later on, when needed, the compressed air is released and sent to turbines to produce electricity. Initially, the conversion occurs from electrical to mechanical. Then, mechanical energy is converted into electricity. Therefore, the conversion efficiencies are high. This technique is considered suitable when an appreciable part of the power load is carried by power stations and where suitable finished salt caverns make it easy to build the compressed gas reservoirs [7].

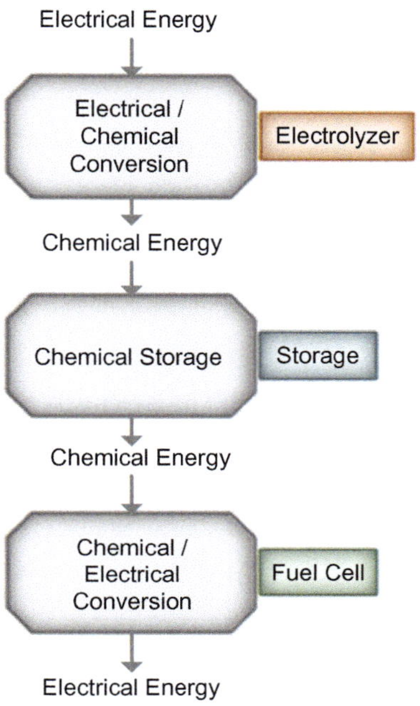

FIG. 3.10 Main steps of electrochemical energy storage using reversible fuel cells.

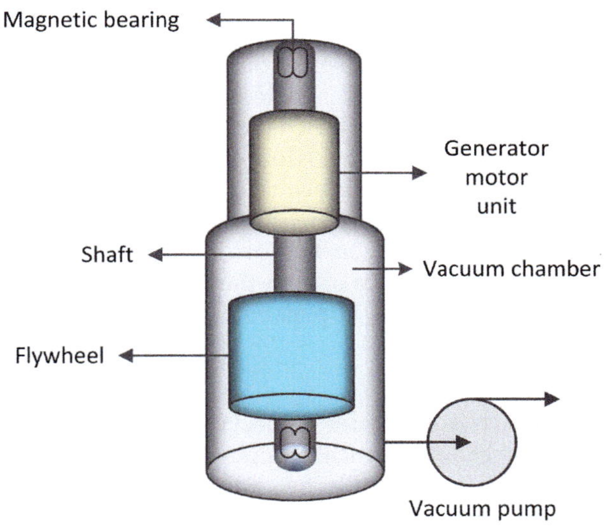

FIG. 3.11 Schematic illustration of a flywheel energy storage unit.

3.2.6.5 Biological storage

This technique consists of the storage of energy in chemical form by means of biological processes. Photosynthesis can be considered in this category since this process converts sunlight into chemical product, which is then used at the feedstock. Lipids, hypodermic fats, glycogen, and starch are means of biological energy storage. Biological methanation is a

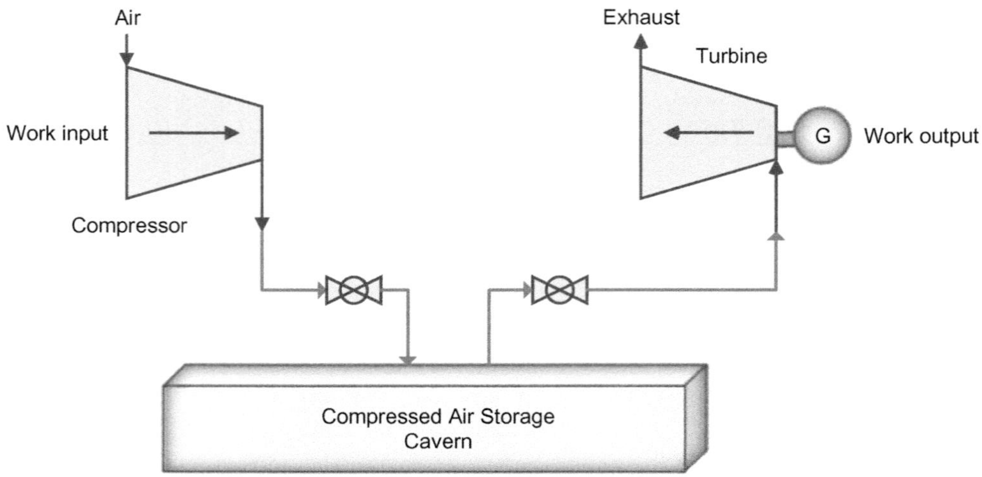

FIG. 3.12 Basic schematic illustration of compressed air storage.

process in which hydrogen and carbon dioxide are converted into methanol via biological processes. Similarly, ethanol can be produced via dark fermentation driven by groups of bacteria in biological processes.

3.2.6.6 *Electromagnetic storage*

Energy can also be stored in a magnetic field such as in large electromagnets. Employing superconducting materials is also considered for future applications. Electromagnetic storage methods include capacitors, supercapacitors and superconductors. Supercapacitors ideally store electric energy in the electrostatic field of the electrochemical double layer, rather than performing any chemistry, as shown in Fig. 3.13. Thus they can be cycled millions of times and have a much longer lifetime compared to batteries. In addition, they can achieve higher power densities compared to batteries due to the short time constant of charging; however, they provide lower energy densities due to the limited surface area of the electrode. One of their advantages is being suitable for very short term storage applications.

3.2.6.7 *Chemical storage*

Chemical storage is also an attractive model considered for, especially, solar thermal applications. The energy is stored by utilizing the heat of chemical reactions. Many chemical reactions occur with thermal energy input since they are endothermic reactions. These reactions can be driven by thermal energy produced in combustion, fuel cell, etc. In addition, chemical storage includes conversion of an energy source into chemical media such as hydrogen and ammonia. For instance, this can be achieved by electrolysis of water for hydrogen generation. Then, hydrogen can be stored. There are various types of hydrogen storage options available in the literature. One of the options is to convert hydrogen into ammonia by reacting with nitrogen from air. In this case, ammonia can be stored in compressed tanks in gas form or in insulated tanks in liquid form.

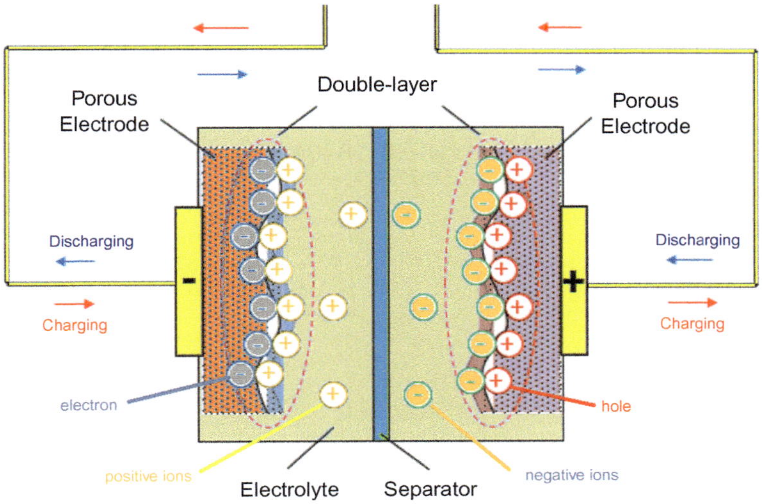

FIG. 3.13 Working principle of a sample supercapacitor for energy storage. *Modified from Nikolaidis P, Poullikkas A. A comparative review of electrical energy storage systems for better sustainability. J Power Technol 2017; 97*

3.2.6.8 *Thermal energy storage (TES)*

Thermal energy storage systems have a huge potential for more efficient use of thermal energy equipment and for simplifying large-scale energy substitutions from the economic perspective [7]. The use of the TES method for thermal applications such as heating, hot water, cooling, air-conditioning, etc., becomes an attractive option, because thermal energy can be directly utilized in absorption coolers, power cycles, heating applications [8] as depicted in Fig. 3.14. TES is a temporary storage process of high- or low-temperature thermal energy for later use. It can be used to store solar thermal energy for overnight heating, summer heating for winter use, winter ice for space cooling in summer, and heating or cooling generated electrically during on-peak hours for use during subsequent peak hours. Even cooling loads, which nearly coincide with maximum levels of solar radiation, often are present after sunset.

TES can be an important mechanism to overcome this mismatch between supply and demand. Thermal energy can be stored as sensible heat and latent heat. Sensible heat storage systems commonly use rocks or water as the storage medium [9]. The heat is stored by increasing the storage medium temperature. Latent heat storage systems store energy in phase change materials (PCMs). The heat is stored when the material changes phase from a solid to a liquid. The PCMs are also considered as an effective energy storage mechanism, especially for small-, medium-scale applications.

3.2.6.9 *Liquid air energy storage*

Liquid air energy storage uses electricity to cool air until it liquefies, stores the liquid air in a tank, brings the liquid air back to a gaseous state (by exposure to ambient air or with waste heat from an industrial process) and uses that gas to turn a turbine and generate electricity, as the system layout shows in Fig. 3.15. It can be integrated with waste heat for better overall efficiency.

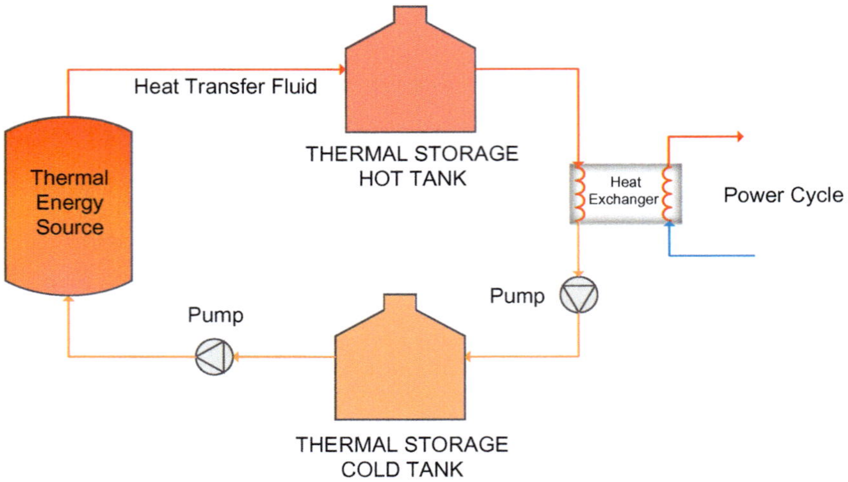

FIG. 3.14 Schematic illustration of a thermal energy storage system.

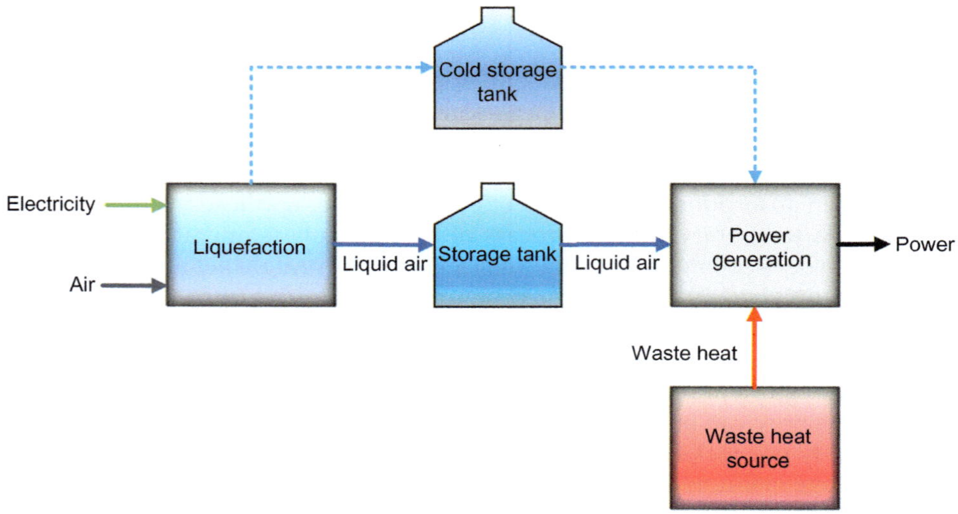

FIG. 3.15 Schematic illustration of liquid air energy storage with back-up waste heat source.

3.2.6.10 *Thermochemical storage*

Thermal energy can be used for thermochemical energy storage systems. Waste heat as well as solar energy are two of the significant candidates for this purpose. A chemical heat pipe can be used for storage and transportation of thermal energy. In this case, high-temperature solar process heat is used to run the endothermic reversible reaction A $\rightarrow$ B as shown in Fig. 3.16. The product B can be kept over the long term and long-range transferred to the location where the energy is required. At that location, the exothermic reaction, which is the reverse of the previous

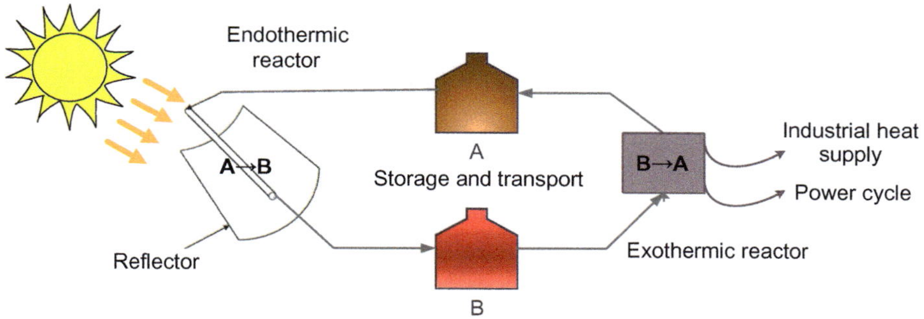

FIG. 3.16 Schematic illustration of thermochemical energy storage using solar energy.

reaction B → A, occurs and yields high-temperature process heat. This heat can be used for power generation cycles or process heat in the industry as well as for other purposes. The chemical substance A at the end of the reverse reaction is reverted to the thermal energy input reactor for reprocessing. One of the significant candidates for this type of energy storage technique is ammonia, which has a reversible synthesis and decomposition process. The dissociation and synthesis of ammonia could be used in a chemical heat pipe for storing and transporting solar energy [10]. The endothermic ammonia decomposition reaction happens thermocatalytically at high-pressure levels of 50–200 bar and at temperature levels above 700 K. The reverse synthesis of hydrogen/nitrogen is an exothermic reaction and can supply heat to any industrial process or to power generation cycles such as the Rankine cycle.

3.2.7 Intelligization

Intelligization is defined as a method in which artificial intelligence methods are employed to better simulate and model, implement, optimize, industrialize and control, and manage and meter [6]. Intelligization necessitates appropriate integration of all modules of the energy market from generation to delivery and distribution, storage, and end usage. Globally, there is a necessity for energy security, economic growth and climate change mitigation. Innovative technologies such as smart energy storage systems are quickly providing the technological readiness and maturity levels required to meet these challenges.

3.2.8 Greenization

Greenization is a term referring to the process of transforming traditional/conventional systems causing greater environmental damage and having less efficiency, more cost, etc. into more effective, more cost-efficient, and more environmentally friendly systems, as shown in Fig. 3.17 [5]. When it comes to traditional energy systems, it is necessary to come up with transitional solutions to improve their energetic, exergetic and environmental standings for sustainable development. In this regard, there is a need to greenize them in the best possible way by considering various criteria.

Dincer and Zamfirescu [11] presented a new thermodynamic conception of greenizing energy systems and suggest a new greenization factor, offering multiple case studies for various

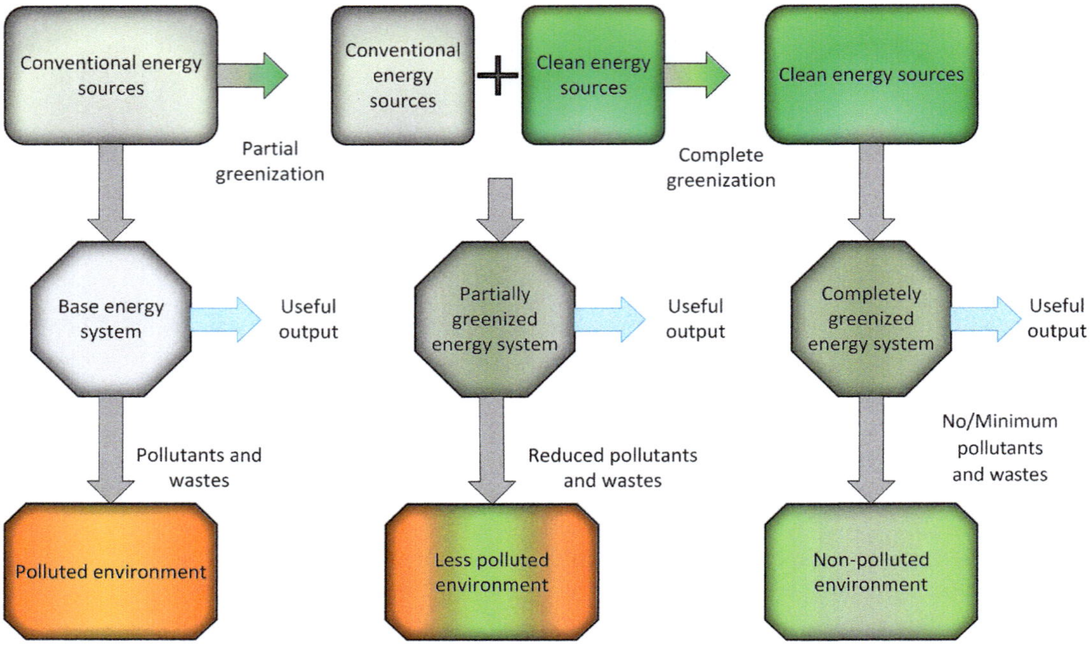

FIG. 3.17 Illustration of greenization process for energy systems. *Modified from Dincer I, Zamfirescu C. Potential options to greenize energy systems. Energy 2012;46:5–15. https://doi.org/10.1016/J.ENERGY.2011.11.061*

greenization criteria. There are numerous expressions for green and greener practices, which refer to the processes to make things more environmentally friendly. An exact application of greenization methods for sustainable energy applications necessitates modification, adaptation, retrofitting, or replacement of existing methodologies, as shown in Fig. 3.17, which are typically based on conventional energy sources, such as fossil fuels, oil, coal, and natural gas. Any energy system is accompanied by environmental consequences through air pollution, water pollution, and solid wastes, which can characterize the magnitude of environmental impact. A system that has been greenized must have a lower environmental impact factor than the original case.

The greenization factor can be defined based on an environmental impact indicator. If one denotes EI the environmental impact, the greenization factor for an energy system is then introduced by [5] as:

$$GF = \frac{EI_{ref} - EI}{EI_{ref}} \tag{3.1}$$

where EI is the environmental impact factor for the greenized system. In this equation, the environmental impact (EI) is defined for two cases: the reference system and the greenized system. Based on the specific problem type, various environmental impact factors can be expressed. For many energy systems, specific greenhouse gas (GHG) emissions can be used as environmental impact factors. It is significant to remark that the greenization factor varies from 0 to 1. A greenization factor of 0 shows that the system is not greenized. If the system is fully greenized then the greenization factor is 1. Completely greenized systems rely on sustainable energy resources, which do not have any or minimal environmental impact at the utilization phase. However, during system construction, there are some associated environmental impacts.

The greenization factor can also be expressed through exergy destruction as

$$GF = \frac{Ex_{d,\,ref} - Ex_d}{Ex_{d,\,ref}} \tag{3.2}$$

where Ex_d is the lifecycle exergy destruction of the concerned system and $Ex_{d,\,ref}$ is the exergy destruction of the reference system.

Both the system and the reference system make identical quantities of product. However, the energy, exergy, ecological and cost parameters of the two systems vary. It is observed from the greenization factor formulas that if $Ex_d \rightarrow 0$ then $GF \rightarrow 1$ implying that the system tends to be greener. If $Ex_d = Ex_{d,ref}$, no greenization effect is observed, as both the reference and the studied system have similar environmental impact.

Another measure to evaluate green energy systems is the improvement potential (IP) defined by Van Gool [12] as

$$IP = D_p\left(1 - \frac{Ex_{out}}{Ex_{in}}\right) \tag{3.3}$$

where D_p is the depletion number, expressed as follows:

$$D_p = \frac{1}{SI} \tag{3.4}$$

A sustainability index (SI) can be presented as the reciprocal of the depletion number and can be determined for any type of energy system. Hence, SI is the sustainability index discussed in detail by Dincer and Rosen [13]:

$$SI = 1 - \eta_{ex} \tag{3.5}$$

where η_{ex} is the exergy efficiency of the system.

The exergy method can be applied to an industrial ecology analysis by computing the exergy flows of every stream of matter and energy and relating them to depletion factors for every independent technology or process. Furthermore, the depletion number of separate and combined technologies can be considered and compared, as shown in Fig. 3.18, for evaluating the advantages of technology combinations from both an energetic and ecologic point of view.

3.2.8.1 Illustrative example: Greenization by fuel switching

In this illustrative example, we briefly describe the emission reduction potential of switching from coal to initially propane and then to natural gas for power generation. Table 3.1 presents the CO_2 emissions from various hydrocarbon fuels due to combustion in the power plants. It is noted that these values occur only during burning, so other life cycle stages are not included. Another important note is that the values are only for CO_2 rather than CO_2 equivalent, implying that other GHG emissions are not included in these numbers.

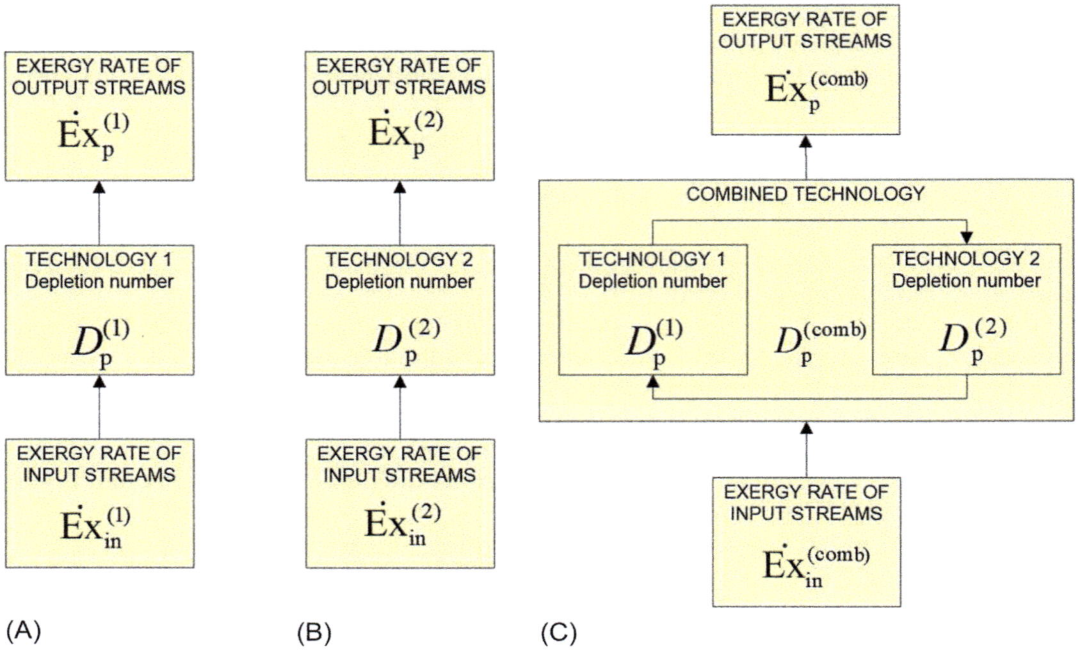

FIG. 3.18 Depletion factor of separate and combined energy systems.

TABLE 3.1 Carbon dioxide emissions due to combustion in the power plants.

Fuel type	Carbon dioxide emission (kg CO_2/kWh)
Coal (anthracite)	0.354
Coal (bituminous)	0.318
Coal (lignite)	0.333
Coal (subbituminous)	0.332
Diesel fuel and heating oil	0.250
Gasoline (without ethanol)	0.243
Propane	0.215
Natural gas	0.181

The values represent only CO_2 emissions, not all GHG emissions.
Source: *U.S. EIA. How much carbon dioxide is produced when different fuels are burned?—FAQ—U.S. Energy Information Administration (EIA). US Energy Inf Adm 2018. https://www.eia.gov/tools/faqs/faq.php?id=73&t=11 (Accessed 26 December 2018).*

Consider a 100 MW coal-fired power plant using anthracite type coal for power generation with a capacity factor of 60%. This means that the power plant runs about 5256 h annually. In this case, the power plant can produce about 525,600 MWh electricity. Based on the emission factors given in Table 3.1, this power plant will emit about 186,000 metric tons of CO_2 annually.

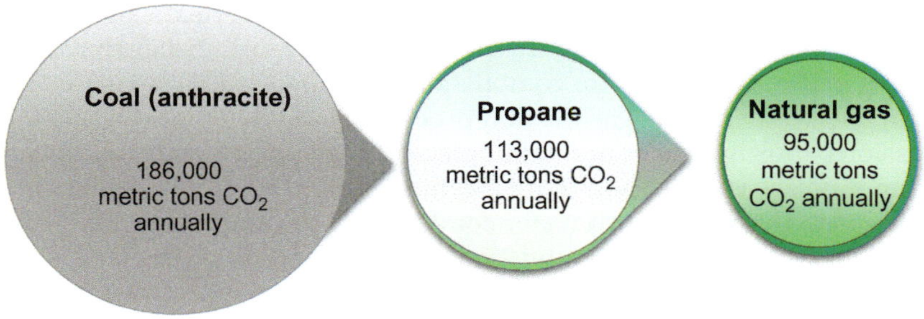

FIG. 3.19 Greenization of power production by fuel switching from coal to propane and natural gas.

Let us consider switching to propane for power generation with the same amount of energy production and capacity factor. In this case, the CO_2 emissions will be reduced to 113,000 metric tons in a year. Similarly, we can consider natural gas, which is the lowest CO_2 emitting common hydrocarbon during combustion. Assuming the same energy production and capacity factors, the CO_2 emissions from a natural-gas fired power plant will be about 95,000 metric tons per year. The CO_2 emission reduction potentials by switching from coal to natural gas as fuels in the power plants are illustrated in Fig. 3.19. There is almost 50% CO_2 emission reduction if natural gas is used instead of coal. Using the greenization factor equation written in the previous section:

$$GF = \frac{EI_{ref} - EI}{EI_{ref}} = \frac{186{,}000 - 95{,}000}{186{,}000} = 0.49 \tag{3.6}$$

One can achieve a greenization factor of about 0.49 for this case.

3.2.8.2 Case study: Greenization of an actual coal-fired power plant

In this case study, a coal-fired power plant is taken as the reference energy system. The plant encompasses a few power generation elements of 500 MW each. It works under a steam reheating Rankine cycle with three stages of turbines (high, intermediate and low pressure). This case study is adapted from Dincer and Zamfirescu [11].

In this case study, there are alternative routes to greenize the power generation plant, namely increasing the overall plant efficiency or increasing the heating value of the fuel.

- Replacement of steam generator with an advanced gasifier (Option 1)

  A coal gasification unit can be used to minimize greenhouse gas emissions of a coal power plant. This system includes an auto-thermal gasifier with the supply of steam, coal and air. After the gasifier, there is a hydrogen-selective palladium membrane system to improve hydrogen conversion and ensure the purity of hydrogen in the stream. For facilitating the hydrogen diffusion through membrane, air is employed as sweep gas such that hydrogen concentration is preserved at low levels. A catalytic combustion process happens on the membrane surface, which finishes the first portion of hydrogen combustion. Hydrogen is additionally combusted with preheated air and the resulting combustion gases are provided to the steam generator. A small part of the combustion gas, including steam, oxygen and nitrogen, is returned to the gasification unit. Here, steam and oxygen gases are reactants. The gases are still hot after the steam generator unit, hence they can be utilized for preheating of the air required for combustion. The combustion gases are also utilized to preheat coal and water for the gasification process. Combustion of hydrogen causes only steam emissions, thus is cleaner and takes place at higher temperatures, hence utilization of the heat in this case is easier and more meaningful. In the system shown in Fig. 3.20, the combustion gases are cooled until water condenses, where it is detached and recovered. Since the flue gases of hydrogen combustion can be cooled more, the coal is utilized more effectively, and consequently for the same useful output, fewer emissions are released to the ambient environment.

- Gasification with biomass co-fueling (Option 2)

  In this option, biomass is introduced into the system to such an extent that it only compensates the reaction heat needed for coal gasification. Biomass is considered to have a neutral GHG, because it is removed from the atmosphere by the plants. The previous system, which was shown in Fig. 3.20, remains the same with an addition of a fuel mixture of

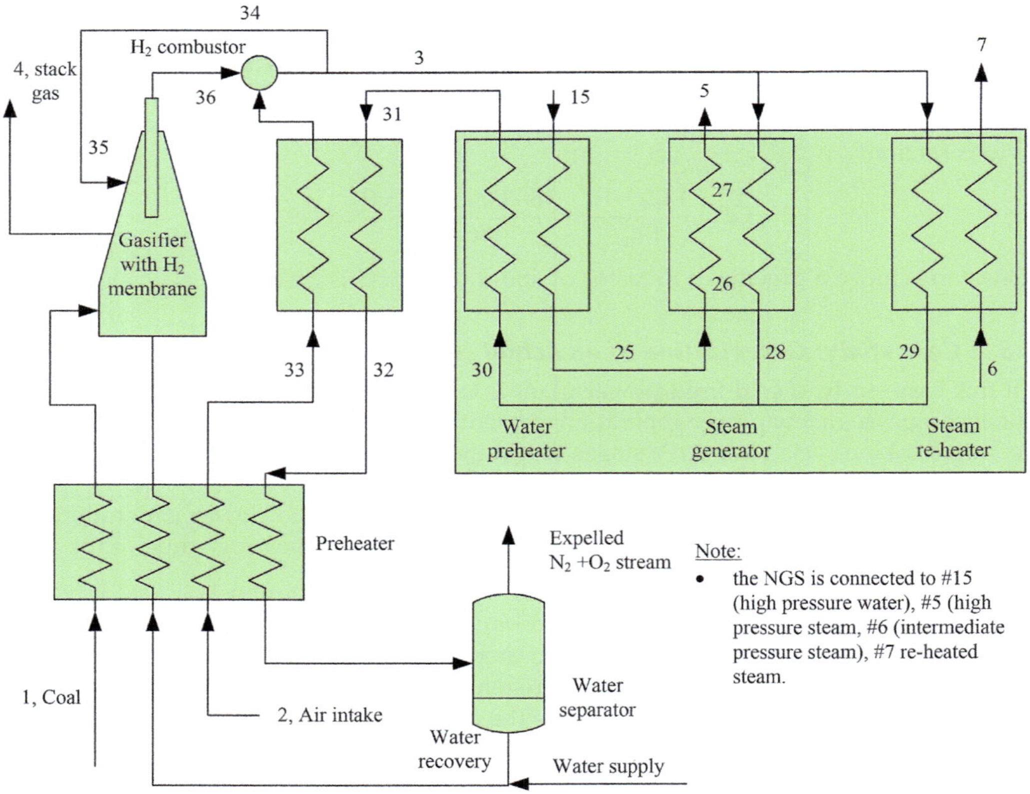

FIG. 3.20 Auto-thermal gasifier and combustor system to replace the coal-fired steam generator (Option 1). *Adapted from Dincer I, Zamfirescu C. Potential options to greenize energy systems. Energy 2012;46:5–15. https://doi.org/10.1016/J.ENERGY.2011.11.061*

coal and biomass. The temperature of hydrogen is about 1000 °C in this case. If biomass is not fed, then the system works in the same way as the previous one (auto-thermal). If biomass is introduced gradually, the amount of hydrogen produced per amount of carbon increases. As the amount of coal consumed reduces, its negative effects on the environment diminish.

- Integration to solid oxide fuel cell and gas turbine system (Option 3)

  In this option, the solid oxide fuel cell (SOFC) consumes hydrogen from the gasifier whereas the ejected gases are combusted with extra air and sent to a gas turbine as the system diagram shows in Fig. 3.21. The compressed air is fed into the fuel cell system and as well into the gasifier. The gases ejected by the SOFC include unreacted hydrogen, nitrogen and oxygen. Consequently, there is no necessity for extra air for combusting. After the gas turbine, the hot gas is redirected toward the steam generator and re-heater. The size of the gasifier is enlarged to produce more fuel than the former version, such that an enhanced power production is achieved.

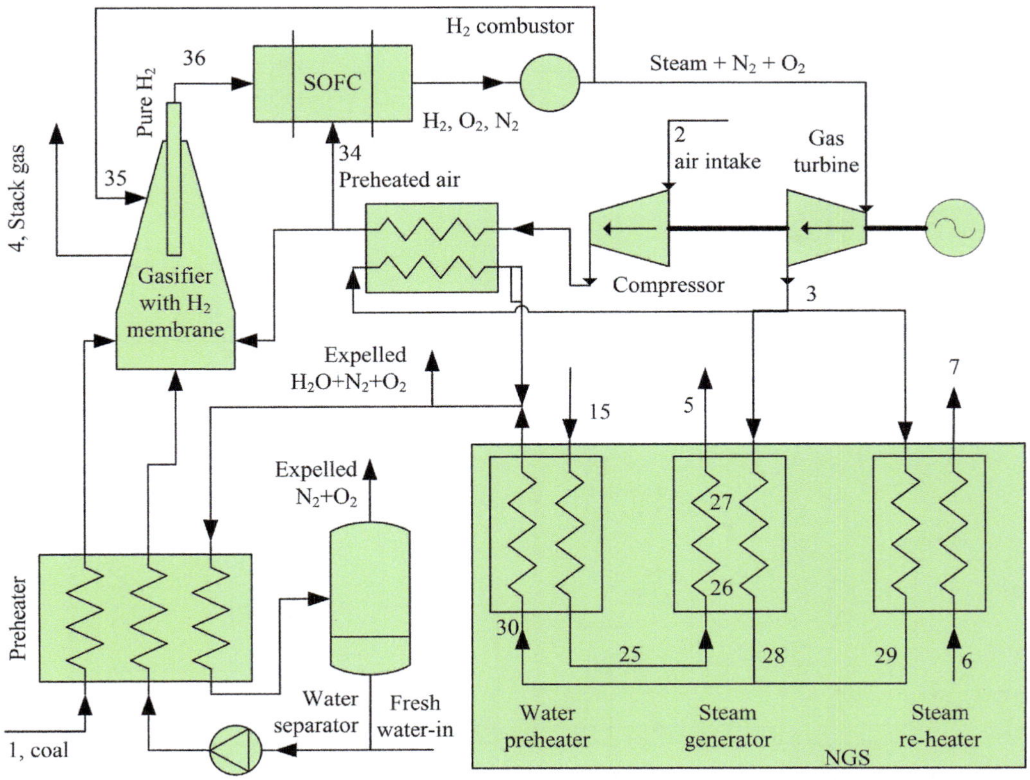

FIG. 3.21 Integration of system from option 2 with SOFC and gas turbine. *Adapted from Dincer I, Zamfirescu C. Potential options to greenize energy systems. Energy 2012;46:5–15. https://doi.org/10.1016/J.ENERGY.2011.11.061*

- Cogeneration with integrated NH_3-H_2O trilateral bottoming cycle (Option 4)

  There are often some possibilities for better heat recovery on the lower temperature side of the integrated power plant. The recovery can be performed by integrating an ammonia-water trilateral flash cycle as a bottoming cycle as shown in Fig. 3.22. The benefit of this cycle is to achieve an outstanding thermal match of temperature profiles of the hot gases derived from the gas turbine as well as enabling cogeneration opportunities. It employs an ammonia concentration of 30% in the ammonia-water working fluid, which permits a temperature rise at the sink side from 40 °C to 80 °C. Thus the cycle can be combined with a heat exchanger to cogenerate hot water with temperatures up to 65 °C.
- Trigeneration (Option 5)

  Power, heating and cooling production via an integrated system is capable of increasing the overall efficiency and reducing the environmental impact. In this trigeneration option, the system is modified by the addition of heating and cooling with a bottoming cycle working on ammonia-water. This cycle uses the low-grade heat by heat recovery in the air compressors (inter-stage cooling) and flows of the hot gases as shown in Fig. 3.23. It includes a heat exchanger for pre-heating the ammonia solution, two-phase expansion unit having two-phases, a heat exchanger that acts as a desorber, a two-phase compressor that provides a two-phase mixture at high pressure and finally a resorber for yielding a liquid aqueous ammonia solution. In this case, there is no net power generation.

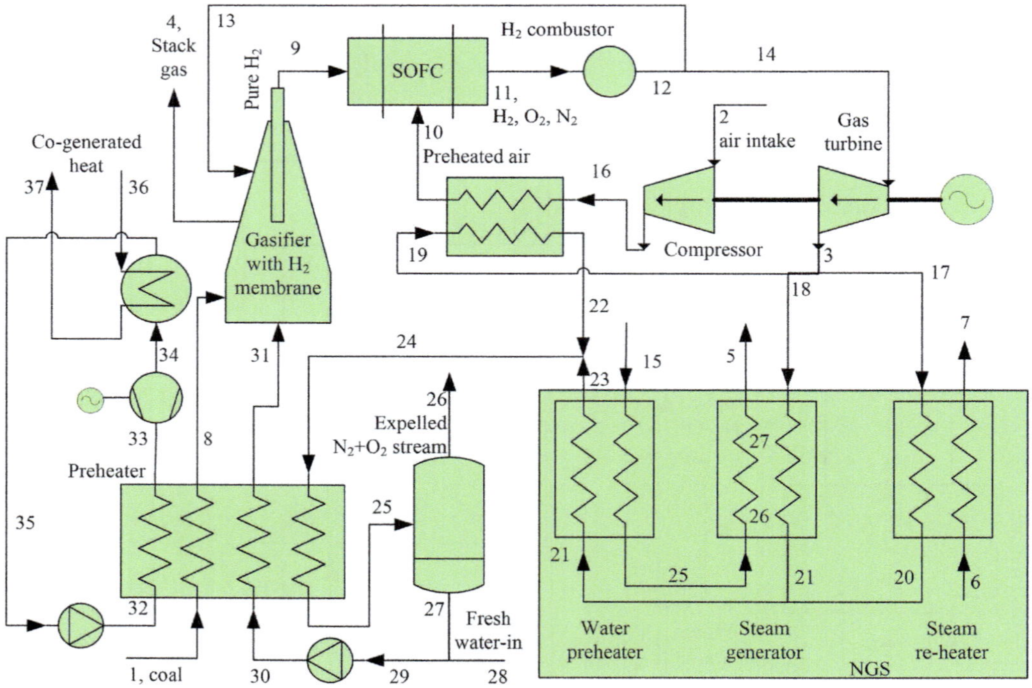

FIG. 3.22 System from option 4 with integrated trilateral bottoming cycle with ammonia water. *Adapted from Dincer I, Zamfirescu C. Potential options to greenize energy systems. Energy 2012;46:5–15. https://doi.org/10.1016/J.ENERGY.2011.11.061*

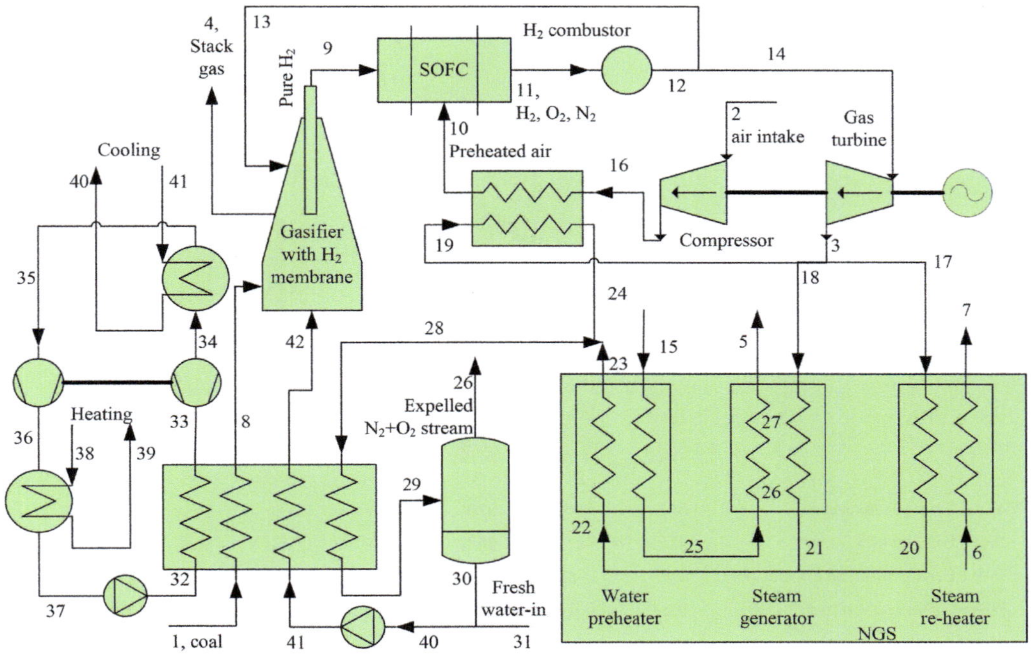

FIG. 3.23 Trigeneration system with biomass assisted coal gasification and ammonia-water bottoming cycle (Option 5). *Adapted from Dincer I, Zamfirescu C. Potential options to greenize energy systems. Energy 2012;46:5–15. https://doi.org/10.1016/J.ENERGY.2011.11.061*

- Carbon capture for full greenization (Option 6)

  Another promising method for enhancing the environmental friendliness of the system is capturing carbon along with sequestration or utilization. Since the installation has a membrane for the gasification unit, the greenhouse gas concentration in the combustion gas is high. This necessitates filtering, gas cleaning and CO_2 removal. Pressurized absorption systems can be applied and improved heat recovery can be achieved. The stack gas can be cooled to a temperature close to room temperature. It is assumed that because of the heat recovery in the stack gas, the coal utilization rate is increased by 30% compared to the previous version. Almost 43% of the rejected heat extracted from the plant is recovered and used for heating and cooling purposes. This option combines carbon capture with increased heat recovery. Moreover, the utilization of captured CO_2 is also possible. For example, captured CO_2 can be used for enhanced oil recovery, methanol synthesis or co-electrolysis.

3.2.8.2.1 Overall comparison of greenization options

In this case study, six different greenization options are presented. The comparative results for all alternatives are depicted in Fig. 3.24. The greenization factor rises from a low value, based on the alternative considered, and reaches unity when the carbon capture is employed. The power plant includes six generating units having about 3.1 GW power production capacity. The primary greenization method is to adjust the coal-fired boiler with advanced

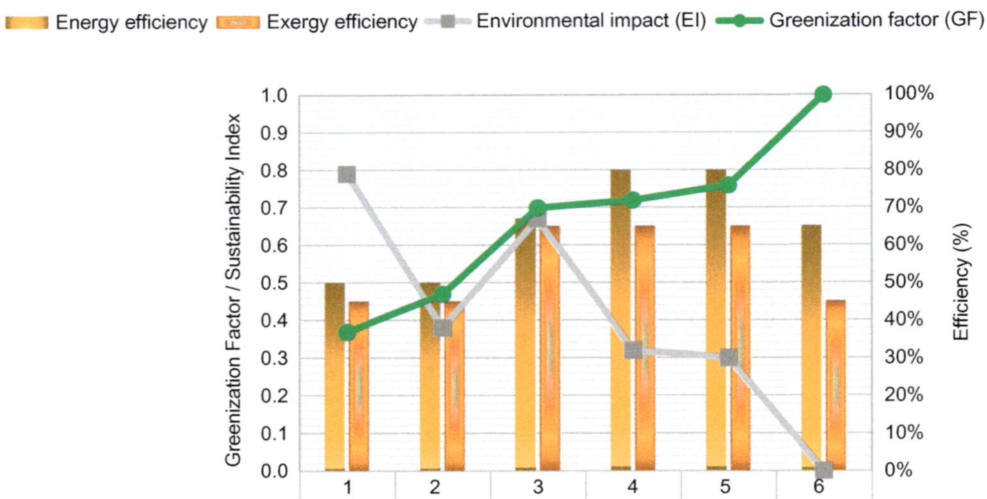

	1	2	3	4	5	6
Energy efficiency	50%	50%	67%	80%	80%	65%
Exergy efficiency	45%	45%	65%	65%	65%	45%
Environmental impact (EI)	0.79	0.38	0.67	0.32	0.30	0.00
Greenization factor (GF)	0.37	0.47	0.70	0.72	0.76	1.00

Options

FIG. 3.24 Comparison of considered greenization methods for a coal-fired power plant. *Data from Dincer I, Zamfirescu C. Potential options to greenize energy systems. Energy 2012;46:5–15. https://doi.org/10.1016/J.ENERGY.2011.11.061*

membrane gasifiers. The greenization factor is 0.37 and the exergy efficiency of the plant rises from 36% to 47% because the heat exchange at the boilers is enhanced as it enhances the coal consumption economy. The second method is to gasify the coal with a biomass blend. The other approach is to replace one of the power generating units with an SOFC-Gas Turbine (GT). The greenization factor increases to 0.7 while the exergy efficiency increases to 65%. Cogeneration and trigeneration are utilized to reach a maximum coal utilization factor of 80%. Lastly, the carbon capture process reduces the efficiency by 20%; nevertheless, the plant releases no CO_2 to the environment [11].

3.3 Integration and multigeneration

There are actually more advantages once a system is designed to yield multiple useful outputs from several kinds of energy resources. In this situation, various differing technologies need to be integrated in a synergistic manner to obtain useful commodities. There are multiple options to convert a single generation system to multigeneration. Some of the main possibilities of integration for advanced energy systems are multistaging, combining, hybridizing, and cascading, as illustrated in Fig. 3.25.

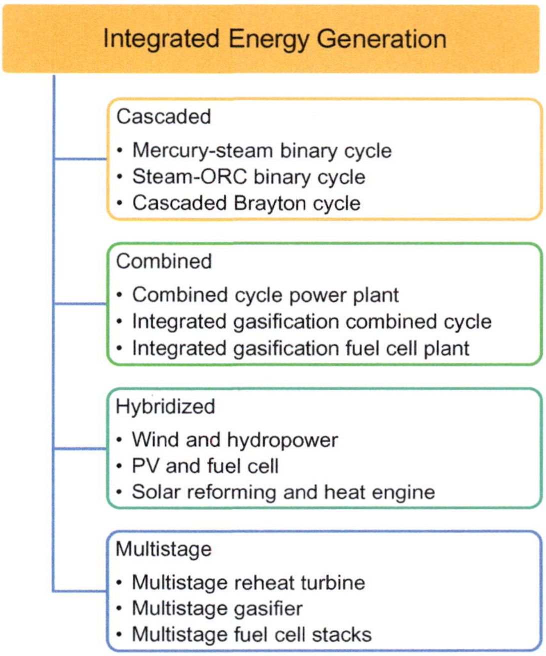

FIG. 3.25 Integrated energy production options via cascading, combining, hybridizing and multistaging.

3.3.1 Multistaged systems

For several power cycles, the employment of multistage processes yields an improvement in energy production and decrease of destructions. Thus, multistaging is a good option to be employed in many methods. One of the most common methods is the usage of multiple expansion-reheat stages as shown in Fig. 3.26.

Using more than three expansion stages does not bring plentiful benefits since the improvement of produced power is too minor to compensate for the further investment [14]. Similar to multistage expansion, there is multistage compression. In fact, it is a way of decreasing the back work ratio in Brayton cycles. Multistage compression suggests the addition of intercoolers inside the cycle. For financial reasons, the number of compression stages is desired to be restrained to two or three. It is noted that an upsurge in the number of stages causes extra irreversibilities inside the power cycle. Nevertheless, for most cases, the overall irreversibility related to a single-stage process is higher than for multistage processes.

Other ways to multistage within Rankine cycles involve the preheating processes and steam generation. Actually, a preheating process in stages employs numerous feedwater heaters (either open or closed types), numerous pumps, and several outlets of steam extraction from the turbine. Multistaging within a Rankine cycle may increase the energy efficiency by a few percentage points, but the improvement in exergy efficiency is much more substantial due to reduced irreversibilities [14].

Multistaging can also be employed in steam (or vapor) generation processes. When the steam production is performed in a single stage, it may cause huge irreversibilities because

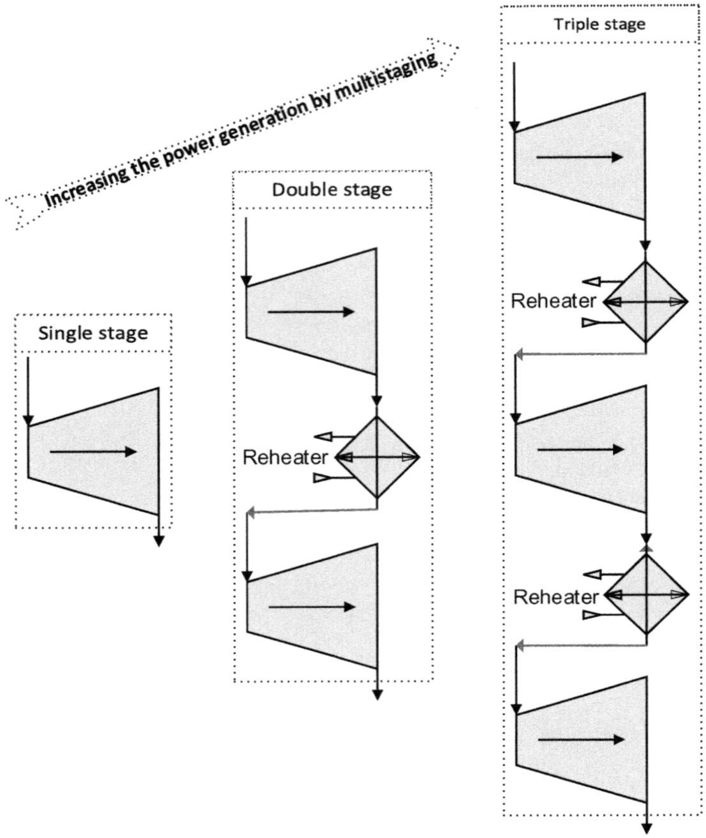

FIG. 3.26 Multistaging for expansion of the turbines.

of the heat transfer at a large temperature difference. Therefore, performing it in multiple steps rather than single stage offers benefits.

Gasification is a different process in which multistaging can be applied to obtain better syngas production with lowered tar. For instance, for a two-stage gasification process, the pyrolysis and gasification regions are obviously detached and there is a likelihood of providing secondary air between the regions. Therefore, the temperature in the gasification zone becomes higher, with more favorable kinetics for tar decomposition and syngas production.

Multistaging in fuel cells is another approach employed to achieve better fuel utilization ratios. In fact, this approach is similar to multistaging in turbines of Brayton and Rankine cycles with intercooling and interheating. In fuel cell systems, the fuel needs to be fed in excess amounts, because of lower cell efficiency. However, highly excessive amounts of fuel can cause a low fuel utilization ratio, and henceforth higher losses. In some examples, the requirement of having a moderately low fuel utilization ratio for higher efficiency of the fuel cell is recompensed by cascading the stack with a gas turbine. In that case, the exhaust, unused fuel, is additionally combusted. Consequently, an alternate solution to this is the staging of fuel cell stacks as shown in Fig. 3.27. In the two-stage arrangement, two stacks are utilized rather than

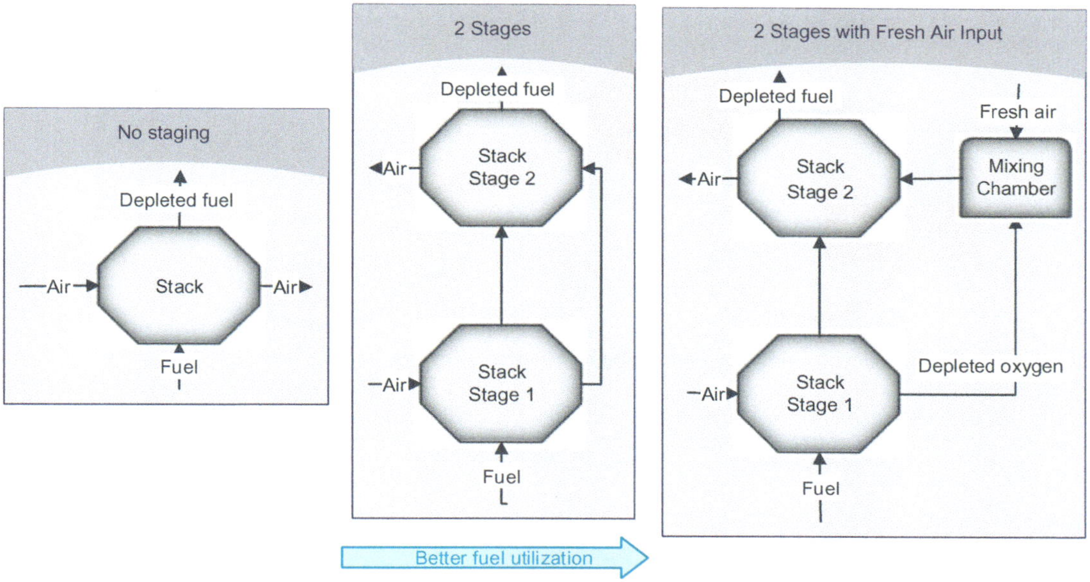

FIG. 3.27 Multistaging in fuel cells for better fuel utilization ratio.

one. The system can be configured to deliver the same amount of power as the un-staged system [14]. However, with the staged system, there is an option of regulating the load individually for each of the stages. As an initial route of staging, both the anodes and cathodes are linked in series. This suggests that the second stage will be provided with the depleted fuel and consumed oxidant from the initial stage. Fresh air can be additionally supplied as well.

3.3.2 Cascaded systems

Cascading is defined as the case where one energy production subsystem accepts or recovers energy from another subsystem to yield an extra useful commodity or power [14]. It is one of the common system integration options, increasing the overall efficiency. A basic illustration of cascading is presented in Fig. 3.28 with two power production cycles, of the same kind. For example, two cascaded Brayton cycles or two cascaded Rankine cycles can be considered. The first one is called the topping cycle, and the second cycle is called the bottoming cycle.

In this case, the heat source for the bottoming cycle is, in fact, a heat sink for the topping cycle. There is power generation in both cycles. More than two systems can also be cascaded but this may complicate the overall system with only minor improvements. As an example, we can mention binary cycles, which are cascaded systems in which the condenser of the topping cycle is the boiler of the bottoming cycle, as shown in Fig. 3.29. When the rejected

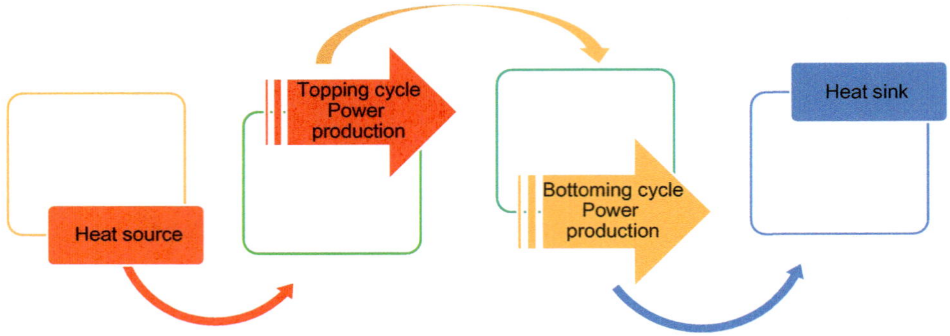

FIG. 3.28 Cascaded energy system with topping and bottoming cycles.

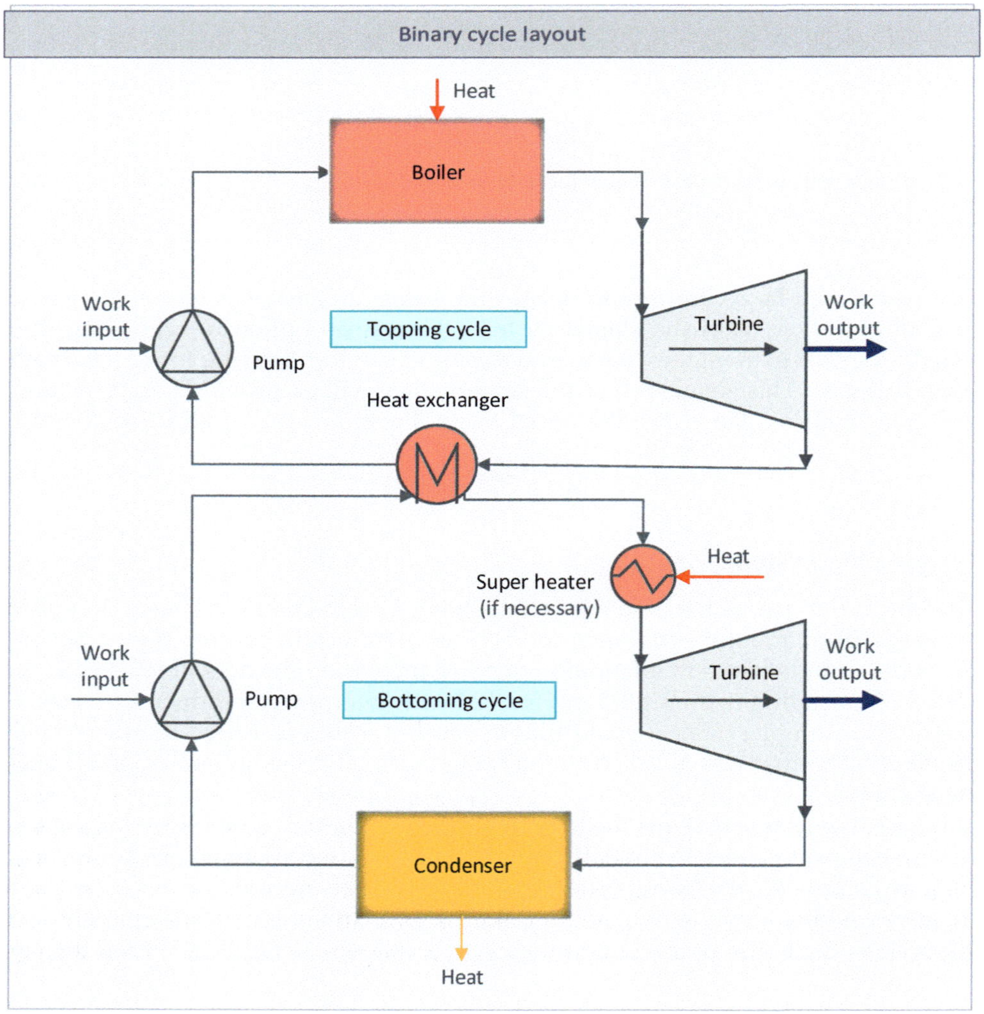

FIG. 3.29 The layout of a sample binary cycle plant.

heat from the topping system is recuperated and utilized in a bottoming system that functions between an intermediate temperature level and the ambient heat sink, more energy is produced compared to a single system. Cascading requires adjusting the working fluid to the external circumstances of the power cycle. Optimum circumstances at source and sink may be attained if two Rankine cycles are well cascaded.

Another aspect of employing binary cycles is to aid recovering a portion of the discharged heat from steam Rankine cycles in a cascaded manner. Using this option, a bottoming cycle is cascaded to the condenser of the main steam Rankine cycle. The system denotes a low-temperature binary cycle to differentiate it from the high-temperature binary cycle, in which a topping system is combined to obtain higher source temperatures. For low-temperature binary cycles, the bottoming system can function with an organic fluid, usually with refrigerants or ammonia.

3.3.3 Combined systems

Integration of power production techniques by means of merging two or more thermodynamic cycles or systems is defined as a combined system. In fact, currently there are many existing combined cycle power plants (CCPP), which are already well-known technologies. These systems combine gas turbines with the Rankine cycle. There are many other combined systems, which are briefly mentioned here.

The CCPP runs with gaseous or liquid fuels since it contains a gas turbine system. On the other hand, there is a possibility of using solid fuels such as coal and biomass by employing an integrated gasification unit. In this case, the overall system is called an integrated gasification combined cycle (IGCC) as shown in Fig. 3.30. The system integration of these combined cycles is especially useful due to the following aspects [14]:

- Coal or biomass gasification can be performed with medium/high pressures using a solid fuel. It can be a slurry water mixture in which water acts as a gasification agent.
- Pure oxygen (oxy-gasification) can be supplied to the gasifier rather than air by employing an air separation plant, yielding numerous advantages.
- Heat recovery from the gasification unit can be utilized for preheating the water in the steam generation unit to reduce the external heat input in the boiler/steam generator.
- Carbon dioxide could be detached from product gas and then sequestrated. In this way, the greenhouse gas emissions could be lowered.
- Coal or biomass can be dried using low-grade heat recovered from flue gases, which enhances the gasification or combustion process.
- For some regions of the world, solid fuels such as coal and biomass are more feasible, easily accessible, and more attractive than natural gas, which implies better energy security.

The heat recovery option from a gasification and gas-cleaning unit is evidently recognized. This recovered heat in addition to the heat recovered from the hot combustion gases emitted by the gas turbine section is quite useful for the heat recovery steam generation unit. The introduction of nitrogen to the gas turbine section is done at an intermediate pressure. The temperature of nitrogen is rather decreased after the air separation plant. Consequently, it cools down the compression process between compression stages. The existence of nitrogen gas in

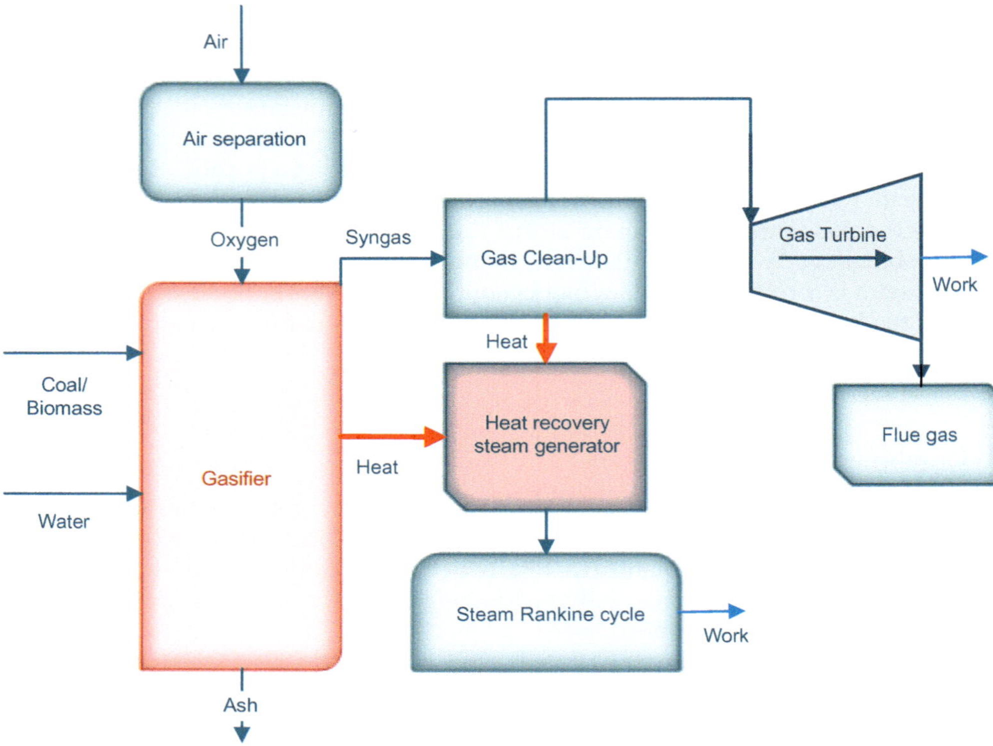

FIG. 3.30 Schematic illustration of an integrated gasification combined cycle.

the entire system drops the combustion temperature and hence decreases the NOx emissions. The usage of a steam turbine rather than a whole Rankine plant significantly increases the efficiency of the system since, instead of transforming low-grade heat into electricity similar to the bottoming cycle of the IGCC, the lower-grade thermal energy is utilized primarily to produce steam that in turn is transformed into an improved gaseous fuel produced with lower exergy destructions [14].

Topping internal combustion engines could be integrated with a bottoming gas turbine as illustrated in Fig. 3.31. Solid oxide fuel cells (SOFCs) can be integrated with molten carbonate fuel cells (MCFCs) that work at a lower temperature and may burn the combustion gases from the topping SOFC. It is possible that both MCFC and SOFC type fuel cells are integrable with bottoming gas turbines to combust the unburned gases and yield more electricity, as shown in Fig. 3.32. These types of high-temperature fuel cells can also be integrated with Rankine cycles. A steam Rankine cycle can also be combined or cascaded with a bottoming organic Rankine cycle (ORC), as shown in Fig. 3.33.

For instance, an SOFC can be combined with a gas turbine and a Rankine cycle for three-stage power production, in which the SOFC acts as the topping cycle, the gas turbine as the intermediate cycle, and the Rankine cycle as the bottoming cycle.

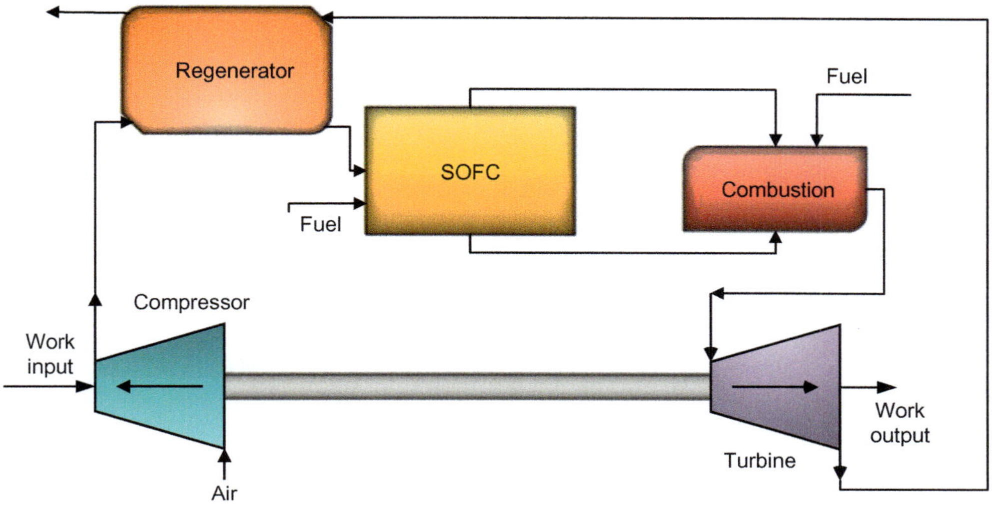

FIG. 3.31 Schematic diagram of a combined SOFC-Gas turbine plant.

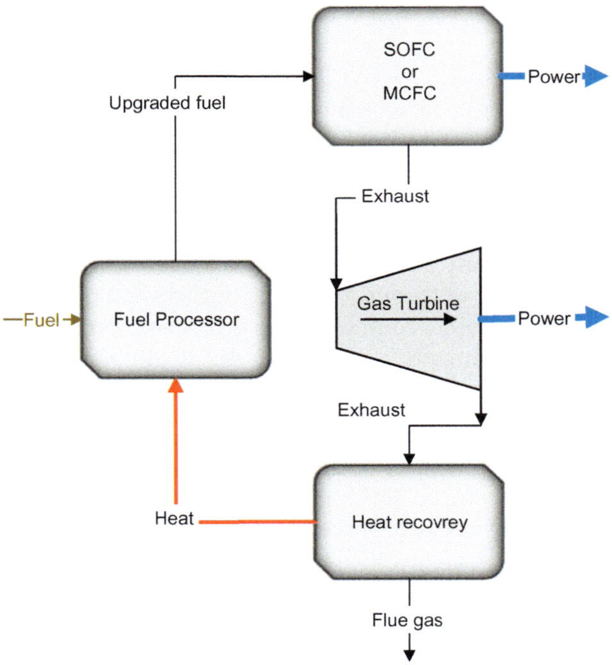

FIG. 3.32 Schematic diagram of a sample high-temperature fuel cell-based combined energy system.

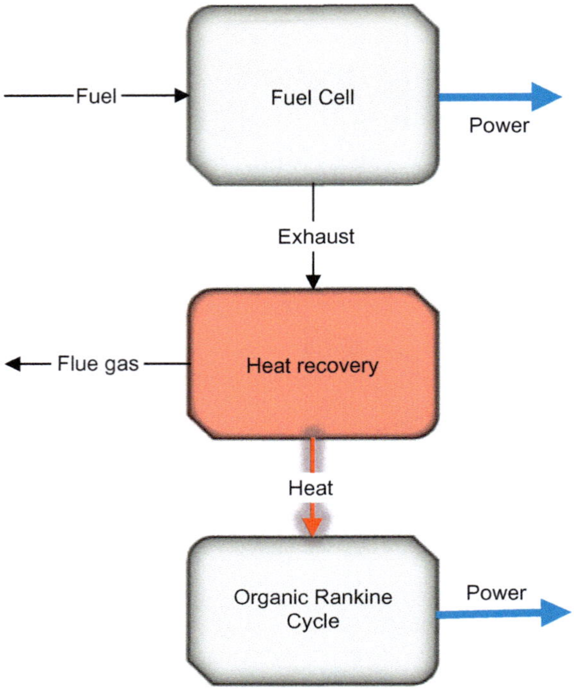

FIG. 3.33 Schematic diagram of a sample medium-low temperature fuel cell-based combined energy system.

3.3.4 Hybrid systems

In hybrid systems, two or multiple different energy systems function synergistically while converting energy from one source or a selection of sources in a harmonized manner [14]. The advantage is that hybridization is likely to combine very different types of energy resources or systems, such as photovoltaic modules and wind turbines, hydro energy and wind turbines, solar and natural gas, coal and concentrated solar power, solar photovoltaics and hydrogen fuel cells, biomass and coal, geothermal and solar, and ocean thermal and solar.

For example, hybridized systems including power and energy storage options can provide nuclear power plants with greater capacity factors and promote better grid load leveling.

A few of the possible routes for hybridized systems are as follows [14]:

- integrated wind and photovoltaic systems,
- integrated systems producing electricity and thermal energy from geothermal and solar resources or from biomass combustion and concentrated solar energy,
- concentrated solar collectors integrated with advanced Rankine power cycles and with natural gas steam generation units for obtaining backup power.

In general, hybrid energy systems provide better energy security, as they can employ various different energy sources in a hybridized fashion. Hybrid systems are especially preferred for renewable energy sources to overcome the challenge of their fluctuating and intermittent nature. Furthermore, hybrid renewable energy systems are preferred for meeting the peak demand, especially if there are also conventional energy systems providing electricity to the grid. It is possible to achieve this hybridization using two options [14]:

- The hybrid system can primarily use renewable energy sources and can also utilize a conventional energy production subsystem like diesel generators for the times when there is a peak demand or unsatisfactory availability of the main resource.
- The hybrid system can primarily use conventional energy systems and can also utilize some types of renewable energy systems to provide additional power generation or to save fuel, particularly in the off-peak periods.

For example, for a hybridized photovoltaic and wind energy system, the produced electricity is provided to the same grid (inverter). The two systems can share an electricity storage unit. They can share the grid connection system as well. For the moment, the usual approach is using battery banks for electricity storage, although the energy density is quite low compared to other types of energy storage techniques. By utilizing both solar and wind energy sources, the capacity factor of the integrated system is increased compared to independent systems. Moreover, the capital cost can also be lowered since they use a shared system for energy conversion, storage, and distribution. Fig. 3.34 displays a basic diagram of a hybrid PV-wind turbine power plant. The system can contain an engine-driven generator like a diesel generator, which functions in harmony with demand as well as wind and solar energy availability.

As mentioned earlier, pumped hydro energy storage is an alternative to utilize excess electricity, particularly for hybrid systems having both wind and hydro energy. Such systems can employ wind or hydropower to pump water to a higher altitude in an artificial lake. Later on, hydraulic turbines and generators can be employed to generate electricity on demand. The systems are also linked to the regional grid. This type of hybridization can smooth the flow

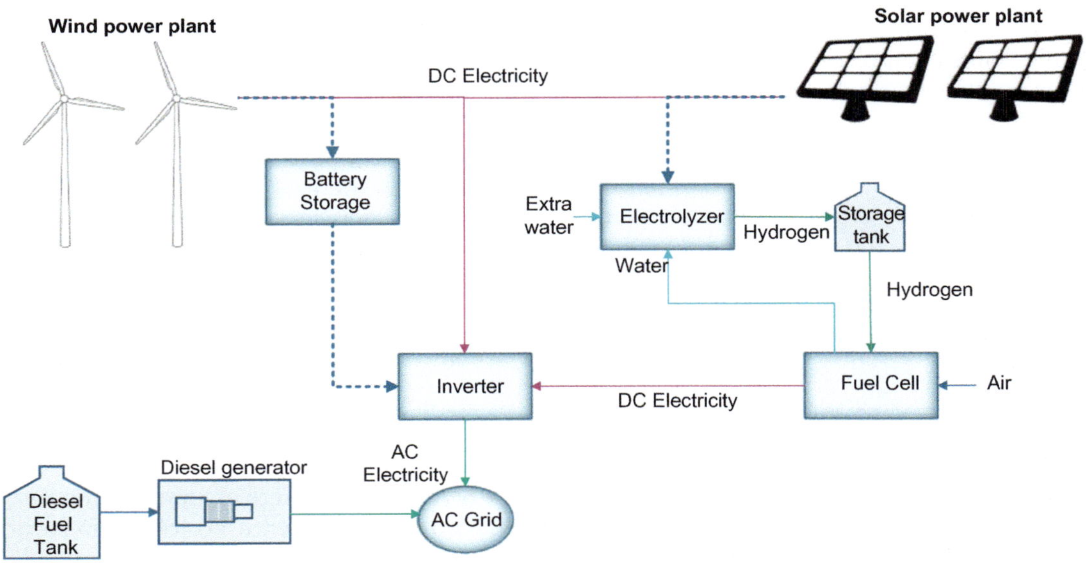

FIG. 3.34 Schematic diagram of a hybridized energy system including solar, wind and diesel power plants.

of energy production by balancing the variable nature of wind energy. The conversion efficiencies of these types of systems are quite high due to the fact that there is only one step of energy conversion, which is mechanical to electrical energy.

Diesel generators are also used to compensate the intermittency and storage challenges of renewables. For example, wind and diesel hybrid energy production units have been realized for grid independent systems of rather small range for small remote communities. In these cases, the two subsystems utilize the same electricity substructure, entailing inverters, controllers, systems for connection to a small local grid, and battery units for energy storage. Actually, the chief objective of the wind and diesel hybridized option is to decrease the substantial use of fossil fuel, which is diesel fuel in this case. The main share of supplied electricity is desired to be from wind power rather than diesel. Fig. 3.34 shows a sample hybridized power-producing system integrating wind, PV, and diesel generators. It is noted that diesel generators can also run on alternative fuels such as ammonia with some modifications. In this case, the harmful emissions can be considerably reduced.

The energy storage challenge of wind and photovoltaic energy systems is one of the difficulties for wide implementation in remote areas, because of storing huge quantities of electricity using chemical batteries. In this case, reversible fuel cells play an important role in hybridization. Hybrid wind, PV and hydrogen fuel cell systems signify one desirable alternative for continuous energy supply to the power grid. In this option, the wind turbines or photovoltaic modules drive an electrolysis subsystem for generation of hydrogen whenever there is renewable energy. A portion of the generated electricity can be directly utilized for the end use. Hydrogen is stored and then utilized in a fuel cell system for energy production when required. In addition, diesel generators can be further added for system hybridization. In this way, emergency cases can be compensated by backup power from diesel generators.

Another wind energy-based hybridized energy system employs a compressed air energy storage unit, as shown in Fig. 3.35. Compressed air storage is, in fact, a thermomechanical process in which energy is charged and recovered through mechanical work and heat, respectively. Air is compressed whenever there is wind and stored in huge underground reservoirs. These reservoirs can be naturally available caverns, salt domes, abandoned mine shafts, depleted gas and oil fields, or manmade caverns. Huge tanks are also possible but would require a large footprint on the ground.

Much has been learned from experience in designing large-scale storage reservoirs in underground salt caverns. Such caverns can be mined using aqueous solutions that dissolve salt and generate the desired shape of the reservoir. The other method involves the accumulation of large amounts of compressed air in aquifers, above the water level, if the geomorphological structure is such that above the aquifer there is hard rock blocking any air leakages. Conversion of compressed air into electricity is typically performed by gas turbine systems having a fast start-up, because the availability of compressed air will have a considerably smaller back work ratio [14]. One disadvantage of compressed air storage is non-transportability of the stored energy.

Many more options are available for integrating energy systems in a hybridized method, based on the applications. For instance, on farms, where there are huge amounts of manure and residual crop/biomass, new hybridized systems could be considered to receive biogas or biomass as input materials. In this case, the fundamental modules that could be integrated to form a hybridized energy production system for farm applications could include wind turbines, PV arrays, diesel or propane generators, micro-hydro systems, bio-digesters (biogas

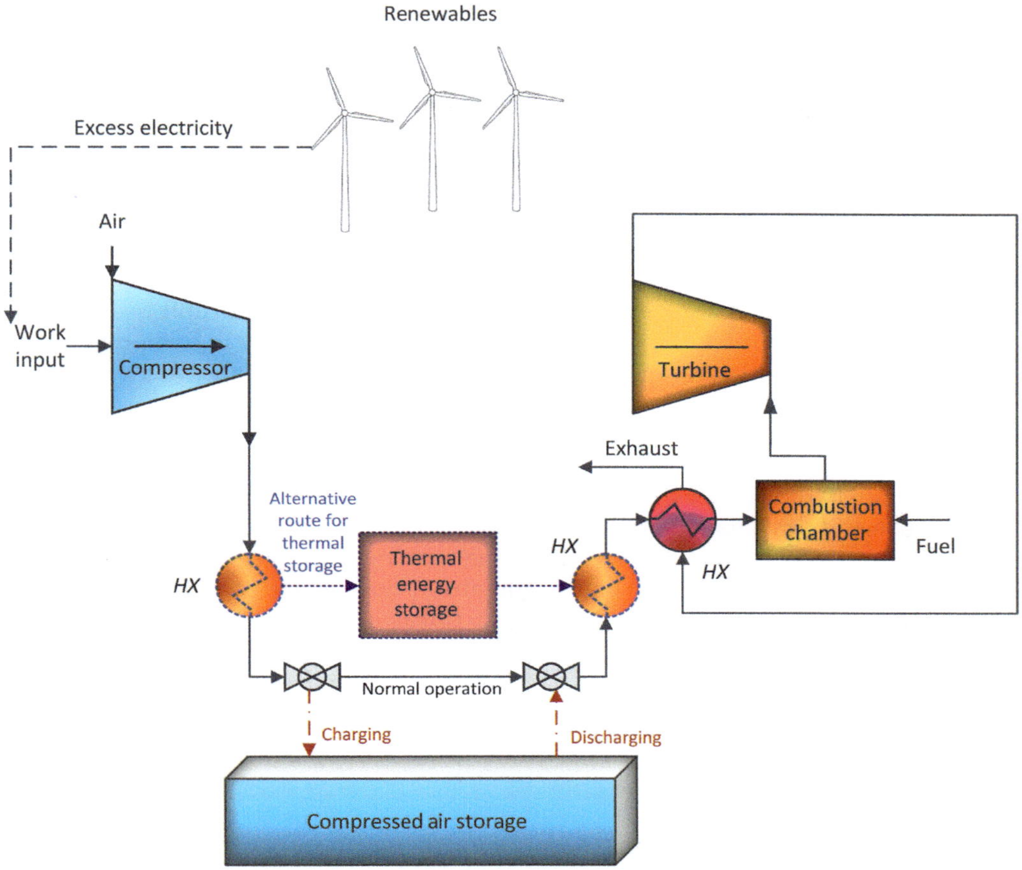

FIG. 3.35 Schematic diagram of a hybrid wind, compressed air storage and conventional system.

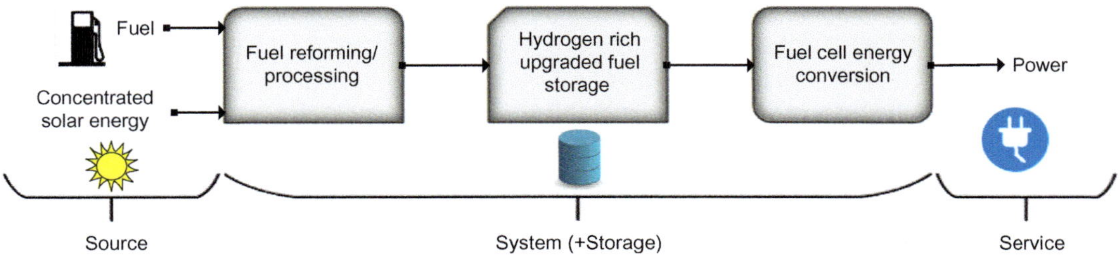

FIG. 3.36 Hybridized solar and fossil fuel generation integrated to solar reforming and fuel cell.

producer), solar-assisted biomass drier-gasifiers, concentrated solar power, fuel cells, etc. Another option is to reform or process a conventional fuel (e.g., coal, natural gas) and obtain a hydrogen-rich fuel as a storage medium, as illustrated in Fig. 3.36. In this case, the fuel is stored either short or long term. Then, using fuel cell conversion technologies, power can be generated on demand with higher efficiencies.

3.3.5 Multigeneration

Multigeneration systems are in fact an extension of the integrated systems. Multigeneration systems are integrated systems that combine various cycles and processes to convert diverse energy sources into multiple useful commodities [14]. Cogeneration and trigeneration systems can be classified under this category, such as combined heat and power (CHP) systems. In cogeneration systems, the energy utilization factor can increase from 15% to 30% in single generation to about 50% to 80% in cogeneration. The systems shown in Fig. 3.37 represent a basic layout of trigeneration and quadruple generation systems. The trigeneration system recovers more heat and provides heating at two different levels: water heating at 45 °C and space heating at 25 °C. On the other hand, the quadruple generation system uses a part of heat recovery to obtain cooling with an absorption refrigerator. In this case, the heat supply to absorption refrigeration is around 80–120 °C, which is essential to run the absorption refrigeration system.

When the number of useful outputs is increased to more than four, the systems become more complicated but also possibly more efficient. The multigeneration systems shown in Fig. 3.38 generate five and six useful outputs: power, space heating, water heating, refrigeration, hydrogen and desalination. In this case, cascaded systems can be employed for system integration. Power generation is produced by the topping cycle. The bottoming cycles use heat recovered from the topping system to run absorption refrigeration and a heat engine such as ORC, which generates additional power to run an electrolyzer and to produce hydrogen. The rejected heat from the bottoming energy conversion cycle can be used for low-grade thermal applications such as hot water production at about 45 °C and space heating at about 25 °C [14].

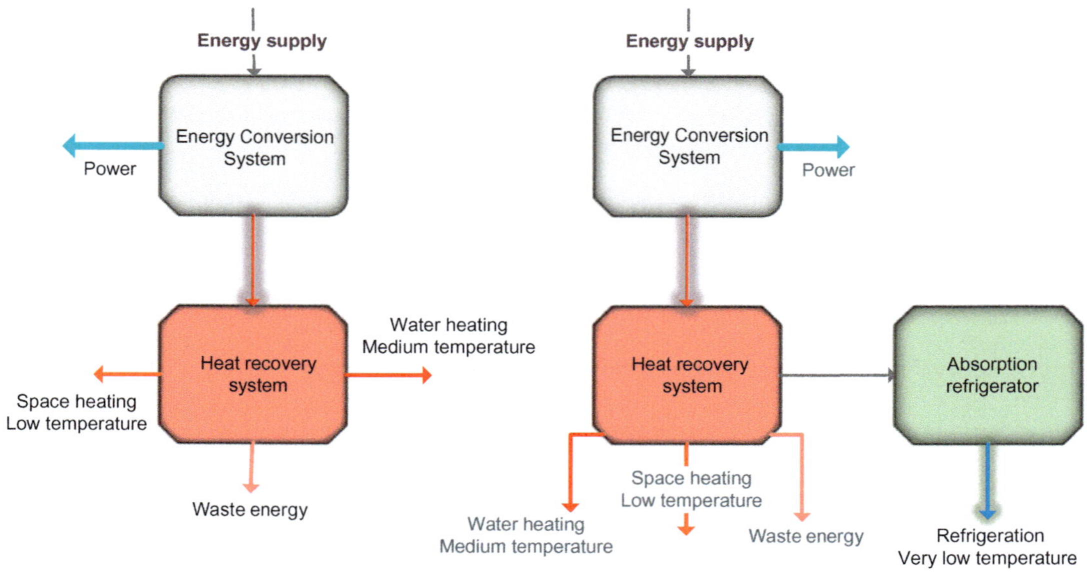

FIG. 3.37 Trigeneration and quadruple generation for improved system efficiencies.

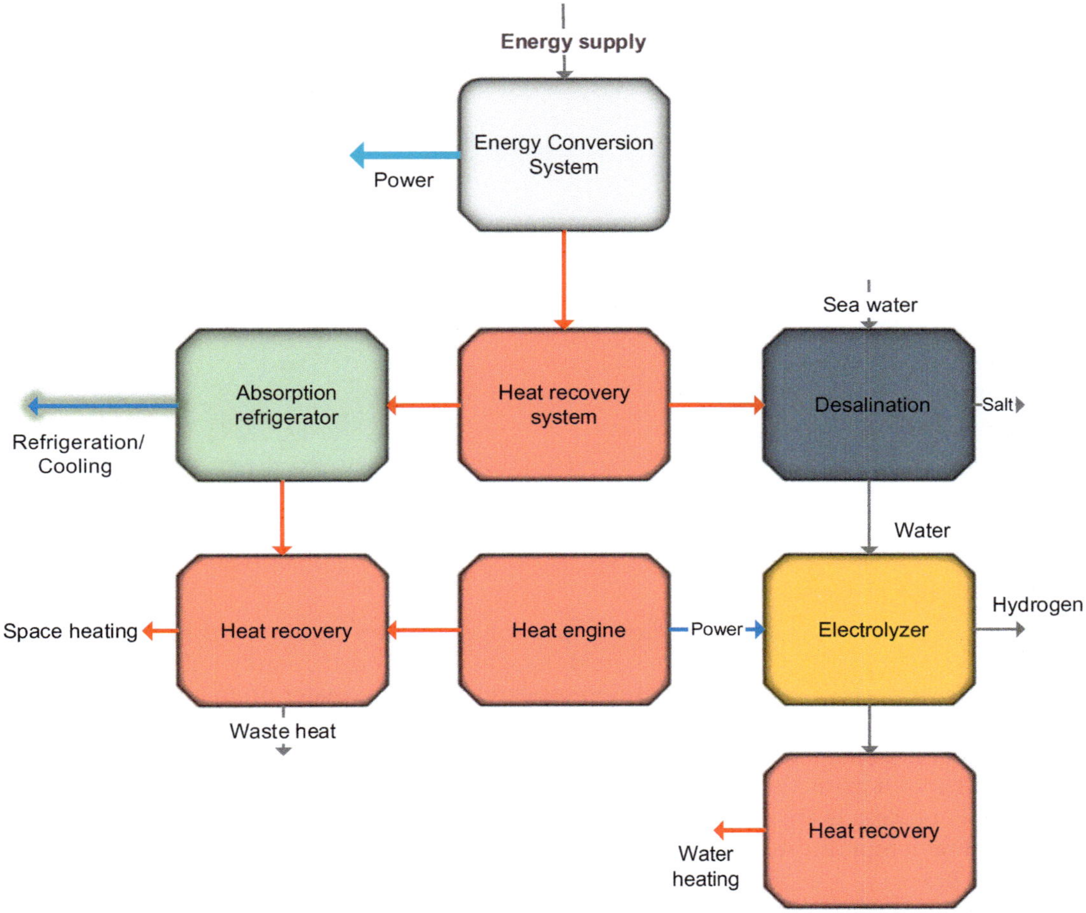

FIG. 3.38 System layout of a multigeneration system including desalination, electrolyzer, heat engine, absorption coolers and heat recovery system.

The number of useful outputs in multigeneration systems can be increased based on the needs of a community as shown in Fig. 3.39. It is suggested that from one or multiple energy sources, many diverse outputs can be produced. The system structure shall be defined by the experienced engineer/researcher for optimizing and maximizing the outputs of the multigeneration system. The system can entail a number of subsystems.

Some possible options for system integration in multigeneration systems are listed here [14]:

- Combustion and gasification (of coal and of biomass)

  Especially when the quality of coal is low, fluidized bed combustion systems are an alternative option for integration with multigeneration systems. In addition, co-gasification of biomass and coal is another example of integration that takes advantage of the high

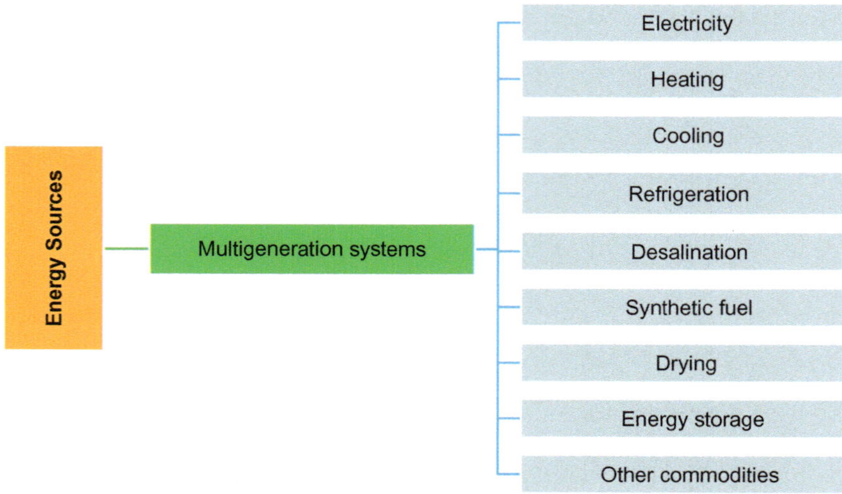

FIG. 3.39 Multigeneration systems for production of multiple commodities.

oxygen content of biomass that allows for a lower oxygen requirement in the pyrolysis phase. Gasifying the coal or biomass yields syngas consisting of mainly CO and H_2 as well as other gases. Therefore, the harmful gases can be captured later on in an easier way by using new technologies, e.g., membrane separation, etc.

- Direct coal liquefaction and plastics-to-fuel conversion

  Direct coal liquefaction is a process that directly converts coal to a liquid hydrocarbon fuel. This process is appealing for multigeneration applications because it occurs at a relatively low temperature, in the range of 400–500 °C, although the pressure is high, about 150–600 bar.

- Indirect liquefaction of coal and biomass

  Indirect liquefaction of coal and biomass can be performed by cascading gasification with a Fischer-Tropsch (FT) process. This process receives syngas as input and transforms it thermocatalytically to a liquid hydrocarbon fuel with a high cetane number. There is also conversion of waste plastics into diesel fuel in which wasted plastic materials from various industries are converted to diesel by thermocatalytic cracking. Furthermore, with recent developments, waste plastics can also be used as feedstock for thermal cracking and for production of hydrogen.

- Heat recovery steam (or vapor) generation

  This is the most common and basic form of using waste heat, which generates steam for industrial processes or combined cycles.

- Methanol synthesis and methanol reforming

  Methanol synthesis is considered useful process with outstanding options for integration with multigeneration systems. Methanol can be produced from syngas under a thermocatalytic reaction, which occurs at about 75 bar and 250 °C. Methanol is used as a fuel and hydrogen carrier. It can also be reformed to hydrogen using steam reforming techniques. Furthermore, methanol can be produced synthetically from green hydrogen, capturing CO_2 in a cleaner way, significantly reducing the overall environmental impacts.

- Ammonia synthesis and ammonia cracking

  Ammonia synthesis is considered an excellent option for energy storage for integration with multigeneration systems. Ammonia can be synthesized from hydrogen and nitrogen at about 200–300 bar and at about 450 to 500 °C though a conventional Haber-Bosch process. It can be stored over longer periods than hydrogen with considerably less loss. Ammonia can be cracked thermocatalytically to generate hydrogen. These are excellent options to store renewable energy within integrated multigeneration systems. In addition, there are new electrochemical ammonia synthesis processes, which are capable of functioning at lower pressures and temperatures. These systems are still being advanced to achieve higher production rates per active electrode area.
- Seawater desalination

  Seawater desalination is performed mainly for freshwater production in many regions of the world that have shortages of freshwater sources. Desalination can be integrated with power generation systems in many ways due to several methods of desalinating seawater, such as (i) multistage-flash distillation (MSF), (ii) multi-effect distillation (MED), (iii) vapor compression distillation (VCD), (iv) reverse osmosis (RO), (v) electrodialysis (ED) and (vi) nano-filtration. Most of the desalination methods require low-grade heat, and hence can be combined with power cycles with partial recovery of the rejected heat.
- Solar light processing (concentration, collimation, and spectral splitting)

  Sunlight can be used to directly drive or supplement photonic processes such as photocatalysis, photoelectrochemical processes, photobiological processes, etc. Solar spectra can be concentrated for high-temperature processes and also can be split for effective utilization of sunlight.
- Biochemical processes (digestion, bio-photolysis)

  A biochemical platform in a biorefinery signifies a set of methods by which biomass can be used to produce various products and fuels and can be integrated with fuel combustion facilities and power generators. This can include lignin production, enzymatic hydrolysis of cellulose, and fermentative processes that eventually lead to ethanol synthesis and other biochemical by-products.

3.3.5.1 Design criteria of multigeneration systems

Multigeneration systems shall be designed in such a way that the extra investment for production of multiple outputs in addition to power is balanced and compensated by higher revenues, a shorter payback period, lower levelized electricity cost, or lower environmental impact. For that reason, evaluation criteria are needed for multigeneration systems to assess their benefits from multiple points of view and compare them with other possible systems.

The main benefit of multigeneration energy systems is the fact that they can provide several useful outputs at once. This is primarily true for energy efficiency, renewable energy use, and the multigeneration of electricity, heating, cooling, chemical fuels such as hydrogen and ammonia, which can be accompanied by many advantages such as economic development, introduction of new jobs, improved energy security, better health and environmental conditions, and climate change mitigation [15].

In this book, we utilize the 3S (Source-System-Service) concept for system integration and multigeneration. This concept was initially introduced by Dincer [5] as shown in Fig. 3.40. Furthermore, energy storage should also be integrated between source and system as well as between system and service.

Sustainable energy systems are desired to meet all the mentioned standards from the source to end use. It is vital to select the most appropriate energy source for a sustainable energy system. Various key criteria exist that must be taken into account for choosing the suitable source of energy. These criteria can be briefly listed as abundance, local availability, affordability, reliability, safety, and environmental friendliness. Some renewables can satisfy most of these criteria whereas others are still in the developmental phase. In addition to renewables, there are some cleaner ways of using fossil fuels (hydrocarbon cracking, gasification, carbon capture, etc.). Nuclear energy is also a credible candidate for sustainable energy systems. Fig. 3.41 shows several improvement methods for an energy system. Process improvement can reduce the losses and maximize the amount of desired output. Efficiency increase can be performed by minimization of irreversibilities. Waste minimization is another important option, using the waste energy or materials for other useful outputs. System integration can enhance the system reliability and increase production rates. Multigeneration can achieve greater quantities of useful products from the same energy input. Energy recovery can reduce the waste energy and can increase the total energy production rate. Reduction of losses can be assured by proper design as well as improved processes.

In the 3S approach, the service step represents the end use phase. Thus, it is crucial to reduce waste, losses, irreversibilities, etc. in this step. Since utilization of fossil fuels in the final use step causes fossil fuel depletion as well as climate change and a wide range of emissions-related impacts, it is desired to use renewables for heating, transportation, metal refining, and the production of manufactured goods.

Heating, ventilation, air conditioning (HVAC), household and industrial appliances, industrial machines, transportation, IT services, lighting, etc. are some of the cases of end-use which are desired to be clean, effective, affordable, practical, and safe. It is also important to note that energy storage is a vital element of the 3S approach, as shown in Fig. 3.40, and is necessary between (i) source and system and (ii) system and service in order to provide continuity, reliability, and availability, as modified by Dincer and Acar [1].

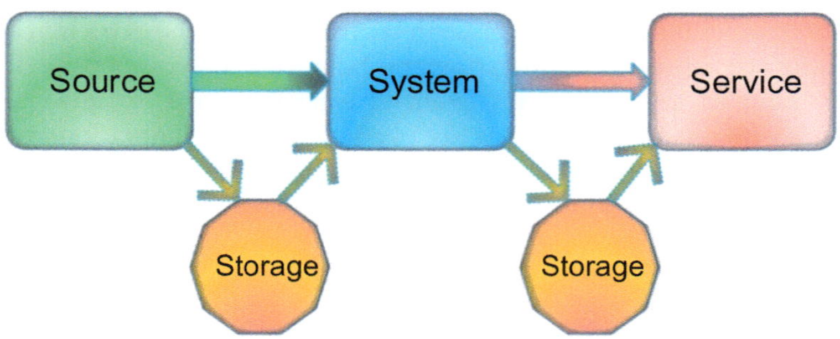

FIG. 3.40 Source-System-Service (3S) approach for energy systems including storage options.

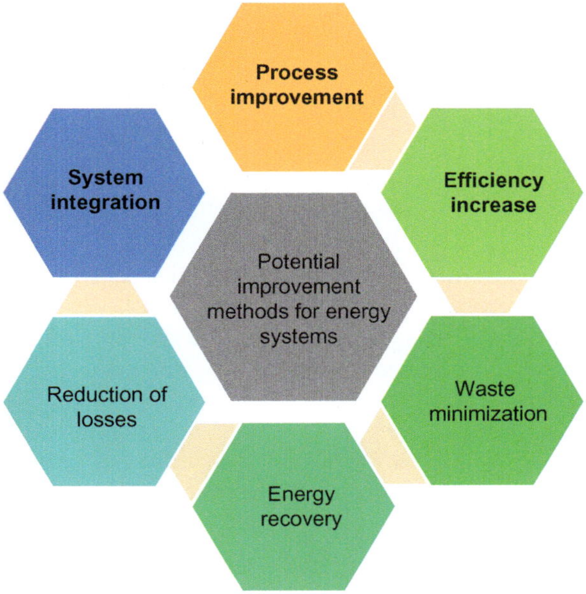

FIG. 3.41 Potential improvement methods for energy systems.

In this book, we present many case studies for integrated energy systems. The chapters are divided with respect to the main energy sources used, namely conventional, nuclear and renewable.

The conventional energy sources include:

- Coal
- Petroleum and derivatives
- Natural gas

The nuclear energy sources include:

- Nuclear fusion
- Nuclear fission

The renewable energy sources include:

- Solar
- Biomass
- Wind
- Geothermal
- Ocean energy
- Hydro energy

These sources can be used in a single or combined manner. One of the fundamental benefits of integrating the energy sources is to eliminate the deficiencies and shortcomings of the individual source. For example, solar energy is known to be an intermittent source, which fluctuates over the day. In order to provide continuous energy, it can be combined with a continual supply such as nuclear or natural gas or geothermal, etc.

The systems are quite diverse, ranging from photovoltaics to nuclear reactors. Based on the sources selected, the suitable systems are assigned and employed in the integrated energy system.

Some of the conversion systems are listed here:

- Rankine cycle
- Brayton cycle
- Absorption cooling system
- Photoelectrochemical energy conversion
- Electrochemical energy conversion
- Gasification
- Concentrated solar collectors
- Electrolysis
- Fuel cells
- Methanation
- Desalination

Services are in fact the useful outputs obtained from the integrated energy system. The determination of the services, which are the useful outputs, should focus on the industrial/residential demand. Although there are numerous useful products, we can list some of them here:

- Electricity
- Cooling
- Heating
- Hydrogen
- Ammonia
- Methanol
- Syngas
- Thermal energy storage
- Freshwater
- Conditioned air

3.3.6 Illustrative example: Hydrogen production

This example shows how to approach a system design by using the 3S (Source-System-Service) method in the context of hydrogen production.

3.3.6.1 Source

Hydrogen can be formed from diverse material sources such as ammonia, biomass, coal, ethanol, H_2S, methane, methanol, water, etc. Currently, steam reforming of natural gas is considered cost-effective, thus it is the most dominant option. However, this choice neither lowers/removes releases nor tackles the limited fossil fuel supply problem. In addition, coal is another supply where carbon capture and storage (CCS) has to be included to achieve greenhouse gas emissions reduction [16].

There is a clear necessity to define reliable, affordable, abundant, and clean sources for H_2 generation. For example, renewable sources deliver the best option for decreasing GHG emissions and reliance on fossil fuels. To further decrease emissions and energy usage, extensive research is being conducted globally to improve and refine systems that generate H_2 in more economical ways from renewable resources. Technological and economic resource capacities are mainly significant for guessing H_2 production potential [2].

Fossil fuels (e.g., coal, natural gas, oil, etc.), nuclear, biomass, and renewables (e.g., geothermal, hydro, ocean, solar, tidal, wind, etc.) can be used as energy sources to yield H_2. An innovative system should support the key pillars of sustainability: better efficiency, better cost-effectiveness, better resources use, better design and analysis, better energy security, and better environment, as previously explained. Henceforth, emissions, efficiencies, cost, renewability, and multigeneration capacity of sources are important for comparison. Renewables have a higher potential to support sustainable production systems. Continuously replenished resources, zero or low emissions, and multigeneration options hold the key to a sustainable future.

3.3.6.2 System

From a broader viewpoint, a hydrogen economy can be considered as a system that uses H_2 as the main carrier in the energy supply cycle. H_2 economy suggests new methods of energy usage that are sustainable and environmentally friendly [2].

There is a robust need to define, develop and apply effective systems to yield H_2 in an environmentally friendly way. These systems must address all key features of design, construction, operation, and maintenance. Agreement with the requirements decreases the system risk to a safe level. A comprehensive array of systems have been proposed for efficient, affordable, reliable, and clean H_2 generation. Nonetheless, currently there are only two main commercially available ones, which have been leading the way for a period of time [2]:

- *Steam methane reforming (SMR)*: This is thermal decomposition of methane with steam to produce H_2. The by-products of this option are CO and/or CO_2.
- *Electrolysis*: In this technique, electricity is applied to split H_2O into H_2 and O_2.

  In addition to these conventional techniques, there are numerous innovative H_2 generation systems that deliver safe, reliable, affordable, and clean H_2. Renewable energy based, integrated, and hybrid systems generate H_2 from numerous renewable resources in a sustainable way. For example, there are novel techniques, such as photoelectrochemical cells, that deliver H_2 with minimum or zero emissions [17]. H_2 production systems can be examined under three options [2]:
- *Thermal*: An example of thermal systems is the conversion of source energy (e.g., fossil fuels, biomass, etc.) to H_2. A different type uses heat together with closed chemical loops to produce H_2 from feedstocks, e.g., water, which is referred to "thermochemical" systems. These systems include natural gas reforming, biomass reforming, coal/biomass gasification, thermochemical water splitting, etc.
- *Electrolytic*: Electricity is utilized for water splitting into H_2 and O_2. This route can result in zero greenhouse gas emissions during operation (not complete life cycle), depending on the energy source.

- *Photonic*: Light energy is utilized for water splitting into H_2 and O_2. These processes show long-term potential for sustainable production with low environmental impact. Photoelectrochemical H_2 and photobiological systems are some samples of photonic H_2 production systems.

3.3.6.3 *Service*

As an energy carrier, H_2 is able to support new relations between energy supply and demand. This can be done using either centralized or decentralized methods, probably refining overall system flexibility. By connecting different energy system components, low-carbon energy sources can be engaged in different end-use preferences, which are challenging to decarbonize, i.e., transport, industry, and buildings. In rural areas with restricted access to the electricity grid, H_2 can advance off-grid access to services with minimum emissions [18].

H_2 has been used in the industry for years. Nevertheless, existing delivery structure and technology are not advanced enough to support broad service and end-use of H_2. As H_2 has quite low volumetric energy density, transportation, storage, and delivery of H_2 to end-use services carry high costs. A cause for this condition is the issues associated with the end-use of H_2 [2]. The main issues of H_2 delivery can be listed as high distribution cost, low efficiency, purity, and leakage. Further research is essential to examine the compromises between production and distribution. Developing the H_2 distribution infrastructure is a significant issue and it is projected to take time to progress and embrace combinations of diverse technologies. Distribution infrastructure sources and demands vary by region and market category (e.g., rural, interstate, urban). Infrastructure alternatives also advance as H_2 demand develops and distribution routes evolve and mature [19].

We need to define the potential application routes where H_2 delivers services and useful commodities (such as fuel) to systems and applications in numerous segments. For example, the use of residential low-carbon H_2 in fuel cells is an encouraging technological route for decreasing both emissions and dependence on imported oil. H_2 can serve in several end-use applications such as fuel cells, multigeneration plants, stoves, Stirling engines, boilers, burners, fans, and numerous other engines in transportation, residential, and industrial fields [20].

Fuel cells are very attractive methods for H_2 service routes, delivering similar service quality compared to conventional cars at lower emissions. A 25% share of fuel cell vehicles in transportation by 2050 could reduce up to 10% of overall transportation-based emissions. Presuming a steady increase in fuel cell vehicle sales, a self-sustaining market could be attained in 15–20 years after the introduction of the first 10,000 fuel cell vehicles [21].

Fruitful conversion to H_2 energy systems relies on the success of the H_2 service sector. Hence, collaboration with industry, academia, research laboratories, and other government institutions is essential to

- Overcome technical problems in research and development of all modules of hydrogen energy systems for multiple services
- Challenge safety issues and help the introduction of more appropriate codes and standards
- Confirm H_2 systems performance in actual service environments
- Train strategic target audiences to promote the near-term use of H_2

3.4 Thermodynamic analysis approach

The first law of thermodynamics is known as the energy conservation law, leading to the statement that energy is neither created nor destroyed, but changes form, for example from heat to work in a thermal power plant and work to heat in a refrigeration/heat pump system. This does not cover losses or inefficiencies or irreversibilities or destructions, and it basically does not reflect the actual operation. If one wants to deal with actual/real systems, one needs to go a further step to the second law of thermodynamics, which is defined as the non-conservation of exergy law. It leads to the statement that exergy is not conserved during any process and is destroyed accordingly based on state properties, operating conditions and interactions with the surroundings. It is a true measure of losses or inefficiencies or irreversibilities or destructions as well as quality and performance. It then brings exergy forward as a potential tool for design, analysis, assessment and improvement [3]. It should be noted that there is a need to define reference state and reference environment in performing exergy analysis by specifying reference temperature, reference pressure and reference chemical composition.

A smart analysis approach has been proposed by Dincer [3]. This approach consists of six critical steps, namely: property, state, process, cycle, balance equations and performance assessment, as illustrated in Fig. 3.42. In this way, one can follow a logical order and make it simple to understand and comprehend the thermodynamic concepts and aspects.

3.4.1 Step 1—Property

Property is defined as a physical characteristic or quantity of a substance which is treated as a key attribute of any thermodynamic system. Any two independent properties usually define the state or condition of the substance, from which all other properties can be derived. There are measurable properties, such as temperature, pressure, mass and volume, and there are also non-measurable properties, such as internal energy, enthalpy, entropy and exergy. Thermodynamic properties are classified as intensive properties (independent of the mass, e.g., pressure, temperature and density) and extensive properties (dependent on the mass, e.g., mass and total volume). Extensive properties per unit mass become intensive properties such as specific volume. Property diagrams of substances are common and generally presented in graphical and tabular forms to summarize the main properties.

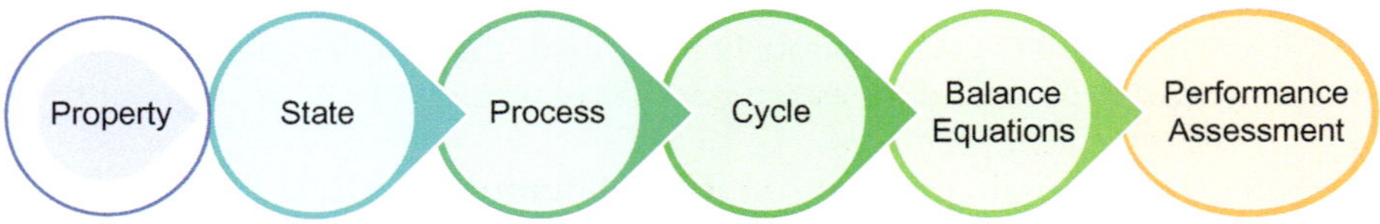

FIG. 3.42 Fundamental thermodynamics analysis approach in six steps. *Adapted from Dincer I. Exergization. Int J Energy Res 2016;40:1887–9. https://doi.org/10.1002/er.3606*

3.4.2 Step 2—State

State is defined as a true condition of any thermodynamic system defined by the thermodynamic properties (particularly, by the measurable properties). In order to obtain the other unknown properties for the subject matter state, there must be at least two known properties. There may be situations where the unknown properties of any state are obtained by using a known property and the state related information, such as saturated liquid, saturated vapor, etc. For instance, in a refrigeration system, there are four common state points, which are saturated vapor after the evaporator, superheated vapor after the compressor, saturated liquid after the condenser and a mixture of liquid and vapor after the throttling valve.

3.4.3 Step 3—Process

Process is defined as a physical or chemical change in the properties of matter or the conversion of energy from one form to another. Several processes are described by the fact that one property remains constant. The prefix iso- is employed to describe a process such as an isothermal process (a constant-temperature process), an isobaric process (a constant-pressure process) and an isochoric process (a constant-volume process). For instance, a refrigeration process is generally expressed by the conditions or properties of the refrigerant at the beginning and end of the process.

3.4.4 Step 4—Cycle

A cycle is a series of thermodynamic processes in which the endpoint conditions or properties of the matter are identical to the initial conditions. For example, in refrigeration, the processes required to produce a cooling effect are arranged to operate in a cyclic manner so that the refrigerant can continuously be reused in a closed system. In thermodynamics, there are cycles, such as closed type and open type. A good example comes from the air Brayton cycle, which has both versions: closed and open.

3.4.5 Step 5—Balance equations

Writing the thermodynamic balance equations for mass, energy, entropy and exergy correctly is the most critical step in thermodynamics. This step is also critical in correctly designing, analyzing and evaluating thermodynamic systems and their components. This should be done carefully and diligently for closed and open systems.

3.4.6 Step 6—Performance assessment

The simplest approach to assess the performance is by the efficiency, which ranges between 0 and 1 or 0% and 100%. As previously explained, the simplest definition of efficiency is given as follows:

$$\text{Efficiency} = \frac{\text{Useful output}}{\text{Total input}} \tag{3.7}$$

If the efficiency is defined under the first law of thermodynamics, it is called first law or thermal or energy efficiency. If it is defined under the second law of thermodynamics, it is then called second law or exergy efficiency. Defining efficiency correctly is recognized as a significant step in thermodynamics. Hence, the energy efficiency under the first law of thermodynamics and exergy efficiency under the second law of thermodynamics can be articulated as follows:

$$\eta_{\text{en}} = \frac{\text{Useful energy output}}{\text{Total energy input}} \tag{3.8}$$

$$\eta_{\text{ex}} = \frac{\text{Useful exergy output}}{\text{Total exergy input}} \tag{3.9}$$

There are situations/applications, where the efficiency values may be >1 or 100%, such as in refrigeration and heat pump systems. In fact, this does not represent the energy conversion process. These devices pump the available heat from a low-temperature medium to a high-temperature medium by an energy input (e.g., compressor work in vapor compression cycles or generator heat in absorption cooling systems). In such cases, we use the coefficient of performance (COP) to assess the performance of refrigeration and heat pump systems. The COP values can be obtained through energy and exergy analyses, and are then called energetic COP and exergetic COP which can be defined as follows:

$$\text{Energetic COP} = \frac{\text{Useful energy output}}{\text{Total energy input}} \tag{3.10}$$

$$\text{Exergetic COP} = \frac{\text{Exergy output}}{\text{Total exergy input}} \tag{3.11}$$

3.4.7 Smart and comprehensive approach

In this section, we provide a smart and comprehensive approach for analyzing integrated energy systems. The smart approach is illustrated in Fig. 3.43.

The fundamental steps to be followed for thermodynamic analysis are listed here:

(1) Define the system/unit type (closed or open)
(2) Sketch the system correctly
(3) Define and sketch the component boundary
(4) Determine and write the conceptually correct assumptions clearly
(5) Show all inputs and outputs (mass, heat and work) on the figure with correct arrow signs
(6) Write all state point numbers on the figure
(7) Define the reference temperature and pressure and write on the corner of figure
(8) Show the known thermodynamic properties (such as temperature and pressure) for each state point on the figure
(9) Calculate the unknown thermodynamic properties using two of the known properties

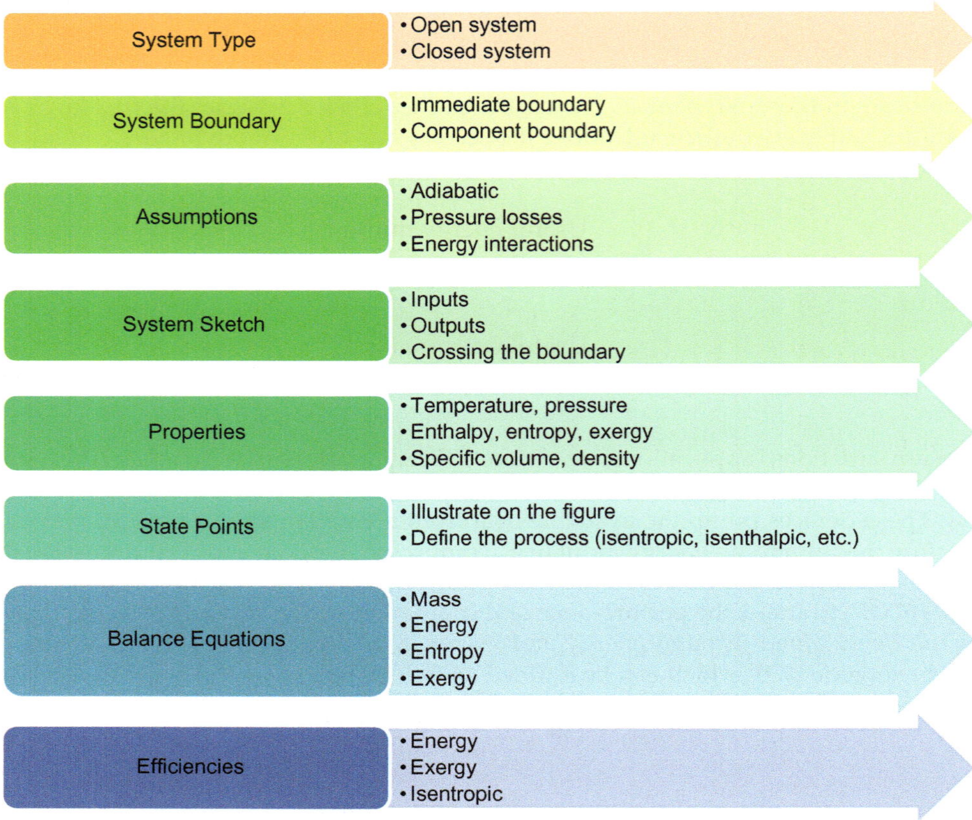

FIG. 3.43 A smart hierarchical approach for thermodynamic analysis.

(10) Write or list all thermodynamic properties for each state point on the figure or table (temperature, pressure, enthalpy, entropy, quality and exergy, etc.)
(11) Write all balance equations (mass, energy, entropy and exergy) based on the figure showing all inputs and outputs
(12) Define and write the energy and exergy efficiencies (useful outputs/total input)
(13) Calculate the respected parameters using balance equations and efficiency definitions
(14) Conduct parametric studies to investigate the effects of important parameters
(15) Compare and validate the attained results with the literature

3.5 Case study: Comparative evaluation of conventional- and renewable-based single generation vs. multigeneration systems

This case study presents a comparative evaluation of single and multigeneration systems from various aspects, namely environmental impact, efficiency and energy source. This case study is adapted from Dincer and Acar [1]. Critical conditions for switching to sustainable energy systems for a sustainable future can be listed as [1]

- energy conservation
- greater use of renewable energies
- smart grids to support renewable energy use
- cleaner technologies
- multigeneration and effective storage of energy carriers and chemicals

3.5.1 Comparison of environmental impacts

One of the objectives of sustainable energy systems is to decrease releases of GHGs. Fig. 3.44 depicts normalized rankings of the specific GHG emissions per kWh for various energy production routes. The data used in this study are taken from the literature [22–25] and they are all most significant releases converted to CO_2 equivalent per kWh based on life-cycle emissions.

In Fig. 3.44, single generation represents only electricity generation, cogeneration represents production of heat and power (CHP), trigeneration is production of heating, cooling and power (CCHP). In addition to CCHP, hydrogen production is also employed in quadgeneration. In the case of wind power or hydro power, it should be noted that there are only two products. Thus, in these options, power and hydrogen are not considered. The ranking is performed using the following equation:

$$Rank_n = \frac{Maximum - Method_n}{Maximum - Minimum} \tag{3.12}$$

Here, *Minimum* and *Maximum* are the minimum and maximum emissions and $Method_n$ represents the emissions of a selected method. This equation allocates a ranking to each

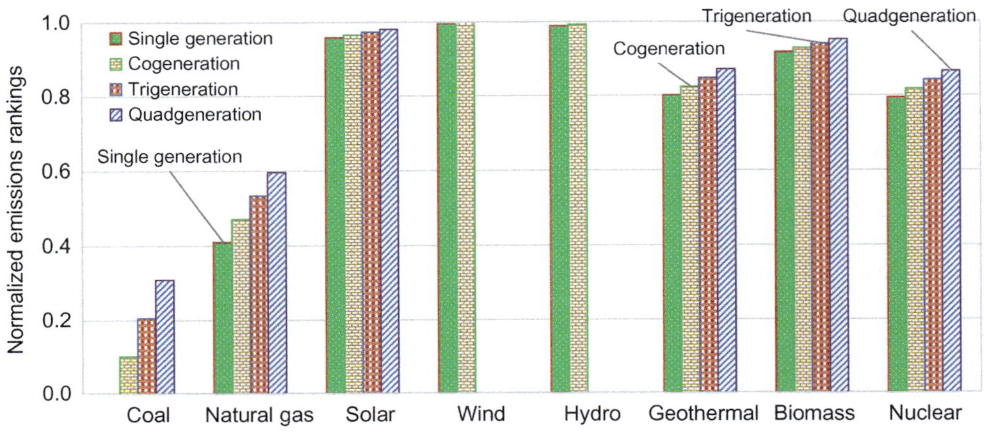

FIG. 3.44 Normalized emissions rankings from various energy sources in different generation technologies (higher values represent better performance). *Data from Dincer I, Acar C. Smart energy systems for a sustainable future. Appl Energy 2017;194:225–35. https://doi.org/10.1016/J.APENERGY.2016.12.058*

technique, which is between 0 and 1, where 0 means the highest quantity of emissions and 1 means the lowest quantity of emissions (cleanest method).

It can be seen that coal has the highest emissions (lowest rankings), followed by natural gas. Hydro and wind have the lowest emissions (highest rankings), followed by solar, biomass, and geothermal. Nuclear energy has fewer emissions compared to fossil fuel-based methods. This is the case even with the fossil fuel-based systems with CCS. Also, regardless of the energy source type, increasing the number of products (switching from single generation to multigeneration) reduces emissions per unit of energy content of the products.

Fossil fuels processing with CCS technology lowers harmful emissions considerably. However, even these technologies have significantly higher emissions than the systems with renewable energies and nuclear. Joint utilization of fossil fuel and biomass with CCS could provide electricity, heat, hydrogen, and industrial commodities with zero or minimal emissions. Quantifying emissions for electricity that is multi generated is complex, since it is not clear how to allocate emissions to different products. The methodology used in this case study is to allocate a fraction discount in emissions equivalent to the emission for the overall multigeneration system divided by the emissions of a reference state (base system) containing separate traditional fossil fuel-based systems with no CCS that separately generate the equal amount of products [1]. It should be noted that even though CCS lowers the emissions related to fossil fuel utilization, it does not address the issue related to limited supply and possible adverse impacts of increasing rates of fossil fuel extraction.

3.5.2 Comparison of energy conversion efficiencies

Fig. 3.45 depicts normalized rankings of the efficiencies for multiple options. The data for conversion efficiencies are the average of the energy and exergy efficiencies for power and hydrogen production from the chosen energy sources. The data are collected from various literature studies [22, 26–28]. Similar to the previous figure, in Fig. 3.45, single generation

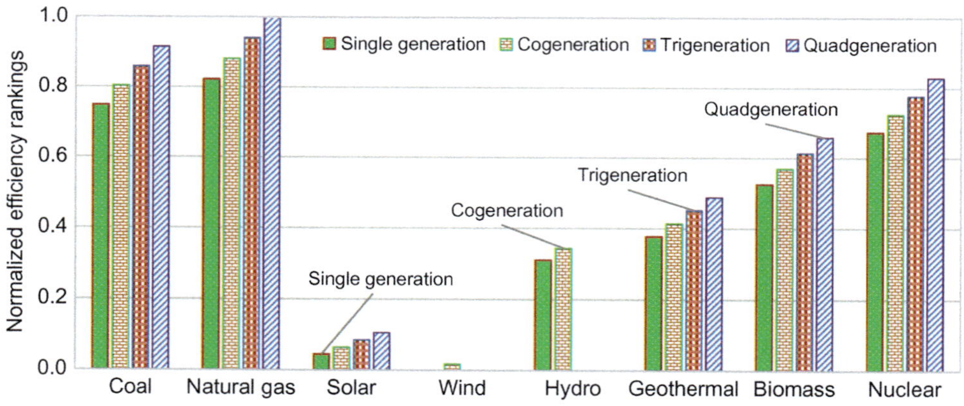

FIG. 3.45 Normalized efficiency rankings from various energy sources in different generation technologies. *Data from Dincer I, Acar C. Smart energy systems for a sustainable future. Appl Energy 2017;194:225–35. https://doi.org/10.1016/J.APENERGY.2016.12.058*

represents only electricity generation, cogeneration represents producion of heat and power (CHP), and trigeneration represents production of heating, cooling and power (CCHP). In addition to CCHP, hydrogen production is also employed in quadgeneration. In the case of wind power or hydro power, it should be noted that there are only two products. Thus, in these options, cogeneration is power and hydrogen whereas trigeneration and quadgeneration are not taken into account. The normalized ranking is conducted by taking the average efficiencies as they are, between 0 and 1 as previously explained.

It is clear from the figure that natural gas has the uppermost efficiencies (highest rankings), followed by coal and nuclear. Solar and wind have the lowest efficiencies (lowest rankings), followed by hydro, biomass, and geothermal. Moreover, irrespective of the energy source, increasing the number of products increases efficiencies, thus increasing energy conservation.

Combining energy sources and increasing the number of products is expected to significantly improve quality and performance of the systems. Although the sun and wind energies seem to have lower rankings compared to other methods, they are supposed to improve when the technology is more developed.

Even though renewable and clean energy sources tend to offer lower efficiencies (mainly because most of them are state-of-the-art or in early development phases) multigeneration increases system efficiencies, as increasing the number of outputs from a single source also increases the overall efficiencies.

3.5.3 Comparison of energy sources

The attributes of renewable energy technologies, which include modularity, flexibility, and low and relatively stable operating costs, are considerably different than those of traditional, fossil-based technologies, whose attributes include large capital investments, long implementation lead times, and uncertain operating costs regarding future fuel prices [1]. The overall benefits of renewable energy technologies are often not well understood and, consequently, these technologies are often evaluated to be not as cost effective as traditional technologies. To assess renewable technologies comprehensively, however, all of their benefits, some of which are often not considered, must be accounted for. One assessment difficulty relates to the fact that renewable energy technologies are sometimes seen as direct substitutes for existing technologies so that their benefits and costs are evaluated in terms of assessment methods developed for the existing technology. For example, solar and other renewable energy technologies can provide small incremental capacity additions to existing energy systems with short lead times. Such power generation units usually provide more flexibility in incremental supply than large, long lead-time units such as nuclear power stations.

Renewable energy resources are vital for sustainable energy systems due to the fact that they are designed as non-depleting resources of energy with enriched quality and minimum or no harmful impact on the environment. Thus, utilization of renewable energy sources with improved energy efficiencies in sustainable energy systems is crucial. The efficiency of any energy conversion technique is limited because of system irreversibilities, for instance heat generation as a consequence of friction. Thus, in conventional single-generation methods, there is always some type of lost or wasted energy. There are numerous instances of energy

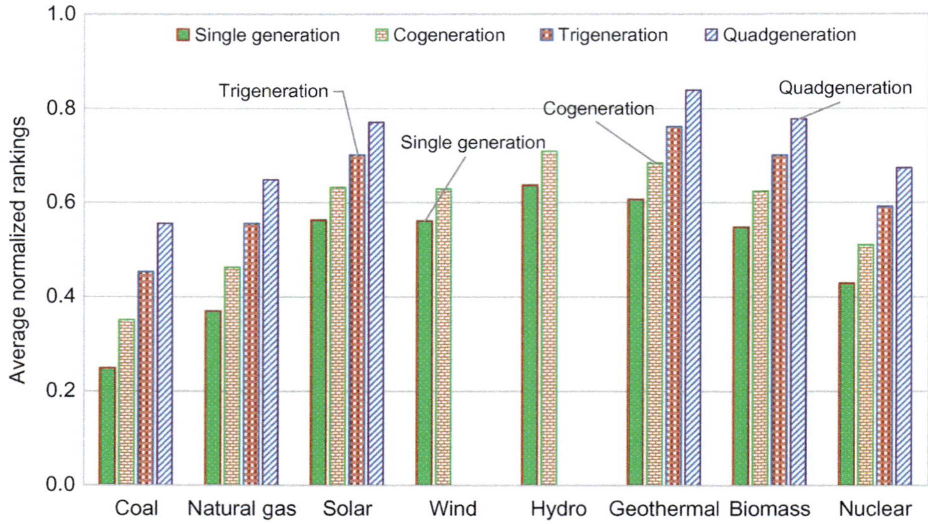

FIG. 3.46 Average normalized overall rankings of various energy sources in different generation technologies (higher values represent better performance). *Adapted from Dincer I, Acar C. Smart energy systems for a sustainable future. Appl Energy 2017;194:225–35. https://doi.org/10.1016/J.APENERGY.2016.12.058*

waste or losses in conventional energy systems, some of them being heat rejected from the heat engines, partial combustion, etc. The main energy usage becomes more effective and cleaner when lost and/or wasted energies are recovered and converted into useful products [29].

The renewable rankings are assigned to each energy source with 0 meaning non-renewable (fossil fuels and nuclear) and 1 meaning renewable, which includes solar, wind, hydro, and geothermal. Biomass is assigned to be 0.5 since its renewability depends on the rate of consumption. If the rate of consumption of the biomass is lower than the rate of biomass regeneration, it can be taken as fully renewable. However, if the rate of consumption exceeds the rate of resource replenishment, then the source is not fully renewable.

The averages of the normalized rankings in terms of emissions, efficiencies, renewability, and a number of anticipated products are taken to find the overall rankings. Fig. 3.46 depicts these average overall rankings of coal, natural gas, solar, wind, hydro, geothermal, biomass, and nuclear.

It is observed that average rankings rise with a rising number of products. Sustainable energy systems are apparently projected to utilize the cleanest resources in the most effective manner at adequate cost ranges. Among the particular sources, geothermal is found to have the highest potential in terms of using cleaner technologies with energy conservation, renewability and the opportunity of multiple anticipated products from the same source. Solar, hydro, and biomass also have huge potentials. On the other hand, fossil fuels are not very desirable for sustainable energy systems even with carbon capture technologies, because of their emissions and non-renewability.

3.6 Conclusions

This chapter provides the fundamental philosophy of system integration for multigeneration systems. Initially, the importance and necessity of integrating energy systems are discussed. It is emphasized that there is a need to transform conventional methods into greener systems by accounting for several aspects such as environmental impact, resource utilization, efficiency, and cost-effectiveness, which will help achieve better sustainability. Thus, while designing and implementing sustainable system integration, various criteria shall be met, such as being energetically secure, environmentally benign, reliable, integrable, economically feasible, commercially viable, and socially acceptable.

Sustainable energy systems need to support sustainability in every stage of their life cycle, from cradle to grave. Hence, the 3S (Source-System-Service) approach is a key element of sustainable energy designs. Switching to clean energy systems for a sustainable future is presently considered as the main driver of innovation and is an ongoing effort. Novel technology solutions, development in materials science, and the introduction of new sources, systems, and services can possibly facilitate substituting the traditional options with sustainable energy systems.

Multigeneration systems offer distinctive benefits, especially when combined with clean energy sources, such as renewables. The multigeneration system can reduce wastes and environmental impacts, while increasing system efficiencies and cost-effectiveness. Furthermore, in comparison to conventional single generation systems, multigeneration systems can deliver multiple outputs concurrently such as electricity, heating, cooling, chemical fuels, value-added products, energy storage, etc.

Energy storage plays an important role in the broader implementation of system integration. Hence, storagization is necessary for continuous operation as well as better demand-supply management. There are several techniques of storing energy such as pumped hydro, electrochemical batteries, reversible fuel cells, compressed air, chemical storage, thermal energy storage, etc.

Greenization is a process of converting conventional systems with higher environmental damage, less efficiency, more cost, etc. into more efficient, more cost-effective, and more environmentally friendly systems. An exact application of greenization method for sustainable energy applications necessitates modification, adaptation, retrofitting, or replacement of the existing ones.

Multigeneration systems shall be designed in such a way that the extra investment for production of multiple outputs in addition to power is balanced and compensated by higher revenues, a shorter payback period, lower levelized electricity cost, or lower environmental impact. System integration can be performed in various ways such as cascading, multistaging, combining, hybridizing for cogeneration, trigeneration and multigeneration.

The useful products obtainable from multigeneration systems include but are not limited to electricity, heating, cooling, refrigeration, desalination, synthetic fuel, drying, energy storage and other commodities.

Sustainable energy systems could attain a sustainable future by tackling challenges and issues related to production, processing, and end use of energy. This chapter emphasized that increasing the number of useful products from the same energy source decreases emissions per unit product and enhances efficiencies.

In order to analyze multigeneration systems thermodynamically, we presented a smart hierarchical approach. This allows conceptually correct analysis as well as providing a systematic approach. The next chapters of this book will present conventional, nuclear and renewable energy-based integrated systems including several case studies.

Nomenclature

D_p	depletion number
EI	environmental impact
Ex	exergy
GF	greenization factor
IP	improvement potential
SI	sustainability index

Acronyms

3S	source-system-service
CCCP	combined cycle power plants
CCHP	combined cooling, heating and power
CCS	carbon capture and storage
CHP	combined heat and power
COP	coefficient of performance
FT	Fischer-Tropsch
GHG	greenhouse gas
GT	gas turbine
HVAC	heating, ventilation, air conditioning
IGCC	integrated gasification combined cycle
MCFC	molten carbonate fuel cells
ORC	organic Rankine cycle
PCM	phase change materials
PV	photovoltaic
SMR	steam methane reforming
SOFC	solid oxide fuel cell
TES	thermal energy storage

Greek letters

η	efficiency

Subscripts and superscripts

ref.	reference
Ex	exergy
d	destruction
out	output
in	input

References

[1] Dincer I, Acar C. Smart energy systems for a sustainable future. Appl Energy 2017;194:225–35. https://doi.org/10.1016/J.APENERGY.2016.12.058.

[2] Dincer I, Acar C. Innovation in hydrogen production. Int J Hydrogen Energy 2017;42:14843–64. https://doi.org/10.1016/j.ijhydene.2017.04.107.

[3] Dincer I. Exergization. Int J Energy Res 2016;40:1887–9. https://doi.org/10.1002/er.3606.

[4] Dincer I. Smart energy solutions. Int J Energy Res 2016;40:1741–2. https://doi.org/10.1002/er.3621.

[5] Dincer I. Greenization. Int J Energy Res 2016;40:2035–7. https://doi.org/10.1002/er.3619.

[6] Dincer I, Acar C. Smart energy solutions with hydrogen options. Int J Hydrogen Energy 2018;43:8579–99.

[7] Dincer I. Evaluation and selection of energy storage systems for solar thermal applications. Int J Energy Res 1999;23:1017–28. https://doi.org/10.1002/(SICI)1099-114X(19991010)23:12<1017::AID-ER535>3.0.CO;2-Q.

[8] Dincer I, Rosen M. Thermal energy storage: Systems and Applications. Wiley; 2011.

[9] Dincer I, Ezan MA. Heat Storage: A Unique Solution For Energy Systems. 1st ed. Cham, Switzerland: Springer International Publishing; 2018. https://doi.org/10.1007/978-3-319-91893-8.

[10] Lovegrove K, Luzzi A, Soldiani I, Kreetz H. Developing ammonia based thermochemical energy storage for dish power plants. Sol Energy 2004;76:331–7. https://doi.org/10.1016/j.solener.2003.07.020.

[11] Dincer I, Zamfirescu C. Potential options to greenize energy systems. Energy 2012;46:5–15. https://doi.org/10.1016/J.ENERGY.2011.11.061.

[12] Van Gool W. Energy Policy: Fairy Tales and Factualities. Innov. Technol.—Strateg. Policies. Dordrecht: Springer Netherlands; 1997. p.93–105. https://doi.org/10.1007/978-0-585-29606-7_6.

[13] Dincer I, Rosen MA. Exergy: Energy, Environment and Sustainable Development. In: Exergy. 2nd ed. New York, NY: Elsevier; 2013. p. 51–73. https://doi.org/10.1016/B978-0-08-097089-9.00004-8.

[14] Dincer I, Zamfirescu C. Chapter 9—Multigeneration systems. In: Zamfirescu ID, editor. Adv. Power Gener. Syst. Boston: Elsevier; 2014. p. 517–73. https://doi.org/10.1016/B978-0-12-383860-5.00009-2.

[15] Mah DN, Wu Y-Y, Ip JC, Hills PR. The role of the state in sustainable energy transitions: a case study of large smart grid demonstration projects in Japan. Energy Policy 2013;63:726–37. https://doi.org/10.1016/j.enpol.2013.07.106.

[16] Dincer I. Green methods for hydrogen production. Int J Hydrogen Energy 2012;37:1954–71. https://doi.org/10.1016/j.ijhydene.2011.03.173.

[17] Bhandari R, Trudewind CA, Zapp P. Life cycle assessment of hydrogen production via electrolysis—a review. J Clean Prod 2014;85:151–63. https://doi.org/10.1016/J.JCLEPRO.2013.07.048.

[18] Koumi Ngoh S, Njomo D. An overview of hydrogen gas production from solar energy. Renew Sustain Energy Rev 2012;16:6782–92. https://doi.org/10.1016/J.RSER.2012.07.027.

[19] Trainham JA, Newman J, Bonino CA, Hoertz PG, Akunuri N. Whither solar fuels? Curr Opin Chem Eng 2012;1:204–10. https://doi.org/10.1016/J.COCHE.2012.04.001.

[20] Farhat K, Reichelstein S. Economic value of flexible hydrogen-based polygeneration energy systems. Appl Energy 2016;164:857–70. https://doi.org/10.1016/J.APENERGY.2015.12.008.

[21] O'Hayre RP. Fuel cell fundamentals. 2nd ed. New York: John Wiley Sons; 2006. p.409. https://doi.org/10.1017/CBO9781107415324.004.

[22] Dincer I, Acar C. A review on clean energy solutions for better sustainability. Int J Energy Res 2015;39:585–606. https://doi.org/10.1002/er.3329.

[23] Acar C, Dincer I. Comparative environmental impact evaluation of hydrogen production methods from renewable and nonrenewable sources. In: Causes, impacts and solutions to global warming. New York, NY: Springer New York; 2013. p. 493–514. https://doi.org/10.1007/978-1-4614-7588-0_28.

[24] Oladokun MG, Odesola IA. Household energy consumption and carbon emissions for sustainable cities—a critical review of modelling approaches. Int J Sustain Built Environ 2015;4:231–47. https://doi.org/10.1016/J.IJSBE.2015.07.005.

[25] Kais S, Sami H. An econometric study of the impact of economic growth and energy use on carbon emissions: Panel data evidence from fifty eight countries. Renew Sustain Energy Rev 2016;59:1101–10. https://doi.org/10.1016/J.RSER.2016.01.054.

[26] Johansson TB, Patwardhan A, Nakićenović N, Gomez-Echeverri L. Global Energy Assessment (GEA). International Institute for Applied Systems Analysis; n.d.

[27] Christopher K, Dimitrios R. A review on exergy comparison of hydrogen production methods from renewable energy sources. Energ Environ Sci 2012;5:6640–51. https://doi.org/10.1039/C2EE01098D.
[28] Gu W, Wu Z, Bo R, Liu W, Zhou G, Chen W, et al. Modeling, planning and optimal energy management of combined cooling, heating and power microgrid: a review. Int J Electr Power Energy Syst 2014;54:26–37. https://doi.org/10.1016/J.IJEPES.2013.06.028.
[29] Sorknæs P, Lund H, Andersen AN. Future power market and sustainable energy solutions—The treatment of uncertainties in the daily operation of combined heat and power plants. Appl Energy 2015;144:129–38. https://doi.org/10.1016/J.APENERGY.2015.02.041.

Further reading

[30] Nikolaidis P, Poullikkas A. A comparative review of electrical energy storage systems for better sustainability. J Power Technol 2017;97.
[31] US EIA. How much carbon dioxide is produced when different fuels are burned?—FAQ—U.S. Energy Information Administration (EIA). US Energy Inf Adm 2018. https://www.eia.gov/tools/faqs/faq.php?id=73&t=11 (Accessed 26 December 2018).

CHAPTER

4

Integration of conventional energy systems for multigeneration

4.1 Introduction

Conventional energy systems have several drawbacks, such as dependence on fossil fuels, lower efficiencies and wasted heat. In fact, these deficiencies imply that they can be improved considerably. Energy conversion systems are very diverse, including simple applications such as the basic windmill, which converts wind energy to kinetic energy, or more complex systems such as natural gas-fired power plants that convert natural gas into electricity via power cycles. Currently, most power generation plants work on the theory of steam and gas cycles, which utilize coal, oil, natural gas and even nuclear sources. Hence, two of the prime improvement areas are enhancement of gas and steam cycles. This can be accomplished in two ways: reducing the wasted energy and integrating with other systems. The low efficiency levels of conventional systems directed researchers into combined systems. Combined cycles were introduced in order to utilize part of the rejected heat in a bottoming cycle, as discussed in the preceding chapter.

The conventional systems started with single generation in which there is only electricity output. Later on, the waste heat was considered as an asset for further improvement, hence the transition from single to cogeneration systems started after the 1990s. Since the beginning of the 2000s, trigeneration systems have been considered and studied. It started with simple cycles and progressed to combined systems, cascaded, cogeneration and trigeneration systems. Since there are still further improvement potentials of conventional systems, today it is still important to consider, study, design, analyze and implement multigeneration systems. The main target for future energy systems is to achieve integrating multiple systems for multigeneration as illustrated in Fig. 4.1. In order to overcome the main issues with conventional systems, diverse aspects of energy systems need to be considered ranging from technical to environmental.

Conventional power cycles are in fact still the fundamental principle of many systems such as marine propulsion, road vehicle engines, locomotive engines, mechanical drives,

Integrated Energy Systems for Multigeneration
https://doi.org/10.1016/B978-0-12-809943-8.00004-2

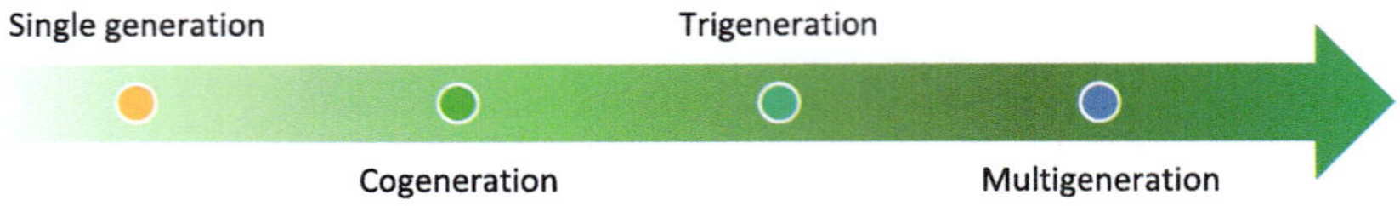

FIG. 4.1 Transition of conventional systems from single generation to multigeneration.

cogeneration plants, and combined cycle power plants. Multigeneration systems provide an opportunity to fulfill multiple useful commodities by incorporating several cycles, which require relatively lower temperature levels.

In this chapter, we provide a general overview of gas and steam cycles and then discuss alternative combined systems for improving the cycle efficiencies. In this regard, initially individual cycles such as Rankine, Brayton, Kalina, etc. are introduced. Then, combined cycles such as Brayton-Kalina, Rankine-Brayton, etc. are described. Several case studies are also presented to demonstrate the improvement potential of conventional energy systems.

4.2 Power cycles

Power cycles are in fact very significant in today's world. The historical background of power cycles goes back to the Industrial Revolution. There are multiple power cycles, which will be further discussed in this section. Commonly, steam and gas power cycles are employed in power generation. Steam power cycles have been applied for commercial purposes in various sectors, ranging from energy production to propulsion in marine vessels, since the early 19th century. The conventional Rankine cycle as introduced by William John Macquorn Rankine used water as its working fluid. The modern steam turbines utilized in industrial applications have evolved to their present formation after going through a process of technological research, improvements and developments that began in the 19th century, where some of the steps can be expressed as follows [1]:

- 1859: The Scottish William Rankine presented a closed thermodynamic cycle converting thermal energy into work. The Rankine cycle systems generally utilize water as the working fluid and even in the 21st century they provide most of the global electricity generation.
- 1896: The American engineer Charles Gordon Curtis developed the first velocity-compound steam engine, reducing the size and weight of the engines by about 90%.
- 1901: The first merchant vessel propelled by steam turbines, TS King Edward, was built and launched in Dumbarton, Scotland.
- 1925: The first high-pressure steam turbine capable of operating under 8274 kPa was introduced by Boston Edison in the United States.
- 1961: Physicist Harry Zvi Tabor and engineer Lucien Bronicki developed an organic Rankine cycle (ORC) using an organic-based working fluid with high molecular mass and relatively lower boiling point. The application of ORCs allows power generation with low-grade heat.

Gas power cycles have been finding applications for commercial purposes in various sectors, ranging from power generation to propulsion in aircraft, since the early 19th century. The gas turbine is a useful device to convert heat energy into mechanical energy, by rotating shaft, with higher capacities and efficiencies. Some of the multiple steps of research and development leading to today's commercial gas power cycles are listed here [1, 2]:

- 1791: The English inventor Josh Barber patented the first gas turbine [2].
- 1872: The American mechanical engineer George Brayton initiated a cycle for gas turbine plants and registered a patent under his name, also known as the Brayton cycle [1].
- 1899: A patent was issued for an American gas turbine. It was sold to General Electric in 1901 [2].
- 1903: Charels Lamale and René Armengaud presented the first successfully working, commercial combustion gas turbine at the Société Turbos Moteurs in Paris using a multistage compressor [1].
- 1953: The first propane-fueled locomotive, having a power rating of 3.6 kW, driven by a gas turbine was constructed and commissioned by the Union Pacific Railroad (UPR) [1].
- 1965: In the mid-1960s, Seippel presented the combined cycle model consisting of a gas turbine system and a steam turbine subsystem [2].

In this section, we present a summary of the main power cycles, namely Rankine, Brayton, Kalina, Stirling, Diesel, Otto and Ericsson cycles.

4.2.1 Rankine cycle

The Rankine cycle is one of the most common and basic vapor power cycles, consisting of four main components: turbine, condenser, boiler/heat exchanger, and pump, as shown in Fig. 4.2. Steam turbines have been widely utilized for commercial applications ranging from energy production to propulsion in marine vehicles [3]. In a simple closed standard Rankine cycle, water is received at low pressures (typically lower than 1 atm) by a pump and

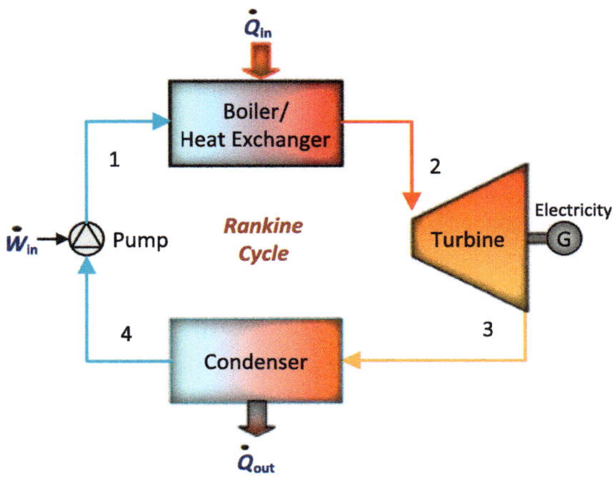

FIG. 4.2 Simple schematic diagram of Rankine cycle.

pressurized to boiler pressure. In a boiler, a phase change process occurs and water is converted into steam at a constant temperature. After this phase change, the steam can also be converted into superheated steam depending on the thermal energy supplied to the boiler, which can increase the temperature. Superheated steam enters a steam turbine, where the steam is expanded to initially produce mechanical work and then electrical work via coupled generators. The steam turbine converts thermal energy into mechanical energy by a rotating shaft.

The ideal Rankine cycle does not involve any internal irreversibilities and comprises four processes as follows (see Fig. 4.2):

- 4-1 Isentropic compression (pressurization) in a pump
- 1-2 Isobaric heat addition in a boiler
- 2-3 Isentropic expansion of the superheated steam in a turbine
- 3-4 Isobaric heat rejection in a condenser

In an ideal Rankine cycle, there are no losses associated with frictional pressure drops in either the condenser or boiler, so the phase change processes occur at constant pressure. Although the ideal Rankine cycle is internally reversible, there are still external irreversibilities because of the temperature change in the heat addition process.

The actual Rankine cycles deviate somewhat from the ideal ones. In actual Rankine cycles, the turbines and pumps are not isentropic. Hence, actual pump work is, in fact, higher than the ideal pump work. Similarly, the produced work in the ideal cycle is higher than the actual cycle. Therefore, we have isentropic efficiency definitions for these types of components. In addition, there can be pressure drops through condensers and boilers in the actual cycle. Because of the friction on bearings between moving parts of the system components, additional losses can form. In an actual Rankine cycle, water is typically subcooled to avoid any cavitation on the pump propeller due to fast vaporization and condensation of the working fluid.

There are several methods used to improve the overall efficiencies of the Rankine cycle, namely: (i) lowering the condenser pressure, (ii) achieving high-temperature superheated steam, (iii) increasing the boiler pressure. The main target of the enhancements is to raise the typical temperature at which heat is transmitted to the working fluid in the boiler, or reduce the typical temperature at which heat is rejected from the working fluid in the condenser. The condensers of steam power plants typically work well below atmospheric pressure. There is a lower limit to this pressure depending on the temperature of the cooling medium. However, lowering the condenser pressure raises the moisture ingredient of the steam at the last stages of the turbine.

For a fixed turbine inlet temperature, the boiler pressure can be increased but the moisture ingredient of steam at the turbine outlet also increases. This can be eliminated by reheating the steam. Hence, there can be two stages of expansion and reheating in between. The temperature of the boiler exit can also be increased to very high temperatures, but it is limited by the material durability of the turbine.

In steam power plants, steam is extracted from the turbine at various points. This steam, which could have produced more work by expanding further in the turbine, is used to heat the feedwater instead. The device that heats the feedwater by regeneration is called a regenerator, or a feedwater heater (FWH). A feedwater heater is essentially a heat exchanger that transfers heat from the steam to the feedwater either by mixing the two fluid streams (open feedwater heaters) or without mixing (closed feedwater heaters). Most steam power plants use a combination of open and closed feedwater heaters.

4.2.1.1 *Organic Rankine cycles*

When the source temperatures are not high enough to obtain superheated steam, other types of working fluids can also be employed in the cycle. In this case, the cycle is called an organic Rankine cycle. The fundamental difference between the steam and organic Rankine cycles is the type of working fluid used, which affects the operating conditions of the cycle. Organic Rankine cycles can work at quite low temperatures, such as 80–100 °C, depending on the selected working fluid [3]. However, most of the studies in the literature employ the organic Rankine cycle when the source temperature is below 250 °C. An organic Rankine cycle (ORC) mostly runs on organic substances such as mixtures of hydrocarbons and refrigerants. In addition, ammonia can be used as a working fluid in an organic Rankine cycle due to its low boiling temperature and decent thermodynamic characteristics. The organic Rankine cycle is more appropriate for small-scale (compared to steam Rankine cycles) power generation applications and waste heat recovery systems. Low-temperature sources suitable for the organic Rankine cycle include waste heat from any industrial process, solar thermal collectors, biomass combustion, engine exhaust gases, geothermal sources, and ocean thermal energy [4].

4.2.1.1.1 Organic Rankine cycle developed at UOIT

A lab-scale organic Rankine cycle setup was developed and built at the University of Ontario Institute of Technology (UOIT) [5]. The system was developed and analyzed as part of a dissertation study conducted by Tarique [5]. The system is a scroll-based organic Rankine cycle focusing on the expansion process. The system is a closed loop structure consisting of an expander, an air-cooled condenser, a vapor generator and auxiliary components. Unlike conventional power generation systems, the work output is obtained by the scroll expander instead of a turbine expander. The scroll expander used in this study is a modified refrigeration scroll compressor operating in the reverse direction. The compressor is a reciprocating type for refrigeration application and suitable for operating under higher pressure ratios and higher discharge pressures. The heater is a radiant electric heater and consists of six heating elements connected in parallel in order to run separately through switches. The heater temperature is set according to the working fluid temperature, which is desired to be a maximum of 200 °C. The operating temperature is selected at this relatively lower level since the designed ORC system was intended for integrating to renewable energy sources or waste heat. This ORC system is suitable for multigeneration applications since it can provide several useful outputs such as power, water/space heating and the cooling effect.

The developed integrated system combines power and cooling cycles where the source heat is utilized to generate power through a scroll expander and a portion of the heat is used in an ejector cooling system. Ammonia-water was selected as the working fluid for both power and cooling cycles. The released heat by the power cycle is captured and utilized for water and space heating. The system integrates a regenerative organic Rankine cycle with an ejector cooling cycle and it was developed for low-capacity power generation applications in the range of kilowatts. Additionally, it can generate heating of service water of tens of kilowatts and generates a couple of kilowatts of refrigeration at around 5 °C. The temperature of the heat source of this system is also set at 100–160 °C since it is considered to be a renewable energy-based system [5].

The power cycle of the system contains an ammonia desorber as vapor generator, a reverse scroll compressor as an expander, and a pump and ammonia resorber as a condenser. The heat is provided to the vapor generator by a heat transfer fluid. After vapor is generated in the ammonia desorber, it drives the expander to generate useful energy. The expanded exhaust stream arrives at the condenser where it rejects its heat and changes its phase. This process is accompanied by ammonia resorption into liquid. The condensate working fluid is sent to the pump and is pressurized to the vapor generator pressure.

The cooling system mainly includes an ejector, an evaporator coil and some piping equipment. The high-pressure ammonia-water vapor channeled from the vapor generator flows through the ejector. An evaporator coil is placed to the ejector in the throat part, the other end being linked to the throttling valve provided with liquid from the receiver located at the bottom of the condenser. Steam flows through the ejector and develops high velocity in the throat area and drops pressure considerably to induce the evaporation of the working fluid in the evaporator. The main vapor stream and the secondary vapor from the ejector thus mix and arrive at the diffuser part to increase the pressure. Eventually, the mixed flows are directed to the condenser and this completes the cycle [5].

Due to the trigeneration feature, the utilization factor of the heat source or fuel energy with this system can exceed 90%. At 120 °C source temperature, the experimental result shows a maximum isentropic efficiency of 67% and an exergy efficiency of about 30%. The experimental result also shows that the concentration of ammonia is a dominant factor in determining the optimum efficiency; it was found that about 40% ammonia concentration is an optimum value, since higher concentration values can significantly reduce the work output. In addition, the cooling and heating can be adjusted without affecting power generation in the trigeneration case.

4.2.2 Brayton cycle

The Brayton cycle is one of the commonly employed cycles for conventional power generation. A simple open-type Brayton cycle consists of three main components, namely a compressor, a combustion chamber, and a gas turbine. The combustion turbines are considered to be one of the most efficient methods of converting the chemical exergy stored in the fuels into electricity [6]. Gas turbine cycles are broadly established systems because of the ability to use numerous types of fuel as the heat source. The working fluid of the open-type Brayton cycle is typically air, which is an abundant material in the environment. There is no phase change process in these cycles, unlike Rankine cycles. Hence, the working fluid remains in the gaseous state during the entire cycle.

There are two type of Brayton cycles based on the working fluid recycling, namely open and closed types. For the closed systems, heat removal after the expansion is provided by a heat exchanger at a constant pressure, where the open systems discharge the combustion gases to the atmosphere after the turbine. Combusting the working fluid causes the feed air requirement in the system. In an actual combustion turbine, pressurized combustion gases are used as the working fluid of the system. Hence, heat addition should also be done by a

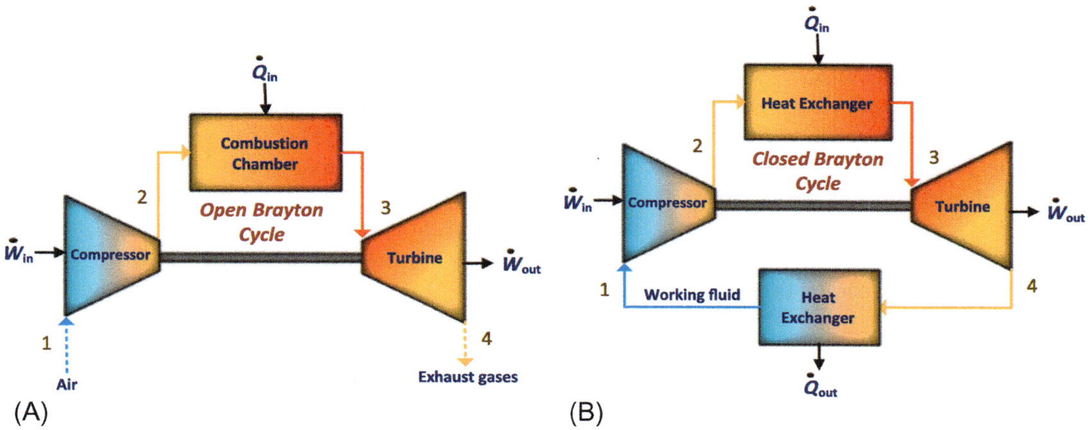

FIG. 4.3 Simple schematic diagram of Brayton cycle (A) open type and (B) closed type.

heat exchanger in the closed systems. Closed systems are commonly favored for the cases when the work output of the cycle needs to be enhanced.

The operation principle of both open- and closed-type ideal Brayton cycles is shown in Fig. 4.3. There are four main processes within a closed-type ideal Brayton cycle:

- 1-2 Isentropic compression in a compressor
- 2-3 Constant-pressure heat addition (by combustion of fuel or heat exchanger)
- 3-4 Isentropic expansion in a turbine
- 4-1 Constant-pressure heat rejection (by heat exchanger in closed-type)

In the open-type Brayton cycle, the last process, which is constant-pressure heat rejection, is not present. The operation principle of the Brayton cycle can be summarized as follows [6]:

- The air is received from the atmosphere by the compressor at state point 1 and then pressurized up to combustion pressure (typical pressure ratios of 5–15). The temperature also rises associated with the compression process.
- The working fluid (air for open-type) enters the combustion chamber or heat exchanger. If it is a combustion chamber, then fuel is also sent to the chamber by a nozzle system based on the specific air/fuel ratio. When the chamber is full with the air-fuel mixture, an igniter starts combustion. While the combustion process takes place in an enclosed medium, both temperature and pressure increase, and reach their maximum values at state point 3.
- When the combustion process is complete, the working fluid flows into the gas turbine for expansion. At state point 4, the temperature and pressure of the combustion gases are significantly reduced to generate useful work. The gas turbine coupled with a generator delivers the useful electrical power output.
- In the open-type Brayton cycle, the exhaust gases are discharged from the gas turbine, whereas in the closed-type cycle, the exhaust gases are sent to a heat exchanger for constant pressure heat rejection and then recycled to the compressor.

Similar to the Rankine cycle, there are some deviations from the ideal Brayton cycle when actual gas turbine cycles are considered. The first difference is that the compression and expansion processes are not isentropic. Consequently, actual compressor work is greater than the ideal compressor work while actual turbine work is lower than the ideal turbine work. Hence, we define isentropic efficiencies of the turbines and compressors. The net work output of a gas turbine cycle is the subtraction of turbine and compressor work. The compressor is connected to the turbine with a shaft, hence uses a portion of the produced mechanical work. Therefore, we define back-work ratio in Brayton cycles to determine how much of the work is recovered and consumed. In order to enhance the net work output, the turbine work output should be improved, or the compressor work input should be reduced. Therefore, multistage turbines can be employed. In this case, an additional heat input to the working fluid occurs between the gas turbine stages. Increasing the number of stages can make the expansion process isothermal.

Since compressing a hotter gas requires more power input in a single stage compressor, a multistage compressor having inter-cooling processes can increase the performance of the compressor and cycle. In this way, the required work input to the compressor is reduced and the heat recovery capacity rises with regeneration [4].

In thermodynamic analysis of the Brayton cycle, air-standard assumptions are commonly considered for simplification [7]. In the air-standard assumptions, the working fluid is air, which continuously circulates in a closed loop and always behaves as an ideal gas. All the processes that make up the cycle are internally reversible. The combustion process is replaced by a heat-addition process from an external source. The exhaust process is replaced by a heat-rejection process that restores the working fluid to its initial state.

4.2.3 Otto cycle

The Otto cycle is the ideal cycle for spark-ignition engines. It is named after Nikolaus A. Otto. He built a successful four-stroke engine in 1876 in Germany using the cycle proposed by Frenchman Beau de Rochas in 1862. The main processes occurring in an Otto cycle are as follows:

- Isentropic compression
- Constant volume heat addition
- Isentropic expansion
- Constant volume heat rejection

Both the intake and the exhaust valves are closed, and the piston is at its lowest position. During the compression stroke, the piston moves upward, compressing the air–fuel mixture. Shortly before the piston arrives at the highest position, the spark plug fires and the mixture ignites, raising the pressure and temperature of the system. The high-pressure gases force the piston down, which sequentially forces the crankshaft to rotate, creating a useful work output through the expansion or power stroke. Close to the end of the expansion stroke, the exhaust valve opens and the combustion gases that are above atmospheric pressure rush out of the cylinder through the open exhaust valve. This process is called exhaust blowdown, and most combustion gases exit the cylinder by the time the piston reaches the lowest position. The cylinder is still filled by the exhaust gases at a lower pressure at the lowest position. The piston moves upward one more time, purging the exhaust gases through the exhaust valve, and down a second time, drawing in fresh air-fuel mixture through the intake valve. It is noted that the pressure in the cylinder is slightly above the atmospheric value during the exhaust stroke and slightly below during the intake stroke.

A four-stroke cycle means that there is 1 cycle, 4 stroke and 2 revolutions, whereas a two-stroke cycle means that there is 1 cycle, 2 stroke and 1 revolution. The two-stroke engines are generally less efficient than their four-stroke counterparts but they are relatively simple and inexpensive, and they have high power-to-weight and power-to-volume ratios.

Similar to other cycles, the actual cycle deviates from the ideal cycle. The ideal Otto cycle contains two strokes equivalent to one mechanical cycle or one crankshaft rotation. On the other hand, the actual engine operation contains four strokes equivalent to two mechanical cycles or two crankshaft rotations. This can be corrected by including intake and exhaust strokes in the ideal Otto cycle.

4.2.4 Diesel cycle

The Diesel cycle is in fact the ideal cycle for compression-ignition engines, consisting of four main processes as listed here:

- Isentropic compression
- Constant-volume heat addition
- Isentropic expansion
- Constant-volume heat rejection.

The compression-ignition engines were first proposed by Rudolph Diesel in the 1890s. The main difference between the Otto cycle and the Diesel cycle is the initiation of combustion. In diesel engines, only air is compressed during the compression stroke, eliminating the possibility of auto ignition. Thus, diesel engines can be designed to operate at much higher compression ratios than spark-ignition engines (usually in the range of 12–24). In spark-ignition engines, the air–fuel mixture is compressed to a temperature that is below the auto-ignition temperature of the fuel, and the combustion process is initiated by firing a spark plug. Hence, the spark plug is replaced by a fuel injector in diesel engines.

The only process differing from the Otto cycle is the heat addition. The fuel injection process in diesel engines starts when the piston approaches to the highest position and continues during the first part of the power stroke. Thus, the combustion process in these engines occurs over a longer interval. Due to this longer period of time, the combustion process in the ideal Diesel cycle can be approximated as a constant-pressure heat-addition process.

The higher efficiency and lower fuel costs of diesel engines make them attractive in applications requiring relatively large amounts of power, such as in locomotive engines, emergency power generation units, large ships, and heavy trucks.

4.2.5 Kalina cycle

The Kalina cycle is an advanced thermodynamic cycle, which can be used for converting thermal energy from a comparatively low-temperature heat source to mechanical energy. This cycle was developed by Aleksandr Kalina in the late 1970s and early 1980s [8]. After that time, there have been several modifications based on the particular application. The Kalina

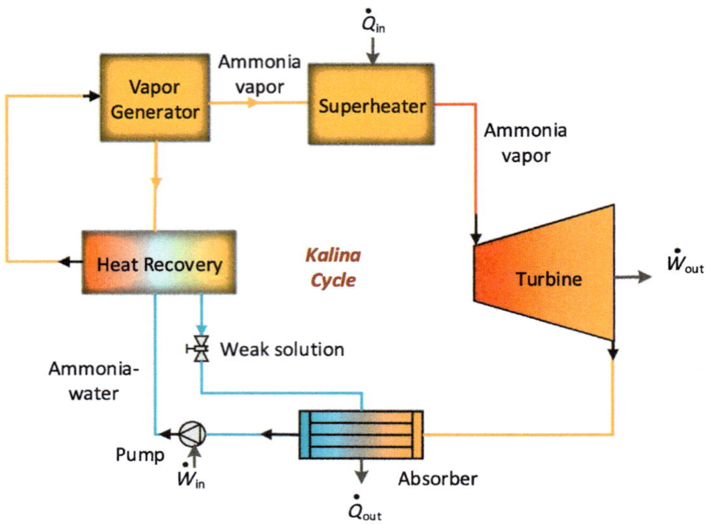

FIG. 4.4 Simple schematic diagram of Kalina cycle.

cycle is principally a modified Rankine cycle, which utilizes the mixture of two different working fluids (water and ammonia) as illustrated in Fig. 4.4. As mentioned earlier, a conventional Rankine cycle utilizes water as a working fluid. Kalina et al. [9] stated that an enhancement of 10–20% in the exergy efficiency of the cycle can be possible in comparison with the conventional Rankine cycle. Since thermodynamic properties of ammonia are attractive, the concentration of ammonia in the working fluid enhances the thermodynamic reversibility leading to higher performances in the cycle.

The operating principle of the Kalina cycle can be summarized as follows [10]:

- The mixture of ammonia and water enters the heat recovery unit before the vapor generator. A concentrated vapor of ammonia is generated in the vapor generator at high pressure and temperature.
- The concentrated vapor is then sent to a superheater, where the temperature of ammonia vapor is further improved.
- After leaving the superheater, the ammonia vapor enters the turbine for expansion. As it expands in the turbine, the available thermal energy in the working fluid is transformed into mechanical energy and then to electricity via a coupled generator.
- After leaving the turbine, the low temperature and pressure vapor enters the absorber. In the absorber, the weak solution and the strong solution streams are mixed at a low temperature. Therefore, a concentrated solution is obtained at low temperature, and it is pumped to the vapor generator after passing through the heat recovery unit.

4.2.6 Stirling cycle

The Stirling engines can function as a work producing device or work consuming device [11]. This brings irreversible operation flexibility. The schematic diagram of the Stirling cycle

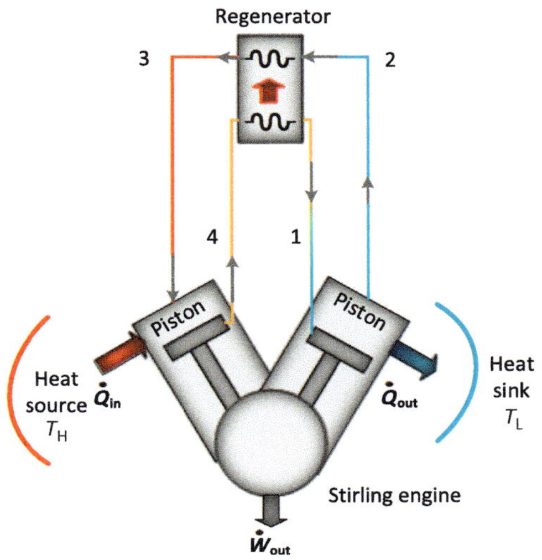

FIG. 4.5 Simple schematic diagram of the Stirling cycle.

operation is shown in Fig. 4.5. The source of high-temperature heat can be concentrated solar, biomass combustion or any waste heat. They are suitable devices for concentrated parabolic dish applications, where the thermal energy can be converted into mechanical energy directly.

The Stirling cycle bears a double-effect piston and cylinder arrangement. A regenerator porous matrix is mounted inside the arrangement. The working fluid can be air, helium, nitrogen, hydrogen, CO_2, etc. The main processes within the Stirling cycle are written as follows [4]:

- Process 1-2: Isothermal compression: The working fluid is compressed isothermally while space discharges the heat to the heat sink. Therefore, the temperature of the heat sink is increased.
- Process 2-3: Isochoric regeneration (heat addition): Heating occurs at the regenerator under constant volume. The temperature of the working fluid increases from T_L to T_H.
- Process 3-4: Isothermal expansion: The working fluid expands isothermally while space is heated externally by the heat source. At this stage, the engine produces useful work.
- Process 4-1: Isobaric regeneration (heat rejection): Cooling occurs at regeneration component at constant volume. Regenerator absorbs heat from the working fluid. The temperature of the working fluid reduces from T_H to T_L.

4.2.7 Ericsson cycle

The Ericsson cycle is very similar to the Stirling cycle. The main difference is that the two constant-volume processes are replaced by two constant-pressure processes. There are two other cycles containing an isothermal heat-addition process (Stirling cycle) and an isothermal heat-rejection process (Ericsson cycle). They differ from the Carnot cycle in that the two isentropic processes are replaced by two constant-volume regeneration processes in the Stirling cycle and by two constant-pressure regeneration processes in the Ericsson cycle. Both cycles utilize regeneration, a process during which heat is transferred to a thermal energy storage device during one phase of the cycle and is transferred back to the working fluid during another phase of the cycle.

Both the Stirling and the Ericsson engines are external combustion engines. Thus, the fuel is combusted outside the cylinder, which is the opposite of the Diesel and Otto cycles. There are several advantages of external combustion, namely (i) flexibility of diverse fuels, (ii) more time for combustion, and hence more complete combustion, (iii) less air pollution and more energy extraction from the fuel, (iv) operation on closed cycles, thus selection of better characteristics of the working fluid.

Due to heat transfer limitations in these cycles, the practicability is a bit low. However, they bear potential for higher efficiency and better emission control, which are significant advantages for especially transportation applications. Hence, there is recent research and interest in these types of cycles.

4.3 Combined cycles

In order to enhance the overall performance of energy production, the above-mentioned cycles can be combined. The combined cycles consist of two different systems having a single source and aiming at the same useful output (e.g., power). In this section, we elaborate on several combined cycles to emphasize the importance of system integration of conventional cycles.

4.3.1 Brayton-Rankine cycle

The Brayton-Rankine combined cycle is the most established application of the combined energy conversion systems. A combined cycle power plant typically employs both a gas and a steam cycle. The main idea of the combination is to utilize the exhaust gases of the Brayton cycle since the exit temperatures of the gas turbines are significantly higher than the atmosphere temperature. Therefore, in the Brayton-Rankine combined cycle, the high-temperature exhaust gases are used (via a heat exchanger) to generate required steam in a Rankine cycle as illustrated in Fig. 4.6. In this way, the overall efficiency of energy conversion is considerably enhanced compared to the individual gas turbine and steam cycles. The gas and steam turbines working in the combined cycle can be evaluated as heat engines since they receive thermal energy from a heat source such as fuel combustion and reject it to a heat sink for generating useful work [10]. The topping cycle of the Brayton-Rankine combined system is mostly a simple Brayton cycle, as shown in Fig. 4.6.

4.3.2 Rankine-Rankine cycle

Two Rankine cycles can also be combined with a heat exchanger depending on the temperature levels of the cycles. As depicted in Fig. 4.7, the exhaust stream of the topping Rankine cycle enters a heat exchanger where it exchanges the thermal energy to the working fluid of the bottoming Rankine cycle. In this case, the heat exchanger functions as a condenser of the

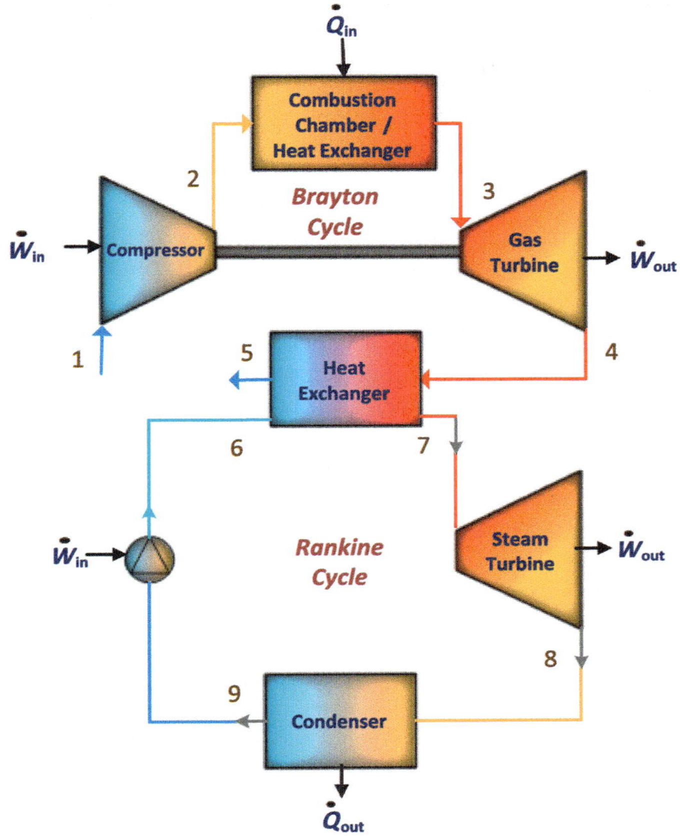

FIG. 4.6 Simple schematic diagram of a combined cycle power plant (combined Brayton-Rankine cycle).

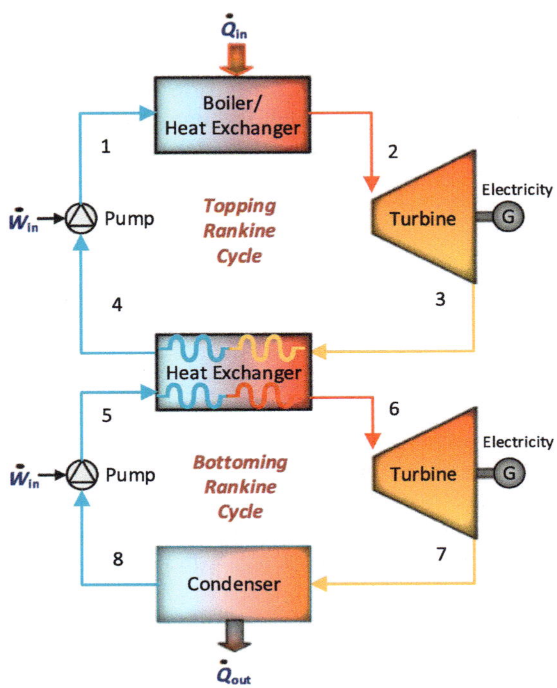

FIG. 4.7 Simple schematic diagram of a combined Rankine-Rankine cycle.

topping Rankine cycle and boiler of the bottoming Rankine cycle. The working fluid is expanded in the secondary turbine to generate more power. The temperature level of the source plays a critical role in this system. If the exhaust temperature of the topping Rankine cycle at state point 3 is at low levels, an organic Rankine cycle can also be combined, having a more appropriate working fluid.

4.3.3 Brayton-Kalina cycle

A simple Kalina cycle includes a waste heat recovery vapor generator, a turbine, and the distillation condensation system. In the distillation condensation subsystem, first the flow incoming from the turbine is cooled by the heater (recuperator), and then the stream is mixed with a lean solution of ammonia to increase the condensation temperature of the working fluid. Finally, the basic solution is condensed in the absorber. The condensed solution is brought to the heater under pressure. A portion of the stream is directed to dilute the ammonia-rich stream coming from the separator. The primary flow passes the recuperator and then it is flashed in the separator. The vapor is mixed with the basic solution. The vapor is condensed and then pressurized by the pump before it flows to the vapor generator [12]. In the Brayton-Kalina combined cycle, the required heat input to the Kalina cycle is provided by the exhaust of the Brayton cycle as illustrated in Fig. 4.8. In this case, a heat exchanger is utilized where the exhaust gas of the gas turbine is sent to deliver thermal energy to the bottoming Kalina cycle.

4.3.4 Brayton-Stirling cycle

In a Brayton-Stirling engine, the required heat for the Stirling engine can be provided from the exhaust of the Brayton cycle through a heat exchanger. In this case, the heat source of the Stirling engine can become the exhaust stream of the gas turbine at state point 4, as shown in Fig. 4.9. As an alternative solution, the combustion chamber itself can also be a heat source when the Stirling engine is placed next to the chamber (at state point 3). The arrangement is set by the ideal operating conditions of the cycle. There is another limitation of cycle configuration due to the type of materials utilized in the head of the Stirling heater [10].

4.3.5 Brayton-fuel cell cycle

Recent developments in fuel cell technologies enable better system integration with conventional systems. A fuel cell system converts chemical energy directly into electricity through an electrochemical process. Most fuel cells also require air/oxygen and fuel input for conversion similar to combustion. However, the conversion efficiencies are not limited to Carnot efficiency, so greater performances are achievable compared to combustion. Due to similarities of inputs, the fuel cell and Brayton cycle can be combined to utilize the excess gases of the fuel cell in a combustion chamber as illustrated in Fig. 4.10. In addition, the exhaust of the gas turbine can be used in a heat recovery steam generator to produce steam. The combined Brayton-fuel cell cycles can achieve efficiencies of about 70% [13]. The Brayton fuel cell cycle bears one of the highest conversion efficiencies compared to other combined cycles. Hence, it is evaluated as a promising combined cycle for future power plants [14].

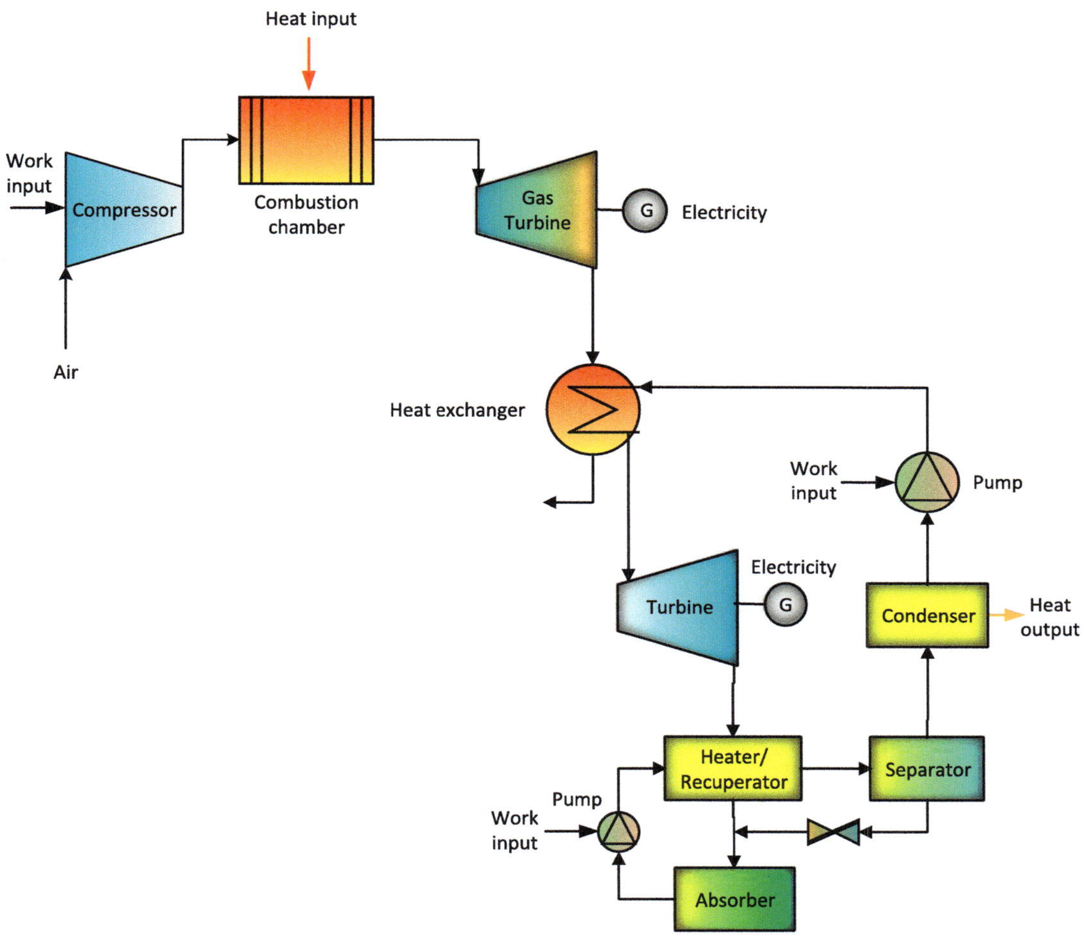

FIG. 4.8 Simple schematic diagram of combined Brayton-Kalina cycle.

4.3.6 Integrated gasification combined cycle

Coal combustion causes several environmental issues. There have been numerous improvements for reducing harmful emissions to the atmosphere, such as setting up of flue gas scrubbers and electrostatic precipitators in conventional coal-fired power plants. Compared to combustion, gasification of coal can yield syngas (mainly H_2 and CO) with some other impurities. The yielding impurities such as sulfur, tar and particulates can be removed using syngas clean-up systems. After coal is gasified, the obtained syngas can be used in combustion chambers to drive a conventional Brayton cycle. In this way, the combustion can be cleaner compared to coal burning. In order to utilize the exhaust of the Brayton cycle, this configuration is later combined with the conventional Rankine cycle acting as a Brayton-Rankine combined cycle. This complete plant is called an integrated gasification combined

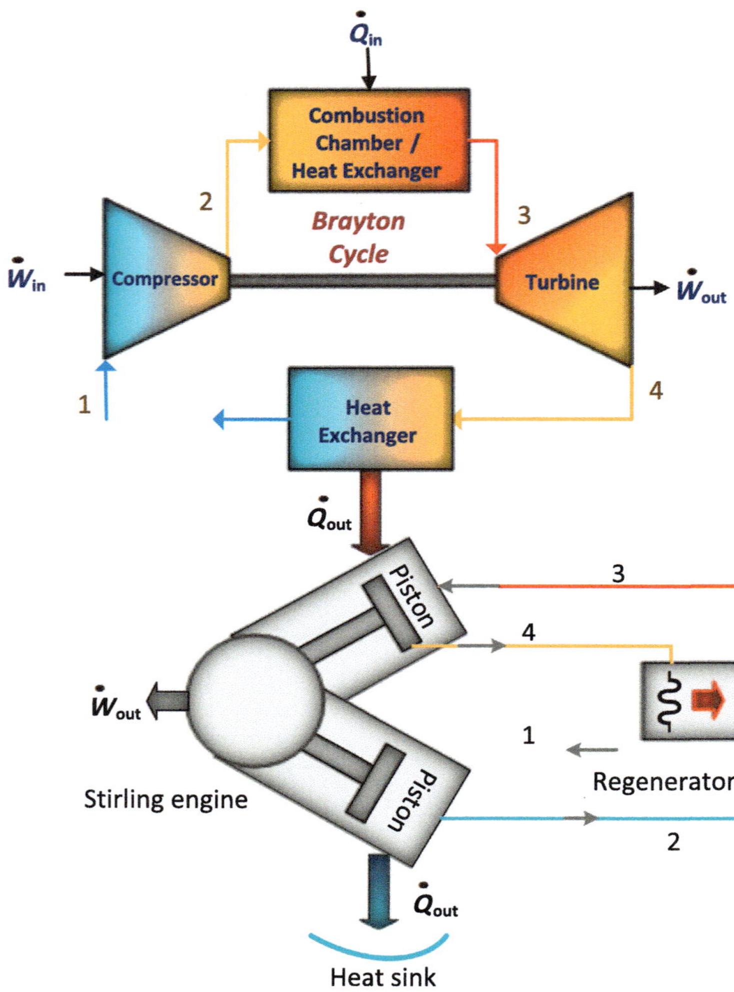

FIG. 4.9 Simple schematic diagram of a combined Brayton-Stirling cycle.

cycle as illustrated in Fig. 4.11. The topping cycle is the Brayton cycle, which supplies the required heat to the bottoming cycle, which is the Rankine cycle. The required steam for the gasification process can be taken from the bottoming Rankine cycle at state point 19 or 20 depending on the temperature. In this case, there should be feed water addition to the Rankine cycle at state point 21.

In the gasifier, gasification is performed as the result of the controlled input of coal, oxygen or air and steam. The end product also includes solid waste in addition to syngas. Oxygen can be directly provided to the gasifier by employing an air separation unit. The syngas produced by the gasifier contains mainly CO and H_2. One of the benefits of integrated gasification combined cycles is that the pollutants such as sulfur and mercury are cleaned before the

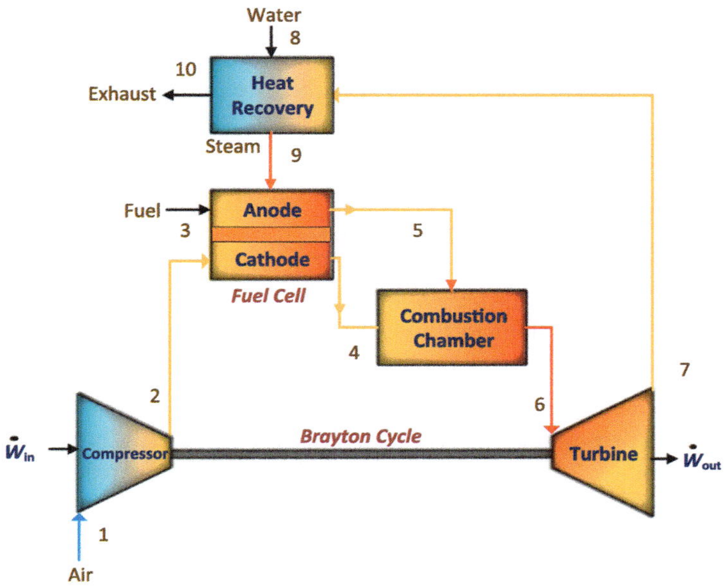

FIG. 4.10 Simple schematic diagram of the Brayton-fuel cell cycle.

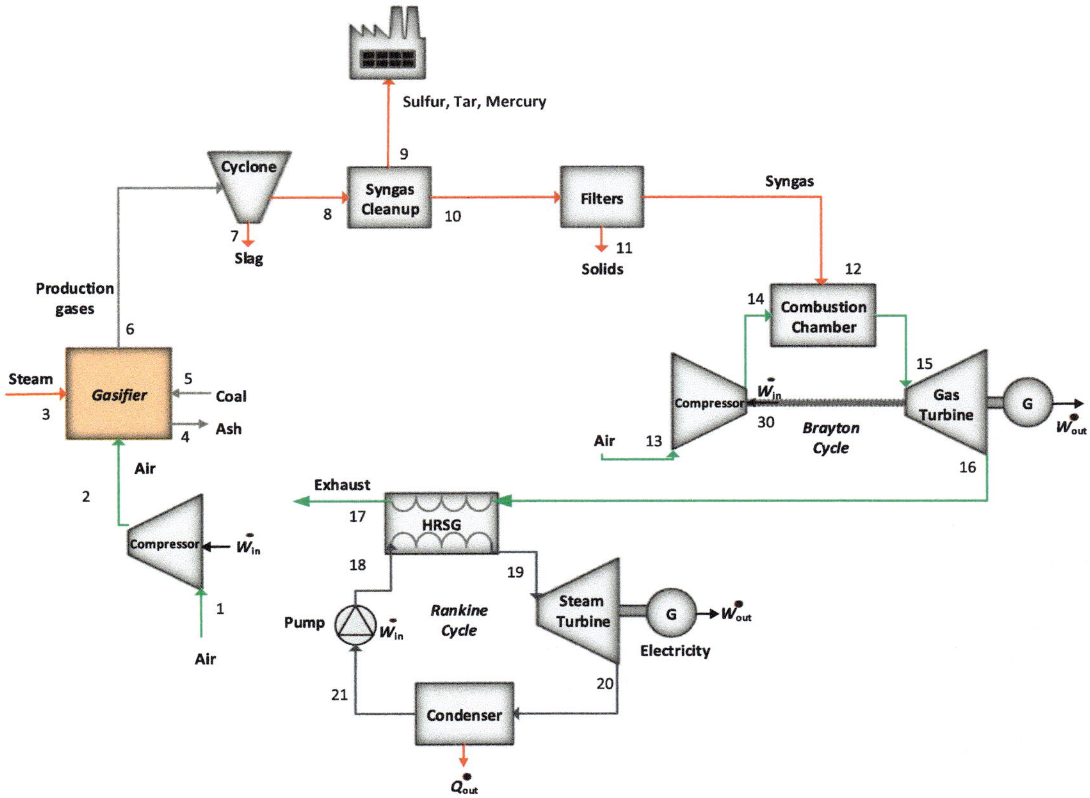

FIG. 4.11 Simple schematic diagram of an integrated gasification combined cycle.

combustion process, unlike the conventional coal-fired power plants [15]. Although integrated gasification combined cycles emit lower SO_2, NO, sulfur dioxide, nitric oxide, mercury, and particulate emissions than similar conventional coal-fired power plants, the yielding of solid waste still requires careful management [16].

4.4 Brainstorming for system integration

Conventional energy systems have mostly well-developed and mature technologies with still further room available for system improvement and integration, as illustrated in Fig. 4.12.

- *Underground coal gasification (UCG)*: Underground coal gasification requires air/oxygen and medium/high-temperature steam to be injected underneath the Earth. The required steam can be sourced via waste heat from any conventional power plant or can be generated on-site using concentrated solar energy or other suitable steam generation techniques.
- *Enhanced oil recovery using captured CO_2*: Fossil fuel processing plants emit significant amounts of CO_2 to the atmosphere. Once it is captured, it can be used for enhancing the oil recovery of the wells. Any CO_2 emitting energy system can also be integrated into oil recovery plants by employing appropriate CO_2 capture technologies.
- *Bitumen extraction using waste steam*: Bitumen extraction from oil sands requires medium-/high-temperature steam to be injected underground (e.g., steam-assisted gravity drainage).

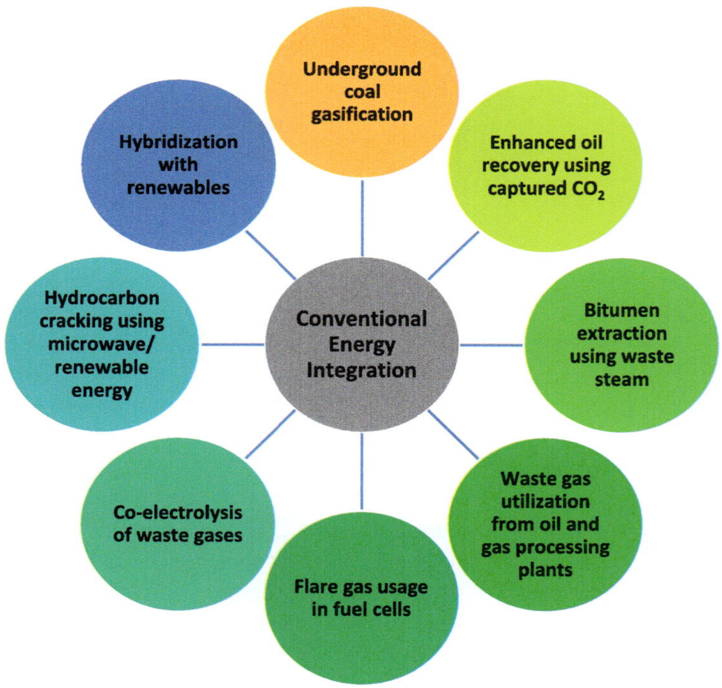

FIG. 4.12 Brainstorming on system integration for conventional energy systems.

Therefore, any energy system having waste steam can be integrated into oil sands to recover bitumen. In addition, renewable sources such as concentrated solar or biomass can also be used to obtain the required steam on-site for cleaner operations.

- *Waste gas utilization from oil and gas processing plants*: CO, CO_2, H_2S, NH_3, CH_4 are among the gases being wasted in oil and gas processing plants during various stages (e.g., flaring, incineration, tail gas). These gases can be used in gas-to-power technologies such as employing fuel cells or other electrochemical routes.
- *Flare gas usage in solid oxide fuel cells*: A large portion of flare gas in natural gas processing plants consists of CH_4, implying that it can be a feed gas for solid oxide fuel cells once treated appropriately. The wasted energy in the flaring gas can be recovered and the overall plant efficiencies can be enhanced in addition to greenhouse gas reduction to the atmosphere.
- *Co-electrolysis of waste gases*: Waste gases such as CO_2 and CH_4 can be electrolyzed by applying excess electricity from the plants with power-to-gas technologies. In this way, storage of energy can be possible as chemical compounds rather than batteries or thermal energy.
- *Hydrocarbon conversion using microwave energy and renewable sourced-thermal energy*: Renewable sources can be used to produce electricity, which in turn can drive a microwave generator. Microwave energy can be used to drive high-temperature processes such as hydrocarbon cracking. Once hydrocarbons are converted into useful gases such as H_2 and CO, they can be used for power generation, synthetic fuel production, etc. Furthermore, concentrated solar energy or biomass combustion can also drive hydrocarbon conversion processes (e.g., hydrocarbon cracking, reforming, gasification), which can make the complete process more environmentally friendly.
- *Hybridization with renewable sources*: Any conventional system can be hybridized to power the grid electricity. Recent developments in the energy storage area can even convert the intermittent types of renewables into base load suppliers.

4.5 Case study 1: Compressed air energy storage (CAES) for an integrated gas turbine power plant

In this case study, a compressed air energy storage for an integrated gas turbine power plant is proposed and analyzed. In this way, excess electricity can be stored underground as compressed air and later converted into electricity on demand. This case study is adapted from DinAli and Dincer [17] and hence presents the main findings of a recent work conducted by DinAli and Dincer [17].

Demand and supply management of any power plant is a significant factor for sustainable power generation, especially for the grid operation. The requirement for load shifting caused by the changes is needed in the load profiles of the utility. The deviations in load behavior are because of many diverse factors, such as (i) higher peak loads on workdays compared to weekends and holidays, (ii) heating and cooling load changes due to daily temperature changes. Therefore, storing the excess electricity is a highly beneficial and desired technique. However, there are not many operational compressed air storage units in the world. One of

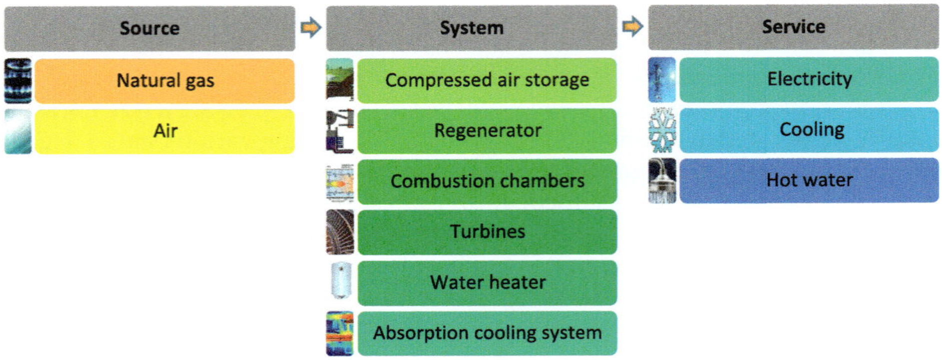

FIG. 4.13 3S approach for case study 1.

the operational plants is the Huntorf plant that is situated in Germany having a maximum capacity of 290 MW [18] and round-trip efficiency of 41% [19]. Another similar plant, called the McIntosh plant, has a capacity of 110 MW [20] and roundtrip efficiency of 54%. Although there are limited practical applications, there have been many research studies for CAES. For example, Buffa et al. [21] performed exergy analyses for a CAES ground-based system dependent on multistage compression and expansion with heat recovery from the intercoolers and regenerative reheat. Their work showed that most of the exergy destruction occurs in compressors, intercoolers, and expanders. They found an overall exergy efficiency of about 52%. In this case study, the sources, systems and services (3S) considered during the design stage are shown in Fig. 4.13.

4.5.1 System description

The system consists of a gas turbine plant integrated with CAES, LiBr absorption chiller, and water heater. The compressors do not function continuously, so it is difficult to integrate other systems. However, possible options for further integration with proper techniques can be an organic Rankine cycle and desalination unit. The schematic illustration of the system is shown in Fig. 4.14. The system contains two-stage compressors, motor, generator, compressed air storage cavern, regenerator, combustion chambers, high pressure and low-pressure turbines, water heater and absorption cooling system. The useful outputs from the system are electricity, cooling, and hot water. The electricity is produced via the gas turbine that is run by the compressed air coming from the storage unit. Initially, the air is preheated by the regenerator and then heated to higher temperature levels by the combustion chambers in which natural gas is used as fuel. In this system, the cooling is obtained using an absorption cooler, where the working fluid is LiBr. The required heat source is taken from the waste heat recovered from compression after the cooling process. The hot water is produced by recovering waste heat from compression intercooling.

Constant volume air storage can cause fluctuating pressure ratio throughout the compression process, which can cause reduced efficiency of the compressor because of deviation from

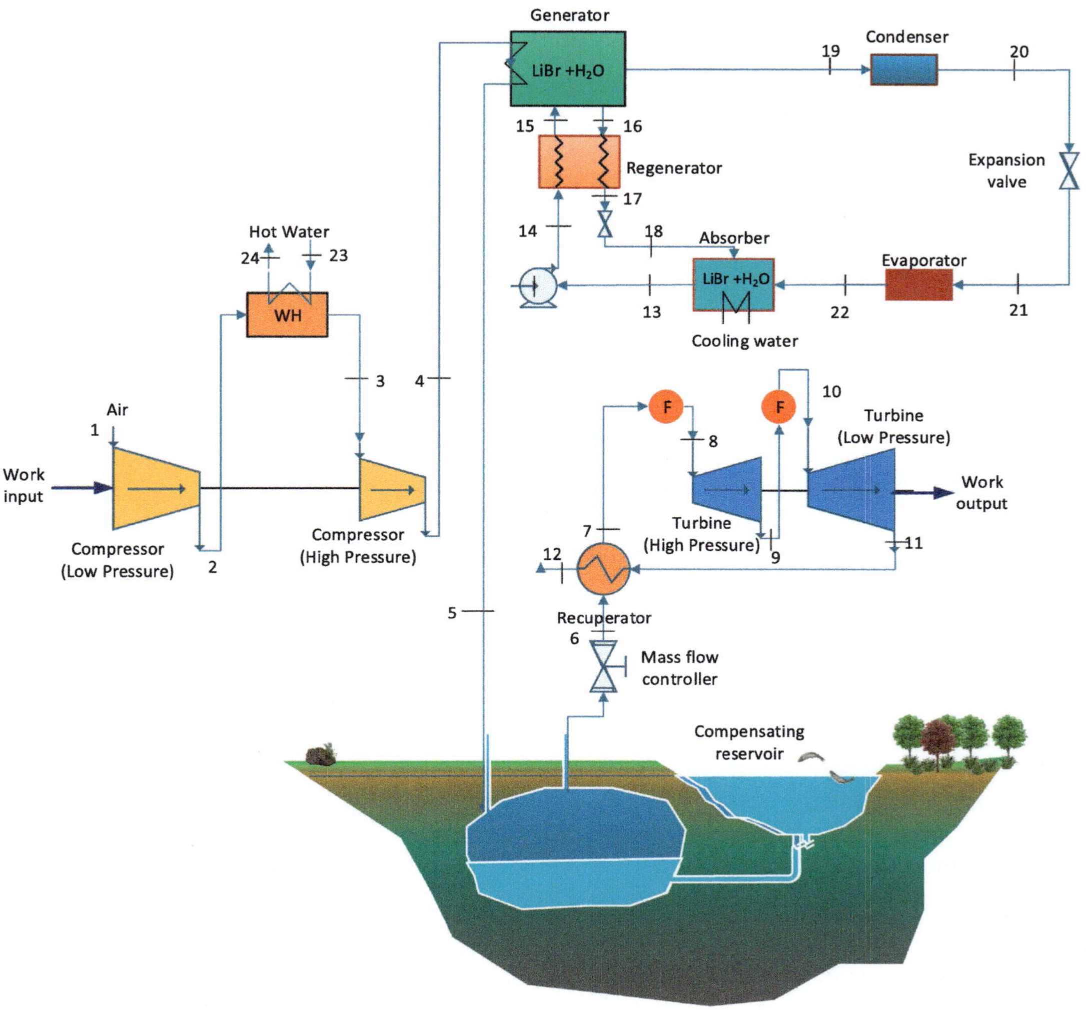

FIG. 4.14 Schematic diagram of the overall system for compressed air energy storage. *(Modified from DinAli MN, Dincer I. Development and analysis of an integrated gas turbine system with compressed air energy storage for load leveling and energy management. Energy 2018; 15;163:604–17. https://doi.org/10.1016/j.energy.2018.08.094).*

the compressor design criteria [22]. Therefore, constant pressure air storage preserved by the water compensation column is considered in this case study. The operation pressure of the storing unit is wholly associated with the height of the compensation water column. In the charging model, air is introduced into the storage from the compressor relocating the water from the cavern via a compensation column. Conversely, the discharging mode operates by discharging the compressed air via the valve and water relocates the extracted air, preserving constant pressure inside the storage.

4.5.2 Assumptions

In order to perform thermodynamic analysis of this multigeneration system, there is a need to make conceptually correct assumptions. Here, the main assumptions undertaken in this case study are listed:

- The reference conditions are $T_0 = 298$ K and $P_0 = 100$ kPa.
- The throttling processes are considered as adiabatic.
- The pressure losses throughout the system are neglected.
- The isentropic efficiency of the turbines and pumps are taken as 80% whereas the compressors have an isentropic efficiency of 90%.
- The pressure of the cavern is presumed to be constant due to constant pressure storage.
- There are no heat losses through the compressors, turbines and pumps.
- The changes in potential and kinetic energy terms are neglected.

4.5.3 Analysis

The energy and exergy analyses are performed for three modes of operations, which are charging, storing, discharging. The main balance equations for the energy storage processes are given in this section. Furthermore, the exergy efficiency definitions of the main processes are also described.

For charging mode, the mass balance equation is:

$$m_1 + \dot{m}_5 \Delta t_{\text{charging}} = m_2 \tag{4.1}$$

Here, $\Delta t_{\text{charging}}$ is the charging duration given in seconds for the air storage where 1 and 2 are the state points before and after the discharging.

The energy and exergy balance equations throughout the charging process for the compressed air storage are defined as follows:

$$m_1 u_1 + \dot{m}_5 h_5 \Delta t_{\text{charging}} = m_2 u_2 + W_b \tag{4.2}$$

$$m_1 \text{ex}_1 + \dot{m}_5 \text{ex}_5 \Delta t_{\text{charging}} = m_2 \text{ex}_2 + W_b + \text{Ex}_{\text{d,charging}} \tag{4.3}$$

Here, W_b denotes the boundary work performed by the dislocation of water.

In the storing phase, the mass, energy and exergy balance equations are defined as follows:

$$m_2 = m_3 \tag{4.4}$$

$$m_2 u_2 = m_3 u_3 \tag{4.5}$$

$$m_2 \text{ex}_2 = m_3 \text{ex}_3 + \text{Ex}_{\text{d,storing}} \tag{4.6}$$

During discharging, the mass, energy and exergy balance equations are defined as follows:

$$m_3 = m_4 + \dot{m}_6 \Delta t_{\text{discharging}} \tag{4.7}$$

$$m_3 u_3 + W_b = m_4 u_4 + \dot{m}_6 h_6 \Delta t_{\text{discharging}} \tag{4.8}$$

$$m_3 ex_3 + W_b = m_4 ex_4 + \dot{m}_6 ex_6 \Delta t_{discharging} + Ex_{d,discharging} \tag{4.9}$$

where $\Delta t_{discharging}$ is the duration of discharging in seconds.

The energy and exergy efficiency of the complete cycle is defined as the ratio of useful output to the required input. The useful outputs are the turbine work, cooling capacity and hot water whereas the required input is the compressor work and fuel energy in the combustion chamber:

$$\eta_{en,overall} = \frac{\dot{W}_{GT1} + \dot{W}_{GT2} + \dot{Q}_{EV} + \dot{Q}_{Heater}}{\dot{Q}_{CC1} + \dot{Q}_{CC2} + \dfrac{\dot{W}_{C2}}{\eta_{motor}} + \dfrac{\dot{W}_{C2}}{\eta_{motor}}} \tag{4.10}$$

$$\eta_{ex,overall} = \frac{\dot{W}_{GT1} + \dot{W}_{GT2} + \dot{E}x^{Q_{EV}} + \dot{E}x^{Q_{Heater}}}{\dot{E}x^{Q_{CC1}} + \dot{E}x^{Q_{CC2}} + \dfrac{\dot{W}_{C2}}{\eta_{motor}} + \dfrac{\dot{W}_{C2}}{\eta_{motor}}} \tag{4.11}$$

The exergy efficiencies of several components in the system are defined as follows:

High-Pressure Gas Turbine:

$$\eta_{ex,GT1} = \frac{\dot{W}_{GT1}}{\dot{E}x_8 - \dot{E}x_9} \tag{4.12}$$

Low-Pressure Gas Turbine:

$$\eta_{ex,GT2} = \frac{\dot{W}_{GT2}}{\dot{E}x_{10} - \dot{E}x_{11}} \tag{4.13}$$

High-Pressure Compressor:

$$\eta_{ex,Comp1} = \frac{\dot{E}x_2 - \dot{E}x_1}{\dot{W}_{C1}} \tag{4.14}$$

Low-Pressure Compressor:

$$\eta_{ex,Comp2} = \frac{\dot{E}x_4 - \dot{E}x_3}{\dot{W}_{C2}} \tag{4.15}$$

Combustion Chamber 1:

$$\eta_{ex,CC1} = \frac{\dot{E}x_8 - \dot{E}x_7}{\dot{Q}_{CC1}\left(1 - \dfrac{T_o}{T_{cc1}}\right)} \tag{4.16}$$

Combustion Chamber 2:

$$\eta_{ex,CC2} = \frac{\dot{E}x_{10} - \dot{E}x_9}{\dot{Q}_{CC_2}\left(1 - \frac{T_o}{T_{cc2}}\right)} \tag{4.17}$$

Recuperator:

$$\eta_{ex,Recuperator} = \frac{\dot{E}x_7 - \dot{E}x_6}{\dot{E}x_{11} - \dot{E}x_{12}} \tag{4.18}$$

Compressed Air Storage:

$$\eta_{ex,CAES} = 1 - \frac{\dot{E}x_{destruction,CAES}}{\dot{E}x_{in,CAES}} \tag{4.19}$$

The fundamental parameters used in the analysis are listed in Table 4.1.

4.5.4 Results and discussion

The exergy destruction rates of the fundamental units are presented in Fig. 4.15. The maximum exergy destruction rate arises at the turbines, implying that the entropy generation is high. The overall energy efficiency of the system is found as 53%, whereas the overall exergy efficiency is found as 41.7%. The energetic and exergetic coefficient of performance for the

TABLE 4.1 Main input parameters

Parameter	Value
Isentropic efficiency of compressor (%)	90
Isentropic efficiency of turbine (%)	80
Compression ratio of compressor	8
Inlet temperature of high pressure turbine (°C)	550
Inlet temperature low pressure turbine (°C)	825
Charging mass flow rate (kg/s)	108
Charging time (h)	8
Discharging mass flow rate (kg/s)	432
Discharging time (h)	2
Volume of storage (m^3)	350,000

Source: *DinAli MN, Dincer I. Development and analysis of an integrated gas turbine system with compressed air energy storage for load leveling and energy management. Energy 2018; 15;163:604–17. https://doi.org/10.1016/j.energy.2018.08.094.*

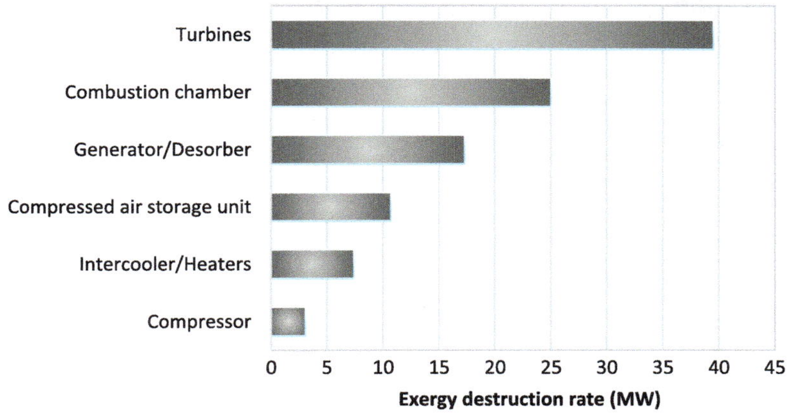

FIG. 4.15 Exergy destruction rates of fundamental system components. *(Data from DinAli MN, Dincer I. Development and analysis of an integrated gas turbine system with compressed air energy storage for load leveling and energy management. Energy 2018; 15;163:604–17. https://doi.org/10.1016/j.energy.2018.08.094.)*

TABLE 4.2 Main mass flow rate values of the integrated system.

Component	Mass flow rate (kg/s)
Combustion chamber 1 fuel rate	1.16
Combustion chamber 2 fuel rate	5.29

absorption cooling system are found to be 0.7825 and 0.206, respectively. In these conditions, the mass flow rate of the obtained hot water is found to be 165.5 kg/s.

Table 4.2 displays the main fuel input rates to the combustion chambers of the integrated system.

The work rate obtained from the high and low pressure turbines as well as power consumption rates of the compressor are illustrated in Fig. 4.16. This figure also shows the obtained cooling and heating from the system. The obtained cooling is about 26 MW, whereas the obtained heating is about 24 MW.

The exergy destruction for air storage increases as the operating pressure inside the storage unit rises. Nevertheless, it has a more significant impact on the discharging stage as depicted in Fig. 4.17.

The exergy destruction rates of the turbines upsurge considerably because of the high entropy generation in the turbines, as shown in Fig. 4.18. This causes a decrease in the exergy efficiency of the turbines.

Increasing the temperature of HP turbine inlet from 500 °C to 1000 °C raises the overall energy efficiency by nearly 5% as presented in Fig. 4.19. The discharging time is a significant element in CAES. It is inversely associated with the power out of the turbine. Based on the peak hours and peak demands, the power output of CAES could be projected. Fig. 4.20 displays the link between the power output of CAES and the discharging time. The data in the graph are based on a specific circumstance with a pressure of 6400 kPa, a storage volume of 350,000 m^3 and charging time.

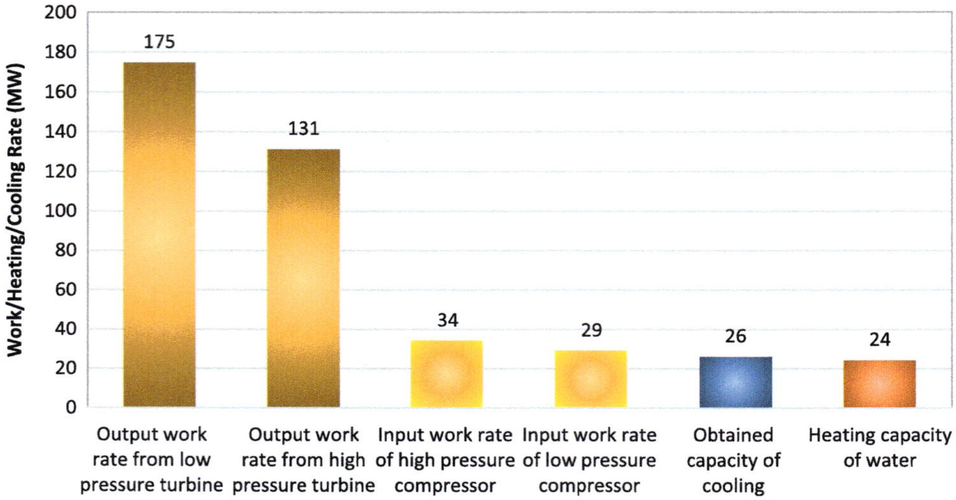

FIG. 4.16 The work, heating and cooling rates of the multigeneration system components. *(Data from DinAli MN, Dincer I. Development and analysis of an integrated gas turbine system with compressed air energy storage for load leveling and energy management. Energy 2018; 15;163:604–17. https://doi.org/10.1016/j.energy.2018.08.094).*

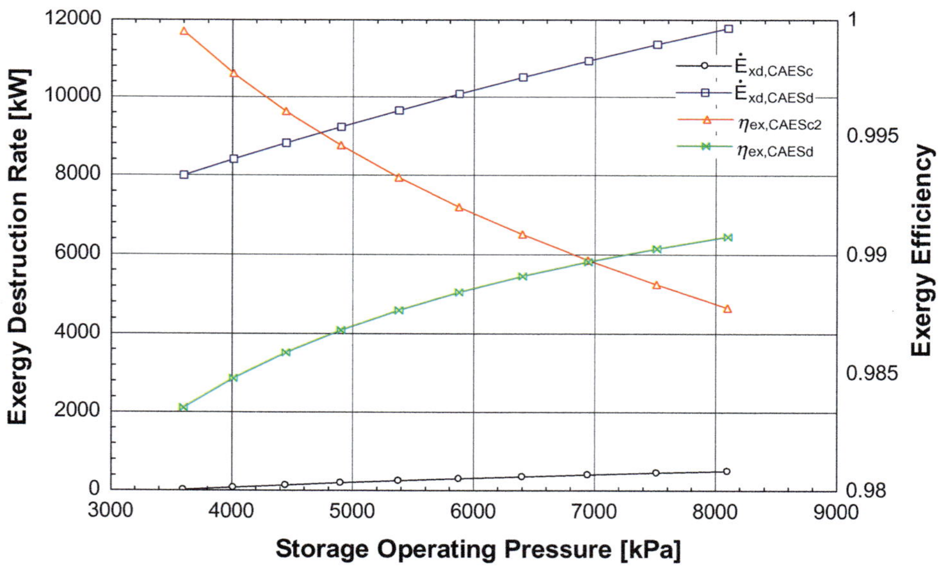

FIG. 4.17 The effects of operating pressure on the exergy destruction value and exergy efficiency of compressed air storage unit. *(Adapted from DinAli MN, Dincer I. Development and analysis of an integrated gas turbine system with compressed air energy storage for load leveling and energy management. Energy 2018; 15;163:604–17. https://doi.org/10.1016/j.energy.2018.08.094).*

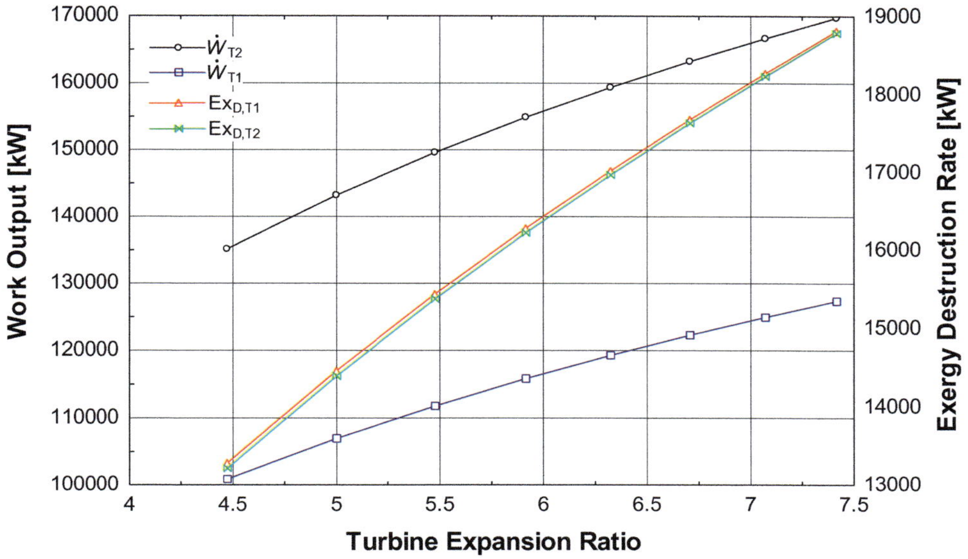

FIG. 4.18 Effects of turbine expansion ratio on turbine work output and exergy destruction. *(Adapted from DinAli MN, Dincer I. Development and analysis of an integrated gas turbine system with compressed air energy storage for load leveling and energy management. Energy 2018; 15;163:604–17. https://doi.org/10.1016/j.energy.2018.08.094).*

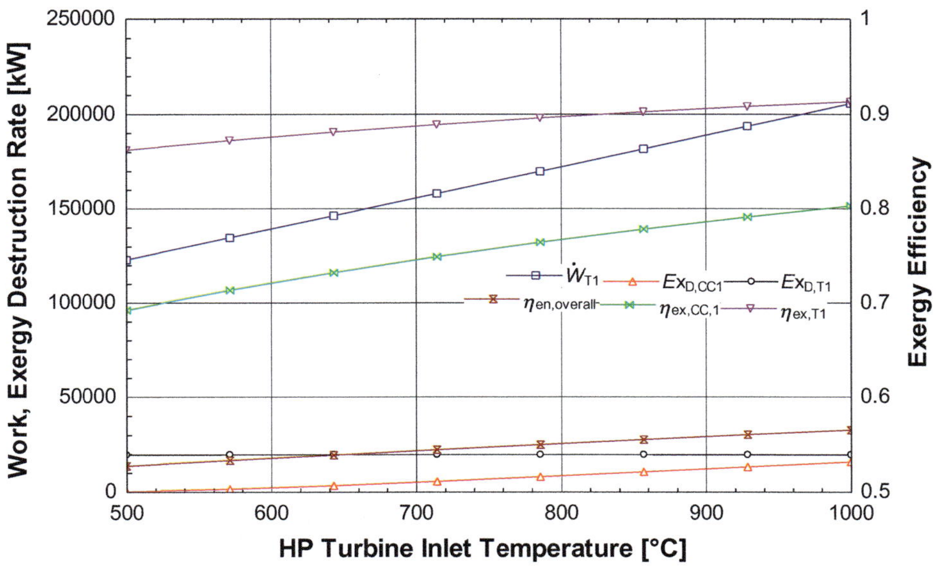

FIG. 4.19 Effects of HP turbine inlet temperature on the overall system and some subsystems. *(Adapted from DinAli MN, Dincer I. Development and analysis of an integrated gas turbine system with compressed air energy storage for load leveling and energy management. Energy 2018; 15;163:604–17. https://doi.org/10.1016/j.energy.2018.08.094).*

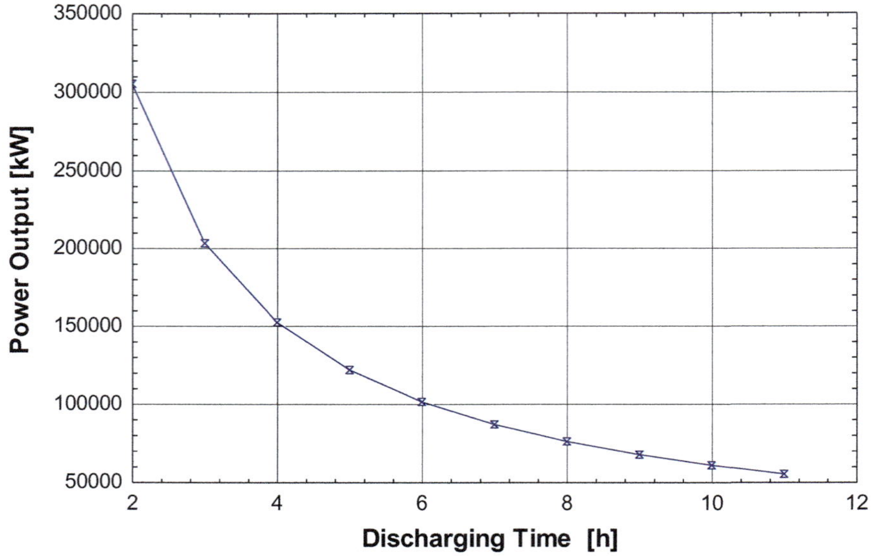

FIG. 4.20 Effects of discharging hours on the power production of the system. *(Adapted from DinAli MN, Dincer I. Development and analysis of an integrated gas turbine system with compressed air energy storage for load leveling and energy management. Energy 2018; 15;163:604–17. https://doi.org/10.1016/j.energy.2018.08.094).*

4.5.5 Concluding remarks

In this case study, a compressed air energy storage based multigeneration system integrated into an absorption cooling system is designed and analyzed to provide electricity, cooling and hot water for municipal use. The useful outputs of the system are about 63.1 MW of turbine work, a cooling capacity of about 26 MW at COP of 0.78 and a municipal hot water output flow of 165.5 kg/s at a temperature of 60 °C. The parametric study results show that the overall energy efficiency rises from 53% to 55% as the turbine inlet temperatures upsurge. The overall energy efficiency does not drop significantly as the compression pressure ratio increases, due to heat recovery from the compression process. The multigeneration system attains an improvement in the overall energy efficiency by about 11% compared to the conventional system represented by the Huntorf plant in Germany. The overall exergy efficiency increases by 6.2% compared to a single generation system. Consequently, the results reveal that using the multigeneration system with CAES can improve the efficiency of a conventional CAES single generation system.

4.6 Case study 2: Integrated gasification combined cycle with a water gas shift membrane reactor for hydrogen production

This case study examines the impacts of the flow rate of gasification oxidant, gasification agent and type of coal on the energy efficiency and hydrogen generation rate of a conventional integrated system. The system contains a pressurized entrained flow gasifier combined

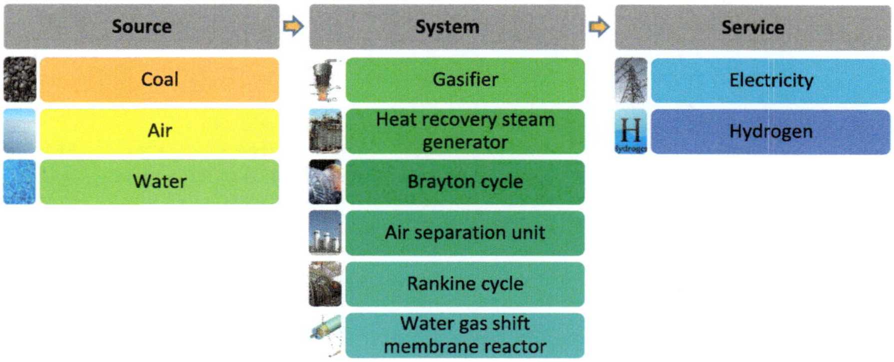

FIG. 4.21 3S approach for case study 2.

with a cryogenic air separation unit, a water gas shift membrane reactor and a combined cycle. This case study is adapted from Al-Zareer et al. [23]. The sources, systems and services in this case study are illustrated in Fig. 4.21.

During development and simulation of the integrated system, Aspen Plus software was utilized. The feed coal is a key element of this type of system. Three types of coal enter the gasification plant; the details are given in Table 4.3. The gasification process model is based

TABLE 4.3 Coal specifications for wet, dry and ash free basis.

		Illinois #6 [24]			Soma [26]			Elbistan [26]		
		Wet basis (wt%)	Dry basis (wt%)	Dry and ash free basis (wt%)	Wet basis (wt%)	Dry basis (wt%)	Dry and ash free basis (wt%)	Wet basis (wt%)	Dry basis (wt%)	Dry and ash free basis (wt%)
Proximate analysis	Moisture	0.20	0.00	0.00	15.10	0.00	0.00	16.30	0.00	0.00
	Fixed carbon	58.01	58.01	68.68	38.48	45.32	52.50	35.73	42.69	50.50
	Volatile matter	26.46	26.46	31.32	34.82	41.01	47.50	35.03	41.85	49.50
	Ash	15.53	15.53	0.00	11.60	13.66	0.00	12.93	15.45	0.00
Ultimate analysis	C	73.90	74.05	87.66	52.48	61.81	71.61	48.76	58.25	68.92
	H	6.24	6.25	7.40	3.80	4.48	5.19	3.25	3.88	4.59
	N	0.71	0.71	0.84	1.32	1.55	1.80	1.27	1.52	1.80
	Cl	0.37	0.37	0.44	0.00	0.00	0.00	0.00	0.00	0.00
	S	1.77	1.77	2.10	2.71	3.19	3.70	3.67	4.39	5.19
	O	1.32	1.32	1.56	12.97	15.28	17.70	13.79	16.48	19.50
	Ash	15.53	15.53	0.00	11.60	13.66	0.00	12.93	14.45	0.00

Continued

TABLE 4.3 Coal specifications for wet, dry and ash free basis.—Cont'd

		Illinois #6 [24]			Soma [26]			Elbistan [26]		
		Wet basis (wt%)	Dry basis (wt%)	Dry and ash free basis (wt%)	Wet basis (wt%)	Dry basis (wt%)	Dry and ash free basis (wt%)	Wet basis (wt%)	Dry basis (wt%)	Dry and ash free basis (wt%)
Sulfur analysis	Pyritic	0.59	0.59	0.70	0.08	0.09	0.10	0.04	0.05	0.06
	Sulfate	0.59	0.59	0.70	0.00	0.00	0.00	0.00	0.00	0.00
	Organic	0.59	0.59	0.70	2.63	3.10	3.60	3.63	4.34	5.13
Heating value	HHV (MJ/kg)	28.1	28.2	33.4	19.9	23.4	27.1	18.6	22.2	26.3
	LHV (MJ/kg)	24.1	27.1	32.5	19.0	22.5	26.2	17.8	21.4	25.5

Source: *Al-Zareer M, Dincer I, Rosen MA. Modeling and performance assessment of a new integrated gasification combined cycle with a water gas shift membrane reactor for hydrogen production. Comput Chem Eng 2017;103:275–92.*

on Gibbs free energy. The Gibbs free energy model was used to evaluate the effects of gasification parameters as it is clearer in terms of input flow rates. Low quality coal gasification has been found to be desirable in terms of energy efficiency. High quality coal is more efficient in terms of combustion efficiency.

4.6.1 System description

The system integrates IGCC with a water gas shift membrane reactor (WGSMR) and CASU. Water, coal and air are the inputs to the system, but the water and air are disregarded from an energy point of view because they are at atmospheric pressure and room temperature. The layout of the integrated system is shown in Figs. 4.22 and 4.23. Furthermore, the layout of the combined cycle part of the IGCC plant is depicted in Fig. 4.24. The key modules of the system are gasification unit, WGSMR unit, CASU and combined cycle. The temperature of the gasifier is typically around 1050 °C for Illinois No. 6 coal [24, 25], which is the temperature at which the pyrolysis processes occur. Based on the coal type given in Table 4.3, this temperature varies. The chemical material balance of the pyrolysis or decomposition reaction is reliant on the chemical composition of the coal, which is fed to the gasification process.

The pyrolysis process yields volatile matter that is vastly combustible and therefore is considered as input to the combustion reactor.

The balanced chemical reactions for volatile matter combustion are written as follows:

$$C_6H_6 + 7.5O_2 \rightarrow 6CO_2 + 3H_2O \tag{4.20}$$

$$H_2 + 0.5O_2 \rightarrow H_2O \tag{4.21}$$

$$CO + 0.5O_2 \rightarrow CO_2 \tag{4.22}$$

$$CH_4 + 2O_2 \rightarrow CO_2 + 2H_2O \tag{4.23}$$

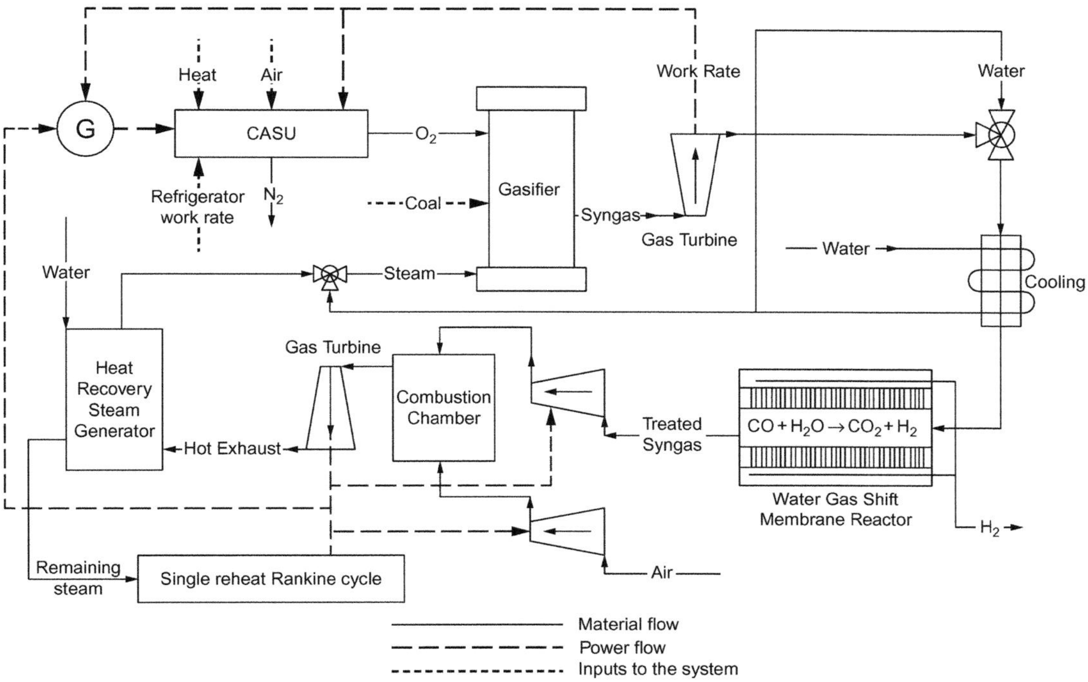

FIG. 4.22 The overall IGCC system including the main subsystems. *(Adapted from Al-Zareer M, Dincer I, Rosen MA. Modeling and performance assessment of a new integrated gasification combined cycle with a water gas shift membrane reactor for hydrogen production. Comput Chem Eng 2017;103:275–92).*

The char decomposition is modeled as a stoichiometric reactor with full conversion of char to the products as follows:

$$\text{Char} \rightarrow C + H_2 + O_2 + N_2 + S + ASH \tag{4.24}$$

After the coal is separated into char and volatile matter, the char is decomposed and the volatile mater is combusted as per the preceding chemical reactions. The resultant chemical groups of the previous reactions are inputs to the final reactor.

The exit of the WGSMR goes into the combustion chamber of the Brayton cycle, which is pressurized to the pressure of the combustion chamber. In this case study, the pressure in the combustion chamber is 23 bar. After the gas turbine in the Brayton cycle, high-temperature exhaust gas comes into the steam generator unit by recovering the available heat. Part of the steam generated by the HRSG is sent to the gasification process, and the remainder arrives at the single reheat supercritical pressure Rankine cycle for electricity generation.

4.6.2 Assumptions

In order to perform thermodynamic analysis of this multigeneration system, there is a need to make conceptually correct assumptions. Here, the main assumptions undertaken in this case study are listed.

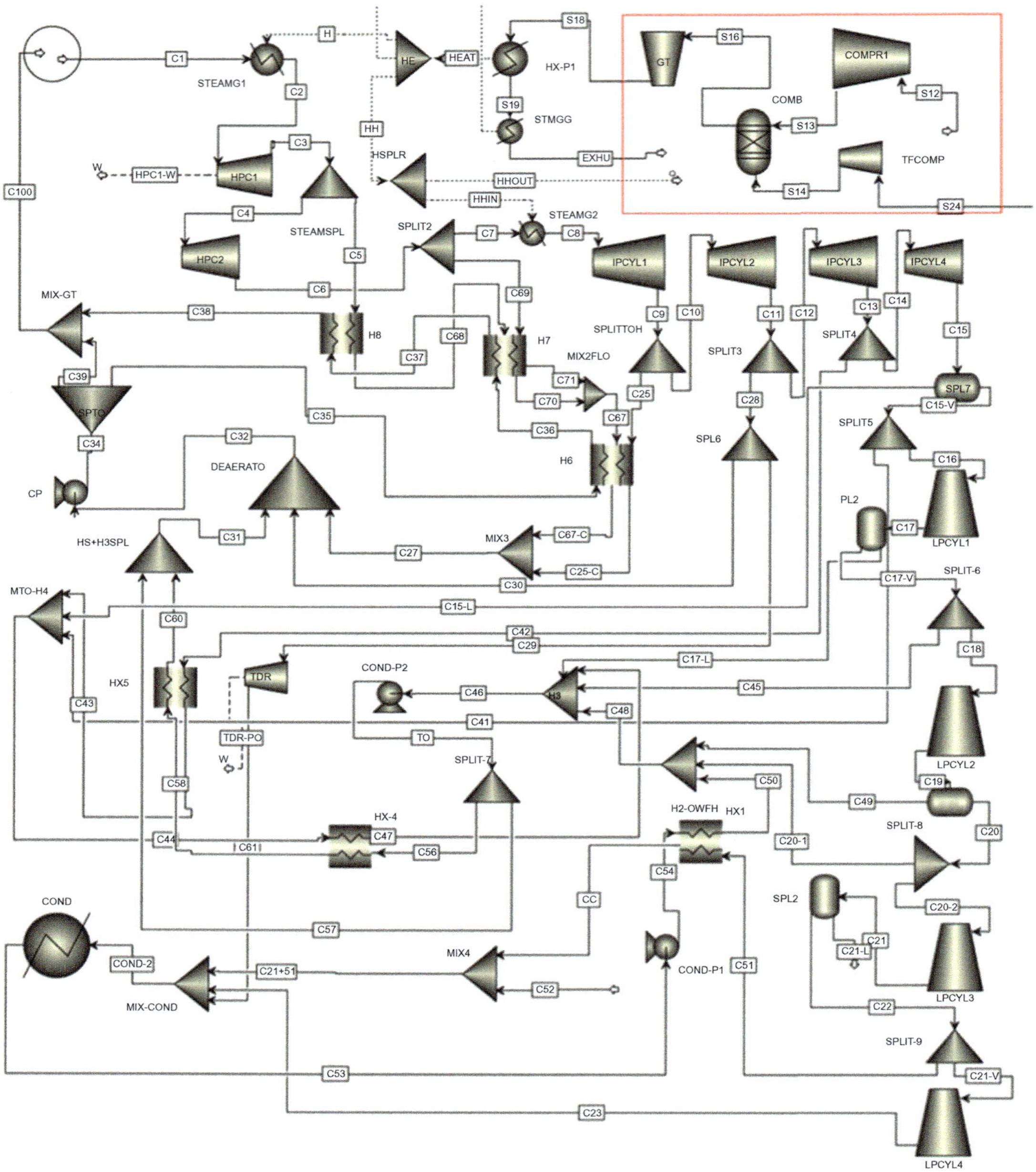

FIG. 4.23 Aspen Plus flow diagram of the combined system. *(Adapted from Al-Zareer M, Dincer I, Rosen MA. Modeling and performance assessment of a new integrated gasification combined cycle with a water gas shift membrane reactor for hydrogen production. Comput Chem Eng 2017;103:275–92).*

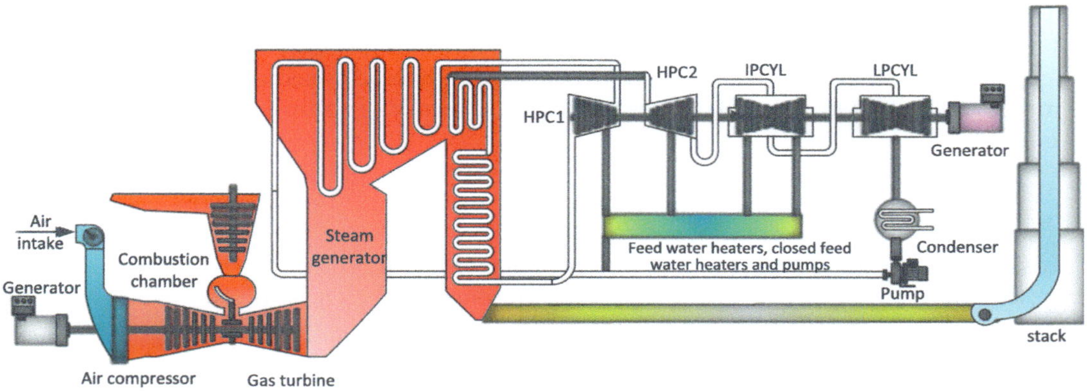

FIG. 4.24 Simplified schematic diagram of the combined cycle part within the IGCC plant including the Brayton and the single reheat supercritical pressure Rankine cycle. *(Modified from Al-Zareer M, Dincer I, Rosen MA. Modeling and performance assessment of a new integrated gasification combined cycle with a water gas shift membrane reactor for hydrogen production. Comput Chem Eng 2017;103:275–92).*

The general assumptions are expressed as follows [23]:

- The system works at steady state conditions.
- The changes in kinetic and potential energies are neglected.
- All fluids flowing in the system are treated as actual.

The following assumptions are considered for the gasification:

- All gas and solid mixtures in the system are homogeneous.
- The pressure drops are insignificant in the gasifier.
- The coal feed is in the form of powder with a fine spherical shape.
- The deviations in temperature in the coal particles and the gasifier are neglected.
- The main syngas products are considered to be CO, H_2, and CO_2 for the analysis.

The following assumptions are made for combustion chambers and the heat exchanger unit:

- There is a complete combustion.
- The heat losses in the HRSG are taken as 10% of the total exchanged heat.
- The heat loss is 10% of the total exchanged heat in the other heat exchangers in the system.

The following assumptions are made for the steam and gas turbines:

- The steam and gas turbines are taken as adiabatic.
- The isentropic efficiencies of all steam and gas turbines are taken as 72%.

The following assumptions are made for the CASU:

- The condensing fluid in the distillation column is taken at a temperature of −179 °C.
- In the distillation column, the condenser operating temperature is very low, requiring a refrigeration system to keep the fluid at low temperature for condensation.

- The coefficient of performance (COP) of the refrigeration system is taken as 1. A COP higher than 1 will lead to small variations in the maximum energy efficiency of the IGCC system. In addition, when a simple refrigerator unit is used based on a Carnot COP, it is around 0.46. A COP of 1 will deliver the same coal order regarding the maximum energy efficiency.

4.6.3 Analysis

Based on the fundamental thermodynamic balance equations for mass, entropy, and energy, the following efficiencies are determined. The energy efficiencies of the gasifier, Brayton cycle and Rankine cycle are defined as follows:

$$\eta_{en,\,gasifier} = \frac{\dot{m}_{syngas}\mathrm{LHV}_{syngas}}{\dot{m}_{coal}\mathrm{LHV}_{coal} + \dot{W}_{pump1} + \dot{W}_{pump2} + \dot{W}_{compressor} + \dot{Q}_{steam} + \dot{Q}_{water}} \tag{4.25}$$

$$\eta_{en,Brayton} = \frac{\dot{W}_{GT} - \dot{W}_{Compressor1}}{\dot{m}_{S26}LHV_{S26}} \tag{4.26}$$

$$\eta_{RC} = \frac{\begin{array}{c}\dot{W}_{HPC1} + \dot{W}_{HPC2} + \dot{W}_{IPC1} + \dot{W}_{IPC2} + \dot{W}_{IPC3} + \dot{W}_{IPC4} + \dot{W}_{LPC1} \\ + \dot{W}_{LPC2} + \dot{W}_{LPC3} + \dot{W}_{LPC4} + \dot{W}_{TDR} - \dot{W}_{CONDP1} - \dot{W}_{CONDP2} - \dot{W}_{CP}\end{array}}{\dot{Q}_{HX,P1}} \tag{4.27}$$

The overall energy efficiency of the system can be written as follows:

$$\eta_{en,overall} = \frac{\dot{W}_{HPC1} + \dot{W}_{HPC1} + \dot{m}_{H_2}\mathrm{LHV}_{H_2} - \dot{W}_{net_{air\ separation}} - \dot{W}_{net,\,gasifier,utilities}}{\dot{m}_{coal}\mathrm{LHV}_{coal} + \dot{Q}_{Brayton} + \dot{W}_{RFG_{Air\ separation}}} \tag{4.28}$$

The Brayton cycle, gasifier, and overall energy efficiencies are calculated using the LHV of coal. This means that the efficiency value for each type of coal signifies a comparison between gasification of coal and combustion of coal.

4.6.4 Results and discussion

Figs. 4.25–4.27 describe the hydrogen generation rate of the integrated system, including its entrained flow gasifier, CASU, WGSMR, and a combined cycle. Illinois No. 6, Soma and Elbistan type of coals were used in the gasification unit. The WGSMR unit captures the hydrogen by selective membranes. While the flow rate of steam in the gasification unit rises starting from 20.3 kg/s, the maximum production rate of the system is reduced and the air flow to the CASU increases in order to reach the maximum hydrogen production rate. However, mass flow rates of steam below 20 kg/s produce hydrogen below 20 kg/s, which means that the maximum hydrogen production rate has an optimal point. The maximum hydrogen production rate increases dramatically while the steam mass flow rises to a maximum. Later, it begins to decrease by reducing hydrogen production, which means it has a poor sensitivity to the steam mass flow. For the three types of coals tested in the IGCC, the maximum generation in hydrogen occurs in the range of 1 kg/s to 21 kg/s.

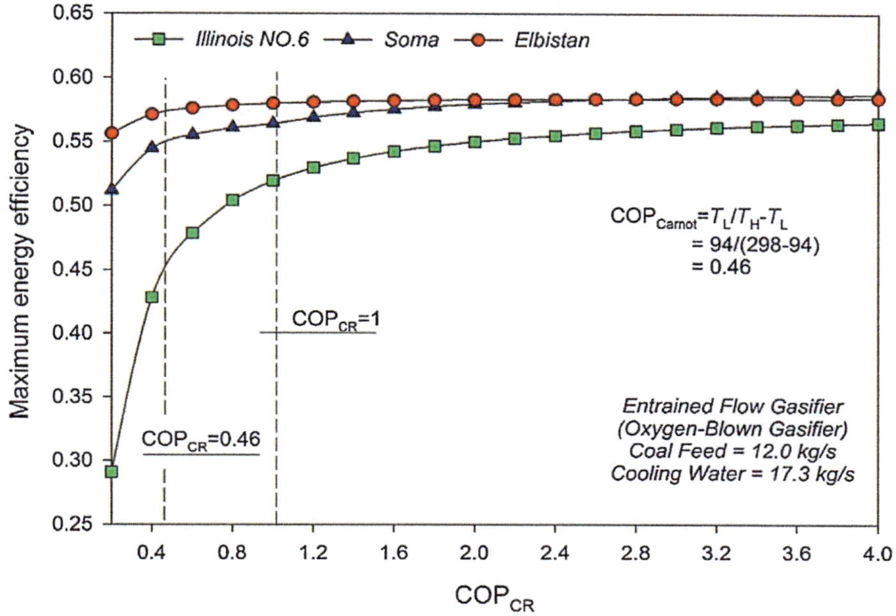

FIG. 4.25 The changes of the maximum energy efficiency of the IGCC system for various coal feeds. *(Adapted from Al-Zareer M, Dincer I, Rosen MA. Modeling and performance assessment of a new integrated gasification combined cycle with a water gas shift membrane reactor for hydrogen production. Comput Chem Eng 2017;103:275–92).*

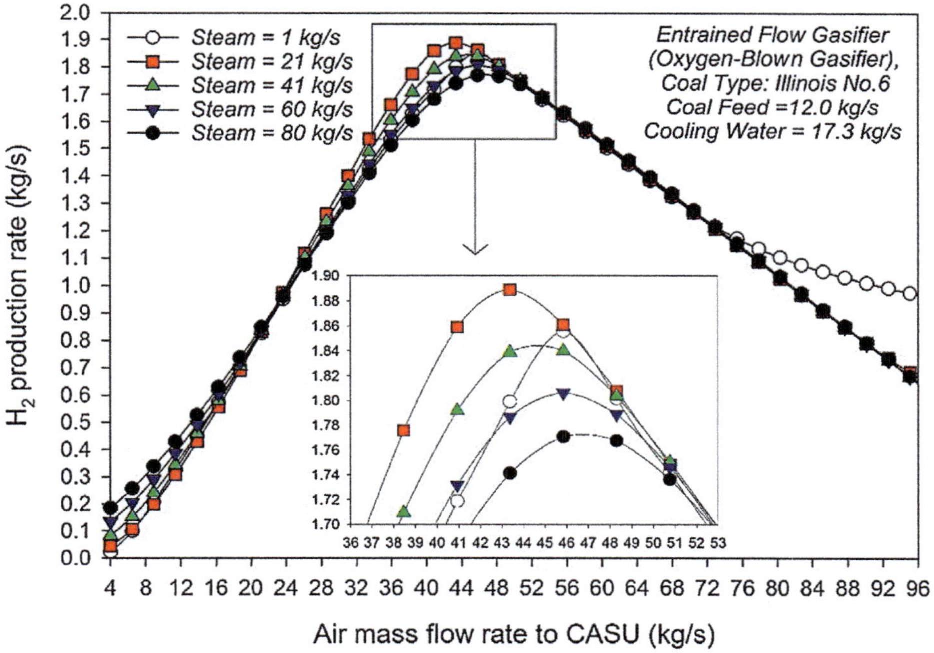

FIG. 4.26 The changes in hydrogen generation rate of the IGCC system for Illinois No. 6 coal type. *(Adapted from Al-Zareer M, Dincer I, Rosen MA. Modeling and performance assessment of a new integrated gasification combined cycle with a water gas shift membrane reactor for hydrogen production. Comput Chem Eng 2017;103:275–92).*

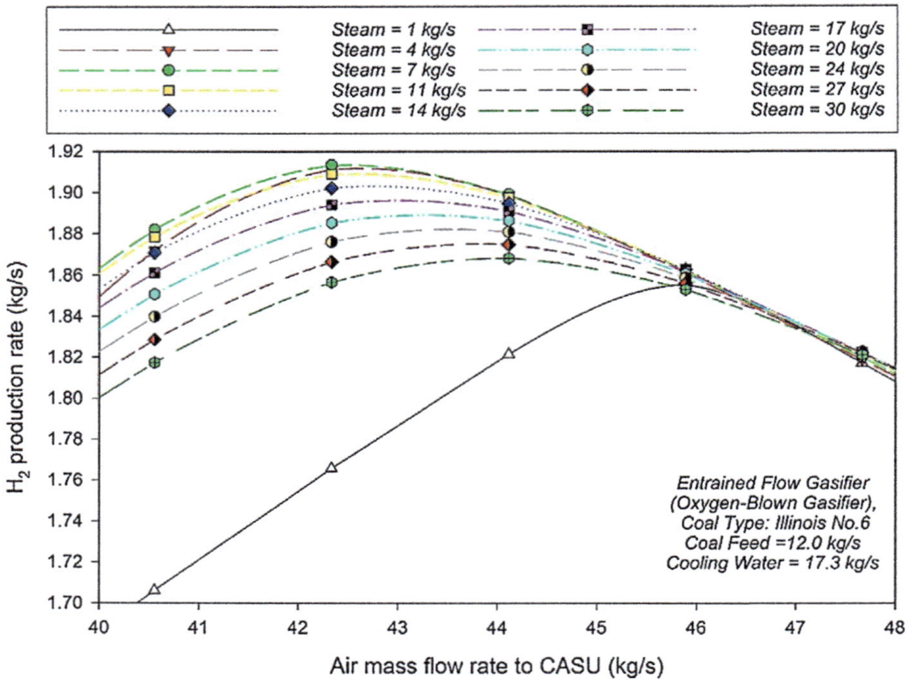

FIG. 4.27 The changes in hydrogen generation rate of the IGCC system for Illinois No. 6 coal type. *(Adapted from Al-Zareer M, Dincer I, Rosen MA. Modeling and performance assessment of a new integrated gasification combined cycle with a water gas shift membrane reactor for hydrogen production. Comput Chem Eng 2017;103:275–92).*

Fig. 4.27 presents the variations in the hydrogen production rate depending on the feed air and steam mass flow rates. It clearly shows at which point the shifting happens. The shift between the rising and declining behavior of the hydrogen generation rate takes place at a steam mass flow rate of approximately 7 kg/s of steam to the gasification unit when Illinois No. 6 coal type is used.

The air flow rate to CASU producing the maximum hydrogen generation rate per unit steam mass flow rate rises as the steam mass flow rate upsurges, excluding the steam mass flow rates of 1 kg/s to 20 kg/s. When the content of carbon in coal drops, an inferior air mass flow rate to the CASU is required to yield the maximum hydrogen generation rate. This is due to the carbon content of coal: the smaller this quantity, the lower the demand for oxygen. As listed in Tables 4.4 and 4.5, the airflow in CASU differs significantly between Illinois No. 6, Soma and Elbistan type coals. Nevertheless, the change in the air mass flow rate to the CASU between the Soma and Elbistan coal types to attain a maximum hydrogen generation rate is much less compared to Illinois No. 6 coal type. The situation is largely due to the carbon content of coal, because it decreases proportionally. The results indicate that the carbon content of coal is one of the key features affecting an integrated hydrogen production system.

Table 4.5 lists the gasification factors at which the maximum energy efficiency can be attained in addition to the net work rate, CASU refrigerator work rate, and the supplementary

TABLE 4.4 The gasification parameters for maximum hydrogen generation rate in the integrated IGCC system.

Type of coal	Air mass flow rate to CASU[a] (kg/s)	Steam mass flow rate to gasifier[b] (kg/s)	Hydrogen production rate[c] (kg/s)	Net work rate produced by IGCC (MW)	Electrical power required by refrigerator cooling CASU condenser (MW)	Extra heat rate required by proposed IGCC system (MW)	Energy efficiency (%)
Illinois No. 6[d]	40.92	20.75	1.89	11.61	22.04	51.19	48.9[e]
Soma[d]	11.38	20.75	1.60	20.99	6.13	57.29	55.9[e]
Elbistan[d]	4.00	20.75	1.58	24.01	2.15	57.87	58.0[f]

[a] *Generated oxygen arrives at the gasifier around 490 °C.*
[b] *Steam temperature arriving at the gasifier is around 420 °C.*
[c] *Cooling water has pressure of about 14.4 bar and a temperature of about 260 °C.*
[d] *Feed rate of coal is about 12 kg/s.*
[e] *Not maximum energy efficiency.*
[f] *Maximum energy efficiency.*
Source: *Al-Zareer M, Dincer I, Rosen MA. Modeling and performance assessment of a new integrated gasification combined cycle with a water gas shift membrane reactor for hydrogen production. Comput Chem Eng 2017;103:275–92.*

TABLE 4.5 The gasification parameters for maximum energy efficiency in the IGCC system.

Coal type	Air mass flow rate to CASU (kg/s)	Steam mass flow rate to gasifier (kg/s)	Hydrogen generation mass flow rate (kg/s)	Net work rate generated by IGCC (MW)	Electrical power essential by refrigerator cooling CASU condenser (MW)	Additional heat rate necessary for proposed IGCC (MW)	Energy efficiency (%)
Illinois No. 6	45.85	1.00	1.86	13.08	24.69	14.99	51.93
Soma	8.92	20.75	1.58	24.44	4.81	55.10	56.38
Elbistan	4.00	20.75	1.58	24.01	2.15	57.87	57.97

Source: *Al-Zareer M, Dincer I, Rosen MA. Modeling and performance assessment of a new integrated gasification combined cycle with a water gas shift membrane reactor for hydrogen production. Comput Chem Eng 2017;103:275–92.*

heat needed by the system. If Illinois No. 6 coal type is used in the gasification, the maximum achievable energy efficiency does not match up to the maximum hydrogen generation. The difference between the efficiency of the energy and the quantity of hydrogen production products is largely due to the high difference of the CASU air flow and the flow rate of steam. The major difference in CASU air flow implies that more energy is needed in the form of heat and electricity.

As well, the huge variance in the steam flow rate means that the required work for the pump and required heat for water vaporization is higher. On the other hand, when Soma coal type is used in the gasification, the parameters yielding the maximum energy efficiency matches with maximum hydrogen generation. However, for Soma coal, the variance in the air mass flow rate to the CASU between the parameters for maximum hydrogen generation rate and maximum energy efficiency is considerably less than for Illinois No. 6 coal type. If Elbistan coal type is used in the gasification, the parameters for maximum hydrogen generation rate and the maximum energy efficiency are the same. Thus, it can be concluded from the results that when the carbon content is lower and the LHV of coal is lower, the moisture content is higher, so the parameters of both the maximum energy efficiency and the generation rate of hydrogen are closer to each other. Thus, in order to achieve the higher energy efficiency of the integrated system, it is better to gasify the coal than burning, as when the moisture content increases, the carbon content decreases and LHV decreases.

The produced oxygen enters the gasifier at 490 °C, whereas the temperature of steam entering the gasifier is 420 °C. The cooling water used is at a pressure of 14.4 bar and temperature of 260 °C. For Soma and Illinois No. 6 type coals, the values in Table 4.5 do not represent the maximum hydrogen generation rate; nevertheless, it is the maximum hydrogen generation rate for Elbistan.

4.6.5 Concluding remarks

This case study integrates a conventional energy source for production of power and hydrogen. The gasification agent, which is steam in this case, the gasification oxidant, which is oxygen in this case, and the coal type are changed to investigate their impacts on the hydrogen generation rates and the overall efficiency of the integrated system. The integrated system includes a pressurized entrained flow gasifier, a cryogenic unit for air separation, a gas power cycle (Brayton cycle), and a supercritical Rankine cycle having single reheating. Some of the main findings from this case study are:

- The highest hydrogen generation rate is obtained as 1.89 kg/s for Illinois No. 6 coal type.
- The highest energy efficiency is achieved as 58% for Elbistan coal type.
- Gasification of low-grade coals is more effective from an energy point of view than combusting them in a coal-fired power plant.
- When the carbon content of coal is reduced, the gasification parameters for the maximum hydrogen generation rate and those for the maximum energy efficiency become closer.
- The hydrogen generation rate hinges on the combined action of both the gasifier and the water gas shift membrane reactor.
- The parameters yielding the maximum hydrogen generation rate through the combined action of both the gasifier and the water gas shift membrane reactor are different from those that generate the maximum hydrogen mass flow rate just from the gasifier.

4.7 Case study 3: Development of an integrated trigeneration system for dimethyl-ether, electricity and fresh water production using waste heat

In this case study, an integrated trigeneration system utilizing the gas turbine exhaust gas for heat recovery is thermodynamically analyzed. The system also has carbon capturing after expansion in the turbine. Additionally, a Cu—Cl thermochemical cycle is incorporated to

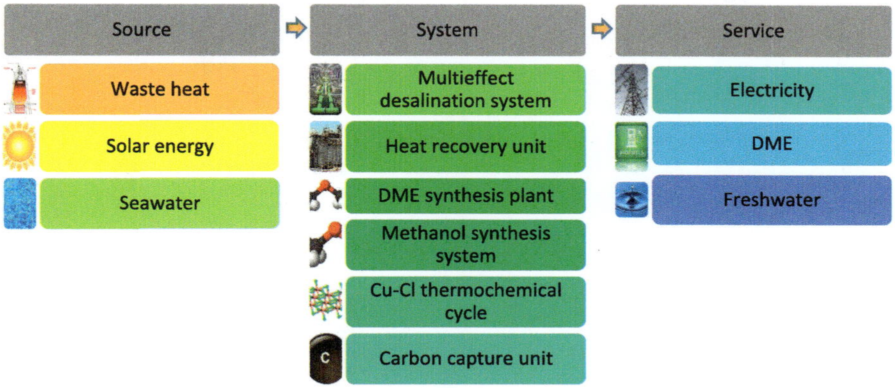

FIG. 4.28 3S approach for case study 3.

generate hydrogen, and a multieffect desalination (MED) is combined to obtain the fresh water required for hydrogen production in the Cu—Cl cycle and for domestic use as well. The required inputs, utilized systems and produced useful outputs are illustrated in Fig. 4.28. The integrated system is modeled and simulated using the Aspen Plus process simulations for chemical processes and the Engineering Equation Solver for the other subsystems. This case study is adapted from DinAli [27].

4.7.1 System description

In the proposed system, gas turbine exhaust gas is the main source of thermal energy. The flue gas comes out of the turbine at a temperature of 840 °C exchanging heat with water in the heat recovery steam generator (HRSG) to produce superheated steam to provide to the thermochemical Cu—Cl cycle with the required heat at a temperature of 550 °C and then to supply the remaining processes, such as methanol synthesis, DME synthesis, and carbon capturing, with the necessary heat. Ultimately, saturated steam is fed to the multieffect desalination (MED). The Cu—Cl thermochemical cycle is the hydrogen production technology incorporated in the system. A four-step Cu—Cl cycle is selected from among the different configurations. The main useful outputs of this system are DME (which is the main focus of this study), electricity, and fresh water. Fig. 4.29 shows the schematic diagram of the integrated trigeneration system.

After the air is compressed, combusted and expanded in the turbine, gas turbine exhausts usually have high thermal energy due to their high temperature, which may reach up to 1000 °C. Heat recovery of gas turbine exhaust gas usually occurs in an HRSG unit producing high temperature steam.

The heat recovery system is providing heat to the different components in a specific order to maintain the proper temperature required in each component. The flue gas conditions and compositions are listed in Table 4.6.

The generated steam passes through the different components in a specific order to maintain the pinch temperature in each component and is then condensed in the first effect of the

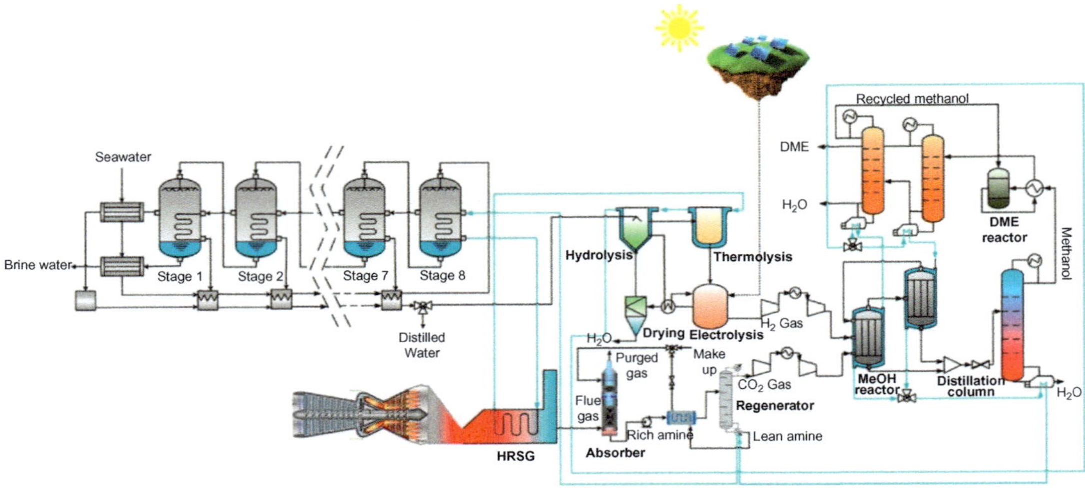

FIG. 4.29 Schematic diagram of integrated trigeneration system for case study 3. *(Modified from DinAli MN. Development and assessment of renewable energy based integrated systems for dimethyl ether production. Master's Thesis. University of Ontario Institute of Technology, 2018).*

TABLE 4.6 Gas turbine flue gas conditions.

Parameter		Value
Temperature		840 °C
Pressure		120 kPa
Total flow		19.46 kg/s
Mole fraction	H_2O	1.6%
	CO_2	23.41%
	N_2	70.28%
	O_2	4.71%

desalination system. For instance, steam first goes to the thermolysis process for providing heat at a temperature of 500 °C. Then, the steam moves to the hydrolysis process to provide heat at a temperature of 400 °C. In addition, the steam moves to lower temperature processes such as the methanol synthesis plant, the DME synthesis plant and finally the carbon capturing plant. The cooled gas turbine exhaust gas passes to the carbon capturing unit. In a chemical absorption system, the exhaust gas passes through an absorber. The solvent reacts with carbon dioxide only and the remaining gases leave the system. In order for the rich amine to reject carbon dioxide gas, the solvent needs to be heated. Following this, the rich amine passes through a heat exchanger and is further heated in the regenerator. The carbon dioxide will then be released at the top of the regenerator, where the lean amine is recycled back to the absorber.

Numerous developmental investigations have been conducted on the experimental Cu—Cl cycles in the Clean Energy Research Laboratory (CERL) at UOIT [28–31]. The Cu—Cl cycle in this system is a four-step cycle that consists of hydrolysis, decomposition, electrolysis, and drying. The first process is hydrolysis, which takes place between copper dichloride and water at a temperature of 400 °C to produce copper oxychloride and hydrochloric acid. The second step is thermolysis, which occurs for copper oxychloride at a temperature of 550 °C to dissociate to oxygen and copper chloride. Then the third step is electrolysis, in which copper chloride reacts with hydrochloric acid from the hydrolysis process at a low temperature to produce hydrogen gas and copper dichloride solution. Hydrogen is then extracted and the copper dichloride solution goes to the fourth step. The last step is drying, in which the copper dichloride solution is dried at 80 °C, and the water is discarded while solid copper dichloride is sent back to hydrolysis. The flowsheet for the carbon capture unit is depicted in Fig. 4.30.

The hydrogen produced through the Cu—Cl thermochemical cycle and carbon dioxide captured from the gas turbine exhaust gas is compressed to 50 bar. The compressed gases are then introduced to the first methanol reactor at a temperature of 235 °C and pressure of 50 bar, producing methanol at a conversion rate of 80%, and the water and some of the gas remain unreacted. The products leave the reactor to the first distillation column, while the unreacted gases leave to a second methanol reactor at the same temperature and pressure of the first reactor. Then, second methanol reactor products have methanol besides unreacted gases, which will be recycled back to the first methanol reactor. The products are then mixed together before entering the distillation column after reducing the pressure to 1 bar to separate water from the produced methanol.

In the DME production subsystem, the produced methanol is dissociated to DME in a process called methanol dehydration. The methanol produced in the previous subsystem is heated to the required temperature, which is 260 °C, through the heat exchanger. Then, the stream is introduced to the DME reactor at a pressure of 13 bar. The products contain

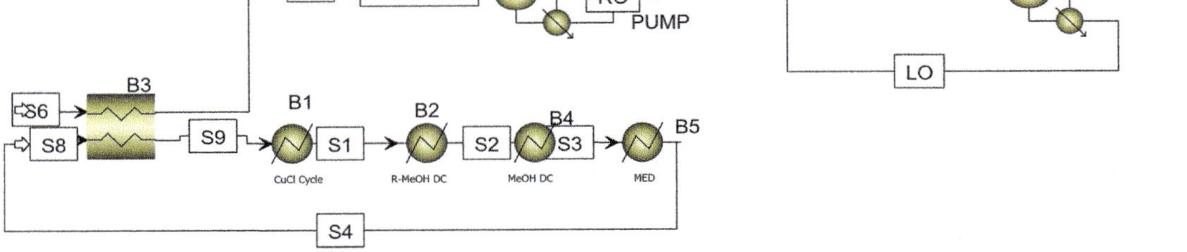

FIG. 4.30 Aspen Plus flowsheet of carbon capture and heat recovery unit. *(Adapted from DinAli MN. Development and assessment of renewable energy based integrated systems for dimethyl ether production. Master's Thesis. University of Ontario Institute of Technology, 2018).*

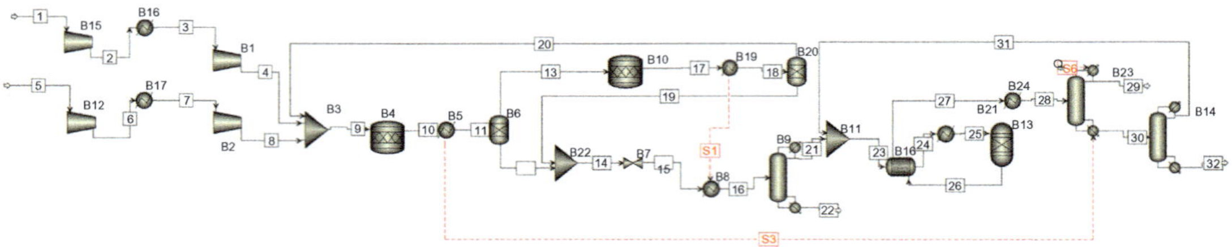

FIG. 4.31 Aspen Plus flowsheet of methanol and DME synthesis system. *(Adapted from DinAli MN. Development and assessment of renewable energy based integrated systems for dimethyl ether production. Master's Thesis. University of Ontario Institute of Technology, 2018).*

the produced DME, water and unreacted methanol. The stream is cooled by providing heat to the feed stream. The stream is then introduced to the second distillation column, which will isolate the produced DME. The remaining mixture will exit the distillation column to enter the third distillation column to isolate water from the unreacted methanol, which will then be recycled back to the DME reactor. Fig. 4.31 shows the Aspen flowsheet for methanol and DME synthesis for the system.

MED is a process of multistep desalination in which the first effect is heated by passing hot steam in tubes and water is sprayed over the tubes, forming vapor that is used as the heating source for the next effect. The main idea of MED is to keep using the same energy in every effect but at lower temperatures and pressures. The configuration considered in this study is a feed forward MED type configuration. In the proposed system, the saturated steam coming from a carbon capturing regenerator enters the first effect of the desalination system to be condensed and delivers the latent heat to it.

4.7.2 Assumptions

In order to perform thermodynamic analysis of this multigeneration system, there is a need to make conceptually correct assumptions. Here, the main assumptions undertaken in this case study are listed. The general assumptions are as follows:

- The system operates under steady state conditions.
- The potential and kinetic energy changes are neglected.
- All compressors and turbines work adiabatically.
- The heat losses and pressure drops are neglected.
- All gas compressors have an isentropic efficiency of 72%.
- The isentropic efficiencies for gas turbine system compressor and turbine are set as 83% and 87%, respectively.
- The pressure drops are neglected in the combustion chamber.
- There are no heat losses in heat exchangers of the Cu—Cl cycle.
- The electrical work required for electrolysis is set as 63 kJ/mol of hydrogen [29].
- The salinity of feed sea water entering to the MED system is taken as 42,000 ppm.
- The salinity of rejected brine water is limited to maximum 70,000 ppm.
- The Aspen Plus property method used for the Cu—Cl thermochemical cycle simulation was set as SOLIDS.

- The Aspen Plus property method used for the methanol and DME synthesis plant simulation is PSRK. The PSRK property method is based on the predictive Redlich-Kwong-Soave equation-of-state model, which is an extension of the Redlich-Kwong-Soave equation of state.
- The property method used in Aspen Plus for carbon capturing is ELECNRTL. The ELECNRTL property method is a useful electrolyte property method. It can accommodate both aqueous and mixed solvent systems.
- The steam table correlations (TEAMNBS) are used for the properties of water in the analysis.

4.7.3 Analysis

The energy efficiency for a gas turbine system is defined as follows:

$$\eta_{en_{GT}} = \frac{\dot{W}_T - \dot{W}_C}{\dot{Q}_{in}} \tag{4.29}$$

where $\dot{W}_T$ is the work generated by the gas turbine, $\dot{W}_C$ is the work consumed by the compressor and $\dot{Q}_{in}$ is the heat added in the form of fuel. The exergy efficiency for a gas turbine system can be expressed as

$$\eta_{ex_{GT}} = \frac{\dot{W}_T - \dot{W}_C}{\dot{Q}_{in}\left(1 - \frac{T_0}{T_{source}}\right)} \tag{4.30}$$

The second subsystem is the Cu—Cl thermochemical cycle. The energy efficiency for the Cu—Cl cycle is expressed as

$$\eta_{ex_{CuCl}} = \frac{\dot{m}_{H_2} LHV_{H_2}}{\dot{Q}_{Total} + \dot{W}_{elec}} \tag{4.31}$$

where $\dot{W}_{elec}$ is the work consumed in the electrolysis step, and $\dot{Q}_{Total}$ can be expressed as

$$\dot{Q}_{Total} = \dot{Q}_{B2} + \dot{Q}_{B4} \tag{4.32}$$

The exergy efficiency for the Cu—Cl cycle can be expressed as follows:

$$\eta_{ex_{CuCl}} = \frac{\dot{m}_{H_2} ex_{H_2}}{\dot{Q}_{Total}\left(1 - \frac{T_o}{T_{source}}\right) + \dot{W}_{elec}} \tag{4.33}$$

The energy efficiency of the DME system can be expressed as

$$\eta_{enDME} = \frac{\dot{m}_{DME} LHV_{DME}}{\dot{Q}_{total} + \dot{W}_{Total}} \tag{4.34}$$

where $\dot{m}_{DME}$ is the mass flowrate of produced DME, and LHV_{DME} is the lower heating value of the DME. The total heat provided during the methanol synthesis process can be written as

$$\dot{Q}_{total} = \dot{Q}_{B9} + \dot{Q}_{B10} + \dot{Q}_{B23} + \dot{Q}_{B14} \tag{4.35}$$

Similarly, the total work can be expressed as follows:

$$\dot{W}_{Total} = \dot{W}_{B1} + \dot{W}_{B2} + \dot{W}_{B12} + \dot{W}_{B15} \tag{4.36}$$

The exergy efficiency of a methanol synthesis system can be expressed as follows:

$$\eta_{ex_{methanol}} = \frac{\dot{m}_{methanol}\, ex_{methanol}}{\dot{Ex}^{Total} + \dot{W}_{Total}} \tag{4.37}$$

where $ex_{methanol}$ is the total specific exergy of methanol, and $\dot{Ex}^{Total}$ is the total exergy provided in the form of heat.

The performance parameter for desalination systems is usually measured either by the gain output ratio (GOR) or the performance ratio (PR). The GOR for the MED system can be written as

$$GOR = \frac{\dot{m}_d h_{fg}}{\dot{Q}_{in}} \tag{4.38}$$

where $\dot{m}_d$ is the mass flow rate of the produced distilled water, h_{fg} is the enthalpy of vaporization and $\dot{Q}_{in}$ is the heat supplied to the system.

In addition, the performance ratio (PR) can be written as

$$PR = \frac{\dot{m}_d}{\dot{m}_s} \tag{4.39}$$

where $\dot{m}_s$ is the salt water entering the desalination system.

The overall energy efficiency for the integrated trigeneration system is expressed as follows:

$$\eta_{en,ov} = \frac{\dot{m}_{DME} LHV_{DME} + \dot{W}_T + \dot{m}_d h_{fg}}{\dot{Q}_{total} + \dot{W}_{Total}} \tag{4.40}$$

where $\dot{Q}_{total}$ can be expressed as

$$\dot{Q}_{total} = \dot{Q}_{stripper} + \dot{Q}_{B9} + \dot{Q}_{B10} + \dot{Q}_{B23} + \dot{Q}_{B14} + \dot{Q}_{B2} + \dot{Q}_{B4} \tag{4.41}$$

Here, $\dot{W}_{Total}$ can be expressed as

$$\dot{W}_{Total} = \dot{W}_{p,ccs} + \dot{W}_{elec} + \dot{W}_{B1} + \dot{W}_{B2} + \dot{W}_{B12} + \dot{W}_{B15} \tag{4.42}$$

The overall exergy efficiency for the proposed system is expressed as follows:

$$\eta_{ex,ov} = \frac{\dot{m}_{DME} ex_{DME} + \dot{W}_T + \dot{E}x_{Desalination}}{\dot{Ex}^{Total} + \dot{W}_{Total}} \tag{4.43}$$

where $\dot{E}x_{Desalination}$ is the exergy recovered in desalination. $\dot{Ex}^{Total}$ is the total exergy provided in the form of heat, which is written as follows:

$$\dot{Ex}^{Total} = \dot{Q}_{EH}\left(1 - \frac{T_0}{T_{source}}\right) + \dot{Q}_{stripper}\left(1 - \frac{T_0}{T_{source}}\right) + \dot{Q}_{B9}\left(1 - \frac{T_0}{T_{source}}\right) + \dot{Q}_{B10}\left(1 - \frac{T_0}{T_{source}}\right) + \dot{Q}_{B23}\left(1 - \frac{T_0}{T_{source}}\right) + \dot{Q}_{B14}\left(1 - \frac{T_0}{T_{source}}\right) + \dot{Q}_{solar}\left(1 - \frac{T_0}{T_{sun}}\right) \tag{4.44}$$

4.7.4 Results and discussion

The system can produce 0.0217 kg/s of DME. If the plant is operating 24 h per day, the daily production of DME will be 1879.15 kg/day. The plant can generate 233.71 MWh of energy per day from the gas turbine and produce 788.14 ton/day of fresh water through MED desalination system. The overall energy efficiency of the trigeneration system is 39.72%. Similarly, the overall exergy efficiency of the trigeneration system is 55.2%.

Fig. 4.32 shows the distribution of exergy destruction between the different subsystems of the trigeneration system. Most of the exergy destruction takes place in the gas turbine subsystem due to the high irreversibility occurring in the combustion process and gas turbine. This is followed by the Cu—Cl cycle due to the high temperature chemical reactions which lead the exergy destruction to increase.

Fig. 4.33 depicts the exergy destruction rates along with exergy efficiency for each step. The overall energy efficiency for the Cu—Cl thermochemical hydrogen production cycle is 41.6%, while the overall energy efficiency for the Cu—Cl thermochemical hydrogen production cycle is 62%.

Fig. 4.34 shows the analysis results of the carbon capture unit. The maximum exergy destruction rate takes place in the regenerator due to the high chemical irreversibility and the amount of heat rejected to the surroundings through the condenser. The maximum exergy efficiency occurs at the heat exchanger, because there is not much exergy loss in this component.

Fig. 4.35 demonstrates the different exergy performance parameters of each component of the methanol and DME synthesis system. The exergy destruction rates are maximum at the reactors due to the chemical irreversibility of the reactors. On the other hand, minimum exergy efficiency takes place at the reactor coolers because of the considerable amount of heat rejected at very high temperature. The energy efficiency of DME synthesis from hydrogen and carbon dioxide becomes 71% whereas the exergy efficiency of DME synthesis is calculated to be 81%.

Table 4.7 shows the operating parameters used and results for the MED subsystem analyses. The detailed mass flows of distilled water, brine water and brine salinity in each stage are plotted in Fig. 4.36. The mass flow of distilled water in each effect is almost the same except that it slightly decreases at the last few effects.

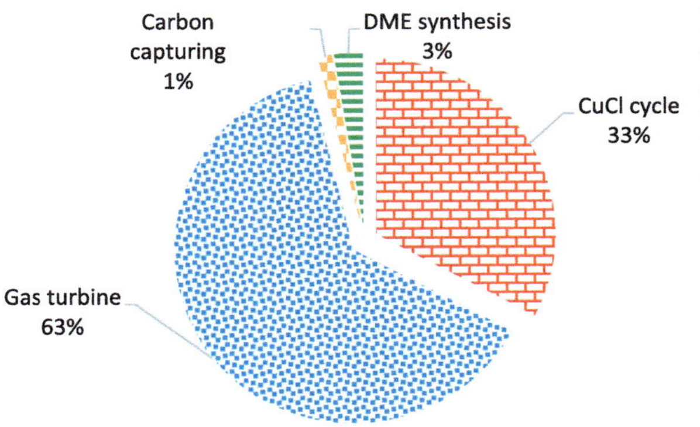

FIG. 4.32 The main exergy destruction ratios of the integrated subsystems. *(Data from DinAli MN. Development and assessment of renewable energy based integrated systems for dimethyl ether production. Master's Thesis. University of Ontario Institute of Technology, 2018).*

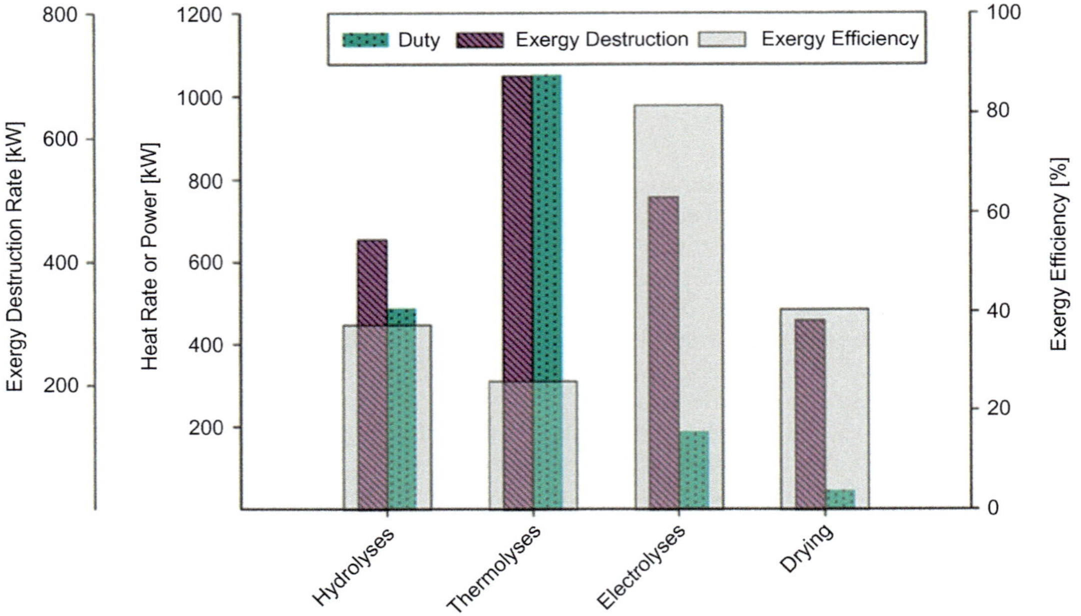

FIG. 4.33 Energy and exergy analyses results for Cu—Cl cycle. *(Adapted from DinAli MN. Development and assessment of renewable energy based integrated systems for dimethyl ether production. Master's Thesis. University of Ontario Institute of Technology, 2018).*

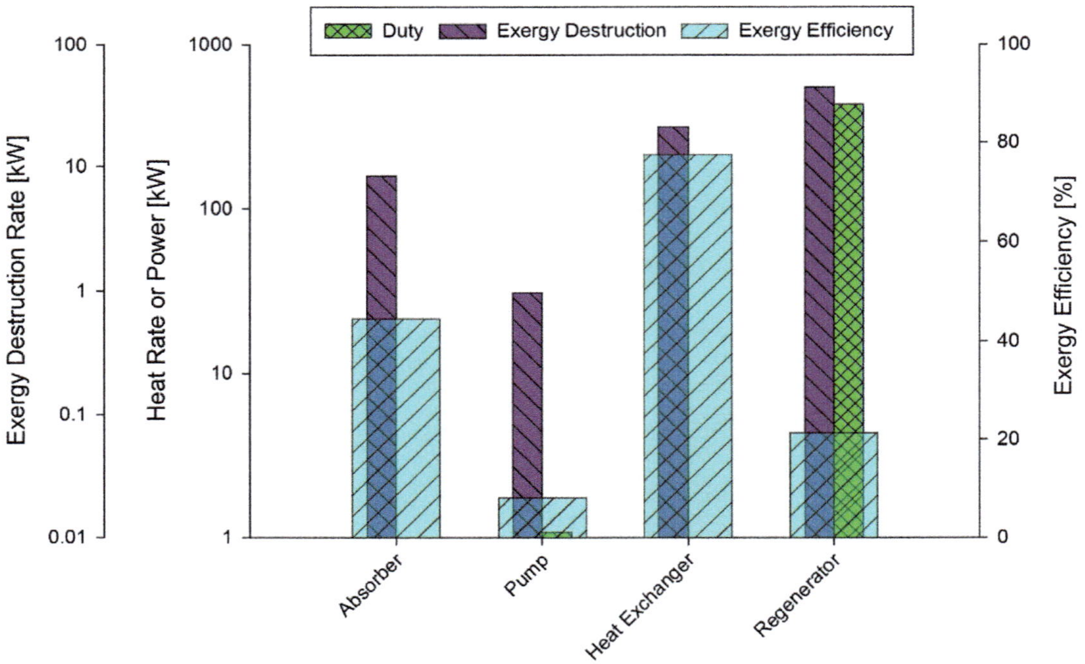

FIG. 4.34 Exergy analyses results for carbon capturing subsystem. *(Adapted from DinAli MN. Development and assessment of renewable energy based integrated systems for dimethyl ether production. Master's Thesis. University of Ontario Institute of Technology, 2018).*

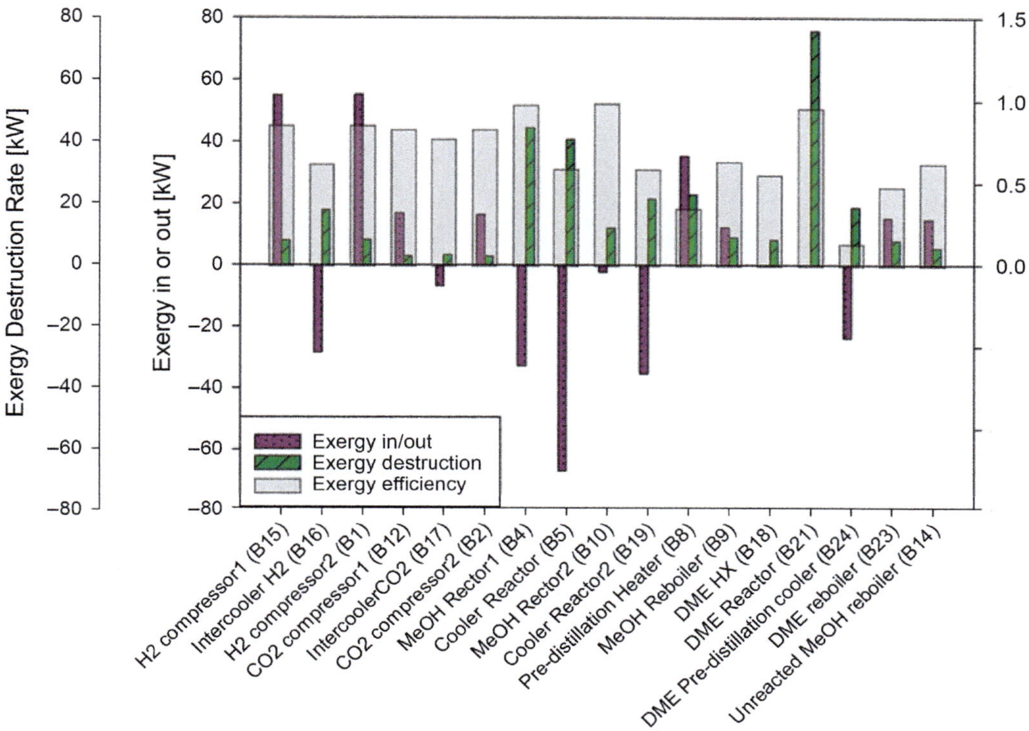

FIG. 4.35 Exergy analyses results for DME synthesis sub system. *(Adapted from DinAli MN. Development and assessment of renewable energy based integrated systems for dimethyl ether production. Master's Thesis. University of Ontario Institute of Technology, 2018).*

TABLE 4.7 MED subsystem operating parameters.

Parameter	Value
Number of stages	8
Feed water temperature	25 °C
Feed water salinity	42,000 ppm
Feed water flowrate	22.94 kg/s
Motive steam flow rate	1 kg/s
Motive steam temperature	140 °C
Brine exit temperature	40 °C
Rejected brine salinity	70,000 ppm
Brine water flow rate	13.76 kg/s
Temperature drop in each stage	2 °C
Distilled water flowrate	9.176 kg/s

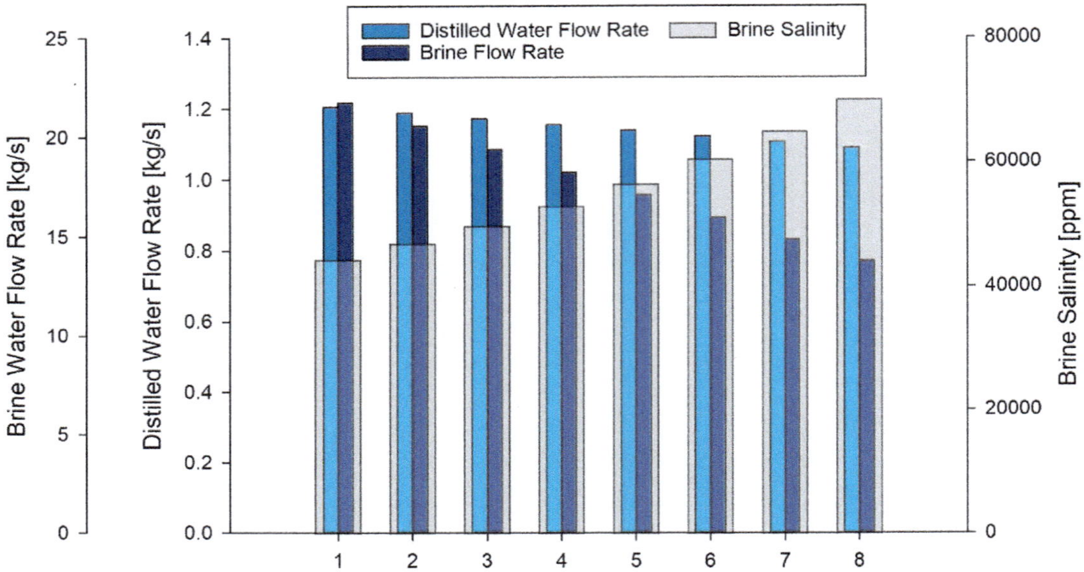

FIG. 4.36 Distilled water, brine water flow over the different desalination effects. *(Adapted from DinAli MN. Development and assessment of renewable energy based integrated systems for dimethyl ether production. Master's Thesis. University of Ontario Institute of Technology, 2018).*

Fig. 4.37 shows the temperature of each effect as well as the temperature drop in each effect. The effect temperature significantly decreases as we go towards the last effects due to the temperature drop.

Fig. 4.38 shows the effect of motive steam temperature on the distilled water flowrate at a different number of stages. The results show that the distilled water flow slightly decreases as the motive steam temperature increases.

Fig. 4.39 shows the effect of increasing motive steam temperature on performance ratio at a different number of stages. The results show that performance ratio slightly decreases as motive steam temperature increases due to the fact that saturated steam has less enthalpy of evaporation as the temperature of the steam increases.

4.7.5 Concluding remarks

This case study integrates a trigeneration system for DME, electricity production and seawater desalination. The integrated system mainly utilizes the turbine exhaust heat. For providing the electricity required in the Cu—Cl thermochemical cycle, solar energy is used to generate electricity via photovoltaic modules. The carbon dioxide is captured from the gas turbine exhaust after recovering the heat to provide the second component for DME production besides hydrogen. A multi-effect desalination unit is integrated with the heat recovery system as well to provide fresh water for both feeding the Cu—Cl cycle and external use. The system is capable of generating about 1.88 tons of DME per day, generating about 233 MWh of electricity per day and producing about 788 tons of fresh water per day. The

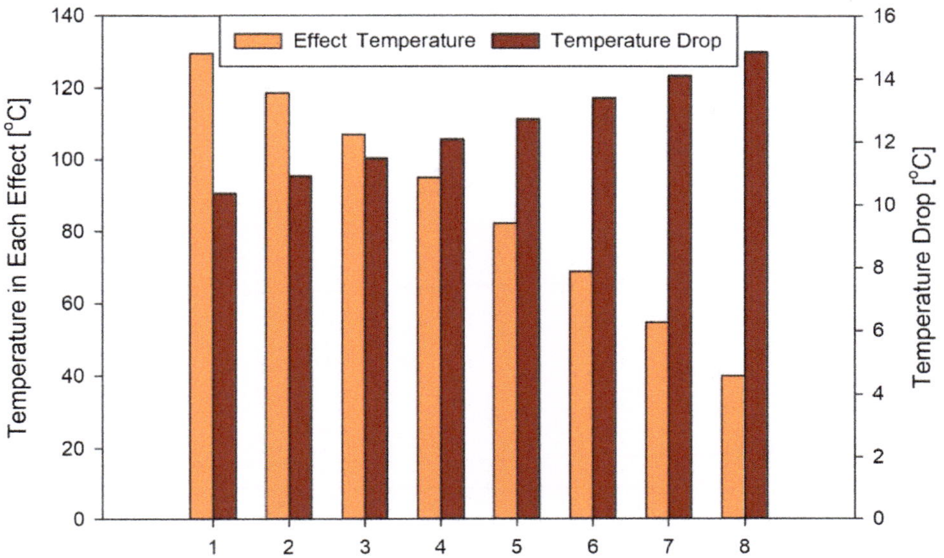

FIG. 4.37 Temperature and temperature drop in each effect of MED plant. *(Adapted from DinAli MN. Development and assessment of renewable energy based integrated systems for dimethyl ether production. Master's Thesis. University of Ontario Institute of Technology, 2018).*

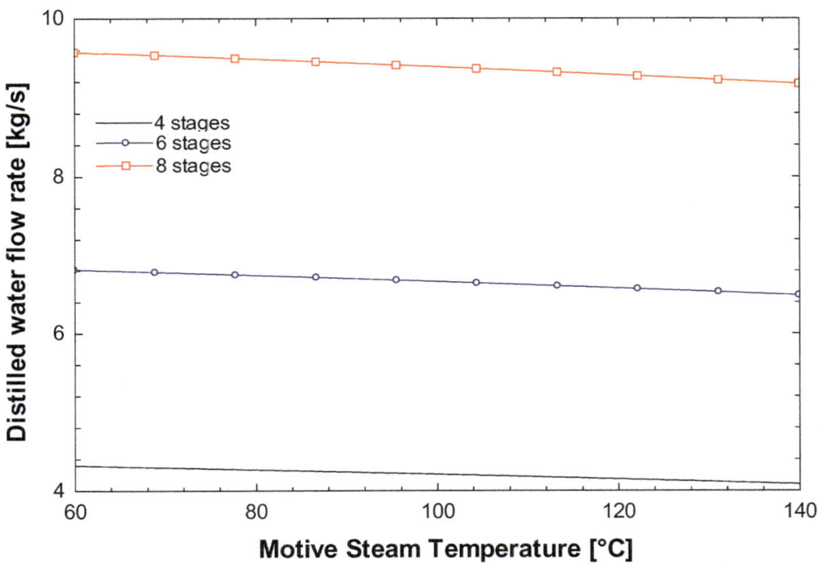

FIG. 4.38 Effect of motive steam temperature on the distilled water flowrate at different number of stages. *(Adapted from DinAli MN. Development and assessment of renewable energy based integrated systems for dimethyl ether production. Master's Thesis. University of Ontario Institute of Technology, 2018).*

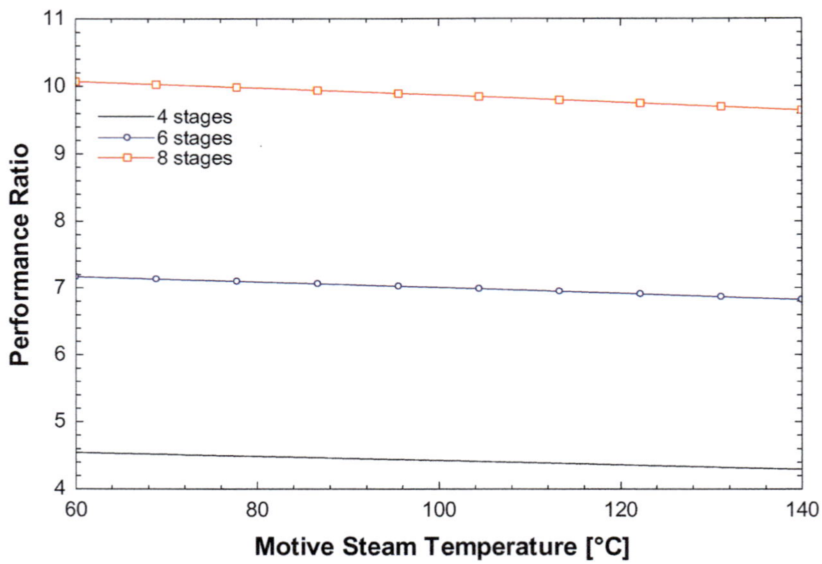

FIG. 4.39 Effect of increasing motive steam temperature on performance ratio at different number of stages. *(Adapted from DinAli MN. Development and assessment of renewable energy based integrated systems for dimethyl ether production. Master's Thesis. University of Ontario Institute of Technology, 2018).*

overall energy efficiency of the trigeneration system is calculated to be about 40%, whereas the overall exergy efficiency is calculated as 55%.

4.8 Case study 4: Integrated solid oxide fuel cell (SOFC) and coal gasification system for cogeneration

This case study examines an integrated gasification and solid oxide fuel cell (SOFC) system with a gas turbine and steam cycle that uses heat recovery of the gas turbine exhaust. The integrated coal system is based on the production of syngas, which is utilized in solid oxide fuel cells (SOFC). The fuel output from the fuel cells is combined with natural gas, and then directed to the combustion chamber of the gas turbine cycle as the bottoming cycle. Heat recovery is utilized to generate the steam for a steam turbine with the release of heat from the exhaust gases of the gas turbine. The required inputs, systems and useful outputs are illustrated in Fig. 4.40. This case study is adapted from El-Emam et al. [26].

4.8.1 System description

In this case study, a modified combined cogeneration system is developed based on the system studied by Ghosh and De [32]. The system entails a gasification unit for coal and an SOFC unit in combination with gas and steam cycles. The effects of the various parameters are studied for a specific type of coal as a fuel source for a gasifier. The properties of coal type

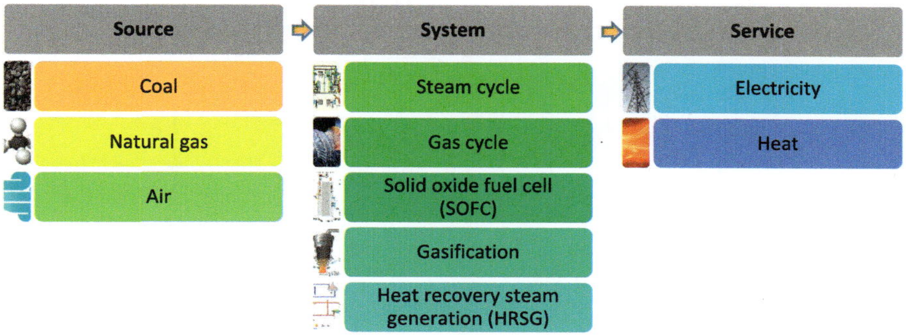

FIG. 4.40 3S approach for case study 4.

TABLE 4.8 Heating values of the coal type in this case study.

Heating value	kJ/kg
LHV	21,920
HHV	23,200

Source: *U.S. Department of Energy. Fuel Cell Handbook. 7th ed. Honolulu, Hawaii: University Press of the Pacific, 2005; Miller BG. Coal energy systems. New York: Elsevier; 2005.*

are shown in Table 4.8. Fig. 4.41 shows a schematic diagram of the combined cogeneration power plant.

Atmospheric air arrives at the air compressor. The compressor provides compressed air to the gasifier, SOFC and the combustion chamber of the gas turbine cycle. Oxygen is separated and sent to the gasifier, and some of the air is sent to the air separation plant. The syngas formed from the gasification process is then cooled, purified and guided to supply the SOFC anode side. Preheated water from the heat recovery steam generator (HRSG) is utilized to cool the syngas. Superheated steam from the cooler is sent to the steam turbine to generate electricity. The compressed air enters the cathode of the fuel cell after being heated via a heat exchanger using the exit of the fuel cell. Outputs from anode and cathode are mixed together and sent into the bottoming cycle for the combustion chamber. The exhaust of the gas turbine is recovered in the HRSG for providing steam to the steam turbine. The additional heat recovered is used to produce saturated steam at low pressure. The coal composition (ultimate analysis) is shown in Fig. 4.42.

4.8.2 Assumptions

In order to perform thermodynamic analysis of this multigeneration system, there is a need to make conceptually correct assumptions. Here, the main assumptions undertaken in this case study are listed as follows:

- Atmospheric air is considered to be an ideal gas with the composition of 21% oxygen and 79% nitrogen.

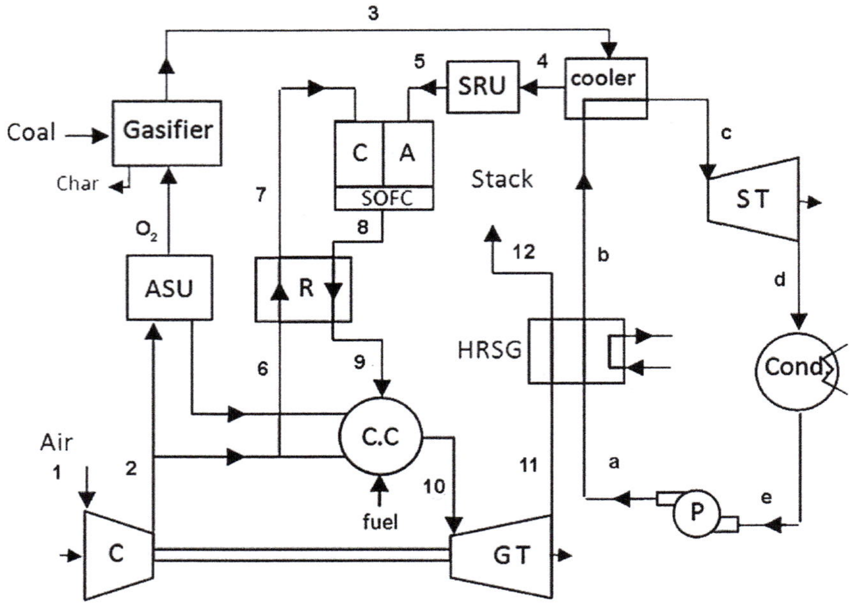

FIG. 4.41 A schematic diagram of the modified integrated system. *(Adapted from El-Emam RS, Dincer I, Naterer GF. Energy and exergy analyses of an integrated SOFC and coal gasification system. Int J Hydrogen Energy 2012;37:1689–97. https://doi.org/10.1016/j.ijhydene.2011.09.139).*

FIG. 4.42 Ultimate analysis of the coal types in this case study. *(Data from [33, 34]).*

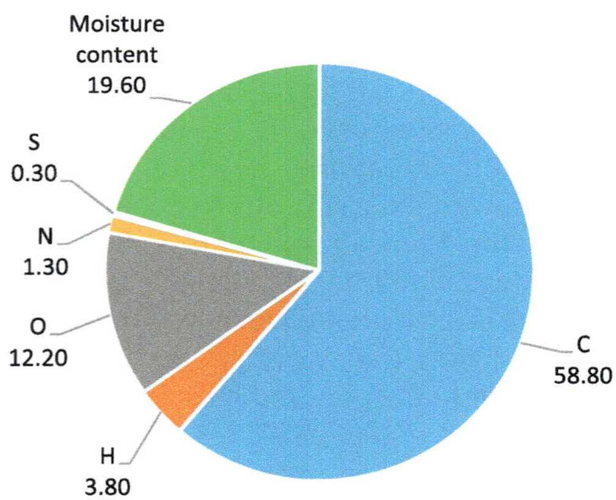

- All flow steams are assumed to be ideal gases.
- There are no pressure losses in the pipelines.
- Ambient temperature and pressure are constant.
- A steady state steady flow operation is assumed.
- Heat exchangers, compressors and the turbine are assumed adiabatic.
- The SOFC is presumed to be internally reforming, and the reforming and shifting reactions maintain equilibrium conditions.

- In the analysis of the solid oxide fuel cell, it is assumed to be insulated [35].
- The pressure drop across the fuel cell is ignored.
- The gas mixture leaves the fuel cell at a chemical equilibrium condition.
- The fuel cell is insulated and there is no heat exchange with the surroundings.
- The mixed gas exiting from the fuel cell is directed to the combustion chamber after it flows through the recuperator.
- The natural gas has the following composition: 93.9% CH_4, 3.2% C_2H_6, 1.1% C_3H_8, 1% CO_2 and 0.8% N_2 [33].
- The separated nitrogen from the air separation unit is also mixed in the combustion chamber for restraining the temperature at the gas turbine inlet to 1250 °C, and to aid in reducing the formation of nitrous oxides.

In addition, the parameters shown in Table 4.9 are the assumed as input conditions.

4.8.3 Analysis

The energy efficiency can be written as a ratio of the produced power to the lower heating value of the fuel

$$\eta_{en} = \frac{\dot{W}}{\dot{m}_f \, LHV_{fuel}} \tag{4.45}$$

The net power production from the integrated system can be defined as

$$\dot{W}_{net} = \dot{W}_{SOFC} + \dot{W}_{GT,net} + \dot{W}_{ST} - \dot{W}_{pump} \tag{4.46}$$

TABLE 4.9 Performance data for the system operation in the base case.

Assumed performance parameters	
Pressure ratio of compressor	15
Inlet temperature of gas turbine (K)	1450
Operating temperature of SOFC (K)	1273
Inlet temperature of steam turbine (K)	823
Inlet pressure of steam turbine (bar)	80
Feeding rate of coal (tons/h)	6.69
Isentropic efficiency of gas turbine (%)	85
Isentropic efficiency of compressor (%)	85
Isentropic efficiency of pump (%)	96
Atmospheric air; O_2 and N_2 (% vol.)	21 and 79

Source: *El-Emam RS, Dincer I, Naterer GF. Energy and exergy analyses of an integrated SOFC and coal gasification system. Int J Hydrogen Energy 2012;37:1689–97. https://doi.org/10.1016/j.ijhydene.2011.09.139.*

The exergy efficiency for each component is determined using the total exergy output to the total exergy input as

$$\eta_{ex} = \frac{\dot{Ex}_{out}}{\dot{Ex}_{in}} \tag{4.47}$$

The total exergy efficiency of the integrated system is determined based on fuel and air exergy as well as total exergy destruction as per the following equation:

$$\eta_{ex} = 1 - \frac{\sum \dot{Ex}_{d,i}}{\sum \dot{Ex}_{fuel} + \dot{Ex}_{air,in}} \tag{4.48}$$

The chemical exergy of the coal is determined based on the following formula, which is used for solid industrial fossil fuels [36]:

$$Ex_{ch,fuel} = (LHV + w\,h_{fg})\,\beta_{dry} + 9{,}417S \tag{4.49}$$

where LHV is the net lower calorific value of the fuel, w is the moisture content of the fuel, h_{fg} is the latent heat of water at T_o and S signifies the mass fraction of sulfur in the fuel. Moreover, β_{dry} can be expressed in terms of the dry organic substances included in the coal as per the following equation [38]:

$$\beta_{dry} = 0.1882\frac{H}{C} + 0.061\frac{O}{C} + 0.0404\frac{N}{C} + 1.0437 \tag{4.50}$$

where C, H, O and N signify the mass fractions of carbon, hydrogen, oxygen and nitrogen in the fuel, respectively. This formula is appropriate with a mass ratio of oxygen to carbon less than 0.667 [37]. The typical syngas composition leaving the gasifier as percentage molar compositions is 10.6% CO_2, 51.6% CO, 0.1% CH_4, 35.1% H_2, and 2.6% N_2 and others [38]. The reaction happens in the gasifier using compressed oxygen split from air in the air separation unit, and it can be written as follows:

$$\begin{aligned}&(y_C C + y_{H_2}H_2 + y_{H_2}H_2 + y_{O_2}O_2 + y_{H_2O}H_2O + y_{N_2}N_2 + y_S S + ash) + X.O_2 \rightarrow\\ &Y.(y_{CO_2}CO_2 + y_{CO}CO + y_{CH_4}CH_4 + y_{H_2}H_2 + y_{N_2}N_2) + y_{H_2O}H_2O + Z.C + ash\end{aligned} \tag{4.51}$$

where y_i is the molar fraction of the composition 'i', and X, Y and Z are the mole fractions of oxygen, syngas and char, which can be computed from the molar balance equations for individual elements in the chemical reaction. The syngas temperature can be defined from the energy balance of the gasifier.

The supply fuel is mixed with circulated gas exiting from the anode side. The steam reforming process of methane and the water gas shift reaction happens in the anode side where the electrochemical reaction also happens. The compressed air is directed to the cathode side, as it acts as an oxidant. The electrons, which are formed at the anode side, are directed to the cathode side and react with the oxygen molecules. The oxide ions generated from this reaction at the cathode side diffuse to the anode through the electrolyte layer.

For the reforming reaction:

$$CH_4 + H_2O \leftrightarrow CO + 3H_2 \tag{4.52}$$

For the shifting reaction:

$$CO + H_2O \leftrightarrow CO_2 + H_2 \tag{4.53}$$

The net electrochemical reaction of the fuel cell is given by:

$$H_2 + \frac{1}{2}O_2 \rightarrow H_2O \tag{4.54}$$

where the equilibrium constants can be defined as a function of the operating temperature.

4.8.4 Results and discussion

Based on an energy analysis of the system for the given coal type, the power generated by the steam turbine is about 5.5 MW. The fuel cell yields about 13 MW of electric power at 55.8% energy efficiency. The gas turbine net work output is about 19 MW. Under the same operational conditions, the overall energy and exergy efficiencies of the system were found as 38.1% and 27%, respectively.

The most important exergy destruction happens in the fuel cell. Substantial exergy destruction can also be observed in the gasifier, combustion chamber and the HRSG as well. Fig. 4.43 displays the effect of varying the reference temperature on the exergy performance of different modules of the system. The performance is calculated by the exergy efficiency of the modules. Three different values of reference temperatures were taken into account, namely 10 °C, 20 °C and 30 °C, respectively.

In Fig. 4.44, the performance of the components is depicted based on the exergy destruction as a percentage of the total exergy destruction within the integrated system, at different reference temperature values. From Fig. 4.44, it is observed that the exergy destruction is influenced by the changes of reference temperature. The exergy destruction rises with temperature for the gasifier, combustion chamber and HRSG, though it declines for the fuel cell.

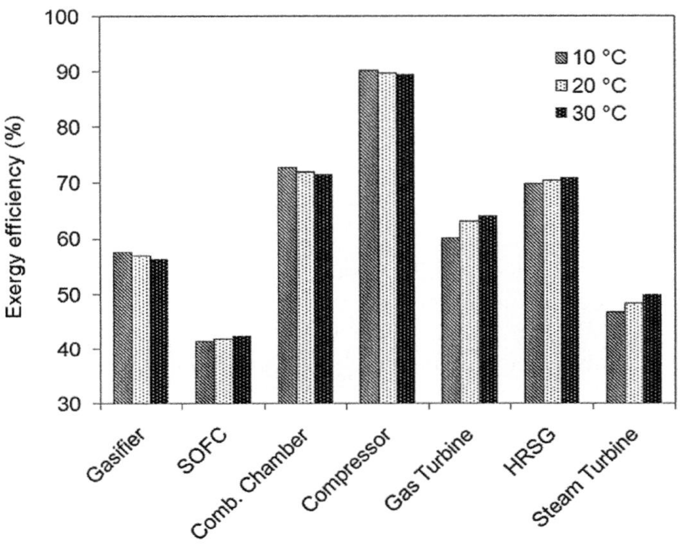

FIG. 4.43 Exergy efficiency of the components within the integrated system. *(Adapted from El-Emam RS, Dincer I, Naterer GF. Energy and exergy analyses of an integrated SOFC and coal gasification system. Int J Hydrogen Energy 2012;37:1689–97. https://doi.org/10.1016/j.ijhydene.2011.09.139).*

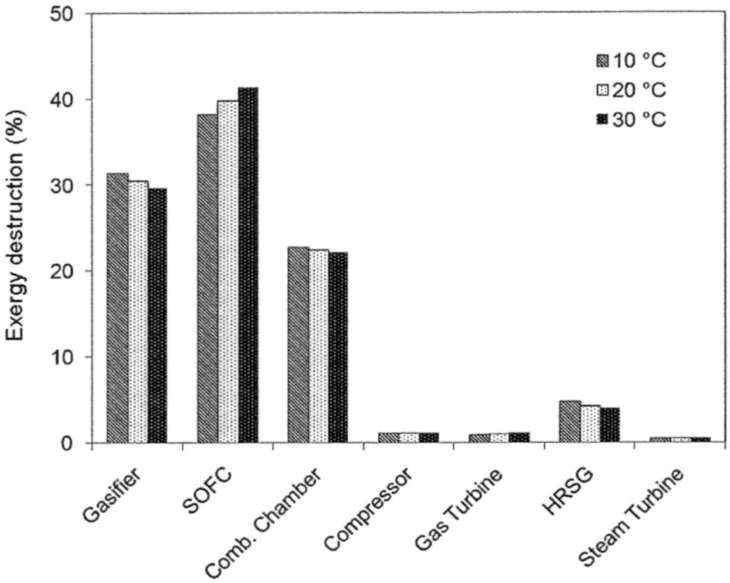

FIG. 4.44 Exergy destruction ratios of the system components as a percentage of total exergy destruction. *(Adapted from El-Emam RS, Dincer I, Naterer GF. Energy and exergy analyses of an integrated SOFC and coal gasification system. Int J Hydrogen Energy 2012;37:1689–97. https://doi.org/10.1016/j.ijhydene.2011.09.139).*

Figs. 4.45–4.47 explain the effects of varying the pressure ratio of the gas turbine compressor on the exergy efficiency as well as exergy destructions of the gasifier, fuel cell and combustion chamber. The exergy destructions are presented as a percentage of the exergy of fuel input. Two different fuel cell operating temperatures are examined for the fuel cell and the combustion chamber. Fig. 4.45 shows that increasing the pressure ratio enhances the exergy

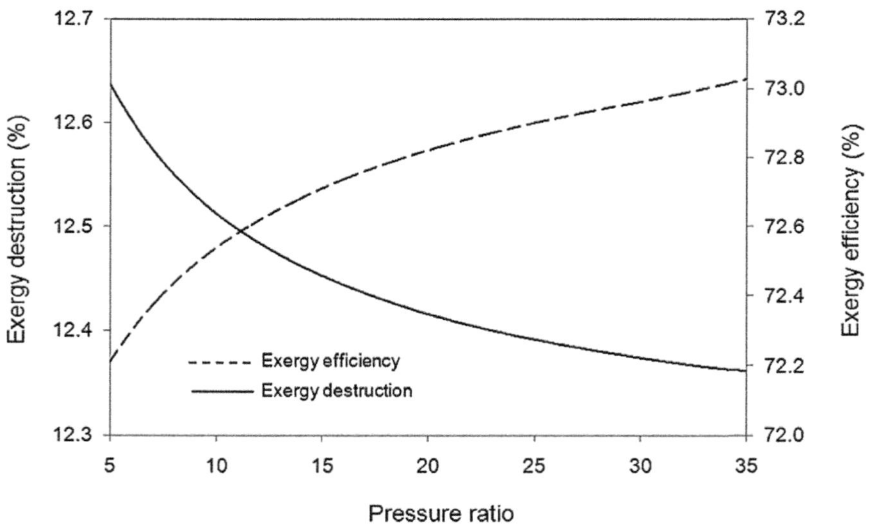

FIG. 4.45 The effects of pressure ratio on the performance of the gasifier and exergy destruction ratio. *(Adapted from El-Emam RS, Dincer I, Naterer GF. Energy and exergy analyses of an integrated SOFC and coal gasification system. Int J Hydrogen Energy 2012;37:1689–97. https://doi.org/10.1016/j.ijhydene.2011.09.139).*

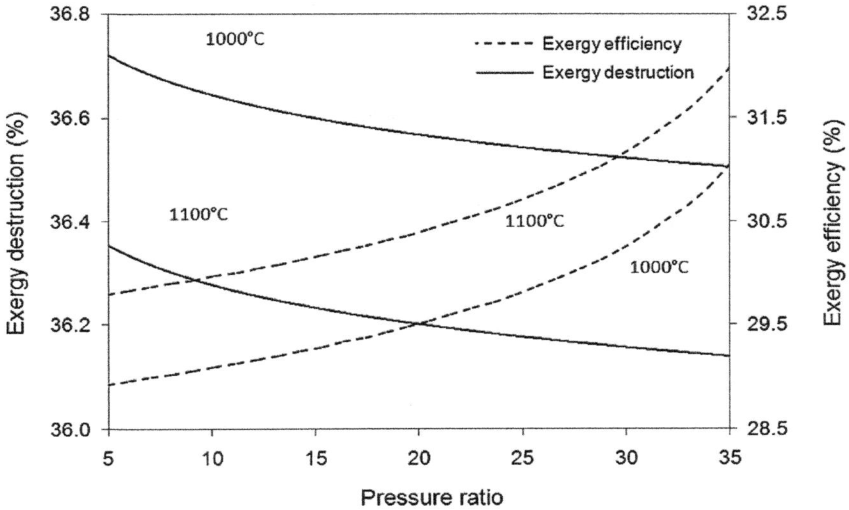

FIG. 4.46 The effects of pressure ratio on the fuel cell performance and exergy destruction ratios at different SOFC temperatures. *(Adapted from El-Emam RS, Dincer I, Naterer GF. Energy and exergy analyses of an integrated SOFC and coal gasification system. Int J Hydrogen Energy 2012;37:1689–97. https://doi.org/10.1016/j.ijhydene.2011.09.139).*

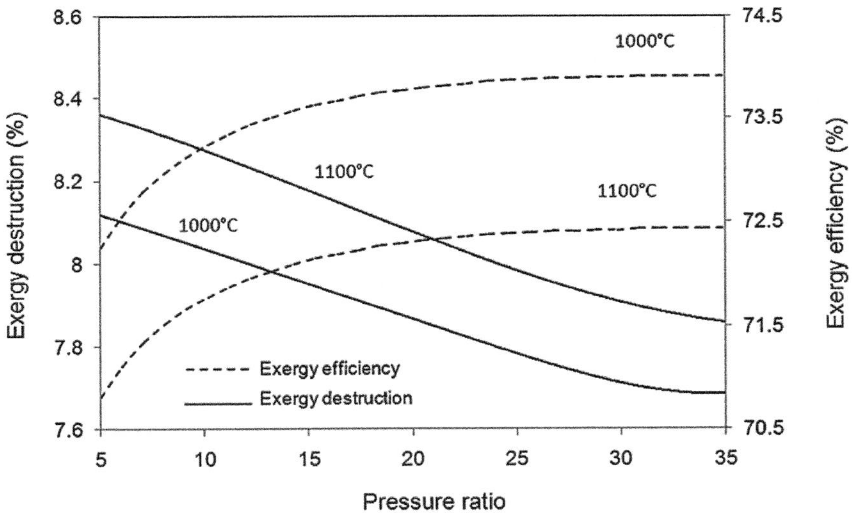

FIG. 4.47 The effects of pressure ratio on the combustion chamber performance and exergy destruction ratio at different SOFC operating temperatures. *(Adapted from El-Emam RS, Dincer I, Naterer GF. Energy and exergy analyses of an integrated SOFC and coal gasification system. Int J Hydrogen Energy 2012;37:1689–97. https://doi.org/10.1016/j.ijhydene.2011.09.139).*

efficiency of the gasifier. The output syngas will be at a higher pressure and temperature, leading to a higher exergy output of the system.

For the SOFC unit as shown in Fig. 4.46, increasing the pressure ratio reduces the exergy destruction. This might be the result of the increase of the fuel cell power with the upsurge of the pressure ratio, causing the exergy destruction to drop and exergy efficiency to rise. Conversely, raising the fuel cell operating temperature reduces the exergy efficiency of the SOFC and hence, the exergy destruction upsurges.

Fig. 4.47 depicts the effects of pressure ratio variation on the combustion chamber. Raising the pressure ratio causes a reduction in the combustible gases in the exhaust of the fuel cell, which is directed to the combustion chamber. This infers a decrease in the extent of combustion, and reduction in the exergy destruction of the combustion chamber. Nevertheless, after a pressure ratio value of 25, the improvements of performance become lower, because of mixing with natural gas fuel in the combustion chamber. Once the fuel cell temperature rises, it means an increase of the SOFC exhaust gas flow, which is supplied to the combustion chamber. This leads to higher exergy destruction in the combustion chamber.

4.8.5 Concluding remarks

This case study demonstrated thermodynamic analyses of an integrated system having coal gasification and solid oxide fuel cell systems. The overall energy and exergy efficiencies of the system are calculated as 38.1% and 27%, respectively. The effects of changing pressure ratio of the gas turbine compressor on different components are also studied. For the gasifier, SOFC, and the combustion chamber, raising the pressure ratio improves the performance of these components. Raising the SOFC operating temperature causes more exergy destruction in the combustion chamber and the SOFC. Three different reference temperature values are applied to study the effect on the exergetic performance of different components.

4.9 Case study 5: Integrated underground coal gasification and steam assisted gravity drainage (SAGD) with SOFC fuel cell system

In this case study, a combined system for syngas, bitumen, hydrogen and power production is developed. This case study is modified from Bicer and Dincer [39, 40]. The proposed system consists of steam assisted gravity drainage (SAGD), underground coal gasification (UCG), solid oxide fuel cell (SOFC), integrated gasification combined cycle (IGCC) and an electrolyzer as illustrated in Fig. 4.48. The developed system proposes important practical benefits for the areas, comprising both coal and oil resources. Coal is gasified using high temperature and pressurized steam in an underground cavity. Formed syngas is used in the IGCC and SOFC systems. By using excess steam in the IGCC, the SAGD process is run and bitumen is removed as an in-situ extraction. Furthermore, a portion of produced power is utilized in an electrolyzer to generate hydrogen.

4.9.1 System description

In this multigeneration system, the useful outputs are bitumen, power, hydrogen and syngas whereas the main inputs are air, water, underground coal and underground high viscous bitumen. A complete diagram of the multigeneration system is depicted in Fig. 4.49.

The system contains a Rankine cycle, hydrogen generation system, Brayton cycle, an underground coal gasifier, SAGD system and fuel cell. The considered multigeneration system is practical for oil sand regions containing both oil sands and coal reserves. Producing the

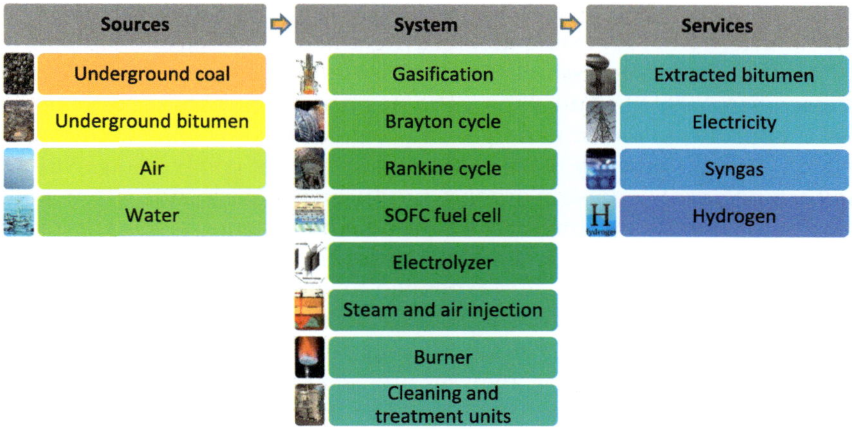

FIG. 4.48 3S approach for case study 5.

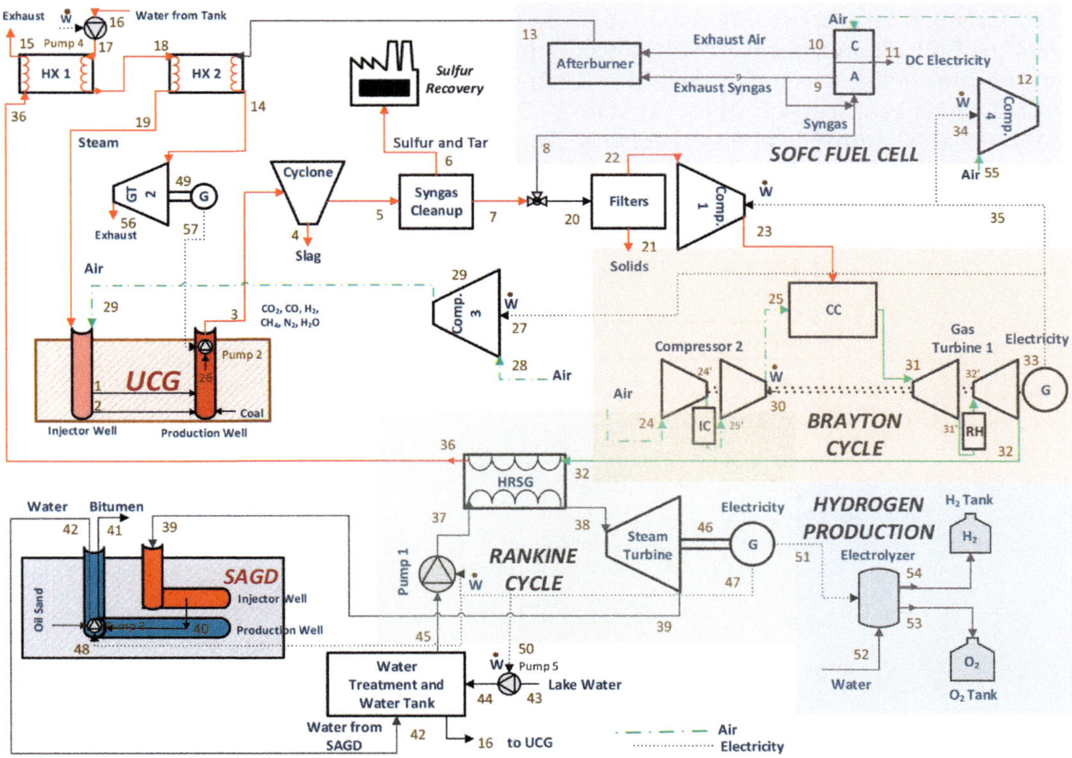

FIG. 4.49 The layout of integration of oil sands into underground coal gasification. *(Modified from Bicer Y, Dincer I. Energy and exergy analyses of an integrated underground coal gasification with SOFC fuel cell system for multigeneration including hydrogen production. Int J Hydrogen Energy 2015;40:13323–37; Bicer Y, Dincer I. Development of a multigeneration system with underground coal gasification integrated to bitumen extraction applications for oil sands. Energ Conver Manage 2015;106:235–48*

steam for SAGD and UCG processes is necessary. In the current unit, steam is produced for UCG through an afterburner for excess air and syngas in the fuel cell and utilizing excess heat after the HRSG unit. Air is likewise delivered to the underground to yield syngas. Compressor 3 compresses air to 2000 kPa and 300 °C. The steam at the inlet of the underground gasifier is about 550 °C and 2400 kPa. The underground coal gasifier has one injection well and one production well drilled vertically. Coal is burned by partial combustion with air, oxygen, steam, or any arrangement of these in order to produce a syngas in-situ. Via an injection well, either oxygen (or air) and steam are introduced to the underground layer. The injection well is linked to the production well with a horizontal well. Nearby the production well, there is a burner to ignite the reaction. The coal is expended towards the production well and an underground cavity is shaped. No ash or slag removal or special handling process is necessary, as sluggish material mainly remains in the underground cavities.

The features of coal utilized in the analyses are listed in Table 4.10. The coal feed rate is assigned to be 30 kg/s at the design conditions. The Illinois 6 type coal, which has an LHV of 25,088 kJ/kg, is used as formerly analyzed in [41, 42]. It has approximately 63% of carbon content. The produced syngas has an LHV of 13,918 kJ/kg according to literature studies [42, 43]. The molar fractions and flow rates of syngas are listed in Table 4.11. While 27% of generated syngas is H_2, almost 43% is CO. The volumetric flow rate of generated syngas is about 10.6 m^3/s. Both air and steam are introduced into the production well and syngas is obtained from the production well with the aid of high pressure inside the cavity. Then, syngas is diverted to syngas clean-up units. At state 3, the syngas outlet temperature and pressure are about 550 °C and 1500 kPa, respectively. The syngas cleaning processes are not detailed in this case study but reasonable assumptions are used based on literature studies as mentioned in the assumptions. Desulfurization temperatures from 400 °C to 650 °C do not affect

TABLE 4.10 Illinois 6 coal properties used in the analysis.

Property	Value
Moisture	0.112
LHV (kJ/kg)	25,088
HHV (kJ/kg)	27,130
C	0.6375
H	0.045
O	0.0688
N	0.0125
S	0.0251
Ash	0.097

Source: *Data from Herdem MS, Farhad S, Dincer I, Hamdullahpur F. Thermodynamic modeling and assessment of a combined coal gasification and alkaline water electrolysis system for hydrogen production. Int J Hydrogen Energy 2014;39:3061–71. doi:10.1016/j.ijhydene.2013.12.068; Ozturk M, Dincer I. Thermodynamic assessment of an integrated solar power tower and coal gasification system for multi-generation purposes. Energ Conver Manage 2013;76:1061–72. https://doi.org/10.1016/j.enconman.2013.08.061.*

TABLE 4.11 The compositions and main properties of the syngas used in the analysis.

Syngas	Molar fraction
H_2	0.2742
CO	0.4331
CO_2	0.0374
N_2	0.1761
H_2O	0.0505
CH_4	0.0216
Mass flow (kg/s)	47.67
Volume flow (m^3/s)	10.6
LHV (kJ/kg)	13,918

Source: *Data from Giuffrida A, Romano MC, Lozza GG. Thermodynamic assessment of IGCC power plants with hot fuel gas desulfurization. Appl Energy 2010;87:3374–83. https://doi.org/10.1016/J.APENERGY.2010.05.020.*

the efficiency remarkably even though power output slightly decreases [44]. Thus, a hot syngas clean-up process is deliberated. After hot syngas is obtained above ground, a cyclone and syngas clean up unit is employed to eliminate particles, sulfur and tar. It is presumed that the syngas temperature drops 5 °C and there are pressure loses at each step of syngas cleaning. Half of the syngas is detached for the fuel cell system, and the other half is directed to the compressor to be used in the Brayton cycle. In comparison with the standard natural gas turbines, in order to evade stall and surge at the compressor and have the same turbine inlet pressure, the air flow needs to be adjusted by variable guide vane settings [44]. The syngas incoming to the combustion chamber is at a pressure of about 2200 kPa and a temperature of about 700 °C. The air needed for the Brayton cycle is compressed by Compressor 2 up to 2200 kPa. The air Brayton cycle has 2 stage compression and expansion with intercooling and reheating. The power is produced via gas turbine. Some portion of the produced power is consumed by the compressors, pumps and auxiliaries. The combustion gases are directed to the HRSG unit and then employed for necessary steam in the UCG. The input temperature and pressure of the steam turbine are about 570 °C and 3500 kPa, respectively. The steam exits the turbine at 380 °C and 1000 kPa, which will be utilized in the SAGD plant. After reducing the viscosity of bitumen underground, bitumen is extracted through a pump arrangement. Almost 80% of water utilized in the SAGD can be recovered. The recovered water is treated and diverted back to the Rankine cycle. Make up water can be delivered from the lake. The required water for electrolysis and the UCG process is stored in water tanks. The volumetric flow rate of bitumen is 0.1591 m^3/s and the LHV of bitumen is about 40,600 kJ/kg. Produced power from the Rankine cycle is consumed by Pump 3 and Pump 5. The remaining power is delivered to the electrolyzer for hydrogen generation. Half of the syngas, which was divided after production, is used in the SOFC. The syngas temperature at the input of the SOFC is about 540 °C. The air inlet is 540 °C and 1300 kPa. The operating temperature of the SOFC is set to 600 °C. Exhaust air and syngas is burned in an afterburner to be utilized in the steam production and excess combustion gases with a temperature of 565 °C is utilized in gas turbine 2 to produce power.

4.9.2 Assumptions

In order to perform thermodynamic analysis of this multigeneration system, there is a need to make conceptually correct assumptions. Here, the main assumptions undertaken in this case study are listed as follows:

- The system operates at steady state.
- The kinetic and potential energy changes during the processes are ignored.
- Ideal gas principles apply for the gases for Brayton cycle.
- There is no pressure loss in the heat exchangers.
- Air is treated as an ideal gas with constant specific heat for the Brayton cycle.
- Compressors are adiabatic.

For the SOFC and steam generation system units, the following assumptions are made:

- The SOFC is internally reforming, and the reforming and shifting reactions sustain equilibrium conditions.
- Pressure drops along the fuel cell are ignored.
- Full combustion happens in the combustor.
- The temperatures at the channel inlets are the same. Furthermore, the temperatures at the channel exits are identical.
- The fuel cell is insulated, and there is no thermal interaction with the environment.
- Contact resistances are neglected. Radiation transfer between solid structure and gas channels is neglected.
- The gas mixture at the fuel channel exit is at chemical equilibrium. Only hydrogen is electrochemically reacted. CO is converted to CO_2 and H_2 by water gas shift reaction.

For the SAGD and Rankine cycle, the following assumptions are employed:

- The steam turbine is adiabatic.
- Bitumen is considered as asphalt in the calculations.
- 80% of water used in the SAGD process is recovered.

For underground coal gasification (UCG), the following assumptions are employed:

- Nitrogen does not react within the gasifier and the syngas produced by the UCG is in chemical equilibrium.
- Syngas consists of the following gas species: CH_4, CO_2, CO, H_2O, H_2, H_2S and N_2. Nitrogen within the product gas is introduced entirely through air injection.
- The syngas mass flow rate is estimated using the mass fraction and the flow rate of nitrogen in the syngas.
- Waste of syngas cleaning filter and cyclone is assumed as solid soil.
- Since removal unit analyses are not detailed, sulfur is assumed as benzene and tar is assumed as H_2S corresponding to 1% of total syngas mass flow rate [45, 46].
- There are no pressure losses in the combustion chamber.

4.9.3 Analysis

The Brayton cycle exergy efficiency is net power output from gas turbine over net exergetic heat input:

$$\eta_{ex,BC} = \frac{\dot{W}_{out,GT,1} - \dot{W}_{in,C2}}{\dot{E}x^{Q_{CC}}} \tag{4.55}$$

The combustion heat is calculated using

$$\dot{Q}_{CC} = \dot{m}_{23} LHV_{syngas} \eta_{comb} \tag{4.56}$$

The exergetic content of heat is calculated as follows:

$$\dot{E}x^{Q_{CC}} = \dot{Q}_{cc}\left(1 - \frac{T_0}{T_{31}}\right) \tag{4.57}$$

Here, $\dot{Q}_{CC}$ is thermal energy and $Ex^{\dot{Q}_{CC}}$ is the thermal exergy delivered to the Brayton cycle. The combustion efficiency is assumed as 90%. The turbine inlet temperature at state 31 is 1330 °C.

The energy efficiency of the Brayton cycle is defined as

$$\eta_{en,BC} = \frac{\dot{W}_{out,GT,1} - \dot{W}_{in,C2}}{\dot{Q}_{CC}} \tag{4.58}$$

The SAGD process efficiency is the exergy difference between extracted bitumen and oil sand over net exergy input by steam as follows:

$$\eta_{ex,SAGD} = \frac{m_{41}\, ex_{41} - m_{oilsand}\, ex_{oilsand}}{m_{39}\, ex_{39} - m_{42}\, ex_{42}} \tag{4.59}$$

Similarly, the energy efficiency of the SAGD process can be written as

$$\eta_{en,SAGD} = \frac{m_{41}\, h_{41} - m_{oilsand}\, h_{oilsand}}{m_{39}\, h_{39} - m_{42}\, h_{42}} \tag{4.60}$$

where $h_{oilsand}$ is the enthalpy of bitumen underground.

The Rankine cycle exergy efficiency is the ratio of net work generated by turbine over total exergy input as follows:

$$\eta_{ex,RC} = \frac{\dot{W}_{out,ST} - \dot{W}_{in,P1}}{(\dot{m}_{32}(ex_{32} - ex_{36})) - (\dot{m}_{38}(ex_{38} - ex_{37}))} \tag{4.61}$$

Similarly, the energy efficiency of Rankine cycle is written as

$$\eta_{en,RC} = \frac{\dot{W}_{out,ST} - \dot{W}_{in,P1}}{(\dot{m}_{32}(h_{32} - h_{36})) - (\dot{m}_{38}(h_{38} - h_{37}))} \tag{4.62}$$

The electrolysis reaction is the opposite of the water formation reaction:

$$H_2O + \text{Electrical Energy} \rightarrow H_2 + \frac{1}{2}O_2 \tag{4.63}$$

The exergy content of produced hydrogen is calculated as follows:

$$\dot{E}x_{H_2} = \dot{m}_{H_2}\left(ex_{H_2,ch} + ex_{H_2,ph}\right) \tag{4.64}$$

where $ex_{H_2,ch} = \frac{236.1 \times 1000}{MW_{H_2}}$ where 236.1 kJ/g mole is taken as the exergy content of hydrogen using standard chemical exergy values of common substances [37] and MW_{H_2} is the molar mass of hydrogen in kg/kmol.

The efficiency of the PEM electrolyzer is taken to be 70% [47–49] as per the following formula:

$$\eta_{en_{electrolyzer}} = \frac{\dot{m}_{H_2} HHV_{H_2}}{\dot{W}_{electrolyzer}} \tag{4.65}$$

where $\dot{W}_{electrolyzer}$ is received from the Rankine cycle net power output. HHV represents the higher heating value of hydrogen.

The electrolysis efficiency of Rankine cycle-based hydrogen production is written as

$$\eta_{en,elec} = \frac{\dot{m}_{H_2} h_{H_2} + \dot{m}_{O_2} h_{O_2}}{\dot{W}_{electrolyzer}} \tag{4.66}$$

and the exergy efficiency can be expressed as

$$\eta_{ex,elec} = \frac{m_{H_2}\, ex_{H_2} + m_{O_2}\, ex_{O_2}}{\dot{W}_{electrolyzer}} \tag{4.67}$$

The SOFC energy efficiency can be determined as the actual voltage over reversible voltage:

$$\eta_{en,SOFC} = \frac{E}{E_{rev}} \tag{4.68}$$

The SOFC exergy efficiency is total power produced by the SOFC over $-\Delta G$:

$$\eta_{ex,SOFC} = \frac{\dot{W}_{SOFC}}{-\Delta G_{H_2O}} \tag{4.69}$$

For the energy efficiency of the UCG process, the useful output is the produced syngas (LHV value) and the required inputs are coal (LHV value), injected steam and air energy contents:

$$\eta_{en,UCG} = \frac{\dot{m}_{syngas}\, LHV_{syngas}}{\dot{m}_{coal} LHV_{coal} + \dot{m}_{19} h_{19} + \dot{m}_{29} h_{29}} \tag{4.70}$$

The exergy efficiency of the UCG process is determined using exergy content of syngas over exergy content of coal, injected steam and air:

$$\eta_{ex,UCG} = \frac{\dot{m}_{syngas} Ex_{syngas}}{\dot{m}_{coal}\, Ex_{coal} + \dot{m}_{19}\, ex_{19} + \dot{m}_{29}\, ex_{29}} \tag{4.71}$$

where $Ex_{coal} = (LHV + w h_{fg})\, \beta_{dry} + 9{,}417S$.

Here, β_{dry} is defined as follows [45]:

$$\beta_{dry} = 0.1882\frac{H}{C} + 0.061\frac{O}{C} + 0.0404\frac{N}{C} + 1.0437 \tag{4.72}$$

where w is the moisture content of the fuel, h_{fg} is the latent heat of water at T_o and S denotes the mass fraction of sulfur in the fuel and C, H, O and N represent the mass fractions of carbon, hydrogen, oxygen and nitrogen in the fuel, respectively [45].

For the exergy efficiency of IGCC, the useful outputs are net work production from Rankine and Brayton cycles and unused syngas exergy (which is half of the produced amount) while the required input is the exergy content of coal:

$$\eta_{ex,IGCC} = \frac{\dot{W}_{net,BC} + \dot{W}_{net,RC} + \frac{\dot{m}_{syngas}}{2} Ex_{syngas}}{\dot{m}_{coal} Ex_{coal}} \tag{4.73}$$

Similarly, the energy efficiency of the IGCC plant can be written as

$$\eta_{en,IGCC} = \frac{\dot{W}_{net,BC} + \dot{W}_{net,RC} + \frac{\dot{m}_{syngas}}{2} LHV_{syngas}}{\dot{m}_{coal} LHV_{coal}} \tag{4.74}$$

Here, the mass flow rate of syngas is divided by two because half of the syngas is utilized in the IGCC whereas the remaining half is sent to the SOFC.

Considering the main inputs and outputs of the system, the energy and exergy efficiencies of the overall multigeneration system can be written, respectively:

$$\eta_{en,ov,sys} = \frac{\dot{W}_{net,Sys} + \dot{m}_{41}h_{41} + \dot{m}_{hydrogen} LHV_{hydrogen}}{\dot{m}_{43}h_{43} + \dot{m}_{55}h_{55} + \dot{m}_{28}h_{28} + \dot{m}_{52}h_{52} + \dot{m}_{oilsand}h_{oilsand} + \dot{m}_{coal} LHV_{coal}} \tag{4.75}$$

$$\eta_{ex,ov,sys} = \frac{\dot{W}_{net,Sys} + \dot{m}_{41}ex_{41} + \dot{m}_{hydrogen} ex_{hydrogen}}{\dot{m}_{43}ex_{43} + \dot{m}_{55}ex_{55} + \dot{m}_{28}ex_{28} + \dot{m}_{52}ex_{52} + \dot{m}_{oilsand}ex_{oilsand} + \dot{m}_{coal} Ex_{coal}} \tag{4.76}$$

4.9.4 Results and discussion

The maximum exergy destruction rates are obtained in heat exchanger 1 and the water tank and treatment unit shown in Fig. 4.50. The causes for an exergy destruction in the heat exchanger is heat transfer over large temperature differences. Based on the results of the exergy destruction rates and exergy loss ratios, heat exchangers and water treatment units could be enhanced. Hence, minimization of exergy destruction rates is important to allow higher performance and lower costs and emissions. This aids in decreasing the associated environmental impacts. The energy and exergy efficiencies of the electrolysis process are found to be 78.1% and 78.2%, respectively. About 1.6 MW of the produced power from the Rankine cycle is consumed in the electrolysis process for hydrogen generation. The remaining electricity is used by pumps and other auxiliaries.

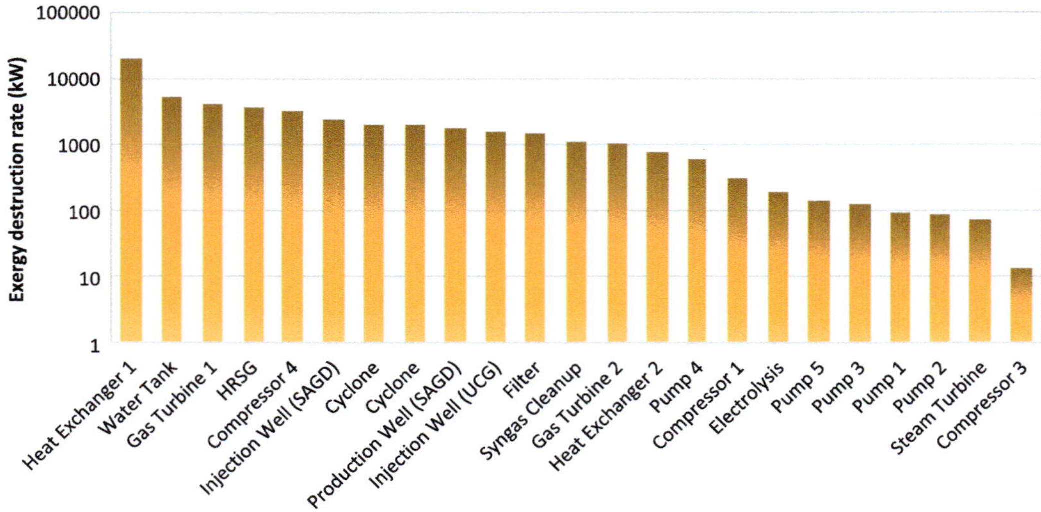

FIG. 4.50 The exergy destruction rates of main components within the multigeneration system.

The overall IGCC system energy efficiency is calculated as 63% even though exergy efficiency is about 27%. The energy and exergy efficiencies of the UCG process are found as 75% and 18% respectively. Even though the UCG process exergy efficiency tends to be low, the exergy efficiency of the IGCC system extends to 27% which is very close to commercial IGCC power plants, meaning that integrated power plants can be more practical and effective. The energy efficiency of the SAGD is found to be 56% whereas the exergy efficiency is about 16%. The exergy destruction rate of the SAGD process is about 4567 kW comprising both production well and injection well. The Brayton cycle energy and exergy efficiencies are found to be 50% and 49%, respectively. The exergy efficiency of the Rankine cycle is found as 32%. The SOFC stack energy efficiency is found as 38% whereas the corresponding exergy efficiency is about 56%. The overall energy and exergy efficiencies of the multigeneration system are found as 19.6% and 17.3% respectively under given conditions at 25 °C and 100 kPa. Furthermore, the exergy efficiencies of all subsystems and components are comparatively shown in Figs. 4.51 and 4.52, respectively. For evaluating the effects of changing parameters on the subsystem and overall efficiencies, the following parametric studies are conducted.

4.9.4.1 *Effect of ambient temperature and mass flow rates*

As revealed in Fig. 4.53 when ambient temperature upsurges from 5 °C to 45 °C, the SAGD process and Rankine cycle exergy efficiencies decline. The reduction rate is lower in the Rankine cycle. This can suggest that system performance will be better in colder sites.

When the water mass flow rate in the Rankine cycle is increased from 5 kg/s to 10 kg/s as presented in Fig. 4.54, the SAGD process efficiency reduces from 17.7% to 8.9%. On the contrary, produced power from the Rankine cycle and injected heat energy through the injection well increase. This can be understood as less water being consumed for the SAGD process to extract bitumen and more water being circulated in the Rankine cycle. Water recovery ratio defines how much water is recovered from the SAGD process. Usually, 80% of water used in

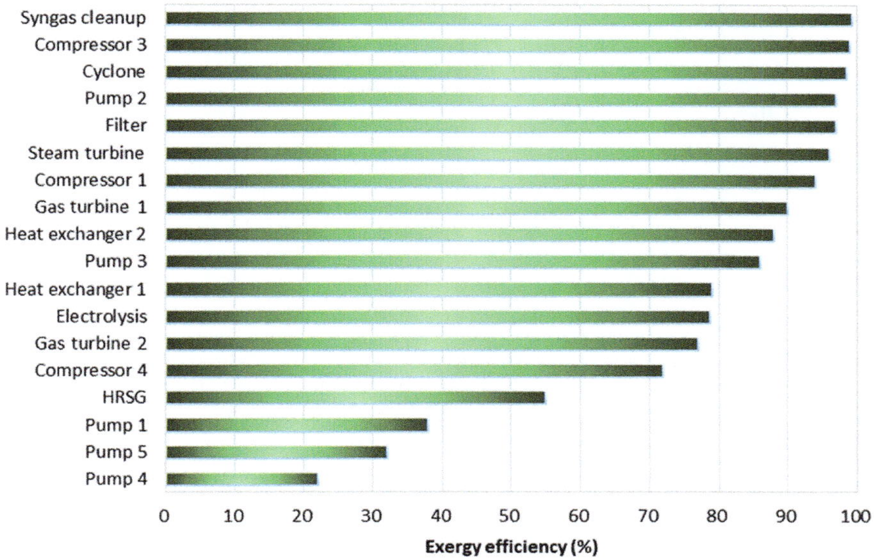

FIG. 4.51 The exergy efficiencies of main components within the multigeneration system.

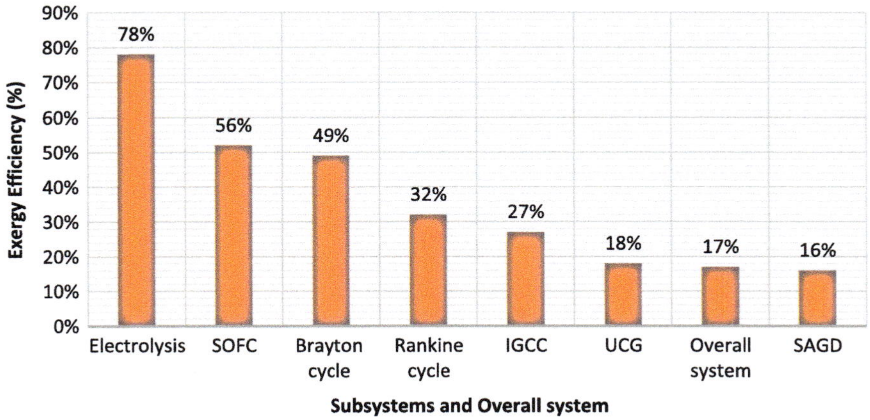

FIG. 4.52 The calculated exergy efficiencies of the subsystems as well as the overall integrated system.

SAGD can be recovered to reproduce steam. When the water recovery ratio rises from 30% to 100%, even though energy efficiency of SAGD upsurges from 54.3% to 57.6%. Nevertheless, the exergy efficiency remains very similar, falling from 16.2% to 16.1% as shown in Fig. 4.55.

4.9.4.2 Effect of syngas mass flow rate and syngas LHV

In Fig. 4.56, as a portion of syngas is consumed in the Brayton cycle, even though produced power upsurges from the increase in syngas mass flow rate, energy and exergy efficiencies of

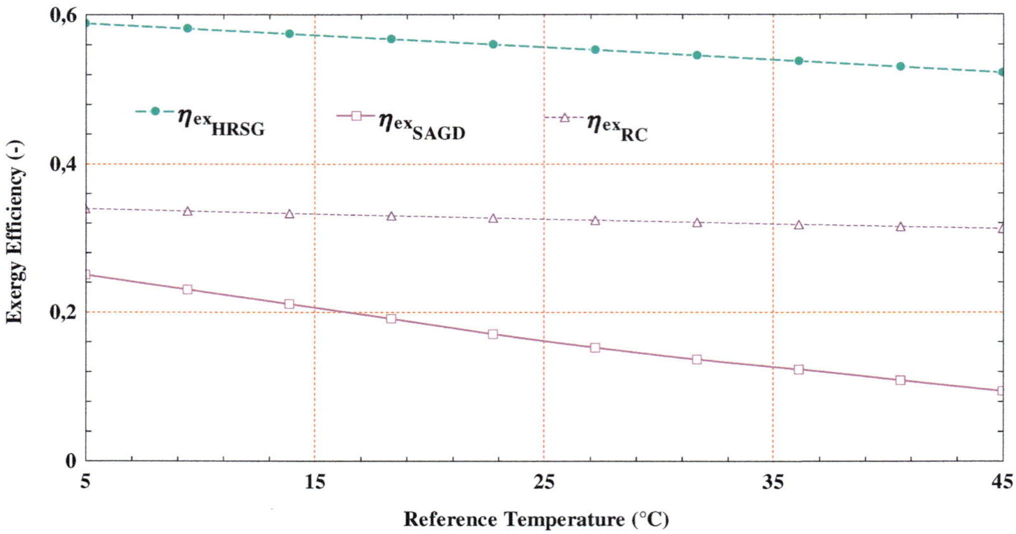

FIG. 4.53 The effects of ambient temperature on the Rankine cycle and HRSG exergy efficiencies.

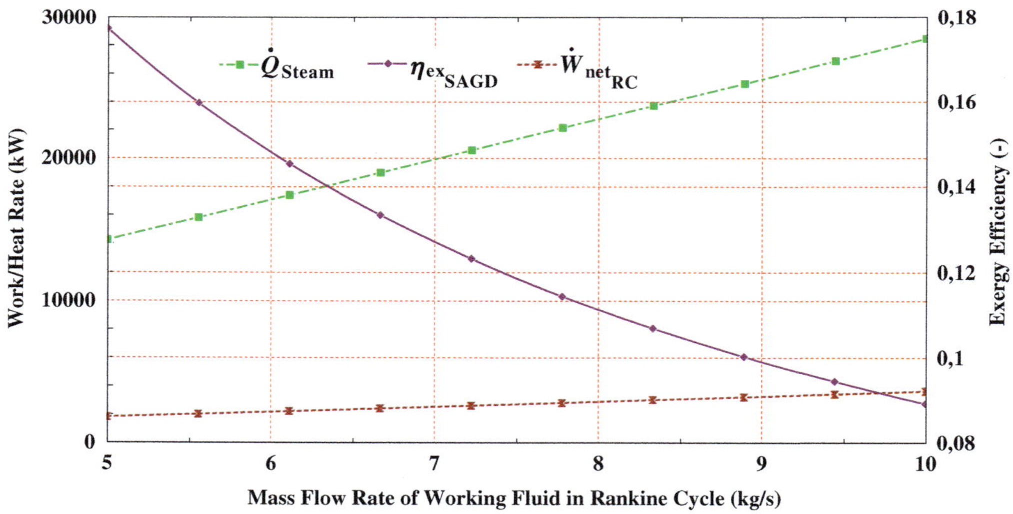

FIG. 4.54 The effects of the water mass flow rate in Rankine cycle on the exergy efficiencies and work/heat rates.

the Brayton cycle decline to about 47.5%, showing that input energy, which is heat in the combustion chamber, upsurges much more than produced power from the gas turbine. Likewise, increasing LHV of formed syngas increased produced power but reduces efficiencies, as depicted in Fig. 4.57. Though LHV upsurges, other parameters, such as coal feed rate and syngas composition, stay the same. Thus, Fig. 4.57 displays the effects of varying LHV on the system performance only when different coal feed rate and syngas compositions are used.

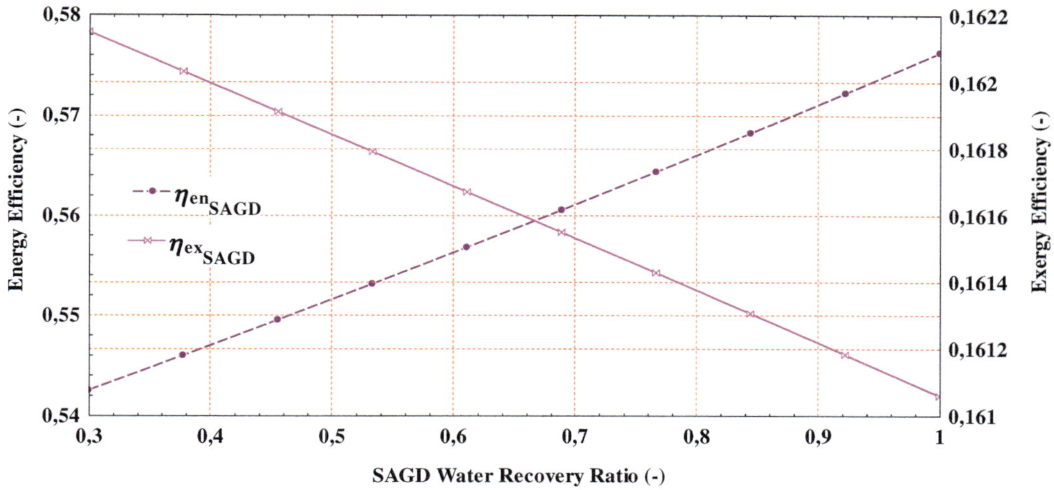

FIG. 4.55 The effects of water recovery ratio on SAGD efficiencies.

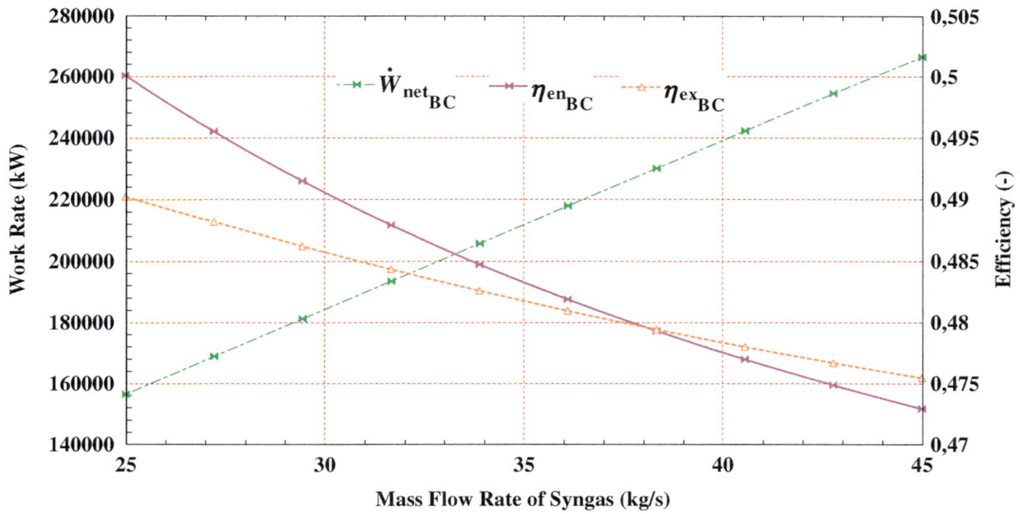

FIG. 4.56 The effects of syngas mass flow rate on the energy and exergy efficiencies of Brayton cycle.

4.9.4.3 Effect of air/fuel pressure and syngas temperature

After syngas is produced from underground coal gasification, the temperature of syngas is important for power generation in SOFC as shown in Fig. 4.58. When syngas temperature rises to 700 °C, SOFC exergy efficiency increases to 70% and generated power reaches 200 kW, since higher operating temperatures give better results in SOFC. Similarly, operating

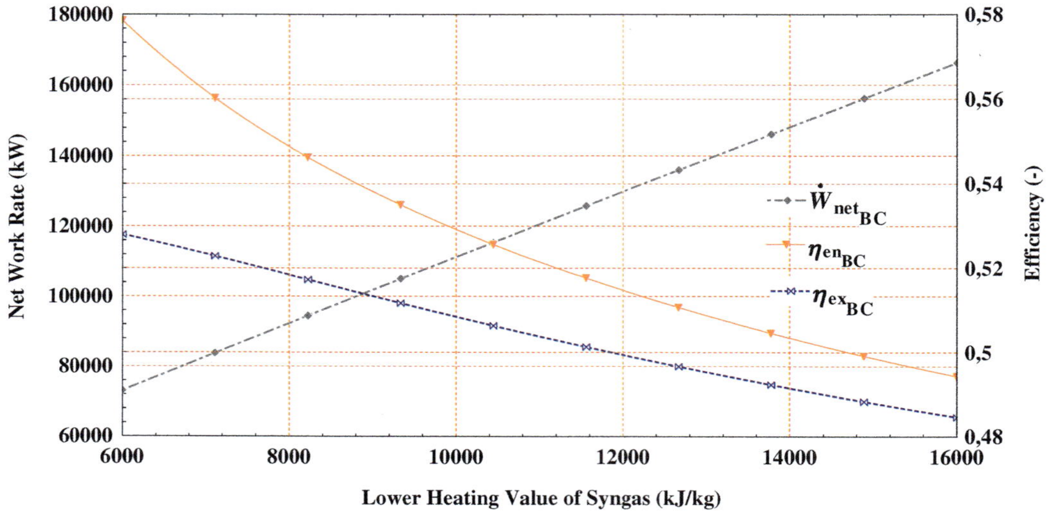

FIG. 4.57 The effects of syngas LHV on the energy and exergy efficiencies of Brayton cycle and net power outputs.

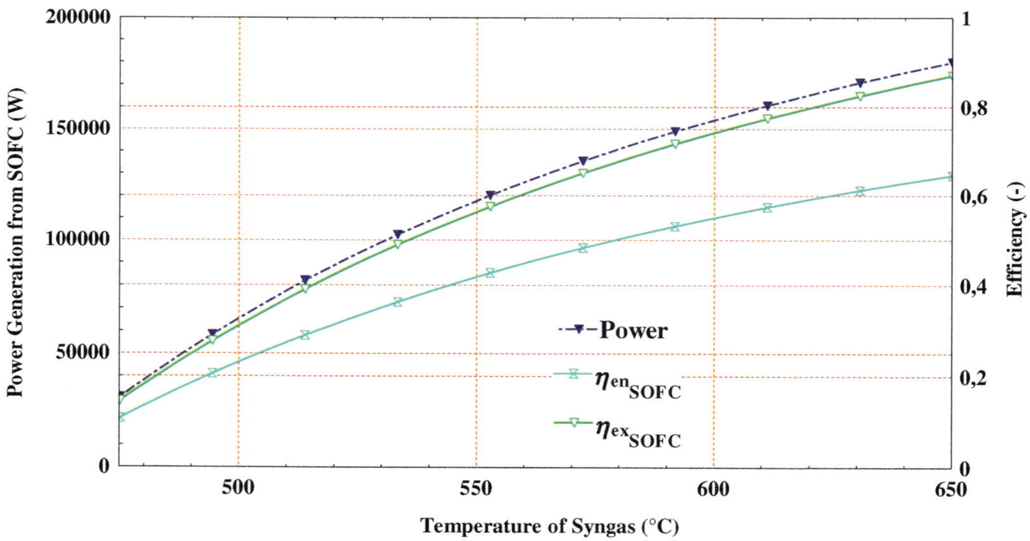

FIG. 4.58 The effects of the syngas temperature on the energy and exergy efficiencies of SOFC and power output.

pressure of SOFC has a positive effect on both energy and exergy efficiency. The exergy efficiency increases to 55% as seen in Fig. 4.59.

4.9.4.4 *Effect of syngas molar fractions*

The CO content of syngas is significant for UCG and electricity production, since efficiency values of UCG and IGCC upsurge particularly as shown in Fig. 4.60. The N_2 quantity in the

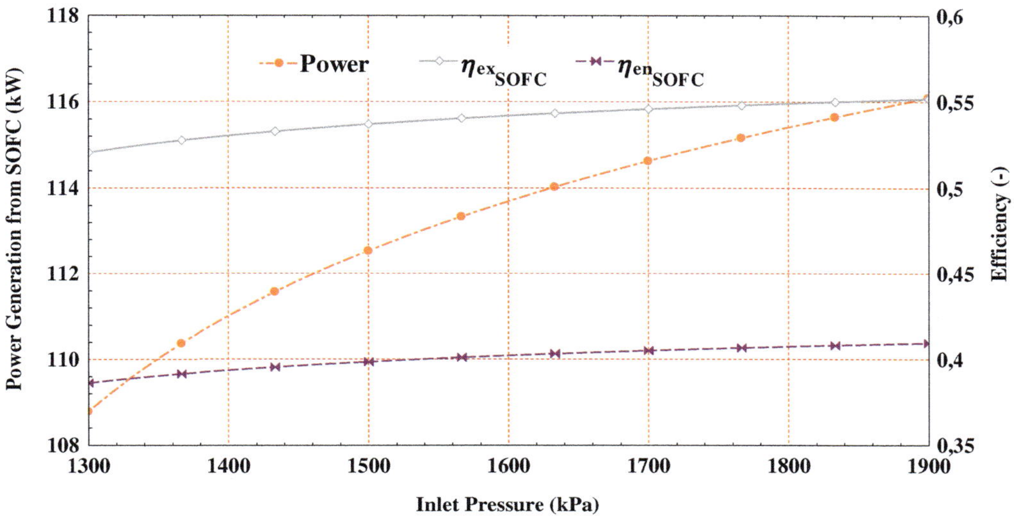

FIG. 4.59 The effects of air and fuel pressures in SOFC on the power output and efficiencies.

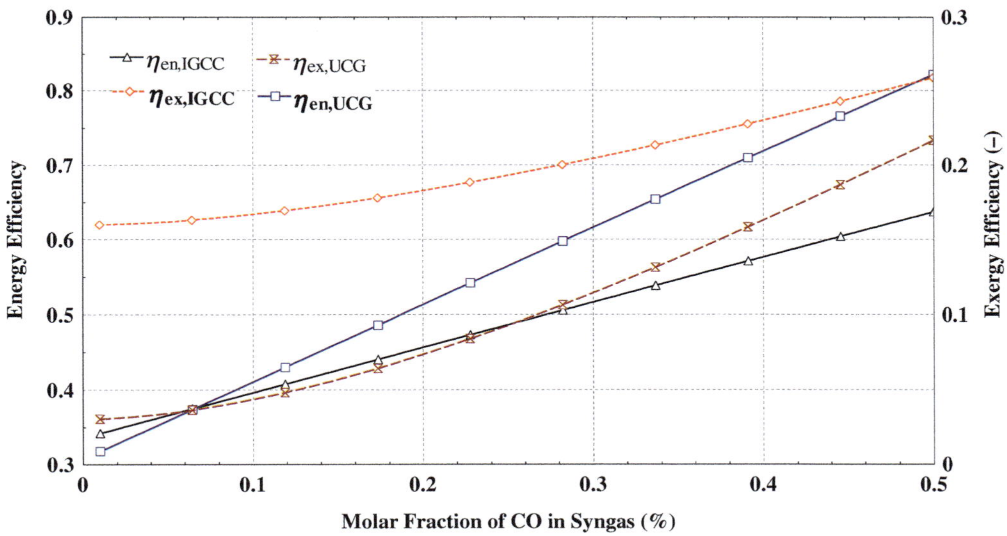

FIG. 4.60 The effects of CO molar fraction in syngas on the energy and exergy efficiencies of IGCC and UCG.

formed syngas is associated with the injected air mass flow rate to the underground gasifier. N_2 does not react in the processes, so N_2 content of syngas is not anticipated; henceforth when N_2 molar fraction upsurges, both IGCC and UCG efficiency drops severely as depicted in Fig. 4.61. This designates probable results of varying air injection mass flow rate and describes that oxygen input would produce greater efficiencies than air input to the gasifier. It is shown

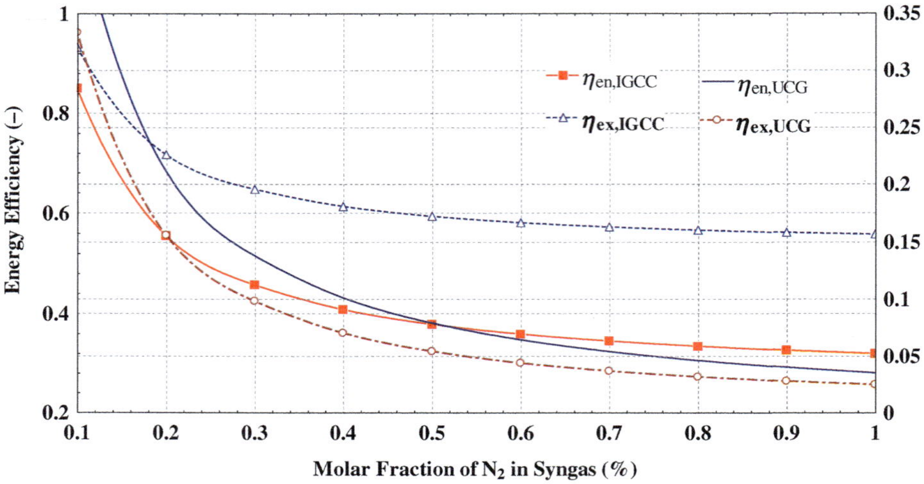

FIG. 4.61 The effects of N_2 molar fraction in syngas on the energy and exergy efficiencies of IGCC and UCG.

that evaluation of exergy efficiency brings more reliable thermodynamic assessment since the exergy rate of the heat varies while transporting it from the reference environment to the working environment.

4.9.4.5 *Effect of the lower heating value of coal and mass flow rate of injected air*

If the LHV of coal upsurges, efficiencies decline as Fig. 4.62 shows, since increased rate of power output is less than the increased rate of input energy. When the mass flow rate of provided air rises from 5 kg/s to 18 kg/s, the energy efficiency of IGCC and UCG processes extends to 62% and 82%, respectively and mass flow rate of generated syngas upsurges linearly

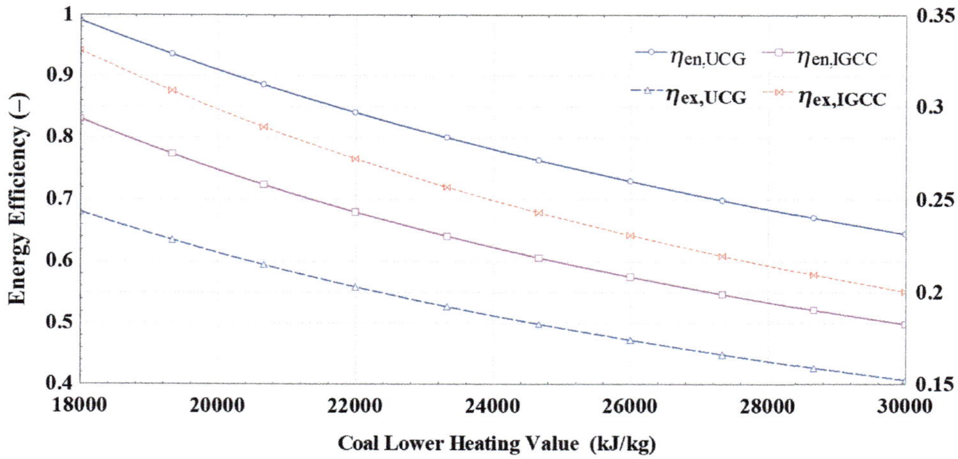

FIG. 4.62 The effects of LHV of coal on the energy and exergy efficiencies of IGCC and UCG.

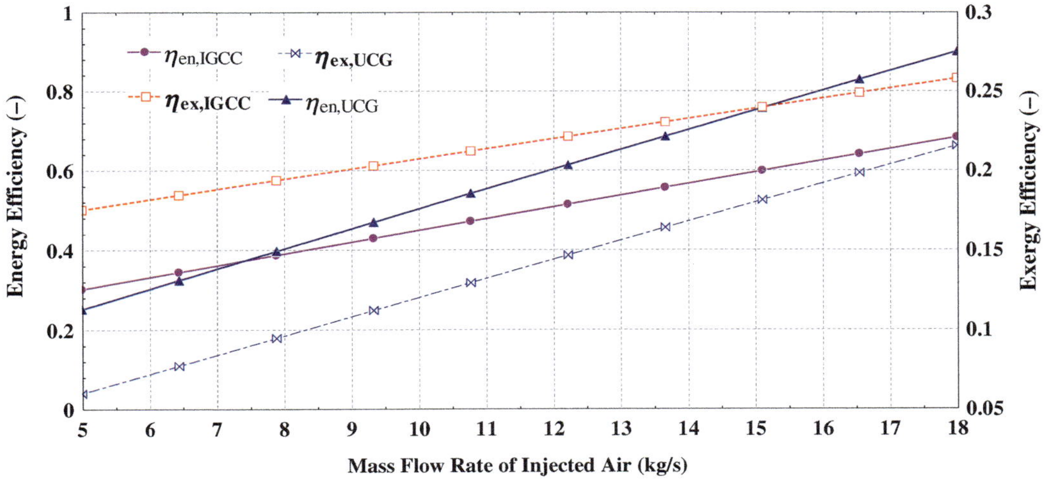

FIG. 4.63 The effects of injected air into UCG on the energy and exergy efficiencies of UCG and IGCC.

to 58 kg/s as depicted in Figs. 4.63 and 4.64. This clarifies that as more air is delivered, more syngas is formed and higher efficiencies are achieved.

Figs. 4.63 and 4.64 signify the parametric study results if the injected air mass flow rate upsurges and other parameters stay constant. The injected mass flow rate may not surpass the selected values in the figures due to constraints in the system.

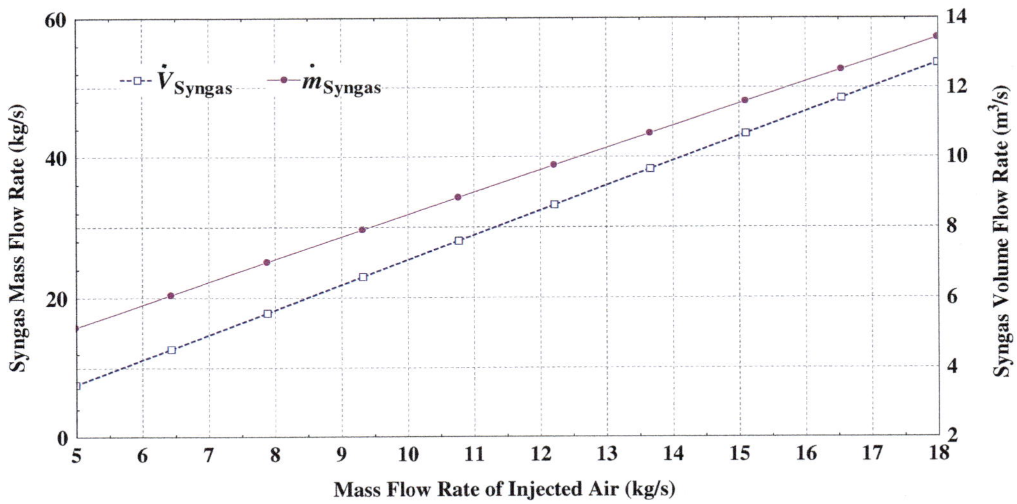

FIG. 4.64 The effects of injected air mass flow rate on the syngas production rates.

4.9.5 Concluding remarks

An integrated system for power, syngas, bitumen and hydrogen generation together with SAGD and UCG systems is designed and thermodynamically analyzed. This multigeneration system highlights the significance of multigeneration systems for the regions containing both oil and coal resources that can be removed by using in-situ technologies. By determination of each state point, component-based irreversibilities are found and their efficiencies comparatively evaluated. The effects of particular working conditions on the subsystems and overall system performance are examined by performing parametric studies. The performance of the system under diverse ambient temperature, pressure and mass flow rate values is observed and the results are analyzed. The energy efficiencies of subsystems are found to be 38% for SOFC, 78% for electrolysis, 50% for the Brayton cycle, 63% for IGCC plant, 75% for the UCG process and 56% for the SAGD process. The energy and exergy efficiencies of the overall system are calculated as 19.6% and 17.3%, respectively under given conditions.

4.10 Conclusions

Most of the energy systems, particularly for power generation, work on conventional systems such as Rankine and Brayton cycles. Although they mostly employ fossil fuels, there are still several methods for improving and greenizing these techniques. This can be accomplished in a couple of ways, such as reducing the wasted energy and integrating with other systems. The potential of improving the conventional systems is noteworthy to study. Hence, there have been many attempts to improve efficiency. A significant option is to employ integrated energy systems, which can include renewables as well as energy storage.

There are several improvement methods to enhance the overall efficiencies of the Rankine cycle, namely (i) lowering the condenser pressure, (ii) achieving high-temperature superheated steam, (iii) increasing the boiler pressure. When the source temperatures are not high enough to obtain superheated steam, other types of working fluids can also be employed in the cycle. In this case, the cycle is called an organic Rankine cycle. A simple open-type Brayton cycle consists of three main components: a compressor, a combustion chamber, and a gas turbine. Since compressing a hotter gas requires more input in a single stage compressor, a multistage compressor having inter-cooling processes can increase the performance of the compressor and cycle. The Kalina cycle is an advanced thermodynamic cycle that can be used for converting thermal energy from a comparatively low-temperature heat source to mechanical energy. The Kalina cycle is principally a modified Rankine cycle, which utilizes the mixture of two different working fluids (water and ammonia). The Stirling engines can function as a work-producing device or a work-consuming device. This brings irreversible operation flexibility. The combination of these cycles can bring higher overall efficiencies as well as multiple outputs in addition to power. A fuel cell system converts chemical energy directly into electricity through an electrochemical process. The similar nature of fuels in the fuel cell and Brayton cycle make them advantageous for system integration. Hence, there are several studies for combined solid oxide fuel cell (SOFC)-gas turbine (GT) cycles.

Combustion of conventional coal has a number of has environmental consequences. Thus, gasification is considered a cleaner approach, mainly yielding syngas with some other impurities. The yielded impurities such as sulfur, tar and particulates can be removed using syngas clean-up systems. The obtained syngas can then be burned in a combustion chamber to run a conventional Brayton cycle, which is therefore called an integrated gasification combined cycle (IGCC).

In this chapter, we presented various case studies starting with compressed air storage integrated to gas turbine and ending with steam assisted gravity drainage integrated to underground coal gasification. Furthermore, several other system integration options are available for conventional energy systems, such as enhanced oil recovery using captured CO_2, bitumen extraction using waste steam, waste gas utilization from oil and gas processing plants, underground coal gasification, flare gas usage in solid oxide fuel cells, co-electrolysis of waste gases, hydrocarbon conversion via renewable sourced-thermal energy and hybridization with renewables.

Nomenclature

$\dot{Ex}$	exergy rate (kW)
$\dot{m}$	mass flow rate (kg/s)
$\dot{Q}$	heat rate (kW)
$\dot{W}$	work rate (kW)
ΔG	Gibbs free energy (J or kJ)
Δt	time difference (s)
Cp	specific heat at constant pressure (kJ/kg)
Cv	specific heat at constant volume (kJ/kg)
E	energy (kJ)
ex	specific exergy (kJ/kg)
Ex	exergy rate (kJ)
g	gravitational acceleration (m/s^2)
h	specific enthalpy (kJ/kg)
m	mass (kg)
MW	molar mass (kg/kmol)
P	pressure (kPa)
s	specific entropy (kJ/kg K)
T	temperature (°C or K)
u	specific internal energy (kJ/kg)
v	specific volume (m^3/kg)
V	velocity of the flow (m/s)
W	work rate (W or kW)
z	elevation from reference point (m)

Acronym

ASU	air separation unit
C	compressor
CAES	compressed air energy storage
CCS	carbon capture storage
COP	coefficient of performance
CSS	cyclic steam stimulation
DME	dimethyl ether
EES	Engineering Equation Solver
FWH	feedwater heater

GOR gain output ratio
HHV higher heating value
HP high pressure
HX heat exchanger
HRSG heat recovery steam generator
IC intercooling
IGCC integrated gasification combined cycle
LCA life cycle analysis
LHV lower heating value
LP low pressure
MED multieffect desalination
PHS pumped hydro storage
ppm parts per million
PR performance ratio
PV photovoltaic
REGEN regenerator
RH reheating
SAGD steam assisted gravity drainage
SOFC solid oxide fuel cell
SRU sulfur removal unit
UCG underground coal gasification
WGSMR water gas shift membrane reactor

Greek letters

η efficiency
β parameter used for hydrocarbon exergy

Subscripts and superscripts

o Reference condition
abs absorber
AC after cooling
ACS absorption cooling system
BC Brayton cycle
C compressor
CC combustion chamber
ch chemical
con condenser
comp compressor
CR compression ratio
d destruction
DC distillation column
el electrolyzer
Elec electrolysis
en energy
EV evaporator
ex exergy
F filter
f fuel
gen generator
GT gas turbine
H high

HRSG heat recovery steam generation
HX heat exchanger
in input
i state number
IC intercooler
IGCC integrated gasification combined cycle
in input
L low
LA lean amine
MeOH methanol
Mix mixture stream
out output
ov overall
P pump
Ph physical
r ratio
RA rich amine
RC Rankine cycle
Reb reboiler
Rec recycle stream
Reg regenerator
rev reversible
s storage
SAGD steam assisted gravity drainage
SOFC solid oxide fuel cell
ST steam turbine
sys system
T turbine
TV throttling valve
UCG underground coal gasification
Unrec unreacted gases stream

References

[1] Cleveland CJ, Morris C. Handbook of energy. In: Chronologies, top ten lists, and word clouds. vol. II. 2014 https://doi.org/10.1016/B978-0-12-397219-4.00009-6.

[2] Hunt RJ. The history of the industrial gas turbine (part 1 the first fifty years 1940–1990). Indep Tech Forum Power Gener 2011;582–2:1–48.

[3] Dincer I, Demir ME. 4.8 Steam and organic Rankine cycles. In: Dincer I, editor. Compr. Energy Syst. 1st ed. Cambridge, United States: Elsevier; 2018. p. 264–311. https://doi.org/10.1016/B978-0-12-809597-3.00410-7.

[4] Dincer I, Zamfirescu C. Advanced power generation systems. Elsevier; 2014.

[5] Tarique MA. Experimental investigation of scroll based organic Rankine systems. University of Ontario Institute of Technology; 2011.

[6] Dincer I, Demir ME. 4.7 Gas turbine cycles. In: Dincer I, editor. Compr. Energy Syst. 1st ed. Cambridge, United States: Elsevier; 2018. p. 209–63. https://doi.org/10.1016/B978-0-12-809597-3.00408-9.

[7] Çengel YA, Boles MA. Thermodynamics: an engineering approach. 8th ed. US: McGraw-Hill Education; 2015.

[8] Zamfirescu C, Dincer I. Thermodynamic analysis of a novel ammonia–water trilateral Rankine cycle. Thermochim Acta 2008;477:7–15. https://doi.org/10.1016/j.tca.2008.08.002.

[9] Kalina A, Tribus M, El-Sayed Y. A theoretical approach to the thermophysical properties of two-miscible-component mixtures for the purpose of power-cycle analysis. In: ASME. Vol. 86. 1989.

[10] Dincer I, Demir ME. 4.9 Combined energy conversion systems. In: Dincer I, editor. Compr. Energy Syst. 1st ed. Cambridge, United States: Elsevier; 2018. p. 312–63. https://doi.org/10.1016/B978-0-12-809597-3.00411-9.

[11] Gheith R, Hachem H, Aloui F, Ben Nasrallah S. 4.6 Stirling engines. In: Dincer I, editor. Compr. Energy Syst. 1st ed. Cambridge, United States: Elsevier; 2018. p. 169–208. https://doi.org/10.1016/B978-0-12-809597-3.00409-0.

[12] Korobitsyn MA. New and advanced conversion technologies: Analysis of cogeneration, combined and integrated cycles. PhD thesis. University of Twente, 1998.

[13] Rao AD, Samuelsen GS, Yi Y. Gas turbine based high-efficiency "Vision 21" natural gas and coal central plants. Proc Inst Mech Eng Part A J Power Energy 2005;219:127–36. https://doi.org/10.1243/095765005X6827.

[14] Poullikkas A. An overview of current and future sustainable gas turbine technologies. Renew Sustain Energy Rev 2005;9:409–43. https://doi.org/10.1016/j.rser.2004.05.009.

[15] Ozturk M, Dincer I. 4.10 integrated gasification combined cycles. In: Dincer I, editor. Compr. Energy Syst. 1st ed. Cambridge, United States: Elsevier; 2018. p. 364–473. https://doi.org/10.1016/B978-0-12-809597-3.00412-0.

[16] Moran MJ, Shapiro HN, Boettner DD, Bailey MB. Fundamentals of engineering thermodynamics. 2010.

[17] Dinali MN, Dincer I. Development and analysis of an integrated gas turbine system with compressed air energy storage for load leveling and energy management. Energy 2018; 15;163:604–17. https://doi.org/10.1016/j.energy.2018.08.094.

[18] Crotogino F, Mohmeyer K-U, Scharf R. Huntorf CAES: More than 20 years of successful operation. Orlando, Florida, USA, 2001.

[19] Castellani B, Presciutti A, Filipponi M, Nicolini A, Rossi F. Experimental investigation on the effect of phase change materials on compressed air expansion in CAES plants. Sustainability 2015;7:1–14.

[20] Lee S-S, Kim Y-M, Park J-K, Moon S-I, Yoon Y-T. Compressed air energy storage units for power generation and DSM in Korea, In: 2007 IEEE Power Eng. Soc. Gen. Meet., IEEE; 2007. p. 1–6. https://doi.org/10.1109/PES.2007.385857.

[21] Buffa F, Kemble S, Manfrida G, Milazzo A, Buffa F, Kemble S, et al. Exergy and exergoeconomic model of a ground-based CAES plant for peak-load energy production. Energies 2013;6:1050–67. https://doi.org/10.3390/en6021050.

[22] Kim YM, Favrat D. Energy and exergy analysis of a micro-compressed air energy storage and air cycle heating and cooling system. Energy 2010;35:213–20. https://doi.org/10.1016/J.ENERGY.2009.09.011.

[23] Al-Zareer M, Dincer I, Rosen MA. Modeling and performance assessment of a new integrated gasification combined cycle with a water gas shift membrane reactor for hydrogen production. Comput Chem Eng 2017;103:275–92.

[24] Wen CY. Noncatalytic heterogeneous solid-fluid reaction models. Ind Eng Chem 1968;60:34–54. https://doi.org/10.1021/ie50705a007.

[25] Wen CY, Chaung TZ. Entrainment coal gasification modeling. Ind Eng Chem Process Des Dev 1979;18:684–95. https://doi.org/10.1021/i260072a020.

[26] El-Emam RS, Dincer I, Naterer GF. Energy and exergy analyses of an integrated SOFC and coal gasification system. Int J Hydrogen Energy 2012;37:1689–97. https://doi.org/10.1016/j.ijhydene.2011.09.139.

[27] DinAli MN. Development and assessment of renewable energy based integrated systems for dimethyl ether production. Master's Thesis. University of Ontario Institute of Technology, 2018.

[28] Naterer G, Suppiah S, Lewis M, Gabriel K, Dincer I, Rosen MA, et al. Recent Canadian advances in nuclear-based hydrogen production and the thermochemical Cu-Cl cycle. Int J Hydrogen Energy 2009;34:2901–17. https://doi.org/10.1016/j.ijhydene.2009.01.090.

[29] Zamfirescu C, Dincer I, Naterer GF. Thermophysical properties of copper compounds in copper-chlorine thermochemical water splitting cycles. Int J Hydrogen Energy 2009;35:4839–52. https://doi.org/10.1016/j.ijhydene.2009.08.101.

[30] Ishaq H, Dincer I, Naterer GF. Industrial heat recovery from a steel furnace for the cogeneration of electricity and hydrogen with the copper-chlorine cycle. Energ Conver Manage 2018;171:384–97. https://doi.org/10.1016/j.enconman.2018.05.039.

[31] Al-Zareer M, Dincer I, Rosen MA. Analysis and assessment of a hydrogen production plant consisting of coal gasification, thermochemical water decomposition and hydrogen compression systems. 2018https://doi.org/10.1016/j.enconman.2017.11.047.

[32] Ghosh S, De S. Energy analysis of a cogeneration plant using coal gasification and solid oxide fuel cell. Energy 2006;31:345–63. https://doi.org/10.1016/j.energy.2005.01.011.

[33] U.S. Department of Energy. Fuel Cell Handbook. 7th ed. Honolulu, Hawaii: University Press of the Pacific; 2005.
[34] Miller BG. Coal energy systems. New York: Elsevier; 2005.
[35] Colpan CO, Dincer I, Hamdullahpur F. Thermodynamic modeling of direct internal reforming solid oxide fuel cells operating with syngas. Int J Hydrogen Energy 2007;32:787–95. https://doi.org/10.1016/j.ijhydene.2006.10.059.
[36] Dincer I, Rosen MA. Exergy, environment and sustainable development. Elsevier; 201351–73. https://doi.org/10.1016/B978-0-08-097089-9.00004-8.
[37] Kotas TJ. The exergy method of thermal plant analysis. 1st ed. Essex, Great Britain: Elsevier; 1985https://doi.org/10.1016/C2013-0-00894-8.
[38] De S, Nag PK. Thermodynamic analysis of a partial gasification pressurized combustion and supercritical steam combined cycle. Proc Inst Mech Eng Part A J Power Energy 2000;214:565–74. https://doi.org/10.1243/0957650001538100.
[39] Bicer Y, Dincer I. Energy and exergy analyses of an integrated underground coal gasification with SOFC fuel cell system for multigeneration including hydrogen production. Int J Hydrogen Energy 2015;40:13323–37. https://doi.org/10.1016/j.ijhydene.2015.08.023.
[40] Bicer Y, Dincer I. Development of a multigeneration system with underground coal gasification integrated to bitumen extraction applications for oil sands. Energ Conver Manage 2015;106:235–48.
[41] Herdem MS, Farhad S, Dincer I, Hamdullahpur F. Thermodynamic modeling and assessment of a combined coal gasification and alkaline water electrolysis system for hydrogen production. Int J Hydrogen Energy 2014;39:3061–71. https://doi.org/10.1016/j.ijhydene.2013.12.068.
[42] Bartone L, White J. Industrial size gasification for syngas, substitute natural gas and power production. US Dep Energy, Natl Energy Technol Lab Morgantown, WV, Pap No DOE/NETL-401/0406072007.
[43] Ozturk M, Dincer I. Thermodynamic assessment of an integrated solar power tower and coal gasification system for multi-generation purposes. Energ Conver Manage 2013;76:1061–72. https://doi.org/10.1016/j.enconman.2013.08.061.
[44] Giuffrida A, Romano MC, Lozza GG. Thermodynamic assessment of IGCC power plants with hot fuel gas desulfurization. Appl Energy 2010;87:3374–83. https://doi.org/10.1016/J.APENERGY.2010.05.020.
[45] El-Emam RS, Dincer I. Thermal modeling and efficiency assessment of an integrated biomass gasification and solid oxide fuel cell system. Int J Hydrogen Energy 2015;40:7694–706.
[46] Jess A. Mechanisms and kinetics of thermal reactions of aromatic hydrocarbons from pyrolysis of solid fuels. Fuel 1996;75:1441–8. https://doi.org/10.1016/0016-2361(96)00136-6.
[47] Roy A, Watson S, Infield D. Comparison of electrical energy efficiency of atmospheric and high-pressure electrolysers. Int J Hydrogen Energy 2006;31:1964–79. https://doi.org/10.1016/j.ijhydene.2006.01.018.
[48] Nieminen J, Dincer I, Naterer G. Comparative performance analysis of PEM and solid oxide steam electrolysers. Int J Hydrogen Energy 2010;35:10842–50. https://doi.org/10.1016/j.ijhydene.2010.06.005.
[49] Artuso P, Gammon R, Orecchini F, Watson SJ. Alkaline electrolysers: Model and real data analysis. Int J Hydrogen Energy 2011;36:7956–62. https://doi.org/10.1016/j.ijhydene.2011.01.094.

CHAPTER

5

Integration of nuclear energy systems for multigeneration

5.1 Introduction

The 19th century was a time of countless scientific discoveries in many fields and one of major discovery was radioactivity. The term radioactivity was proposed by Marie Curie, who received two Nobel Prizes for her discoveries in physics and chemistry. The radioactive element polonium that she discovered in 1898 was named for her native country—Poland.

By definition, radioactivity is the phenomenon of spontaneous emission of nuclear radiation (alpha, beta, gamma, neutron particles, etc.), which is generated either directly by unstable atomic nuclei or by a nuclear reaction. Some of these types of radiation, such as beta and gamma, were known from previous discoveries like that of Roentgen, who was the first to discover X-rays. Gamma-rays and X-rays have the same nature, both being a form of high energy electromagnetic radiation. Gamma-rays are produced by radioactive decay of some nuclei (which will be explained in this chapter), while X-rays are in general produced by special electronic equipment. Radioactivity can occur spontaneously by disintegration of some radioactive nuclei that exist in nature. Natural radioactivity was discovered by Heri Bequerel in 1896.

The process of radioactivity can be induced, in which case it is called artificial radioactivity. Generation of artificial radioactivity was studied first in Marie Curie's group at the University of Paris in the 1920s, with a significant contribution by Curie's student Stefania Maracineanu, who demonstrated the first laboratory experiment proving the possibility of artificially producing nuclear radiation. In her PhD thesis, completed in 1924, Stefania Maracineanu did experiments with lead and polonium and showed that lead, when activated with radioactive polonium, emits radiation. This first laboratory proof was investigated for 10 more years by Frederic and Irene Joliot-Curie, who formulated a theoretical model for artificial radioactivity in aluminum bombarded with alpha particles. The two experts received the Nobel Prize for chemistry in 1935 and acknowledged the initial discovery of Stefania Maracineanu in an article published in "Neues Wiener Journal" (from 5th June 1934).

Integrated Energy Systems for Multigeneration
https://doi.org/10.1016/B978-0-12-809943-8.00005-4

Generating nuclear radiation in a controlled manner was of utmost importance for the progress of nuclear energy engineering. As pointed out by Ernest Rutherford in 1911, the heat generated by radioactive decay is enormous as compared to that generated by any chemical reaction. Another major step in explaining the process of radioactivity is attributed to James Chadwick in 1932, followed by relevant experiments conducted by Enrico Fermi who succeeded in producing nuclear reactions induced by neutron radiation. Fermi demonstrated the first nuclear reactor when he conducted the chained fission reaction of uranium in 1942 in Chicago. Later, a nuclear reactor of 1 MW was built and tested at Oak Ridge Laboratories and the first full-size nuclear reactor of 250 MW thermal installed capacity—called the B reactor—began to operate near Richland, WA.

Intensive research work was pursued in the 1940s in countries engaged in World War II—United States, Germany, United Kingdom and the USSR—to develop nuclear power mainly for military applications. The first nuclear power plant to produce electricity of 100 kW was commissioned in Idaho in the United States in 1951. In the USSR, the first nuclear power plant of 5 MW was constructed and connected to the grid. Atomic Energy Canada Limited (AECL) was established in 1944 in Chalk River (Ontario) with the aim of developing nuclear energy in Canada. In 1945 AECL tested the first nuclear reactor and some years later developed the CANDU (CANadian Deuterium Uranium) nuclear power plant technology.

During the past 40 years many kinds of nuclear reactors have been developed for the purpose of electric power production and marine and space mission applications (where nuclear energy can be used for propulsion, power generation, heating, cooling and any other needs). In recent years the idea of hydrogen production with nuclear energy has gained major interest. Other novel applications of nuclear energy are in desalination, petroleum extraction from oil sands, and generating of sustainable process heat.

Nuclear power technologies were developed considerably and reached maturity in the second half of the 20th century. They are founded on the discoveries and contributions of many renowned scientists like Thompson, Roentgen, Bequerel, Curie, Bohr, Rutherford, Einstein and others and these led to the formation (during a period of about 75 years, between 1875 and 1950) of new bodies of physics such as nuclear physics and quantum mechanics. On one hand, nuclear and quantum physics have provided a thorough understanding of the laws of matter, and on the other hand they have supported the development of various engineering applications, among which the production of controlled nuclear heat in nuclear reactors appears to be one of the most important. Nowadays, nuclear energy is used to produce heat, which powers electricity generation plants and generates power for marine vessel propulsion or extraterrestrial spacecraft.

Although nuclear energy has played an important role in the world's energy supply, there is reluctance worldwide to continue with nuclear power options for overcoming the energy insufficiency. Some countries, such as Germany and South Korea, are good examples of this. It is quite apparent that, for many developing countries, ranging from Turkey to China, it is hard to meet energy demands without a nuclear contribution, which is why they try to develop such technologies to cover their power needs.

Nuclear power plants commonly employ uranium as the feedstock nuclear fuel, which is extracted from uranite mineral resources. Australia has almost 23% of global uranite minerals. The remaining worldwide uranite sources are distributed as follows: Kazakhstan 15%, Russia 10%, South Africa and Canada 8%, the US 6%, Brazil, Namibia, and Niger 5%, Ukraine 4%, Jordan and Uzbekistan 2%, and India 1% [1].

In this chapter, the fundamentals of nuclear energy systems are discussed and integration of nuclear power plants into several systems is presented. Nuclear plants offer large-capacity heat and power production owing to their fission reaction. They are suitable candidates for cogeneration and even multigeneration, which brings wider application in various sectors, ranging from process heat, desalination, hydrogen production, district heating, and synthetic fuels.

5.2 Fundamentals of nuclear energy

In nuclear energy plants, a nuclear reaction is continuously controlled in the reactor. The nuclear fission reaction produces essentially neutron radiation, which is instantly transformed into high-temperature thermal energy. In order to remove the heat generated from the reaction, and to attain pressurized superheated steam, a steam generation unit is commonly used. Based on the reactor type, the design and structure of the steam generation unit can vary. The steam generation unit is part of a steam Rankine cycle, which converts heat into power.

Certain chemical elements have radioactive characteristics. These elements are able to convert themselves into other chemical elements by releasing or capturing radiation via nuclear processes such as radioactive deterioration, or they are also able to undergo fusion or fission reactions. The atom consists of a nucleus and a number of electrons. The nucleus contains protons and neutrons. Under normal situations, the atom is neutral in terms of electric charge, meaning that the number of electrons is equal to the number of protons. The nucleus cannot be separated by chemical reactions, but it is possible for it to be altered in its constituency by nuclear reactions.

The explanations of some significant particles in nuclear physics are briefly presented here [1]:

- Electron (e^-): Subatomic particles that have one negative elementary electric charge and a quite minor rest mass, insignificant compared to the nucleus mass.
- Positron (e^+): The antimatter that annihilates an electron, which has the same rest mass but a positive charge.
- Neutrino (v): It carries zero electric charge and non-zero rest mass; it is smaller than an electron and travels with a speed close to that of light.
- Proton (p^+): Subatomic particle, which has one positive elementary electric charge and rest mass.
- Antineutrino ($\overline{v}$): The antimatter that can annihilate a neutrino.
- Photon (hv): The elementary quanta of electromagnetic radiation (light). The photon simultaneously acts as a wave and particle. Its speed of propagation is the speed of light; hence it has no rest mass. It has no electric charge, but it has dynamic mass, momentum, and spin. Photons can be annihilated by absorption when light interrelates with matter.
- Neutron (n): Subatomic particle, which has no electric charge and a rest mass almost equal to that of a proton. The neutron is classified as a nuclide.

The balance of atomic masses is especially significant for analyzing nuclear reactions. The atomic mass is a factor that classifies chemical elements. Any chemical element is recognized by the number of protons in its nucleus, which is called the atomic number (or proton number). However, the number of neutrons can vary from the number of protons within the nucleus for a chemical element. A chemical element can bear some level of variability in its nucleus resulting in a variation in the number of neutrons. Thus, even though the electric charge of the nucleus of a chemical element is stable, the nucleus mass differs based on the number of neutrons, making a range of subspecies. Nuclei that have the same number of protons and a different number of neutrons are called isotopes. The richness of an isotope characterizes the possibility of defining the specific isotope in nature among all known isotopes of the same chemical element.

Nuclear reactions yield emission or absorption of several kinds of nuclear radiation. Consistent with quantum mechanics, matter absorbs and emits radiative energy in fixed packages (quanta). Generally, nuclear radiation contains a flux of unstable particles. The main forms of radiation observed in nuclear reactions are as follows [1]:

- Alpha (α): It is a fast-moving flux of 4He nuclei (helium nucleus consisting of two protons and two neutrons). Therefore, this is an ionizing radiation of which particles have two elementary charges (+2e). The characteristic kinetic energy is 5 MeV per particle.
- Beta (β^-): Negatively ionized particle radiation, which consists of a flux of fast-moving electrons
- Beta (β^+): Positively ionized particle radiation, which consists of a flux of fast-moving positrons
- Gamma (γ): Electromagnetic radiation in the wavelength spectrum of picometers having the energy of hundreds of keV per photon
- Rontgen (X): Electromagnetic radiation in the spectrum of 0.1–10 nm having an energy of about 1–100 keV per photon
- Neutron (n): It contains a beam of free neutrons. Thus, it is non-ionized radiation. Nevertheless, when interrelating with atoms it yields strong ionizing impacts since photon emission happens, which relocates electrons at higher energy levels and ionizes the atom. The neutron radiation may be characterized based on the kinetic energy of the neutrons. Some subgroups of neutron radiation are thermal neutrons (25 meV), slow neutrons (1–10 eV), and fast neutrons (1–20 MeV).

The release and absorption of energy throughout the nuclear reaction can be defined by the law of conservation of energy, which must take into account the probability of mass-energy conversion in quantum mechanics. Nuclear reconfiguration after the reaction is responsible for the energy discharged or absorbed. The nuclides that make up the nucleus are both attractive and repulsive; Coulomb forces yield proton repulsion and the Yukawa forces attract the nuclides. The balance between attractive and repulsive nuclear forces may be stable, unstable or metastable on the basis of their exact status. If the basic balance is stable, the isotope is stable. If the nuclear balance is unbalanced or metastable, the isotope is radioactive. This means that it interrupts the construction of a stable equilibrium system. The atomic mass elements between magnesium and xenon are relatively stable because the stability factor is moderately flat. For elements heavier than xenon, the size of the nucleus becomes immense, which affects the balance of forces. In this case, the attractive forces become weaker and the repulsive forces get stronger. Thus, the heavy elements resulting from the nuclear fission reaction tend to degrade into lighter elements. The simplest fissionable elements are the isotopes of uranium 235 and plutonium 239. The fissile elements are flooded with neutrons that begin to separate them. When these radioactive isotopes are released, they process new radioactive chemicals and release additional neutrons, causing a chain reaction in the presence of other fissile materials. Although uranium (atomic number 92) is the heaviest natural element, many nuclear accelerators or reactors can add neutrons to the central nuclei to produce many other elements. The most important reaction in nuclear energy systems is currently the fission of the uranium nucleus, which occurs naturally as a fuel. In a fission reaction, a heavy nucleus is bombarded with a "slow" neutron. When capturing the neutron, the nucleus reaches the metastable state.

For reaching equilibrium, the nucleus separates into two parts of about half weight and creates neutron radiation and energy. The only known naturally arising fissionable isotope is $^{235}_{92}U$. However, the richness of this isotope on Earth is only about 0.7%. Two other fissionable artificial isotopes are known: the uranium isotope $^{233}_{92}U$ and the plutonium isotope $^{239}_{94}Pu$. Both these isotopes can be formed starting from more plentiful resources through neutron bombardment. Plutonium-239 is acquired from $^{238}_{92}U$, which occurs about 99.3% in nature over the following overall reaction [1]:

$$^{238}_{92}U + n \rightarrow {}^{239}_{94}Pu + 2\beta^- \tag{5.1}$$

The artificial fissionable uranium isotope (uranium-233) is formed from thorium-232, which is 100% abundant, by the overall reaction [1]:

$$^{232}_{90}Th + n \rightarrow {}^{233}_{92}U + 2\beta^- \tag{5.2}$$

The nuclear fission reaction in a simplified form can be expressed as follows:

$$^{235}_{92}U + n_{th} \rightarrow {}^{236m}_{92}U \rightarrow {}^{A1}_{Z1}E_1 + {}^{236-\kappa-A1}_{92-Z1}E_2 + \kappa n + Q_{out} \tag{5.3}$$

$^{236m}_{92}U$ is initially created. Then, it is split into lighter nuclei, which are $^{A1}_{Z1}E_1$ and $^{236-\kappa-A1}_{92-Z1}E_2$ while emitting κ neutrons. These resulting elements E_1 and E_2 bear metastable nuclei structure, hence split causing emission of gamma rays.

5.2.1 Nuclear fuel

When the fuel is removed from natural deposits, there is a detailed fuel processing sequence before it is ready to be used in nuclear reactors. Currently, most of the main nuclear material utilized in the reactors is uranium, whereas there are some rare cases utilizing thorium. However, thorium-232 is expected to gain a large share and to become the primary material to be used as nuclear fuel [1].

The nuclear fuel cycle contains a front-end phase, a fuel utilization phase, and a back-end phase. The front-end phase includes all the processes, beginning with mining until fuel production. The back-end phase comprises fuel recycling, materials recovery, and disposal. At the mining location, the uranium mineral is extracted, milled, and removed using chemical leaching. The end product is a uranium cake that is the rough commercialized type of uranium fuel encompassing triuranium octaoxide (U_3O_8). Approximately 260 Gt of ore is necessary to ultimately yield 0.1 Gt of fuel [1]. When enhancement is required, then U_3O_8 is transformed into UF_6 and managed via isotope split to increase the amount of 235 U to

3.5%. When enhancement is not required the triuranium octaoxide is transformed into uranium dioxide (UO_2) that can be fed directly to reactors such as CANDU types. At the removal stage of triuranium octaoxide, about 520 Mt of product is produced for 100 Mt of fuel, whereas about 260 Gt of mill tailings are deposited [1]. The transformation of U_3O_8 to UO_2 is over a thermal process. If enrichment is necessary, the uranium dioxide needs to be additionally transformed into uranium hexafluoride.

When uranium oxide is obtained, there is a hydro-fluorination process in which uranium oxide reacts with anhydrous hydrofluoric acid (HF). In another transformation phase, uranium tetrafluoride is combined in the gas phase with fluorine to produce uranium hexafluoride. Product extraction is carried out by distillation and crystallization using a special steel drum. About 770 Mt of UF_6 can produce about 150 Mt of enriched fuel [1].

Additional treatment involves the formation of uranium fuel in the form of pellets. Several options are available for collecting and covering fuel pellets. Treated fertile materials are also incorporated into the fuel pellets. In practical reactors, fuel is used in solid form. However, there are also liquid fuels for developed systems and research, which contain liquid uranyl salts and molten salts. A typical fuel element is sintered and formed in cylindrical strips with a uranium core, which may be uranium metal, alloy or uranium dioxide (UO_2) or uranium carbide (UC). Fuel cladding is another important step. Nuclear fuel has to be incorporated into a layer of corrosion-resistant material to allow adequate heat transfer between the fuel core and the reactor cooling agent. The housing has a smaller section for the absorption of thermal neutrons and blocks the release of radioactive material. The cladding is generally made of zirconium alloy (2.5% niobium and 97.5% zirconium, which highly penetrates into neutrons) or, as in the most recent systems, silicon carbide [1]. One of the problems associated with the fuel cladding is the catalytic effect of water separation and hydrogen production. This problem is often a concern in the event of an accident if the temperature of the layer increases and the catalytic activity improves. The silicon carbide has a much lower water decomposition activity.

The spent fuel is discharged from the air and stored in interim locations. Whenever possible, used uranium is transported to a permanent storage or recovery facility. For breeding reactors, uranium can be continuously recycled. In this case, the fuel is recycled to remove the fertile isotopes and fissionable isotopes to produce a new fuel pellet in the production area. The residual materials after reprocessing can also further be treated by recycling important isotopes for other usages.

Nuclear waste materials include radioactive isotopes and transuranic elements, which have quite a long half-life. A few of these elements are neptunium, plutonium, tectenium and iodine. Maintaining this species over a long period of time creates vitrification or calcination, which makes an amorphous matter, due to which they do not react for a long time. The most important factor is the increase in radiation caused by nuclear waste producing heat, which is about 1 to 1.5 kW per ton of uranium after 10 years [1]. In theory, this level of heat could be used as a source of energy for electricity generation or heating purposes by using heat engines. However, because security issues could arise during use of such a waste system, these systems have not been implemented so far. After completing the treatment, the waste is placed in permanent storage underground and the depth of the tunnels is more than 1 km below the ground. It is noted that there are concerns about the storage of nuclear waste in many countries.

Nuclear wastes can be classified into three categories:

- *Low-level waste*: This includes the remaining radioactive material from medical and industrial sites that use radionuclides for numerous methods employing gamma- or X-ray generation, etc. Some of the radioactive material disposed of by other nuclear facilities is likewise low-level waste.
- *High-level waste*: This is formed throughout the nuclear fuel cycle, containing solid and liquid materials that are very radioactive.
- *Transuranic waste*: This waste comprises radioactive materials with a decay time of longer than 20 years, releasing alpha particles with changing intensity.

5.2.2 Conventional reactors

The first generation of nuclear reactors encompasses four major concepts. However, currently, there is no known operational nuclear power plant from this first generation. The models advanced in the first generation were a boiling water reactor (BWR), gas-cooled reactor, liquid metal fast breeder reactor (LMFBR), and pressurized heavy water reactor (PHWR). Some of the past nuclear power plants from first generation reactors were [1].

- Dresden Generating Station 1, Illinois, USA (Boiling water reactor): 210 MW installed capacity, commissioned in 1960, decommissioned in 1990.
- Fermi 1 in Michigan, USA (Liquid metal fast breeder reactor): 100 MW installed capacity, commissioned in 1956, decommissioned in 1972.
- Mangox 1, Clader Hall, UK (Gas-cooled reactor): 60 MW installed capacity, commissioned in 1956, decommissioned in 2003.
- NPD Rolphton, Ontario, Canada (Pressurized heavy water reactor): 19.5 MW installed capacity, commissioned in 1961, decommissioned in 1987.

One of the commonly utilized kinds of reactors is the pressurized water reactor. The earlier versions of pressurized water reactors used heavy water as moderator and were called pressurized heavy water reactors (PHWR). They were initially advanced by Atomic Energy Canada Ltd. under the name CANDU (CANada Deuterium Uranium). The fuel is positioned in tubes over which PHWR coolant flows. The coolant transmits the nuclear thermal energy to a heat exchanger (at about 350 °C), which is usually a steam generation unit. The moderator contains a heavy water bath in which the tubes are submerged at the exterior side of the pressure tubes. The pressure level in the moderator is about 1 atm. The moderator is constantly cooled and the heat is rejected at about 80 °C. The fuel bars can be exchanged or extracted individually at any time via distinct apparatuses. In the moderator bath, there are chains of graphite bars in order to lessen the neutron flux and hence adjust the reaction rate and criticality. The CANDU reactor has been further developed, with better alternatives and enlarged capacity based on the same pressure tube model and the use of natural uranium and heavy water.

The essential model of the BWR relates to the first generation of nuclear reactors. A simplified version of the BWR reactor is schematically shown in Fig. 5.1. The BWR utilizes water for both moderator and coolant purposes, with the variance that the reactor yields saturated steam. The pressure in the BWR is usually about 75 bar, and the corresponding temperature is

FIG. 5.1 Simplified illustration of a BWR type nuclear reactor.

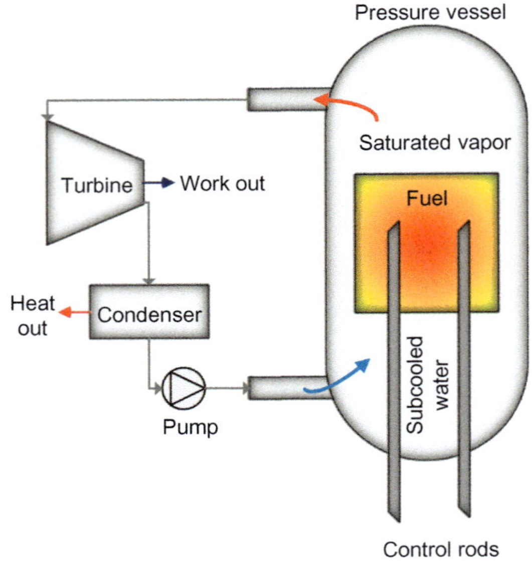

about 285 °C [1]. The reactor has a different design, ensuring a stable boiling process. After water arrives at the reactor, water is initially sent to a preheating process in a downcomer. Then, it upsurges and runs around the vertical fuel bars over the reactor core area. The resulting two-phase steam current, having a vapor mass quality of about 15%, is sent to the cyclone separator and to the steam drying unit in the upper part of the reactor, where the saturated steam is obtained. The reactor is located in a containment structure as per the design conditions, whereas other equipment including a steam turbine, a condenser and a pump are placed on the exterior side.

The first gas-cooled reactor (GCR) was industrialized in the UK (MAGOX Reactor). In the original MAGOX reactor, carbon dioxide having a pressure of approximately 20 bar was utilized as coolant. The main fuel was natural uranium, and the reactor was a fast breeding type. The moderator was made of graphite. The level of temperature of the produced heat is appropriate to be integrated into the steam power cycles. Fig. 5.2 displays the basic illustration of a gas-cooled reactor using carbon dioxide as a coolant.

Light water is an outstanding moderator. For regulating the reaction rate, extra movable bar type moderators are utilized made from boron carbide or Ag–In–Cd. The pressure is preserved over a pressure container equipped with immersed electric heaters intended to sustain a vapor pressure of about 155 bar [1]. The vapor is situated at the top of the pressurizer, while the bottom part is constantly full of subcooled liquid. A heat exchanger is employed to provide the nuclear thermal energy from the primary circuit to a secondary circuit, which may be a steam generation unit expected to drive a turbine that usually turns a generator or drives a marine vehicle. The generated steam is normally at 275 °C and 60 atm [1]. An illustration of a PWR-type nuclear reactor is shown in Fig. 5.3.

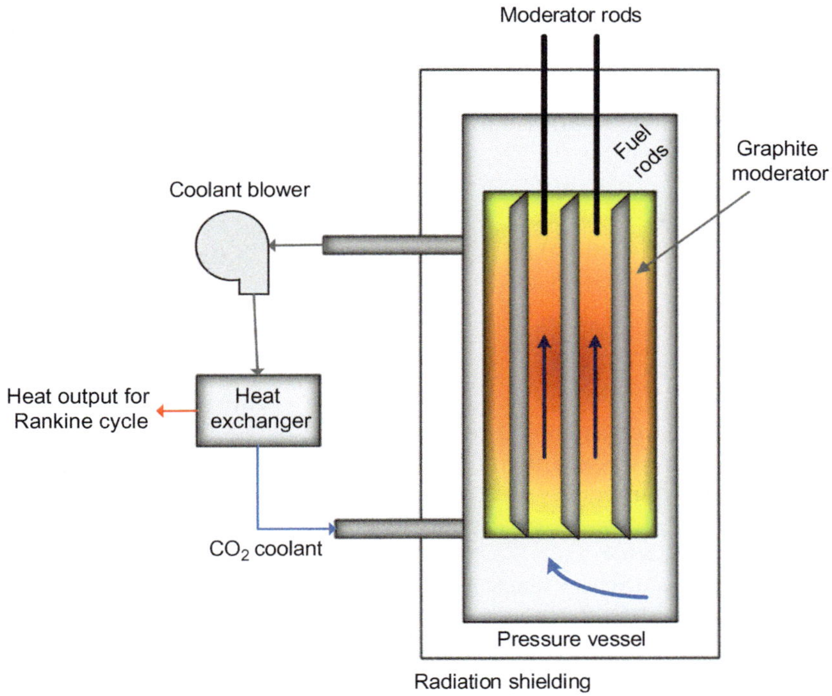

FIG. 5.2 Simplified illustration of first-generation gas-cooled type nuclear reactor.

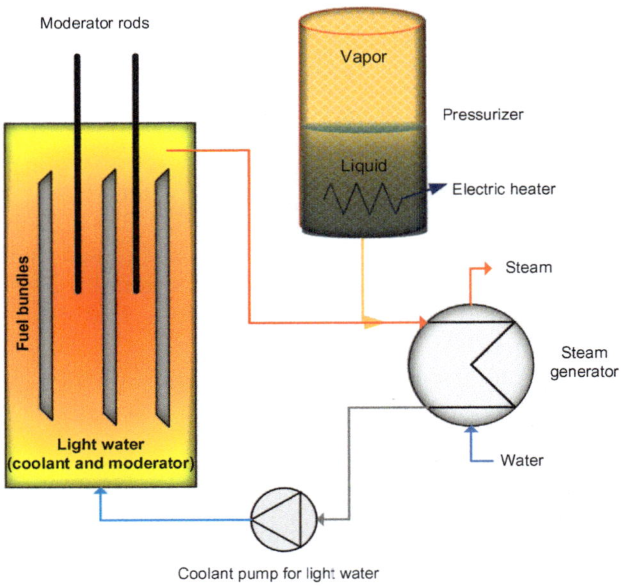

FIG. 5.3 Simplified illustration of PWR-type nuclear reactor.

Numerous advanced nuclear reactors are presently in the design phase, implying that the conceptual step of advancement is now mature and the detailed design is underway. These type of reactors are commonly called Generation III+; some of them are listed here [1]:

- Advanced CANDU (ACR-1000) reactor.
- European pressurized reactor (EPR).
- VVER-1200.
- APWR-IV.
- ESBWR, economic simplified BWR.
- AP-1000 reactor.
- APR-1400.
- EU, APWR.
- IFR, integral fast reactor.
- PBR, pebble bed reactor.
- HTGC, high-temperature gas-cooled reactor.
- SSTAR, small sealed transportable autonomous reactor.
- CAESAR, clean, environmentally safe advanced reactor.
- SR, subcritical reactor.
- TBR, thorium-based reactor.
- AHWR, advanced heavy water reactor.
- KAMINI, Uranium-233 based reactor.

In addition, there are Generation IV nuclear reactors; where some of them are listed here [2]:

- *Gas-cooled fast reactor (GFR)*: It is a fast-neutron-spectrum, helium-cooled reactor and closed fuel cycle. The spent fuel is reprocessed and reused. It produces about 1200 MW and thermal energy at about 850 °C.
- *Very-high-temperature reactor (VHTR)*: It is a graphite-moderated, helium-cooled reactor with a once-through uranium fuel cycle. In this reactor type, spent fuel is not reprocessed. The outlet temperature is estimated at about 1000 °C and the units have power output of about 600 MW.
- *Supercritical-water-cooled reactor (SCWR)*: This is a high-temperature, high-pressure water-cooled reactor that functions beyond the thermodynamic critical point of water. It is intended for 1700 MW of power output and a core outlet temperature of 500–650 °C.
- *Sodium-cooled fast reactor (SFR)*: It is a fast-spectrum, sodium-cooled reactor and closed fuel cycle for effective controlling of actinides and transformation of fertile uranium. Its core outlet temperature is about 550 °C, and it is intended for three different sizes: low (50–150 MW), intermediate (300–600 MW) and large (600–1500 MW).
- *Lead-cooled fast reactor (LFR)*: It is a fast-spectrum lead/bismuth eutectic liquid-metal-cooled reactor and a closed fuel cycle for effective transformation of fertile uranium and management of actinides. It has two different types: (i) small power capacity in the range of 20 MW and about 560–570 °C core outlet temperature and (ii) high power capacity of 600 MW and temperature of about 480 °C.
- *Molten salt reactor (MSR)*: It yields fission power in a circulating molten salt fuel mixture with an epithermal-spectrum reactor and a full actinide recycling fuel cycle. The reference power capacity is about 1000 MW with a coolant exit from the core at 700–800 °C.

5.2.3 Nuclear thermochemical cycles

Thermochemical cycles include a closed sequence of chemical reactions in which there can be an endothermic reaction at high temperature driven thermochemically. The cycle may also include an electrochemical reaction, which is called a hybrid thermochemical cycle. Pure thermochemical cycles commonly run with high-temperature thermal energy. These types of cycles are closed processes mainly used for splitting water into hydrogen and oxygen and recycling all other intermediate chemicals. Hybrid thermochemical cycles require heat at an intermediate temperature level (about 500 °C) and electricity to be driven. One of the favorable options for this temperature level is the CuCl thermochemical cycle. The input energy can be concentrated solar radiation from a solar tower or solar through system or nuclear heat. The heat input is used in part for electricity generation and in part to source the endothermic reaction within the thermochemical cycle. Some multi-step techniques involving recycled chemicals were developed such as two-step, three-step, four-step, five-step, six-step, etc., where "steps" denotes the number of chemical reactions. Thermochemical cycles do not usually need catalysis to energize the chemical reactions. In these options, oxygen separation membranes are not needed. The temperature of the necessary heat is at a reasonable range, from 600 to 1200 K [1]. Hybrid thermochemical cycles require both electrical and thermal to drive the endothermic chemical reactions and electrochemical reactions to reduce operating temperatures.

5.2.3.1 *CuCl cycle*

CuCl thermochemical cycles contain a sequence of chemical reactions and are characterized by a number of main chemical steps. Two-, three-, four-, and five-step CuCl cycles for thermochemical water splitting are discussed in the literature. The main inputs to the CuCl thermochemical water splitting cycle are water, thermal energy, and electricity, while oxygen and hydrogen are the obtained outputs. The five core chemical reactions in the cycle are as follows:

- HCl (g) production (hydrolysis),
- Oxygen production (copper oxychloride decomposition),
- Copper (Cu) production,
- Drying of cupric chloride,
- Hydrogen production.

The four-step cycle combines the third and fourth reactions. The three-step CuCl cycle integrates the H_2 production step and the copper production, and drying of cupric chloride steps to reduce the complexity and equipment needs. In the cycle, some of the reactions are exothermic whereas there are also endothermic reactions. The reaction temperatures are different for various steps. This is, in fact, a challenge for thermal management.

5.2.3.2 *MgCl cycle*

The MgCl thermochemical water splitting cycle is a hybrid process requiring both thermal energy and electricity to decompose water for hydrogen production. The maximum temperature within the cycle is about 550 °C. The MgCl cycle has three main steps in which two steps are thermochemical and the other step is electrochemical. In the hydrolysis step, a solid gas reaction takes place, in which hydrogen chloride (HCl) and magnesia (MgO) are formed during the hydrolysis of magnesium chloride ($MgCl_2$). This reaction is endothermic and has a temperature range of 450–550 °C. The reactants of the hydrolysis step are H_2O and $MgCl_2$. In the chlorination step of the MgCl cycle, MgO (s) and Cl_2 (g) serve as reactants to obtain $MgCl_2$ (s) and oxygen at a reaction temperature of about 400–500 °C. $MgCl_2$ is then sent back to the hydrolysis step to achieve closed cycle recycling of all of the Mg compounds on a continuous basis. Hydrogen can be generated by a thermochemical or an electrochemical reaction in the third step of the MgCl cycle. The hybrid magnesium-chlorine cycle includes three or four reaction steps among which one is electrochemical-based. In this cycle, the electrochemical step is the electrolysis of hydrochloric acid.

5.2.4 Thermodynamic analysis of nuclear power plants

Thermodynamic analysis of nuclear power plants can reveal the key exergy destruction steps within a general nuclear reactor. Fig. 5.4 illustrates a simplified nuclear reactor model to show the fundamental energy flows. Each box within the figure shows the specific processes occurring in the plant. The energy transfer processes such as neutron radiation, conduction heat transfer, combined heat transfer, and fluid flow, are illustrated. The nuclear fission reaction takes place in the fuel pellet core and releases chiefly neutron radiation and gamma radiation or other minor particles. The primary portion of nuclear radiation interacts with fuel pellets at state point 1 and produces heating. Some portion of neutron radiation leaks to the moderator at state point 2 where it is distributed back or partly transforms into thermal energy. Other neutron radiation at state point 3 extends to auxiliary modules of the reactor and is transformed into thermal energy. The hot fuel pellets transmit thermal energy via the heat conduction mechanism toward the pellet surface at state point 4 and further to the cladding surface at state point 5.

The cladding surface interrelates via combined heat transfer mechanisms, in which conduction and convection are dominant at state point 6. Heat transfer fluid is heated from the inlet temperature at state point 7 to the outlet temperature at state point 8. The thermal energy produced within the auxiliary modules of the reactor is transmitted to the

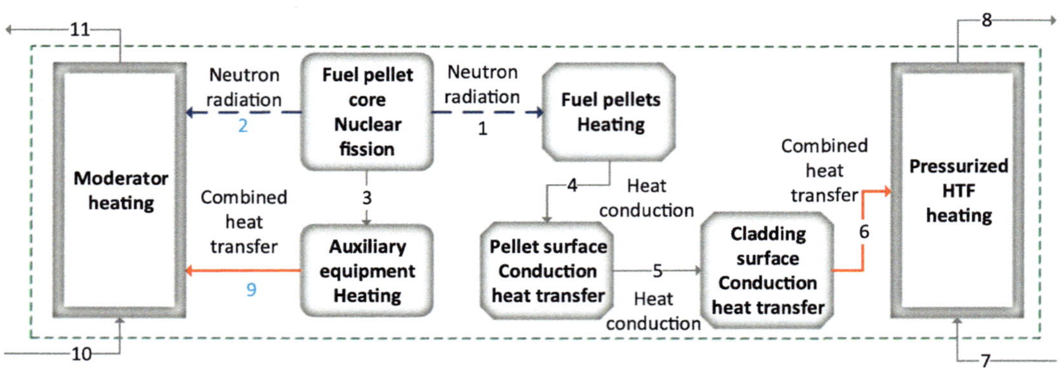

FIG. 5.4 Simplified thermodynamic model of a sample nuclear reactor. *Modified from Dincer I, Zamfirescu C. Chapter 6—Nuclear power generation. Adv. Power Gener. Syst., Boston: Elsevier; 2014, p. 311–68. https://doi.org/10.1016/B978-0-12-383860-5.00006-7*

TABLE 5.1 Fundamental energy and exergy balance equations and efficiencies for a general nuclear reactor

Balance equation and efficiency	Equation	Details
Energy efficiency	$\eta_{en} = \frac{\dot{E}_{htf}}{\dot{E}_{nuclear}}$	Energy rate transferred to heat transfer fluid is the useful output. The energy input is the energy delivered through the fission.
Energy balance equation	$\dot{E}_{nuclear} = \dot{E}_{htf} + \dot{E}_{moderator}$	Here, $\dot{E}_{nuclear}$ is the summation of $\dot{E}_1 + \dot{E}_2 + \dot{E}_3$, which is the nuclear radiation rate obtained via fission reaction. $\dot{E}_{htf}$ is the thermal energy transmitted to heat transfer fluid. It can be defined as $\dot{m}_{htf}(h_8 - h_7)$. $\dot{E}_{moderator}$ is the thermal energy transferred to the moderator fluid that is rejected to the environment. It can be defined as $\dot{m}_{moderator}(h_{11} - h_{10})$.
Exergy efficiency	$\eta_{ex} = \frac{\dot{E}x_{htf}}{\dot{E}x_{nuclear}}$	The exergy transferred to heat transfer fluid is the useful output. The input is the exergy delivered through the fission.
Exergy balance equation	$\dot{E}x_{nuclear} = \dot{E}x_{htf} + \dot{E}x_{moderator} + \dot{E}x_d$	Here, $\dot{E}x_{nuclear}$ is equal to $\dot{E}_{nuclear}$ that is the exergy of nuclear radiation where the Carnot factor is unity and associated temperature is infinity. $\dot{E}x_{htf}$ and $\dot{E}x_{moderator}$ are the exergy transferred to the heat transfer fluid and moderator, respectively. The exergy of moderator is lost. $\dot{E}x_d$ is the exergy destruction over the nuclear reactor.

Source: *Dincer I, Zamfirescu C. Chapter 6—Nuclear power generation. Adv. Power Gener. Syst., Boston: Elsevier; 2014, p. 311–68. https://doi.org/10.1016/B978-0-12-383860-5.00006-7.*

environment moderator at state point 9 by convection or combined convection and conduction. The moderator is usually recycled to a cooling system to remove the moderator heat. It arrives at the reactor colder at state point 10 and leaves warmed at state point 11. The fundamental energy and exergy balance equations for the nuclear reactor as well as associated efficiency equations are defined as listed in Table 5.1.

A comparison of temperature levels and energy densities for nuclear fuels, fossil fuels and biomass is provided in Table 5.2. As shown, nuclear fuels have the highest energy density per unit mass. Nonetheless, the Carnot factor of the combustion methods of biomass and fossil fuels is higher as they provide quite high temperatures compared to the conventional nuclear process. Note that new generation nuclear reactors are capable of generating higher temperatures than conventional nuclear reactors, meaning that Carnot factor will be larger. On the other hand, nuclear power plants have generally higher capacity factors compared to fossil fuels and biomass. There are no direct environmental emissions during operation of nuclear power plants, but indirect emissions are available caused by construction as well as other auxiliary operations.

5.3 Nuclear energy-based multigeneration systems

Nuclear power plants can generate thermal and electrical energy simultaneously, which makes them promising in terms of cogeneration. In most of the thermal power plants, a

TABLE 5.2 Comparison of nuclear fuels, fossil fuels and biomass.

Fuel type	Temperature	Energy density (approximately)	Carnot limit	Heat transfer medium	Greenhouse gas emissions
Biomass	500–1500 °C	0.017 MJ/ton	0.62–0.83	Flue gas	Neutral
Fossil fuel	1000–1500 °C	0.045 MJ/ton	0.76–0.83	Flue gas	CO_2 in flue gas
Nuclear fuel (fission)	300–500 °C	83,000 MJ/ton	0.5–0.6	Liquid	Indirect emissions due to reactor construction
	700–950 °C	16,000 MJ/ton	0.7–0.75	Gas	

Source: *Dincer I, Zamfirescu C. Chapter 6—Nuclear power generation. Adv. Power Gener. Syst., Boston: Elsevier; 2014, p. 311–68. https://doi.org/10.1016/B978-0-12-383860-5.00006-7.*

percentage of the input heat is transformed into electricity with typically 20% to 45% efficiency. Normally, the remaining portion of the heat is released to the surroundings as waste heat. The main advantage of nuclear power plants is the capability of converting this waste heat for generating a high-pressure and high-temperature vapor, which can be used to produce heating, cooling, or fresh water through an integrated system. In this section, we present how to integrate the nuclear plants into other significant industrial processes for multigeneration purposes. The useful outputs from a nuclear multigeneration plant can be extended to heating, fresh water, cooling, hydrogen, process heat, drying, synthetic fuels or other commodities with market value as illustrated in Fig. 5.5.

In district heating or desalination systems, the essential temperature levels are in the range of 60 °C to 120 °C [1]. Utilization of low-grade thermal energy is a suitable approach, which can reduce the overall cost, and the usage of low-grade thermal energy can increase the return on investment in the nuclear power plant. For other applications requiring high-grade

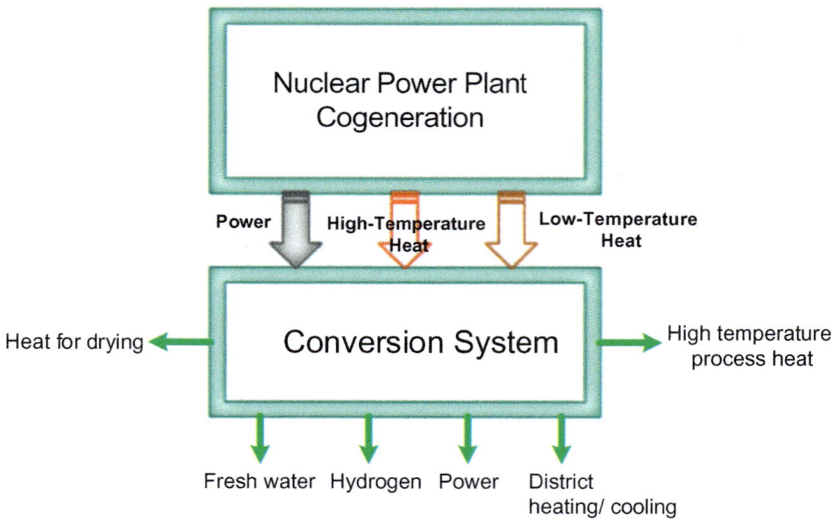

FIG. 5.5 Integrated nuclear multigeneration plant with multiple useful outputs.

thermal energy, the heat recovery process plays an important role. Though there are a few issues, such as dealing with the danger of radioactive pollution, there is evidently huge potential for nuclear heat integration into industrial processes.

Many areas in the world are facing freshwater scarcity due to several reasons, including climate change, contamination of freshwater sources, cost of desalination, etc. The issue is even more problematic when the pollution and salinity of freshwater resources are high. Seawater is an important alternative to maintain current and future sources of fresh water in dry areas near the sea. For desalination plants, a lot of energy is needed to separate the salt from the salty water. The desired energy form can be thermal or mechanical pumping, used for the energy requirement during the pressure increase of seawater through membranes using reverse osmosis (RO). Currently, the main source of thermal energy in this application is fossil fuels, including coal, oil and natural gas. Large-scale desalination plants are commonly located near thermal power plants, which can provide steam and electricity for the process. Nevertheless, the utilization of fossil fuels is undesirable due to its many disadvantages. The major disadvantage is the combustion of fossil fuels releasing huge quantities of greenhouse gases compared to nuclear options. We will cover various alternatives for cleaner desalination processes, in which the required energy is sourced from renewables, in the next chapter. However, here we present utilization of nuclear energy for desalination purposes. Over the long term, it is not desirable to have the desalination process solely dependent on fossil fuels since they have restricted availability, and fresh water is a permanent demand of mankind. Hence, nuclear energy can play an important role in providing a share of desalination worldwide. Nuclear energy offers other advantages for this application, ranging from economic competitiveness to energy supply diversification. One of the leading cogeneration systems using nuclear energy as a topping cycle is employed to yield fresh water through multistage flash desalination methods.

In addition to fresh water, energy storage and energy carriers bear huge significance for a sustainable future. Hydrogen is anticipated to play a substantial role as an energy carrier. It can be used as fuel in virtually every system where fossil fuels are utilized today. Hydrogen can also be a fundamental feedstock for many sectors ranging from petrochemicals to fertilizers. Nuclear energy can be used for splitting of water for hydrogen production via several methods, including nuclear water electrolysis, nuclear thermochemical cycles, nuclear high-temperature steam electrolysis, etc. Furthermore, nuclear-based reforming of fossil fuels is a favorable technique with high viability in the transition toward a fully implemented hydrogen economy, which will decrease the dependence on fossil fuels.

5.4 Brainstorming for system integration

Nuclear energy can be integrated and/or hybridized with many diverse systems, as illustrated in Fig. 5.6.

- *Small modular reactor for distributed power generation*: Small modular reactors are suitable for small-scale power generation purposes and can thus enable decentralized power generation. In addition, with this method, district heating and cooling applications can be easily integrated into small modular reactors.

FIG. 5.6 Brainstorming on system integration for nuclear energy.

- *Nuclear waste heat integration for thermochemical cycles*: Thermochemical cycles require thermal energy, whereas hybrid thermochemical cycles require both heat and electricity. Therefore, nuclear power plants can be easily integrated into hybrid thermochemical cycles such as CuCl and MgCl cycles.
- *Nuclear waste heat for district heating and cooling*: Nuclear waste heat available from the power plants can be used within cities for district heating and cooling purposes. In this way, trigeneration plant efficiency will be higher than single power generation.
- *Nuclear power plant hybridized with renewable power plants*: Nuclear electricity can be integrated with other renewable energy systems for achieving a more sustainable and reliable grid operation.
- *Nuclear based ammonia production*: High-temperature steam electrolysis can be powered by a pressurized water reactor (PWR), a boiling water reactor (BWR), a sodium cooled fast reactor (SFR) or a high-temperature gas-cooled reactor (HTGR) to produce hydrogen. Then, hydrogen is reacted with nitrogen via the Haber-Bosch process or electrochemical ammonia synthesis. Nuclear electricity can drive the electrochemical ammonia synthesis as well as other auxiliary equipment.

- *Nuclear based urea, nitric acid and ammonium nitrate production*: Ammonia produced from nuclear energy can be combined with carbon dioxide to yield urea. In addition, ammonia can be used as feedstock to yield nitric acid via the Ostwald process in which anhydrous ammonia is oxidized to nitric oxide, in the presence of platinum or rhodium gauze catalyst at a high temperature of about 500 K and a pressure of 9 atm. The reaction of nitric acid and ammonia yields ammonium nitrate, which is another fertilizer for agricultural applications.
- *Aluminum production by using nuclear heat*: Aluminum rich bauxite ore (Al_2O_3) must be removed in a mixed suspension with sodium hydroxide at 110–270 °C, where there are intermediate materials of sodium aluminate and aluminum hydroxide. A rotating kiln is employed for calcination of aluminum hydroxide at temperature levels of approximately 1100 °C.
- *Nuclear energy for gasification of coal or biomass*: This process can be carried out in two different ways: i) steam-coal/biomass gasification and (ii) hydrogasification. In the case of hydrogasification, coal/biomass is integrated into hydrogen by an exothermic reaction to produce synthetic natural gas. In this case, nuclear heat can be used to produce synthesis gas and extract hydrogen through a water-gas shift reaction or steam reforming. The process can be integrated with a very high temperature reactor having a temperature of about 950 °C.
- *Nuclear heat for steam-assisted gravity drainage (SAGD) to recover oil*: This process is utilized to extract bituminous oil from oil sands where the required steam is at 10–15 MPa and 200–240 °C. Furthermore, if synthetic crude oil is desired to be produced, hydrogen gas is also needed. In this case, hydrogen can also be produced using nuclear power via electrolysis, thermochemical cycle, high-temperature steam electrolysis, etc.
- *Nuclear-based desalination system*: Nuclear heat can be integrated into desalination plants running on thermally driven processes (e.g., multi-stage flashing, multi-effect distillation). In addition, nuclear electricity can be utilized for freshwater production driven by electrical processes (e.g., reverse osmosis).
- *Nuclear-based synthetic fuel production*: Synthetic fuels such as ethanol, methanol, etc. can be produced using nuclear heat as well as nuclear electricity, as will be discussed in Chapter 7.
- *Nuclear heat for paper and pulp production*: Nuclear heat can be utilized for paper and pulp production where the temperature requirements are in the range of 200–400 °C. This amount of heat can either be supplied by the nuclear reactor or by the steam coming from the steam circuit of the reactor.
- *Nuclear-based vinyl chloride production*: Vinyl chloride can be produced with the help of a nuclear reactor, as it requires a temperature range of 100–200 °C. The electricity and heat requirement for the synthesis of vinyl chloride production can both be provided by a nuclear reactor.

5.5 Case study 1: High-temperature nuclear reactor using thermochemical cycle for electricity and hydrogen production

In this case study, an energy system that runs on nuclear energy for the co-production of electricity and hydrogen as depicted in Fig. 5.7 using a two-step thermochemical cycle is considered. This case study is modified from the detailed thermodynamic study conducted by

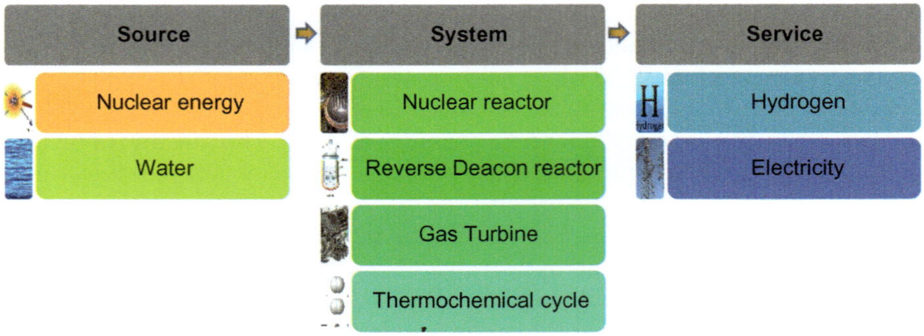

FIG. 5.7 3S approach for case study 1.

Khalid and Bicer [3]. The salient feature of this study is that it discusses the thermodynamic viability of integration of a two-step thermochemical cycle with a nuclear reactor for large-scale hydrogen production.

5.5.1 System description

A schematic of the integration of a two-step thermochemical cycle with high-temperature nuclear reactor for hydrogen and electricity is presented in Fig. 5.8. This system has several components, namely small modular reactor (GT-MHR type), heat exchangers (HEX1, HEX2 and HEX3), compressor, gas turbine, splitter, electrolyzer, and reverse Deacon reactor. At state 1, compressed and hot helium leaves the nuclear reactor, which is utilized to deliver the heat needed by reverse Deacon reactor. At state 2, helium leaves the reverse Deacon reactor and is allowed to pass through heat exchanger 1, so that it can provide the heat and desired temperature required by the steam entering the reverse Deacon reactor. Helium at state 3 enters the gas turbine and is expanded to generate electricity. At state 4, helium at reduced pressure leaves the gas turbine and is further cooled down (state 5) in heat exchanger 2. The temperature of the cooled helium (state 5) is further reduced in heat exchanger 3 (state 7). The cooled helium coming out of heat exchanger 3 gets compressed in the compressor. This compressed helium (state 8) gets heated in heat exchanger 2 and then enters the nuclear reactor core at state 9 by using the fresh stream of water (state 6). At state 14, hydrogen chloride gas enters the electrolyzer (electricity provided by a nuclear reactor) to be split into hydrogen and chlorine gases. At state 15, chlorine gas enters the reverse Deacon where it reacts with high temperature steam (state 12) to produced oxygen (state 16) and hydrogen chloride gases, thus completing the two-step thermochemical cycle. A splitter is used to split the hot steam coming out from the heat exchanger 3, into two parts, namely state 11, and state 12. This is how the whole integrated system works using nuclear energy for co-production of hydrogen and electricity.

5.5.2 Assumptions

In order to perform thermodynamic analysis of this integrated system, there is a need to make conceptually correct assumptions. The following main reasonable assumptions are considered in this case study:

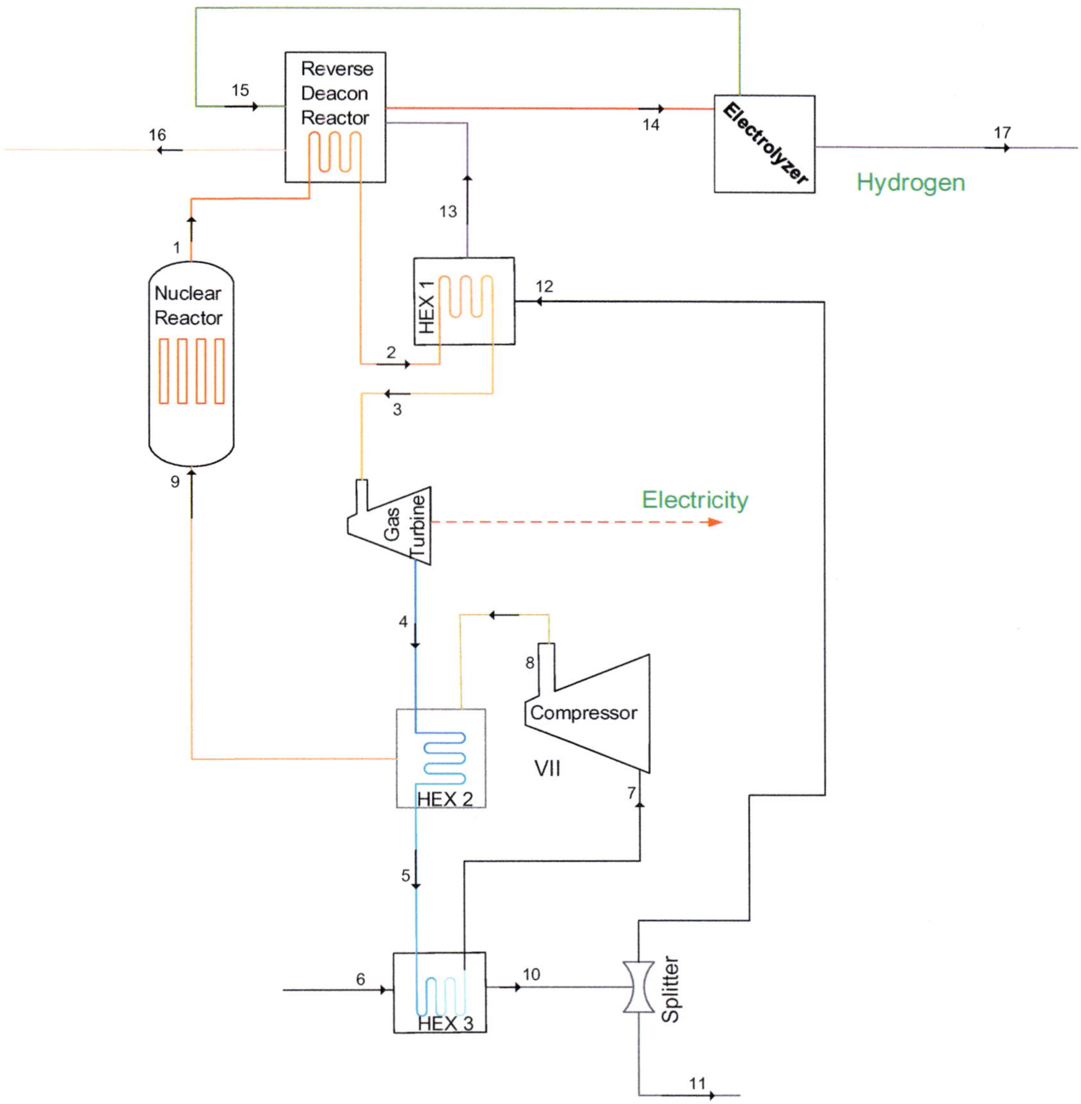

FIG. 5.8 Schematic of the integration of the two-step thermochemical cycle with small modular reactor for co-production of hydrogen and electricity.

- The system works under steady state.
- Full completion for chemical reactions according to their stoichiometry.
- Pressure drops in pipelines and heat exchangers are ignored.
- No heat losses are considered in heat exchangers, pumps and turbines.
- Ambient temperature and pressure are 25 °C and 100 kPa, respectively.

5.5.3 Analysis

The thermochemical cycle and system performance are measured by defining energy and exergy efficiencies.

For the thermochemical cycle, the following expressions of exergy and energy efficiencies are used, respectively:

$$\eta_{ex,cycle} = \frac{\dot{m}_{H_2} ex_{H_2}}{\dot{W}_{elect} + \dot{Ex}^{Q}_{reaction}} \quad (5.4)$$

where the thermal exergy associated with the reverse Deacon reaction is $\dot{Ex}^{Q}_{reaction}$, and $\dot{W}_{elect}$ denotes the electrical power consumed by the electrolyzer.

$$\eta_{en,cycle} = \frac{\dot{m}_{H_2} LHV_{H_2}}{\dot{W}_{elect} + \dot{Q}_{reaction}} \quad (5.5)$$

Likewise, for the overall system performance, the following energy efficiency expression is utilized:

$$\eta_{en,ov} = \frac{\dot{m}_{H_2} LHV_{H_2} + \dot{W}_{net,electric}}{\dot{Q}_{core}} \quad (5.6)$$

where the mass flow rate of hydrogen produced is $\dot{m}_{H_2}$, and the lower heating value of the hydrogen is LHV. Here, $\dot{Q}_{core}$ denotes the heat required by the reactor core. The overall exergy efficiency is calculated by

$$\eta_{ex,ov} = \frac{\dot{m}_{H_2} ex_{H_2} + \dot{W}_{net,electric}}{\dot{Ex}^{Q}_{core}} \quad (5.7)$$

5.5.4 Results and discussion

State points with all their relevant thermodynamic data such as P (pressure in kPa), T (temperature in °C), $\dot{m}$ (mass flow rate in kg/s), ex (specific exergy in kJ/kg) and h (specific enthalpy in kJ/kg) are identified and calculated (see Table 5.3). In addition, Table 5.4 lists the value of all the useful parameters obtained from this study. 67.3 t of hydrogen can be produced daily using the presented integrated system with the system operating at energy and exergy efficiencies of 24.1%, and 23.5%, respectively. The electrolyzer in the presented study operates at 1.7 V for a current density of 1000 A/m^2.

By determining the rates of exergy destruction of various units in the integrated system, key players for losses in the integrated system are identified and have been tabulated in Table 5.5. The exergy destruction losses incurred by the reactor core is 245.5 MW while compressor and electrolyzer contribute 58.4 MW, and 47.3 MW of exergy destruction rates, respectively.

Fig. 5.9 illustrates the variation of energy and exergy efficiencies of the overall system and heat/work rates with the outlet temperature of the nuclear reactor. As the temperature rises from 852 °C to 862 °C, both exergy and energy efficiencies of the system decrease. Fig. 5.9 also shows with the increase in outlet temperature, electrical power consumed by the electrolyzer,

TABLE 5.3 Thermodynamic data for the the cogeneration system.

State no.	Fluid type	*T* (°C)	*P* (kPa)	$\dot{m}$ (kg/s)	*h* (kJ/kg)	ex (kJ/kg)
1	Helium	860.0	7000	439	5910	4911
2	Helium	850.0	7000	439	5858	4872
3	Helium	846.5	7000	439	5840	4859
4	Helium	526.1	2800	439	4164	3139
5	Helium	213.2	2800	439	2540	2283
*6	Water	25	101.3	161.7	−15,866	0
7	Helium	30	2800	439	1589	2063
8	Helium	203.1	7000	439	2501	2842
9	Helium	516.1	7000	439	4125	3685
*10	Water	100	101.3	161.7	−13,282	491.2
*11	Water	100	101.3	147.7	−13,282	491.2
*12	Water	100	101.3	13.93	−13,282	542
*13	Water	630	101.3	13.93	−12,154	1124
*14	Hydrogen Chloride	630	101.3	56.39	−2021	2563
*15	Chlorine	630	101.3	54.83	319.9	1892
*16	Oxygen	630	101.3	12.37	626.1	418.1
*17	Hydrogen	630	101.3	1.559	9109	117,111

[a] *Calculated using NASA properties in EES [38].*

Source: *Data modified from Khalid F, Bicer Y. Energy and exergy analyses of a hybrid small modular reactor and wind turbine system for trigeneration. Energy Sci Eng 2019;00:1–15. https://doi.org/10.1002/ese3.327.*

and heat required by reverse Deacon reaction increases while the electrical power generated by the system decreases.

Fig. 5.10 shows the effect of the outlet temperature of the nuclear reactor on the rate of hydrogen production, and both efficiencies (energy and exergy) of the thermochemical cycle. As the temperature changes from 852 °C to 862 °C, the rate of hydrogen production rises from 0.156 kg/s to 0.935 kg/s while no effect on the efficiencies of the thermochemical cycle is observed.

In Fig. 5.11, the effect of the outlet temperature of the reverse deacon reactor on both efficiencies (exergy and energy) of the thermochemical cycle, and the rate of hydrogen production is plotted. As the temperature increases from 848 °C to 858 °C, no effects are observed on the efficiencies while the rate of hydrogen produced decreases from 0.935 kg/s to 0.156 kg/s. With the reduction in hydrogen produced, the amount of electricity consumed by the electrolyzer also decreases, resulting in no change in efficiencies of the cycle.

As the outlet temperature of the reverse Deacon reactor varies from 848 °C to 858 °C, the system's exergy and energy efficiencies change from 27.4% to 31.6%, and 37.3% to 49.4%, respectively (see Fig. 5.12). Heat rates for the reverse Deacon reaction and reactor core both

TABLE 5.4 Calculated and input values of different parameters used in the thermodynamic study.

Parameter	Value
Heat rate consumed in reverse Deacon reactor (MW)	22.7
Net electricity generated by the system (MW)	95.4
Electric power consumed by the electrolyzer (MW)	255.8
Hydrogen produced daily (tonnes)	67.3
Energy efficiency of the overall system (%)	24.1
Exergy efficiency of the overall system (%)	23.5
Pressure ratio in the gas turbine cycle	2.5
Net electric power produced by the gas turbine cycle (MW)	301.8
Isentropic efficiency of the compressor (%)	90.0
Isentropic efficiency of the gas turbine (%)	92.1
Operating voltage (V)	1.7
Current density (A/m^2)	1000
Energy efficiency of the thermochemical cycle (%)	32.6
Exergy efficiency of the thermochemical cycle (%)	33.6

Source: *Modified from Khalid F, Bicer Y. Energy and exergy analyses of a hybrid small modular reactor and wind turbine system for trigeneration. Energy Sci Eng 2019;00:1–15. https://doi.org/10.1002/ese3.327.*

TABLE 5.5 Rate of exergy destruction of major units in the integrated system.

Component	The rate of exergy destruction[a] (MW)
Reactor core	245.5
Compressor	58.4
Electrolyzer	47.3
Heat exchanger 1	1.7
Gas turbine	19.8
Heat exchanger 2	5.6
Heat exchanger 3	17.1
Reverse Deacon reactor	1.6

[a] *Modified from Khalid F, Bicer Y. Energy and exergy analyses of a hybrid small modular reactor and wind turbine system for trigeneration. Energy Sci Eng 2019;00:1–15. https://doi.org/10.1002/ese3.327.*

decreases with the rise in outlet temperature of the reverse Deacon reactor. This all leads to an increase in exergy and energy efficiencies of the overall system. One should also observe from the figure that with the increase in outlet temperature of the reverse Deacon reactor, the net work rate produced by the gas turbine increases.

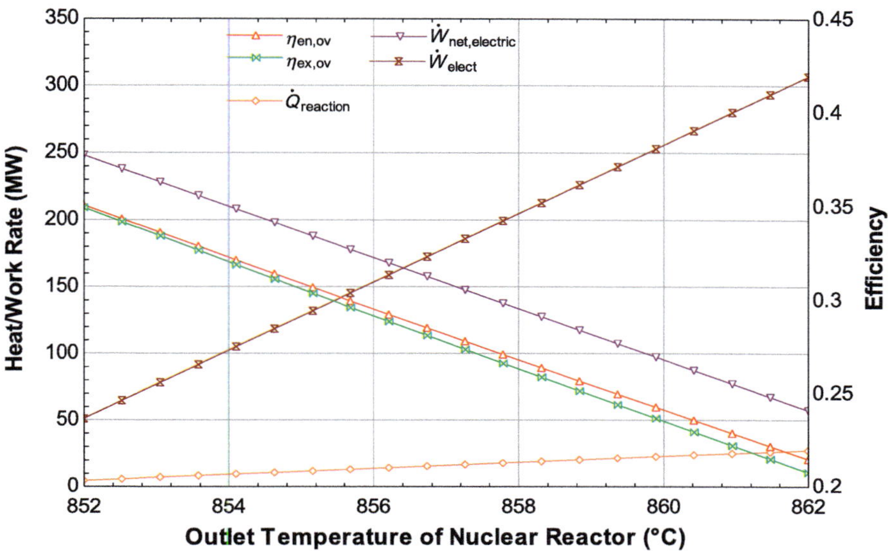

FIG. 5.9 Energy and exergy efficiencies, and heat/work rates variation with the outlet temperature of the nuclear reactor.

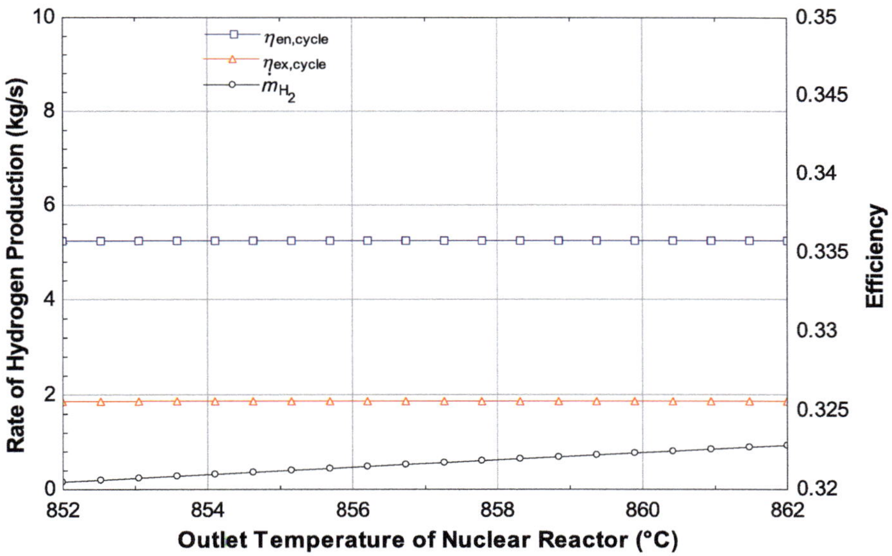

FIG. 5.10 Effects of the outlet temperature of nuclear reactor on the rate of hydrogen production, and energy and exergy efficiencies of the thermochemical cycle.

Operating temperature of the thermochemical cycle plays an important role. Thus, its effect on various efficiencies and work rates is shown in Fig. 5.13. As the temperature of the cycle rises from 650 °C to 800 °C, the exergy efficiency of the thermochemical cycle increases from 39.4% to 40.4% while the overall system exergy efficiency varies from 31.6% to 32.5%.

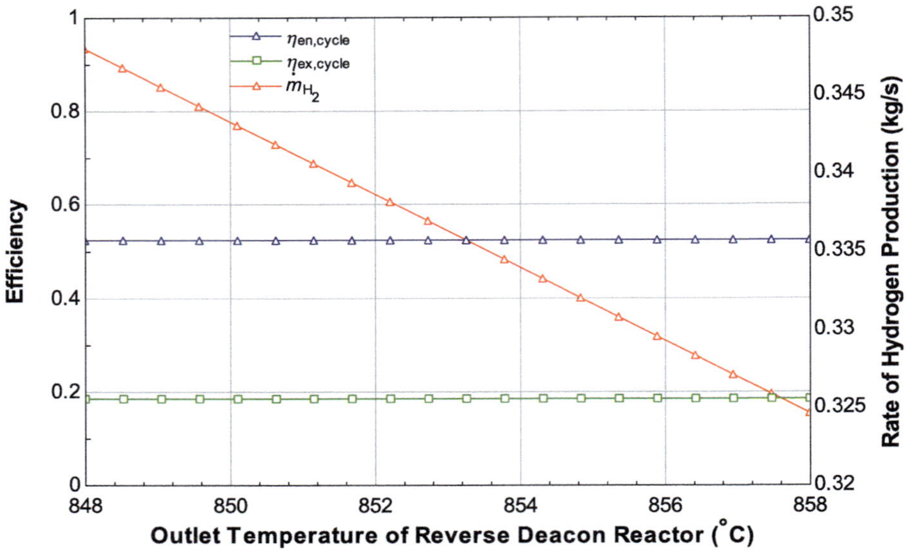

FIG. 5.11 Effects of the outlet temperature of reverse Deacon reactor (T_2) on efficiencies of the thermochemical cycle, and rate of hydrogen produced.

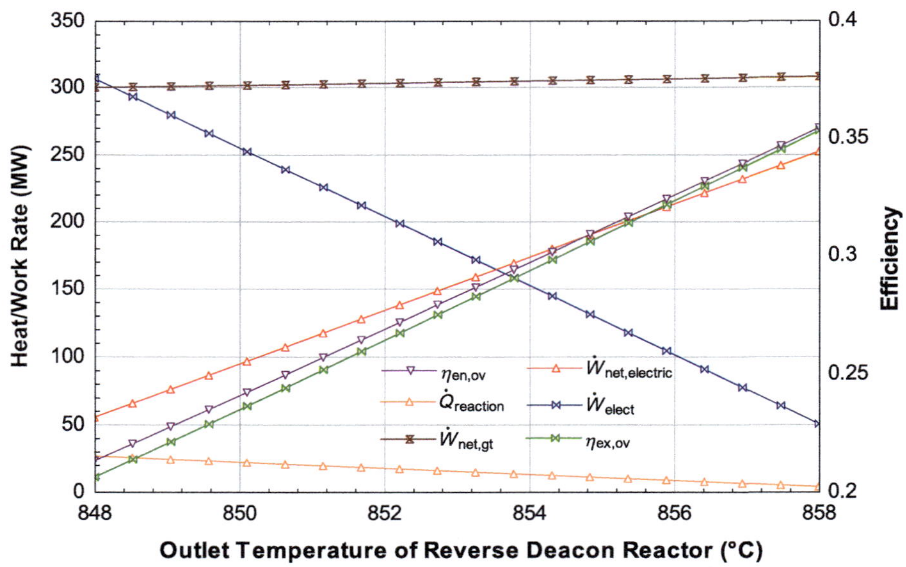

FIG. 5.12 Various efficiencies, heat and work rates variation with outlet temperature of reverse Deacon reactor (T_2).

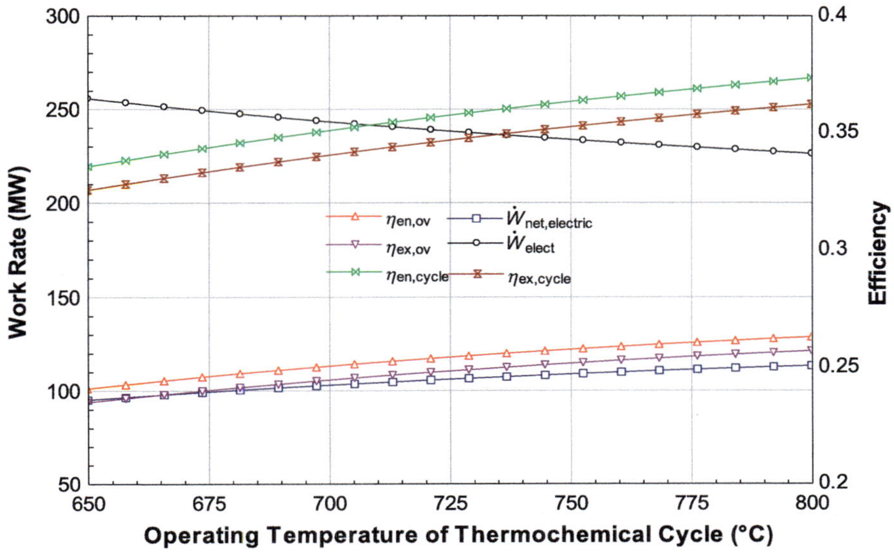

FIG. 5.13 Cycle temperature effects on the various work rates and efficiencies.

It is further observed from the figure that with the rise in operating temperature of the cycle, the work rate required by the electrolyzer decreases from about 256 MW to about 220 MW.

The decrease in electrolyzer work rate can be somewhat explained by Fig. 5.14. As the cycle temperature increases, the voltage required by the electrolyzer decreases from 1.7 V to 1.51 V.

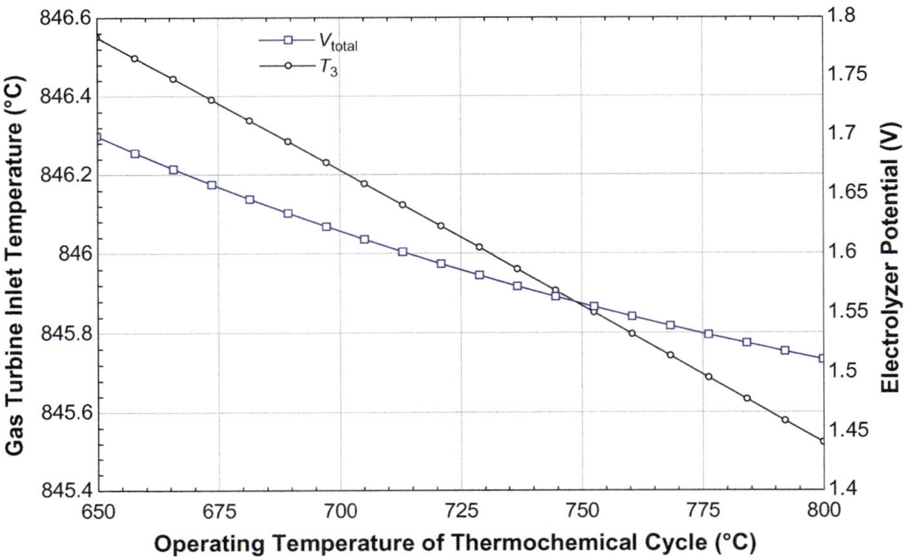

FIG. 5.14 Cycle temperature effects on the gas turbine inlet temperature, and electrolyzer potential.

The decrease in electrolyzer potential is explained by the fact that ohmic losses (due to the conductivity of the membrane) associated with the electrolyzer decreases with increase in electrolyzer temperature (cycle temperature). A drop in the gas turbine inlet temperature is also observed (see Fig. 5.14).

Current density effects on energy efficiencies of the thermochemical cycle and overall system along with exergy efficiencies are shown in Fig. 5.15. The system's energy efficiency changes from 28.4% to 19.8% while the system's exergy efficiency decreases from 27.8% to 19.2% as the current density rises from 500 A/m^2 to 1500 A/m^2. Energy efficiency of the thermochemical cycle changes from 41.6% to 27.2% while its exergy efficiency decreases from 40.5% to 28.2%.

Ohmic losses are directly proportional to current density. As the current density varies from 500 A/m^2 to 3000 A/m^2, the electrolyzer potential and work rate consumed by it increase from 1.34 V to 3.08 V, and 331.6 MW to 370.6 MW, respectively (see Fig. 5.16). The net work rate generated by the integrated system also decreases from 129.4 MW to 61.9 MW.

As the pressure ratio of the gas turbine cycle changes from 2.5 to 4.5, the integrated system's energy efficiency changes from 24.1% to 11.5%, while the integrated system's exergy efficiency declines from 23.5% to 11.1% as shown in Fig. 5.17. The electricity generated by the integrated system shows an interesting trend, i.e., first electricity generated rises with the increase in pressure ratio till it reaches the optimal value, and then starts reducing from there with the further increase in pressure ratio. This is because after the optimal value of pressure ratio, the rate at which compressor work increases is more than the rate at which the turbine produces electric power, leading to the overall decrease in electricity generated by the integrated system.

In Fig. 5.18, the effects of pressure ratio of the gas turbine on the rate of exergy destruction in the reactor core and heat exchanger 3 are illustrated. One can see from the figure that the two entities increase with the rise in pressure ratio. As the pressure ratio rises, a greater

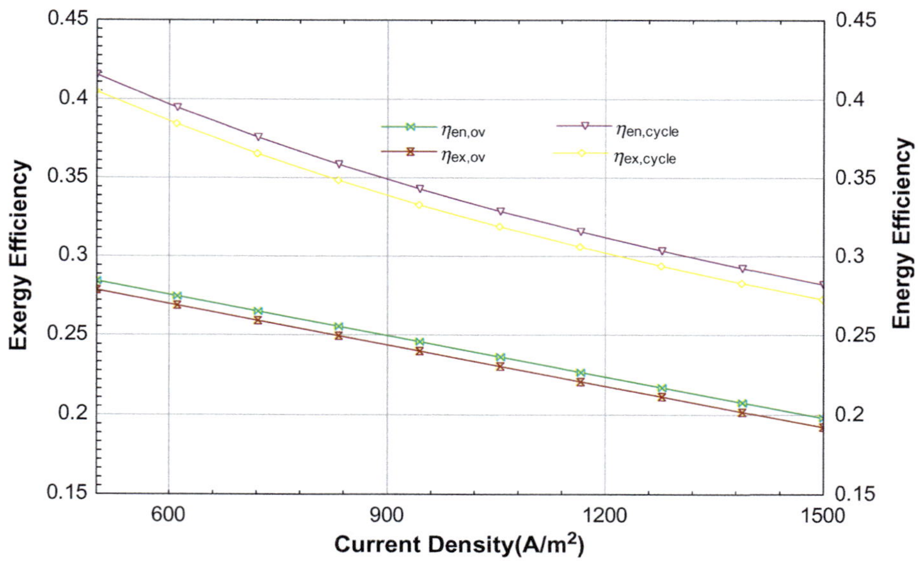

FIG. 5.15 Effect of current density on various efficiencies.

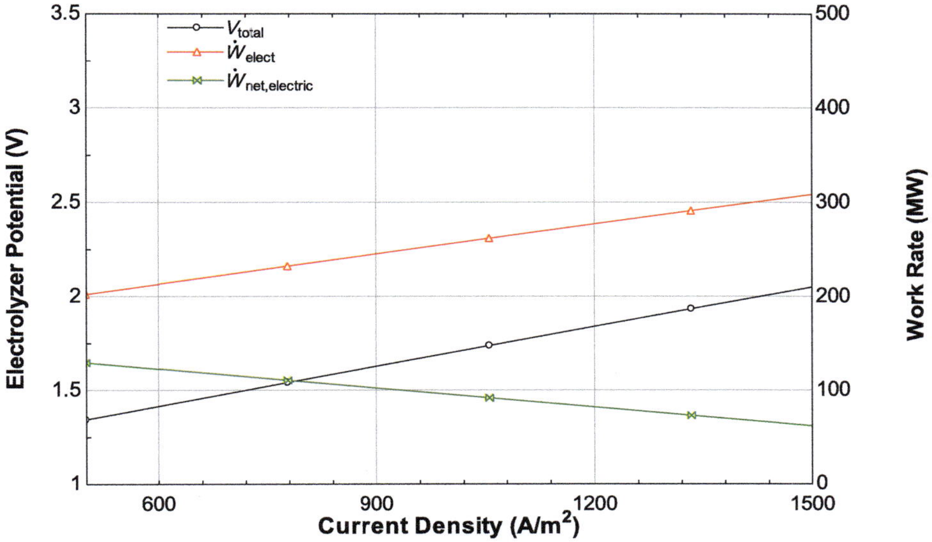

FIG. 5.16 Potential and work rates electrolyzer variation with the current density.

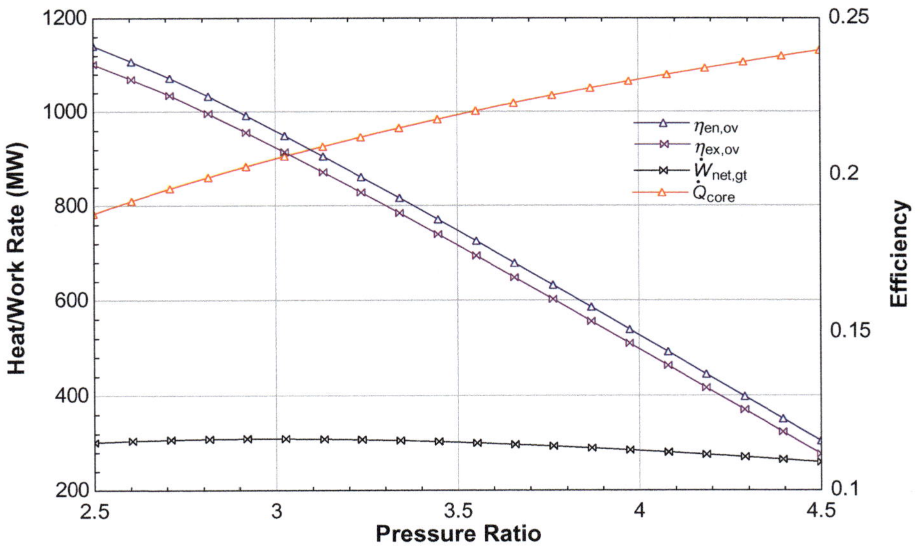

FIG. 5.17 Heat and work rates, and efficiencies variation with pressure ratio.

amount of heat is required from the core to heat the cooling helium coming out from the gas turbine. Since a greater pressure ratio means helium coming out from the turbine will be at much lower pressure, resulting in a decrease in its temperature, thus more heat will be required to again heat the helium to state 1 so that it can provide the required temperature

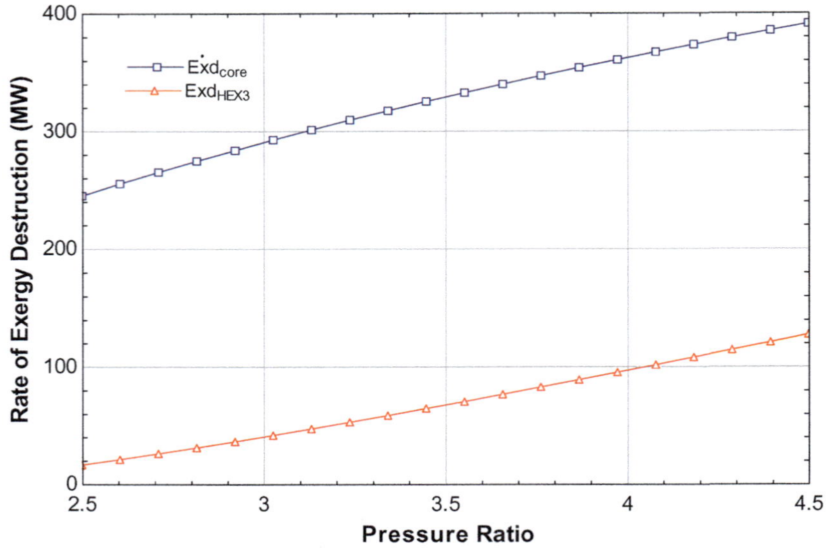

FIG. 5.18 Effects of pressure ratio on exergy destruction rates of the reactor core, and heat exchanger 3.

and heat needed by the reverse Deacon reactor. Hence, with the increase in the heat required by the core, the exergy destruction rate in the core increases.

5.5.5 Closing remarks

In this case study, coupling of a two-step thermochemical cycle with a small modular reactor for co-production of hydrogen and electricity is presented. An exergy assessment with conventional energy analysis is reported here. The results show that hydrogen on a large scale can be produced with the use of the present system, i.e., roughly around 68 t per day. The presented system can achieve an overall exergy efficiency of 23.5% with an energy efficiency value of 24.1%. The results presented in this case study will help researchers consider coupling of thermochemical cycles with small modular reactors for cogeneration or extended to trigeneration/multigeneration depending upon the need.

5.6 Case study 2: Integration of high-temperature nuclear reactor with thermochemical cycle for production of electrical power, hydrogen and hot water

In this case study, a hybrid energy system running on nuclear and wind energies for trigeneration (electrical power, hydrogen, and hot water) using a two-step thermochemical cycle (based on hydrogen chloride) is considered for which the inputs and outputs, as well as systems, are depicted in Fig. 5.19. This case study is modified from the detailed thermodynamic study conducted by Khalid and Bicer [3]. The salient feature of this study is that it presents the thermodynamic viability of integration of wind turbines with a nuclear reactor for large-scale hydrogen production.

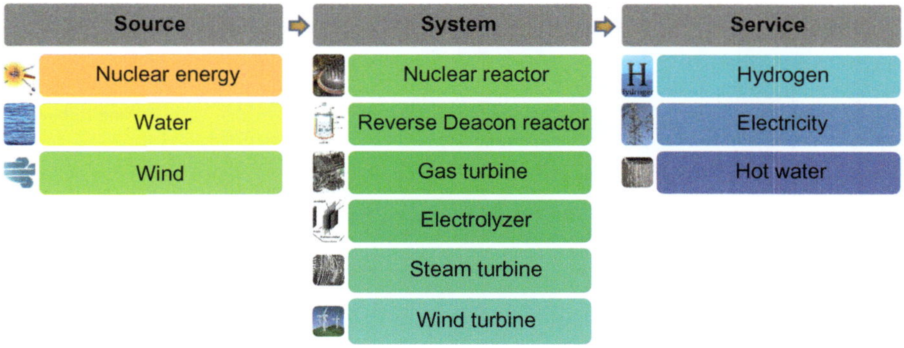

FIG. 5.19 3S approach for case study 2.

5.6.1 System description

The schematic diagram of the integration of wind turbines with a nuclear reactor for hydrogen, electricity and hot water is shown in Fig. 5.20. This system has several components, namely a set of wind turbines, small modular reactor (GT-MHR type), pumps, heat exchangers (HEXs), compressor, gas and steam turbines, mixer, splitter, electrolyzer, and reverse Deacon reactor. At state 1, compressed and hot helium leaves the nuclear reactor and is utilized to deliver the heat needed by the reverse Deacon reactor. At state 2, helium leaves the reverse Deacon reactor and is allowed to pass through heat exchanger 1, so that it can provide the heat and desired temperature required by the water entering the reverse Deacon reactor. Helium at state 3 enters the gas turbine and is expanded to generate electrical power. At state 4, helium at reduced pressure leaves the gas turbine and is further cooled down (state 5) in heat exchanger 2. The temperature of the cooled helium (state 5) is further reduced in heat exchanger 3 (state 7). The cooled helium coming out of heat exchanger 3 gets compressed in the compressor (state 8). This compressed helium gets heated in heat exchanger 2 and then enters the nuclear reactor core at state 9. At state 16, hydrogen chloride gas enters the electrolyzer to be split into hydrogen and chlorine gases. The electrical power consumed by the electrolyzer is generated by the set of wind turbines. At state 17, chlorine gas enters the reverse Deacon where it reacts with high-temperature steam to produce oxygen (state 18) and hydrogen chloride gases, thus completing the two-step thermochemical cycle. At state 12, pressurized water enters the electrolyzer and is heated by the losses incurred from the electrolyzer. This hot pressurized steam leaves the electrolyzer at state 19 and then enters the steam turbine to generate electrical power. The depressurized steam leaves the steam turbine at state 21 and is in heat exchanger 5. The condensate (state 22) obtaind from the heat exchanger 5 is again pressurized in pump 1 to ambient pressure (state 23). Before entering the mixer, this pressurized water gets cooled in heat exchanger 6 to ambient temperature (state 27). In the mixer, the pressurized water coming out from heat exchanger 6 mixes with the freshwater stream (state 25). At state 24, a single mixed stream of water leaves the mixer which enters pump 2 to be pressurized (state 10). This pressurized water gets heated in heat exchanger 3 and leaves at state 11. A splitter is used to split the hot steam coming out from the heat exchanger 3 into two parts, namely state 12 and state 13. The stream at state 13 is utilized

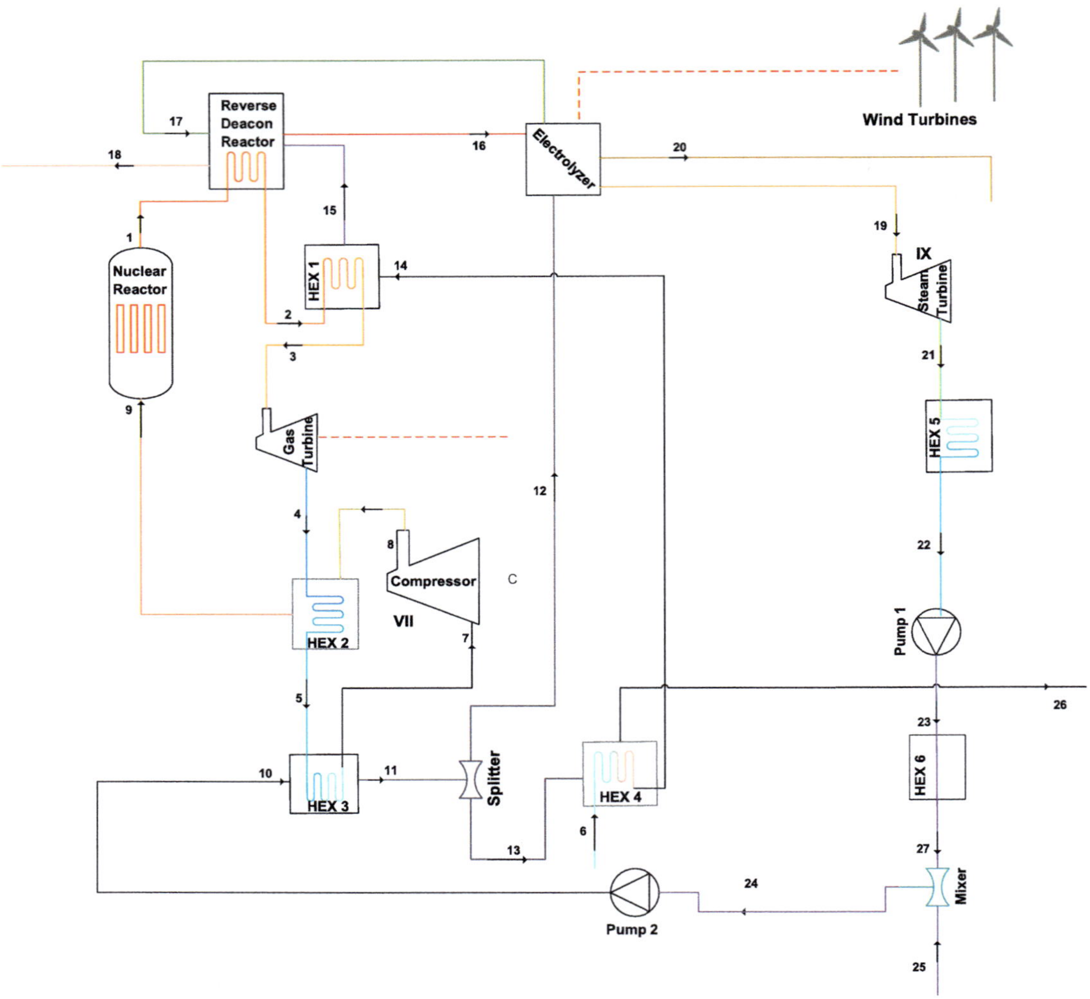

FIG. 5.20 Schematic diagram of the integration of the wind turbines with nuclear reactor for trigeneration.

to convert the ambient water (state 6) into steam at ambient pressure (state 14) and leaves the heat exchanger 4 as hot water (useful commodity). The other stream at state 12 is being utilized in the electrolyzer. This is how the whole integrated system works by utilizing wind and nuclear energies and producing three useful commodities, namely hydrogen, electrical power, and hot water.

5.6.2 Assumptions

In order to perform thermodynamic analysis of this integrated system, there is a need to make conceptually correct assumptions. The following main reasonable assumptions are considered in this case study:

- The system works under steady state.
- Full completion for chemical reactions according to their stoichiometry.
- Pressure drops in pipelines and heat exchangers are ignored.
- No heat losses are considered in heat exchangers, pumps and turbines.
- Ambient temperature and pressure are 25 °C and 100 kPa, respectively.

5.6.3 Analysis

The thermochemical cycle and system performance are measured by defining energy and exergy efficiencies.

For the thermochemical cycle, the following expressions of exergy and energy efficiencies are used, respectively:

$$\eta_{ex,cycle} = \frac{\dot{m}_{H_2} ex_{H_2}}{\dot{W}_{elect} + \dot{Ex}^{Q}_{reaction} + \dot{Ex}^{Q}_{required} - \dot{Ex}^{Q}_{loss,electrolyzer}} \quad (5.8)$$

where the thermal exergies associated with reverse Deacon reaction, heat required, and electrolyzer losses are $\dot{Ex}^{Q}_{reaction}$, $\dot{Ex}^{Q}_{required}$, and $\dot{Ex}^{Q}_{loss,electrolyzer}$, respectively. Here, $\dot{W}_{elect}$ denotes the electrical power consumed by the electrolyzer.

$$\eta_{en,cycle} = \frac{\dot{m}_{H_2} LHV_{H_2}}{\dot{W}_{elect} + \dot{Q}_{reaction} + \dot{Q}_{required} - \dot{Q}_{loss,elect}} \quad (5.9)$$

Likewise, for the overall system performance; the following energy and exergy efficiencies expressions are utilized, respectively:

$$\eta_{en,ov} = \frac{\dot{m}_{H_2} LHV_{H_2} + \dot{W}_{net,electric} + \dot{m}_{26} h_{26} - \dot{m}_{25} h_{25}}{\dot{W}_{wind} + \dot{Q}_{core}} \quad (5.10)$$

where the mass flow rate of hydrogen produced is $\dot{m}_{H_2}$, and the lower heating value of the hydrogen is LHV. Here, $\dot{Q}_{core}$, and $\dot{W}_{wind}$ denote the heat required by the reactor core, and wind energy work rate input, respectively.

$$\eta_{ex,ov} = \frac{\dot{m}_{H_2} ex_{H_2} + \dot{W}_{net,electric} + \dot{m}_{26} ex_{26} - \dot{m}_{25} ex_{25}}{\dot{W}_{wind} + \dot{Ex}^{Q}_{core}} \quad (5.11)$$

5.6.4 Results and discussion

State points with all their relevant thermodynamic data such as P (pressure in kPa), T (temperature in °C), $\dot{m}$ (mass flow rate in kg/s), ex (specific exergy in kJ/kg) and h (specific enthalpy in kJ/kg) are identified and calculated (see Table 5.6). In addition, Table 5.7 lists the values of all the useful parameters obtained from this study. Note that 134.7 t of hydrogen can be produced daily using the presented integrated system with the system operating at energy and exergy efficiencies of 49.4%, and 31.6%, respectively. The electrolyzer in the presented study operates at 1.7 V for a current density of 1000 A/m^2.

TABLE 5.6 Thermodynamic data for the trigeneration system.

State no.	Fluid type	T (°C)	P (kPa)	$\dot{m}$ (kg/s)	h (kJ/kg)	ex (kJ/kg)
1	Helium	860.0	7000	439	5910	4911
2	Helium	840.0	7000	439	5806	4834
3	Helium	833.1	7000	439	5770	4808
4	Helium	516.5	2800	439	4115	3107
5	Helium	213.2	2800	439	2540	2283
6[a]	Water	25	101.3	13.93	−15,866	0
7	Helium	30	2800	439	1589	2063
8	Helium	203.1	7000	439	2501	2842
9	Helium	506.5	7000	439	4076	3654
10	Water	25.9	15,199	1066	122.7	15.1
11	Water	120	15,199	1066	514.4	67.58
12	Water	120	15,199	60.95	514.4	67.58
13	Water	120	15,199	1005	514.4	67.58
14[a]	Water	100	101.3	13.93	−13,282	542
15[a]	Water	630	101.3	13.93	−12,154	1124
16[a]	Hydrogen chloride	630	101.3	56.39	−2021	2563
17[a]	Chlorine	630	101.3	54.83	319.9	1892
18[a]	Oxygen	630	101.3	12.37	626.1	418.1
19	Water	550	15,199	60.95	3448	1511
20[a]	Hydrogen	630	101.3	1.559	9109	117,111
21	Water	45.8	10	60.95	2271	139.4
22	Water	45.8	10	60.95	191.8	2.85
23	Water	45.8	101.3	60.95	191.9	2.94
24	Water	25	101.3	1066	104.9	0
25	Water	25	101.3	1005	104.9	0
26	Water	111.5	15,199	1005	478.6	59.21
27	Water	25	101.3	60.95	104.9	0

[a] *Calculated using NASA properties in EES [38].*

Source: *Data modified from Khalid F, Bicer Y. Energy and exergy analyses of a hybrid small modular reactor and wind turbine system for trigeneration. Energy Sci Eng 2019;00:1–15. https://doi.org/10.1002/ese3.327.*

By determining the rates of exergy destruction of various units in the integrated system, key players for losses in the integrated system are identified and have been tabulated in Table 5.8. The exergy destruction losses incurred by the electrolyzer are 281.4 MW while the reactor core also has a significant contribution to exergy destruction losses (253.8 MW).

TABLE 5.7 Calculated and input values of different parameters used in the thermodynamic study.

Parameter	Value
Heat rate consumed in reverse Deacon reactor (MW)	45.6
Total electricity generated by the system (MW)	341.7
Electric power consumed by the electrolyzer (MW)	511.6
Hydrogen produced daily (tonnes)	134.7
Hot water produced daily (tonnes)	86,850
Energy efficiency of the overall system (%)	49.4
Exergy efficiency of the overall system (%)	31.6
Pressure ratio in the gas turbine cycle	2.5
Net electric power produced by the gas turbine cycle (MW)	294
Isentropic efficiency of the compressor (%)	90.0
Isentropic efficiency of the gas turbine (%)	92.1
Operating voltage (V)	1.7
Current density (A/m^2)	1000

Source: *Modified from Khalid F, Bicer Y. Energy and exergy analyses of a hybrid small modular reactor and wind turbine system for trigeneration. Energy Sci Eng 2019;00:1–15. https://doi.org/10.1002/ese3.327.*

TABLE 5.8 Rate of exergy destruction of major units in the integrated system.

Component	The rate of exergy destruction[a] (MW)
Electrolyzer	281.4
Reactor core	253.8
Compressor	58.4
Heat exchanger 1	3.4
Gas turbine	19.8
Heat exchanger 2	5.5
Heat exchanger 3	40.6
Steam turbine	11.8
Reverse Deacon reactor	3.2

[a] *Modified from Khalid F, Bicer Y. Energy and exergy analyses of a hybrid small modular reactor and wind turbine system for trigeneration. Energy Sci Eng 2019;00:1–15. https://doi.org/10.1002/ese3.327.*

Fig. 5.21 illustrates the variation of energy and exergy efficiencies of the overall system and heat/work rates with the outlet temperature of the nuclear reactor. As the temperature rises from 850 °C to 950 °C, the heat rate of the reverse Deacon increases while both exergy and energy efficiencies of the system decrease. Fig. 5.21 also shows with the increase in outlet

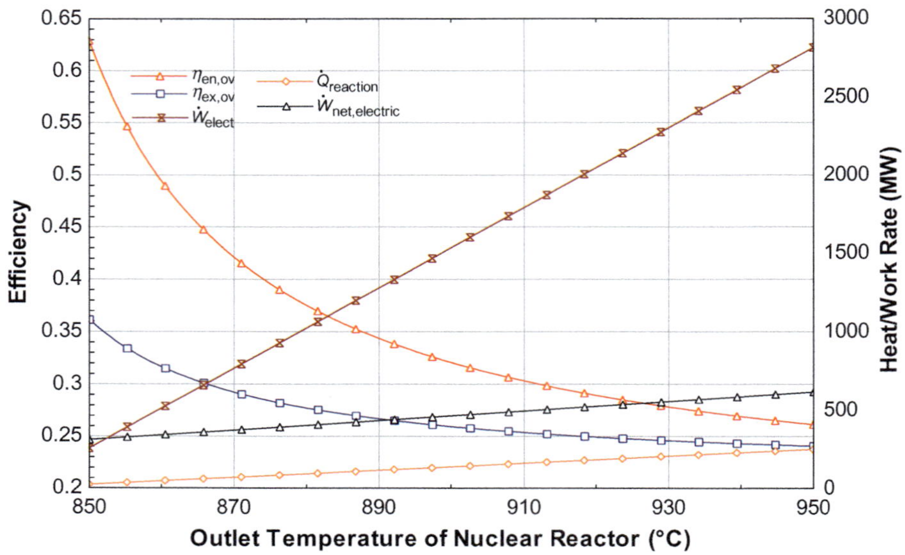

FIG. 5.21 Energy and exergy efficiencies, and heat/work rates variation with the outlet temperature of the nuclear reactor.

temperature, electrical power consumed by the electrolyzer and the electrical power generated by the system also rise.

Fig. 5.22 shows the effect of the outlet temperature of the nuclear reactor on the rate of hydrogen production, and both efficiencies (energy and exergy) of the thermochemical cycle. As the temperature changes from 850 °C to 950 °C, the rate of hydrogen production rises from 0.78 kg/s to 8.57 kg/s while no effect on the efficiencies of the cycle is observed.

In Fig. 5.23, the effect of the outlet temperature of the reverse Deacon reactor on both efficiencies (exergy and energy) of the thermochemical cycle and the rate of hydrogen production is plotted. As the temperature increases from 820 °C to 840 °C, no effects are observed on the efficiencies while the rate of hydrogen produced decreases from 3.12 kg/s to 1.56 kg/s. With the reduction in hydrogen produced, the amount of electricity consumed by the electrolyzer also decreases, resulting in no change in efficiencies of the cycle.

As the outlet temperature of the reverse Deacon reactor varies from 820 °C to 840 °C, the system's exergy and energy efficiencies change from 27.4% to 31.6%, and 37.3% to 49.4%, respectively (see Fig. 5.24). Heat rates for the reverse Deacon reaction and reactor core both decrease with the rise in outlet temperature of the reverse Deacon reactor. This all leads to an increase in exergy and energy efficiencies of the overall system. One should also observe from the figure that, with the increase in outlet temperature of the reverse Deacon reactor, the net work rate produced by the gas turbine increases.

Operating temperature of the thermochemical cycle plays an important role. Thus, its effect on various efficiencies and work rates is shown in Fig. 5.25. As the temperature of the cycle rises from 650 °C to 800 °C, the exergy efficiency of the thermochemical cycle increases from 39.4% to 40.4% while the overall system's exergy efficiency varies from 31.6% to 32.5%. It is further observed from the figure that, with the rise in operating temperature of the cycle, the work rate required by the electrolyzer decreases from 511.6 MW to 452.6 MW.

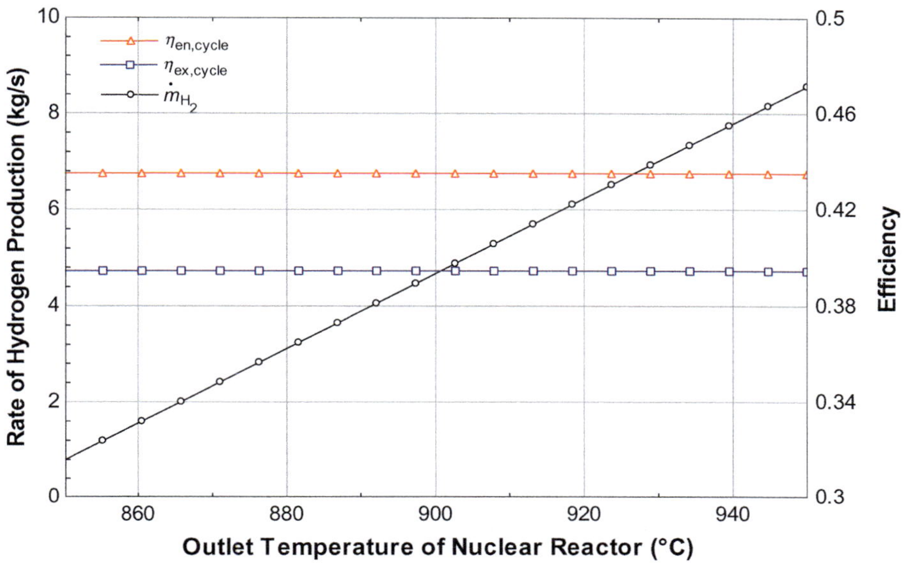

FIG. 5.22 Effects of the outlet temperature of nuclear reactor on the rate of hydrogen production and energy and exergy efficiencies of the thermochemical cycle.

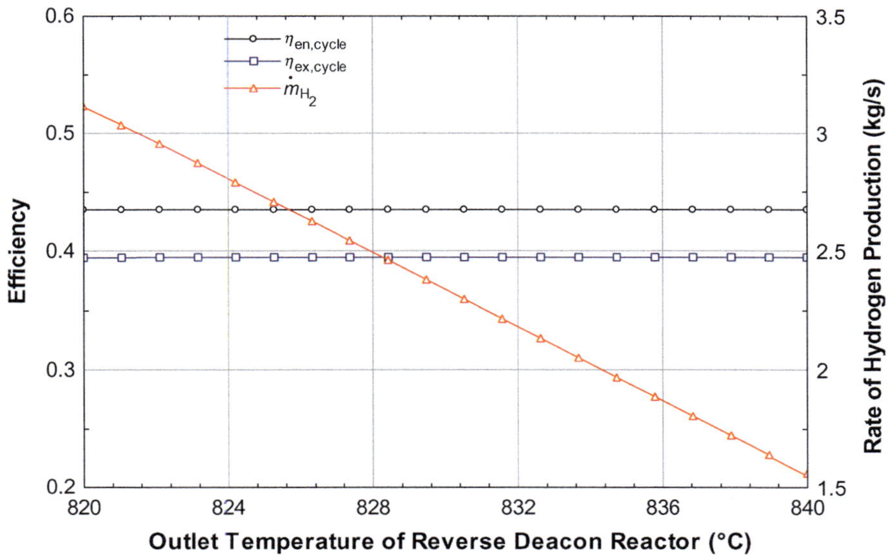

FIG. 5.23 Effects of the outlet temperature of reverse Deacon reactor (T_2) on efficiencies of the thermochemical cycle and rate of hydrogen produced.

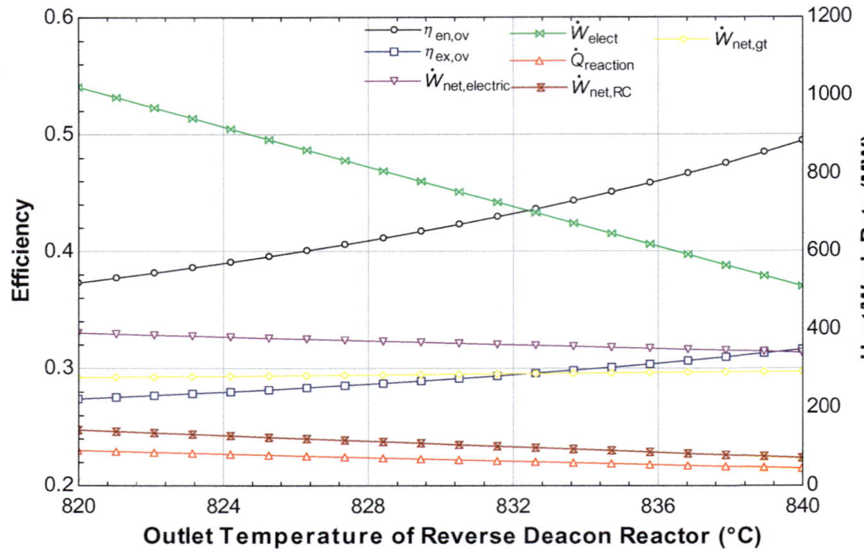

FIG. 5.24 Various efficiencies, heat and work rates variation with the outlet temperature of reverse Deacon reactor (T_2).

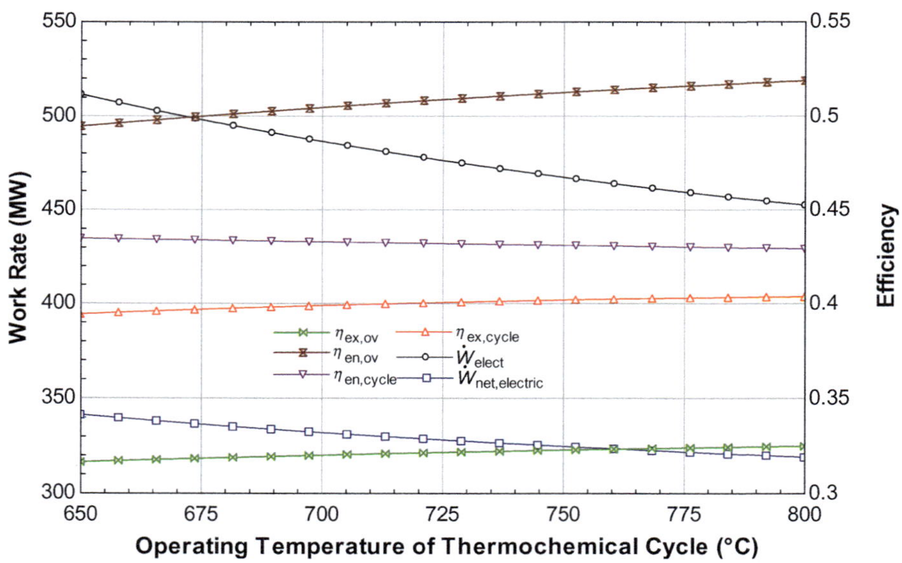

FIG. 5.25 Cycle temperature effects on the various work rates and efficiencies.

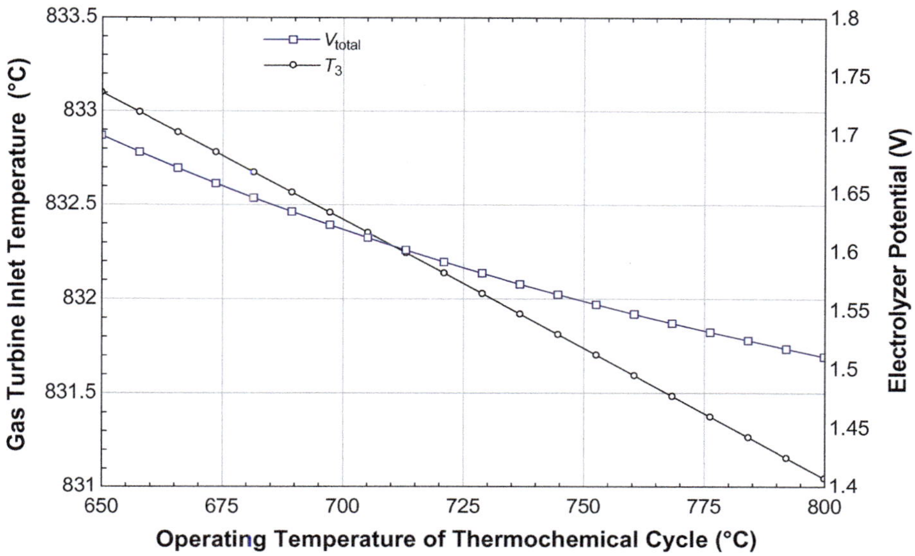

FIG. 5.26 Cycle temperature effects on the gas turbine inlet temperature and electrolyzer potential.

The decrease in electrolyzer work rate can be somewhat explained by Fig. 5.26. As the cycle temperature increases, the voltage required by the electrolyzer decreases from 1.7 V to 1.5 V. The decrease in electrolyzer potential is explained by the fact that ohmic losses (due to the conductivity of the membrane) associated with the electrolyzer decrease with increase in electrolyzer temperature (cycle temperature). A drop in the gas turbine inlet temperature is also observed (see Fig. 5.26).

Current density effects on energy efficiencies of the thermochemical cycle and overall system along with exergy efficiencies are shown in Fig. 5.27. The system's energy efficiency changes from 54.5% to 37.5% while its exergy efficiency reduces from 33.5% to 27.1% as the current density rises from 500 A/m^2 to 3000 A/m^2. However, no effect is observed on the energy efficiency of the thermochemical cycle, while its exergy efficiency decreases from 41.9% to 32.1%. This result shows the advantage of carrying out exergy analysis in addition to energy analysis.

Ohmic losses are directly proportional to current density. As the current density varies from 500 A/m^2 to 3000 A/m^2, the electrolyzer potential and work rate consumed by it both increase from 1.34 V to 3.08 V, and 404.4 MW to 927.8 MW, respectively (see Fig. 5.28). It is also seen that with an increase in current density, the electricity generated by the Rankine cycle also rises, which results in an overall increase in the electricity generated by the integrated system. This phenomenon occurs because, as the electrolyzer losses increase, a greater amount of steam can be heated from these losses, resulting in greater mass flow rate inside the Rankine cycle leading to more power output.

As the pressure ratio of the gas turbine cycle changes from 2.5 to 4.5, the integrated system's exergy efficiency changes from 31.6% to 26.9%, while its energy efficiency increases from 49.4% to 57.5% as shown in Fig. 5.29. The electricity generated by the integrated system

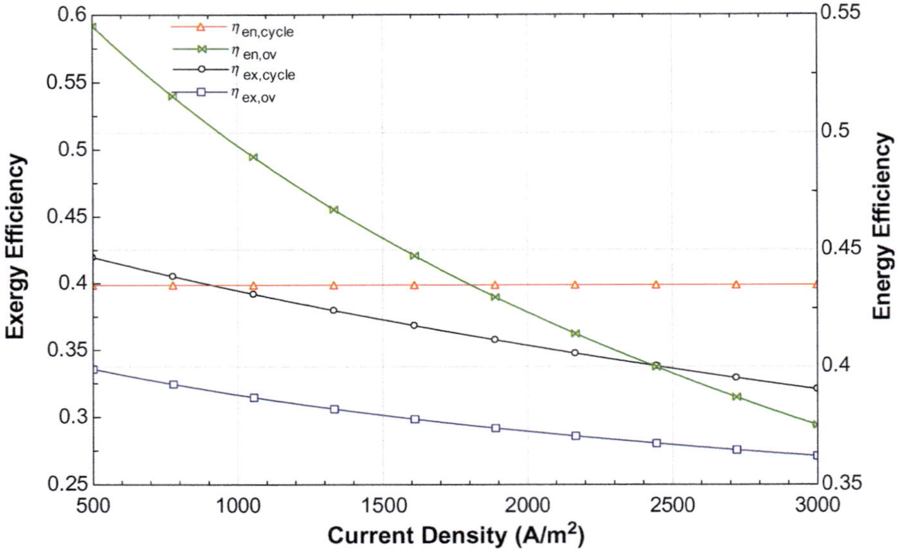

FIG. 5.27 Effect of current density on various efficiencies.

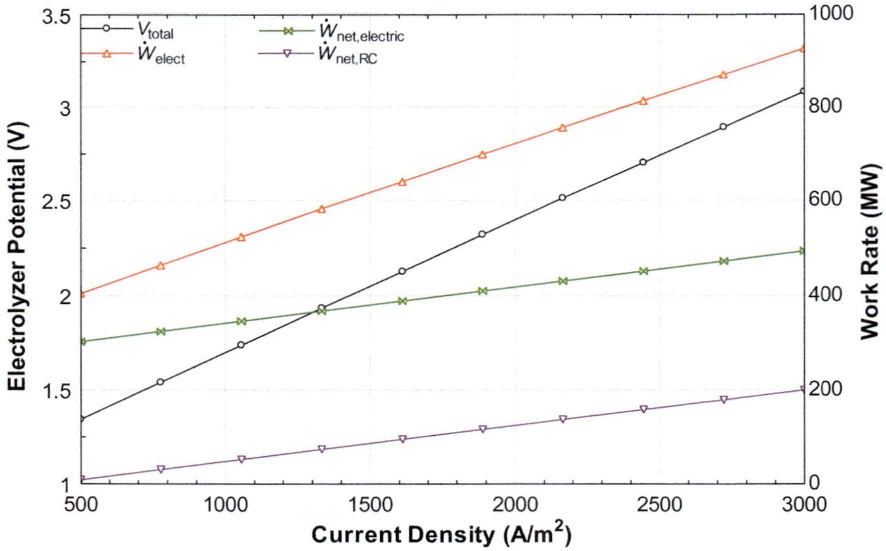

FIG. 5.28 Electrolyzer potential and work rate variation with the current density.

shows an interesting trend, i.e., first electricity generated rises with the increase in pressure ratio till it reaches the optimal value, and then begins to decrease from there with the further increase in pressure ratio. This is because after the optimal value of pressure ratio, the rate at

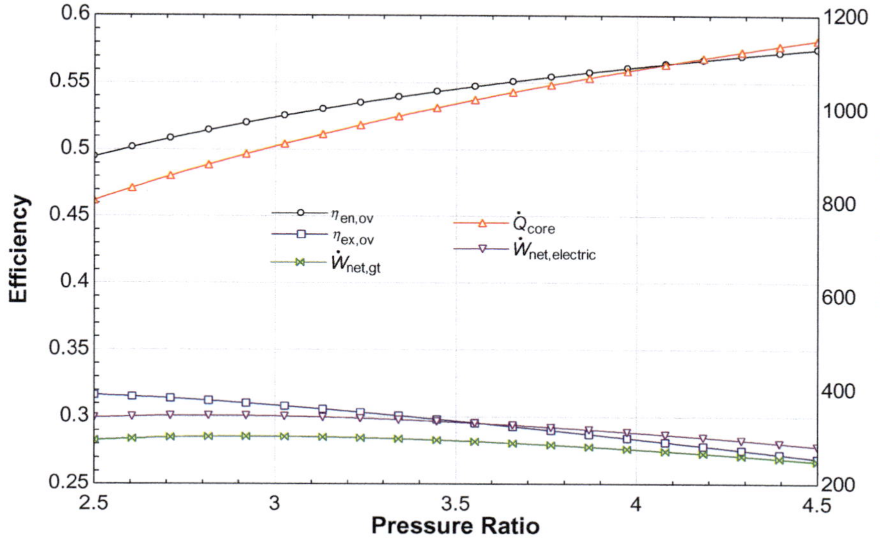

FIG. 5.29 Heat and work rates, and efficiencies variation with pressure ratio.

which compressor work increases is more than the rate at which the turbine produces electric power, leading to the overall decrease in electricity generated by the integrated system.

In Fig. 5.30, the effects of pressure ratio of the gas turbine on the rate of the exergy destruction in the reactor core, heat exchanger 3, and the rate of hot water produced are illustrated.

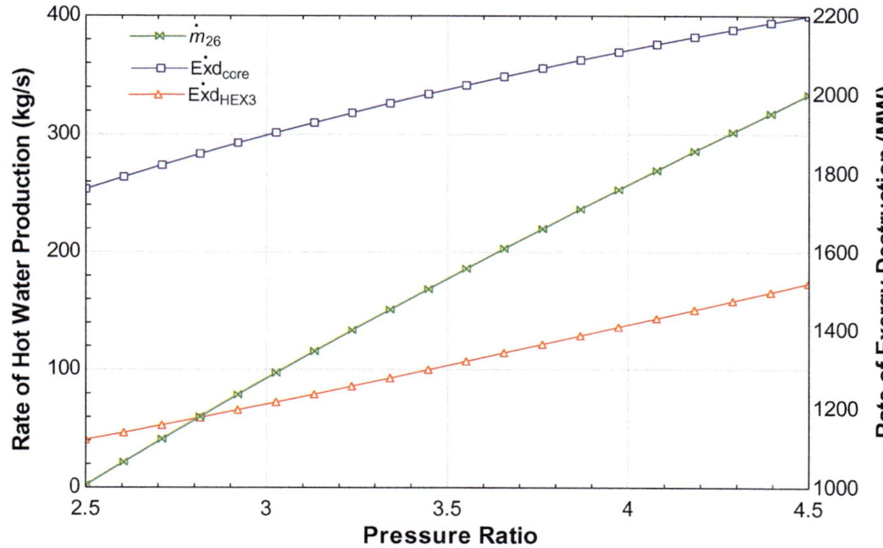

FIG. 5.30 Effects of pressure ratio on the rate of hot water produced, exergy destruction rates of the reactor core, and heat exchanger 3.

One can see from the figure that all the three entities increase with the rise in pressure ratio. As the pressure ratio rises, a greater amount of heat is required from the core to heat the cooling helium coming out from the gas turbine. A greater pressure ratio means helium coming out from the turbine will be at much lower pressure, resulting in a decrease in its temperature. Thus, more heat will be required to again heat the helium to state 1 so that it can provide the required temperature and heat needed by the reverse Deacon reactor. Hence, with the increase in the heat required by the core, the exergy destruction rate in the core increases.

5.6.5 Closing remarks

In this system, hybridization of a set of wind turbines with a small modular nuclear reactor is reported. An exergy assessment with a conventional energy analysis of a novel integrated system that runs on wind and nuclear energies for triple useful commodities, namely electrical power, hot water, and hydrogen, is presented in this case study. Hydrogen on a large scale can be produced with the use of the present system as it utilizes the two-step thermochemical cycle. The results of the current study suggest that the presented system achieves an exergy efficiency of 31.6% and an energy efficiency of 49.4%. This case study will open the avenue for researchers to consider hybridization of renewable energy sources such as wind and solar with small modular nuclear reactors for various useful outputs.

5.7 Case study 3: CANDU 6 and sodium-cooled fast reactors for nuclear desalination and electricity

This case study reports on a comparative assessment of CANDU 6 and sodium-cooled fast reactor-based energy systems for nuclear desalination and electricity production. The reverse osmosis (RO) based desalination method is considered in the case study, where the sources, systems, services are illustrated in Fig. 5.31. This study is based on the investigation conducted by Khalid et al. [4]. The salient features of this study are that, for both the energy systems, the feed of the RO plant is preheated by the condenser heat, and the energy from the pressurized brine is recovered by use of a Pelton turbine.

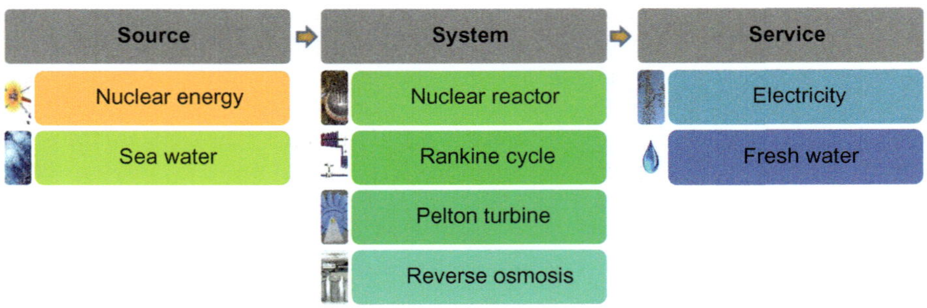

FIG. 5.31 3S approach for case study 3.

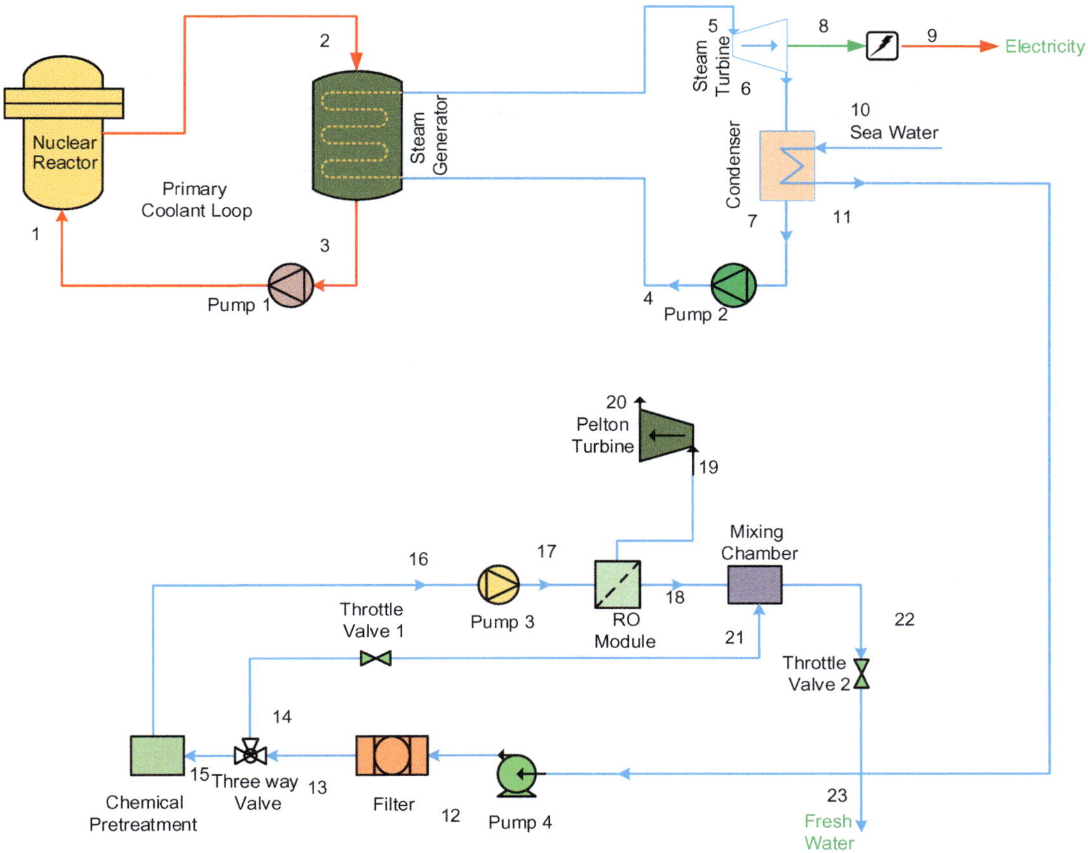

FIG. 5.32 CANDU 6 based energy system for co-production of fresh water and electricity. *Adapted from Khalid F, Dincer I, Rosen MA. Comparative assessment of CANDU 6 and sodium-cooled fast reactors for nuclear desalination. Desalination 2016;379:182–92*

5.7.1 System descriptions

Fig. 5.32 shows the schematic diagram of a CANDU 6 based nuclear power plant for co-production of fresh water and electricity using a reverse osmosis plant. This system has several components, namely a nuclear reactor, steam turbine, steam generator, pumps, Pelton turbine, condenser, throttle valves, RO module, mixing chamber, and filter. At state 2, the primary coolant (pressurized heavy water) leaves the nuclear reactor and enters the steam generator and leaves it at state 3. The pressurized water at state 4 enters the steam generator and leaves it as hot pressurized steam (state 5), which gets expanded in a steam turbine to produce electrical power. The steam leaves the turbine at state 6 and is condensed in the condenser by the inlet seawater coming in at state 10. The condensate leaves the condenser at state 7, and gets pumped through pump 2 to state 4. The coolant coming out from the steam generator (state 3) gets pumped by Pump 1 and enters the nuclear reactor. The heated seawater that comes out from the condenser (state 11) goes to pump 4 where it gets pressurized and enters the filter at state 12. The pressurized seawater leaves the filter at state 13, and splits into two streams, the stream at state 14 and the stream at state 15. The stream at state 15 enters the chemical pre-treatment plant where it gets chemically treated. The treated stream leaves the chemical pre-treatment plant at state 16 and gets pressurized through pump 3 to state 17. The stream at state 17 enters the RO module and produces two outputs, namely fresh water (state 18) and brine (state 19). The brine leaves the RO module at high pressure; thus to recover energy from it, the Pelton turbine is used. The stream at state 14 is throttled by throttle valve 1. This throttle stream (state 21) goes to the mixing chamber where it is mixed with another stream of water (state 18). The mixed streams leave the mixing chambers as one stream (state 22) and enter the throttle valve 2, where the pressure gets reduced (state 23). The stream at 23 is the stream of the fresh water produced by the desalination plant considered in this study.

Fig. 5.33 shows the sodium fast cooled reactor based energy system for co-production of fresh water and electricity. This system also comprises several units, namely a nuclear reactor, primary and secondary heat exchangers, steam generator, steam turbine, condenser, pumps, Pelton turbine, condenser, throttle valves, RO module, mixing chamber, and filter.

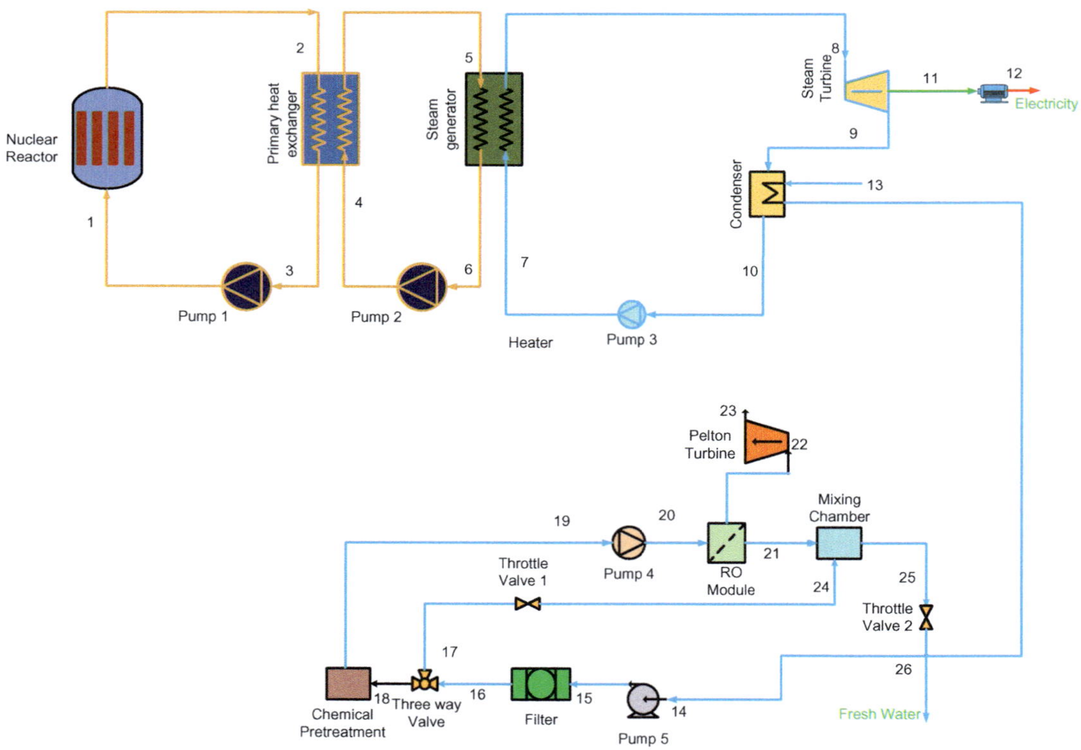

FIG. 5.33 SFR based energy system for co-production of fresh water and electricity. *Modified from Khalid F, Dincer I, Rosen MA. Comparative assessment of CANDU 6 and sodium-cooled fast reactors for nuclear desalination. Desalination 2016;379:182–92*

This system also works in a similar fashion to the system presented in Fig. 5.32. However, in order to make it safer, a primary heat exchanger is additionally used so that no contact of working fluid is possible with the coolant (liquid sodium).

5.7.2 Assumptions

In order to perform thermodynamic analysis of this integrated systems, there is a need to make conceptually correct assumptions. The following main reasonable assumptions are considered in this case study:

- The systems work under steady state.
- Changes in kinetic and potential energies and exergies are negligible.
- The turbines, throttle valves, and pumps work adiabatically.
- 85% isentropic efficiencies of the turbine and pumps are considered.
- Pipelines and heat exchanger pressure losses are neglected.

5.7.3 Analysis

To evaluate the thermodynamic performance of the RO process, electric power generation, and the two energy systems considered in this case study, exergy efficiency is used and is defined as follows:

For the reverse osmosis (RO) plant considered in Fig. 5.32, the exergy efficiency is defined as

$$\eta_{ex,ro1} = \frac{\dot{m}_{23}ex_{23} + \dot{m}_{20}ex_{20} - \dot{m}_{10}ex_{10}}{\dot{W}_{Pump3} + \dot{W}_{Pump4} - \dot{W}_{Pelton}} \tag{5.12}$$

where $\dot{W}_{Pump3} + \dot{W}_{Pump4}$ are the electricity consumed by the pumps and $\dot{W}_{Pelton}$ is the electricity generated by the Pelton turbine.

Similarly, for the RO plant considered in Fig. 5.33, the exergy efficiency is written as follows:

$$\eta_{ex,ro2} = \frac{\dot{m}_{23}ex_{23} + \dot{m}_{26}ex_{26} - \dot{m}_{13}ex_{13}}{\dot{W}_{Pump5} + \dot{W}_{Pump4} - \dot{W}_{Pelton}} \tag{5.13}$$

For the electricity generation efficiency for both systems, the exergy efficiencies are defined, respectively, as

$$\eta_{ex,elec1} = \frac{\dot{W}_{Steam} + \dot{W}_{Pelton} - \dot{W}_{Pump3} - \dot{W}_{Pump4} - \dot{W}_{Pump1} - \dot{W}_{Pump2}}{\dot{E}x^{Q_r}} \tag{5.14}$$

$$\eta_{ex,elec2} = \frac{\dot{W}_{Steam} + \dot{W}_{Pelton} - \dot{W}_{Pump5} - \dot{W}_{Pump3} - \dot{W}_{Pump4} - \dot{W}_{Pump1} - \dot{W}_{Pump2}}{\dot{E}x^{Q_r}} \tag{5.15}$$

The exergy efficiencies for the co-production of fresh water and electricity for both systems are defined as follows:

$$\eta_{ex,cog1} = \frac{\dot{m}_{23}ex_{23} + \dot{m}_{20}ex_{20} - \dot{m}_{10}ex_{10} + \dot{W}_{Steam} + \dot{W}_{Pelton} - \dot{W}_{Pump3} - \dot{W}_{Pump4} - \dot{W}_{Pump1} - \dot{W}_{Pump2}}{\dot{E}x^{Q_r}} \tag{5.16}$$

$$\eta_{ex,cog2} = \frac{\dot{m}_{23}ex_{23} + \dot{m}_{26}ex_{26} - \dot{m}_{13}ex_{13} + \dot{W}_{Steam} + \dot{W}_{Pelton} - \dot{W}_{Pump5} - \dot{W}_{Pump3} - \dot{W}_{Pump4} - \dot{W}_{Pump1} - \dot{W}_{Pump2}}{\dot{E}x^{Q_r}} \quad (5.17)$$

5.7.4 Results and discussion

All the state points are defined with all their relevant thermodynamic data, including P (pressure in kPa), T (temperature in °C), $\dot{m}$ (mass flow rate in kg/s), ex (specific exergy in kJ/kg), and h (specific enthalpy in kJ/kg). Table 5.9 lists the values of some of the useful parameters used in the present case study. It is to be noted from the table that both the energy systems can produce about 5000 kg/s of fresh water. By determining the rates of exergy destruction of various units in the CANDU 6 and SFR energy systems, the key players for losses are identified and have been tabulated in Table 5.10. For the CANDU 6 energy system, steam generator have the maximum exergy

TABLE 5.9 Values of useful parameters used in the present case study.

Parameter	Value
Inlet temperature of the coolant for CANDU 6 reactor (°C)	249.5
Outlet temperature of the coolant for CANDU 6 reactor (°C)	291.9
Mass flow rate of the coolant for CANDU 6 reactor (kg/s)	7724
Mass flow rate of the steam for CANDU 6 reactor (kg/s)	640
Steam turbine inlet pressure for CANDU 6 reactor (kPa)	4250
Steam turbine inlet temperature for CANDU 6 reactor (°C)	254
Condenser operating pressure for CANDU 6 reactor (kPa)	7
Inlet temperature of RO feedwater for CANDU 6 reactor (°C)	36
Amount of fresh water produced for CANDU 6 reactor (kg/s)	5000
Inlet temperature of the primary coolant for SFR (°C)	355.5
Outlet temperature of the primary coolant for SFR (°C)	550
Inlet temperature of the secondary coolant for SFR (°C)	320.5
Outlet temperature of the secondary coolant for SFR (°C)	527
Mass flow rate of the primary coolant for SFR (kg/s)	15,784
Mass flow rate of the secondary coolant for SFR (kg/s)	12,768
Mass flow rate of the steam for SFR (kg/s)	915.3
Steam turbine inlet pressure for SFR (kPa)	17,000
Condenser operating pressure for SFR reactor (kPa)	15
Inlet temperature of RO feedwater for SFR (°C)	40

Source: *Modified from Khalid F, Dincer I, Rosen MA. Comparative assessment of CANDU 6 and sodium-cooled fast reactors for nuclear desalination. Desalination 2016;379:182–92.*

TABLE 5.10 Rate of exergy destruction of major units in the integrated energy systems.

System	Component	The rate of exergy destruction[a] (MW)
CANDU 6	Steam generator	126
	Pump 1	1.6
	Pump 2	0.5
	Pump 3	48
	Pump 4	5
	Condenser	4.5
	Steam turbine	85
SFR	Primary heat exchanger	350
	Steam turbine	160
	Pump 3	2.5
	Pump 4	50
	Pump 5	5
	Condenser	20
	Steam generator	320

[a] *Approximate values based on the data provided by Khalid et al. [4].*

destruction rate (126 MW) while for the SFR energy system, the primary heat exchanger contributes the maximum to the exergy destruction losses (350 MW).

For most thermodynamic systems, ambient temperature plays a vital role in their performance. For example, sometimes system efficiency increases or decreases with the change in ambient temperature. The effect of ambient temperature on the various exergy efficiencies of the CANDU 6 and the SFR-based energy systems is illustrated in Fig. 5.34. It is evident from the figure that, as the ambient temperature rises, the exergy efficiencies for RO plants considered in both the energy systems (i.e., CANDU 6 and SFR) decrease significantly, which eventually leads to the decrease in exergy efficiencies of the cogeneration processes for both the systems. However, it is observed from the figure that the exergy efficiency of the electricity generation process is not affected by the change in ambient temperature.

Variation in the mass flow rate of fresh water on the different exergy efficiencies is depicted in Fig. 5.35. It is seen from the figure that, with the increase in mass flow rate of fresh water, the exergy efficiencies of cogeneration and electricity generation processes for both the energy systems increase. This rise is somewhat explained by the fact that, as the mass flow rate of the fresh water increases, the Pelton turbine generates more work, which eventually results in an increase in the electrical output of the systems considered. The exergy efficiency of the RO process is not affected by the mass flow rate of fresh water (see Fig. 5.35).

The variation of different exergy efficiencies of processes defined for CANDU 6 and SFR-based energy systems with the inlet temperature of RO feedwater are depicted in Figs. 5.36

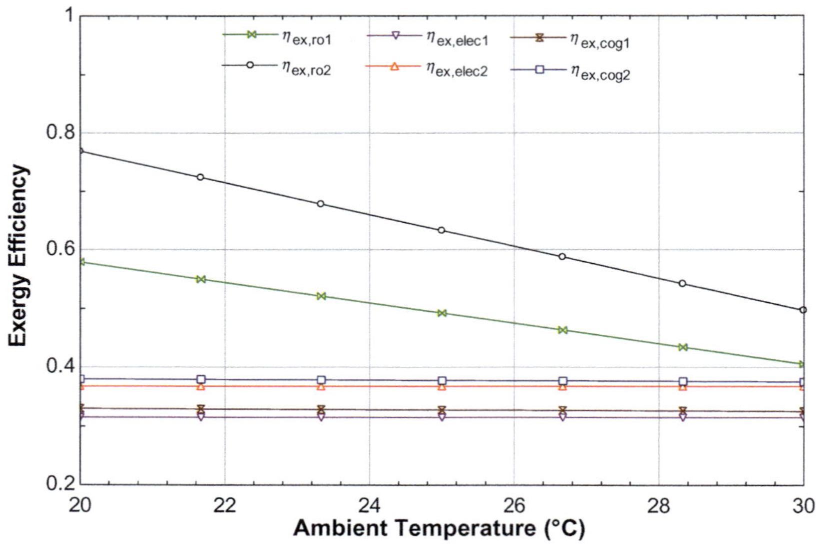

FIG. 5.34 Variation of various exergy efficiencies with ambient temperature.

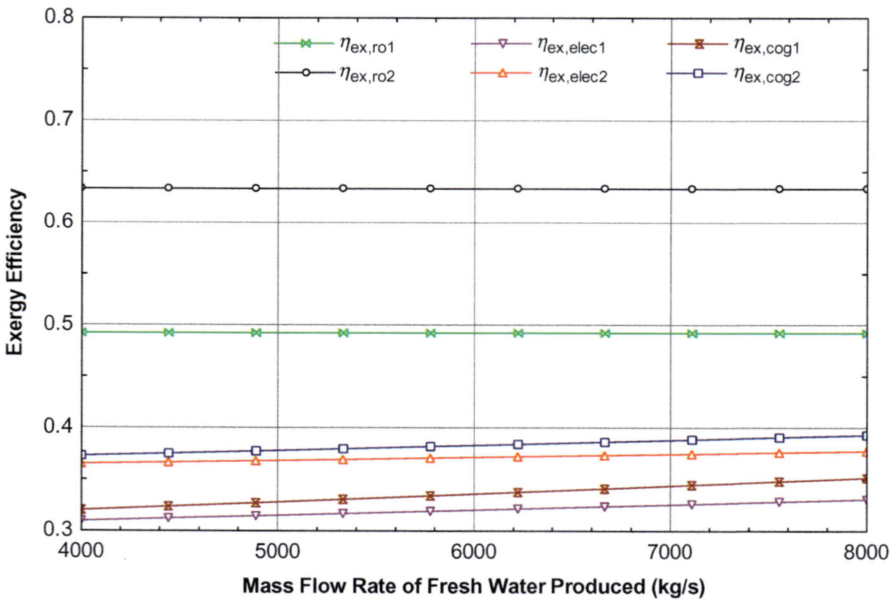

FIG. 5.35 Effect of the mass flow rate of fresh water on exergy efficiencies of various processes.

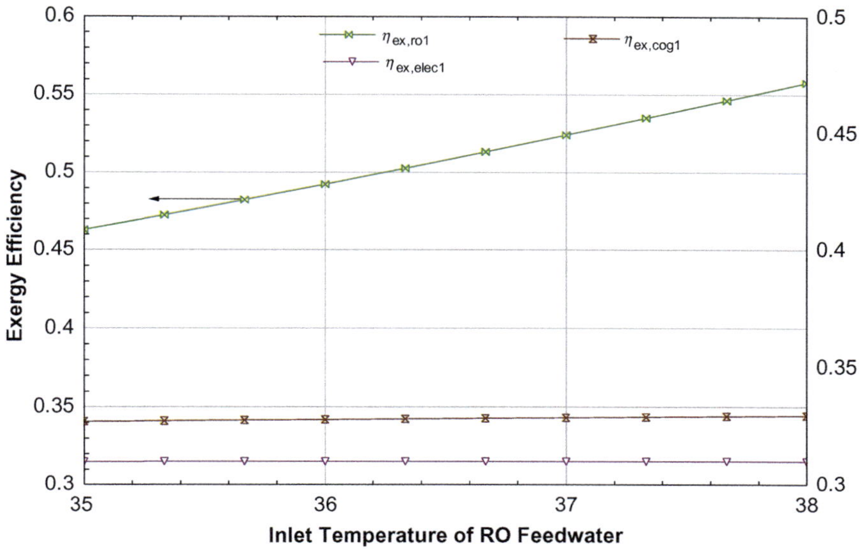

FIG. 5.36 Effect of the inlet temperature of RO feedwater on exergy efficiencies of various processes of CANDU 6 integrated energy system.

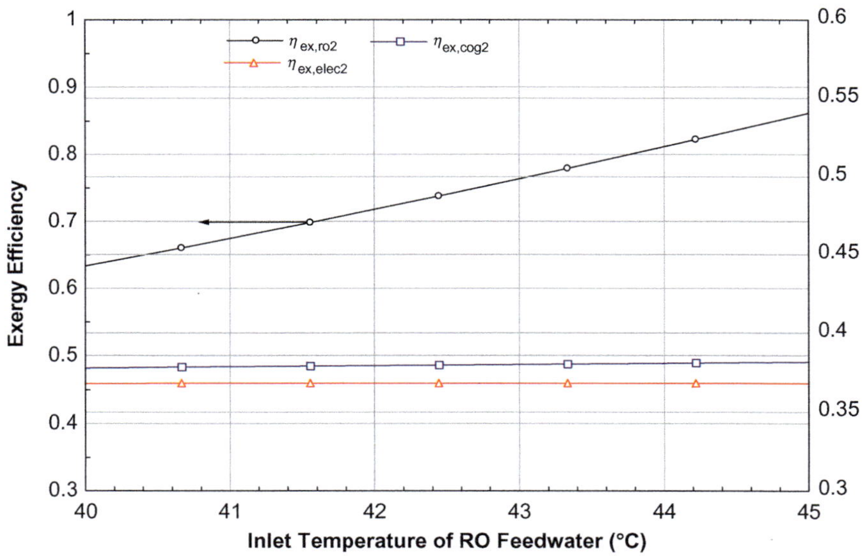

FIG. 5.37 Effect of the inlet temperature of RO feedwater on exergy efficiencies of various processes of SFR integrated energy system.

and 5.37. It is illustrated in the figures that as the inlet temperature of the feed increases, the RO exergy efficiencies increase, resulting in the overall exergy efficiencies of the cogeneration processes for both the energy systems slightly rising. This can be explained by the fact that the seawater viscosity decreases with the rise in feed temperature, permitting the membrane to conduct more ions. However, it is also worth noting that the exergy efficiencies of the electricity generation processes for both the energy systems are not affected by the change in feedwater inlet temperature.

5.7.5 Closing remarks

In the present case study, the exergy assessment of two nuclear energy-based energy systems, namely SFR, and CANDU 6, for co-production of fresh water and electricity is reported. The results show that the SFR-based energy system performs much better compared to the CANDU 6 energy system in terms of exergy efficiency of cogeneration processes. The steam turbine in the SFR system has high inlet pressure compared to the CANDU 6, causing the SFR energy system to perform much better. The other meaningful information one can infer from this case study is that it is beneficial for a reverse osmosis plant to couple with a nuclear power plant condenser for a high inlet temperature of RO feedwater. This case study will help researchers consider coupling of a nuclear power plant with an RO plant for the production of fresh water and electricity.

5.8 Case study 4: High-temperature GT-MHR nuclear reactor for co-production of fresh water and electricity

This case study reports thermodynamic analysis of a high-temperature nuclear reactor for nuclear desalination and electricity. The reverse osmosis (RO) based desalination method is considered in the case study. This study is based on the investigation conducted by Khalid et al. [5]. The salient feature of this case study is that the steam turbine is used to generate electricity in this energy system in order to capture the waste heat. Additionally, the feed of the RO plant is preheated, and to recover the energy from the pressurized brine, a Pelton turbine is used. The system inputs, outputs and subsystems are shown in Fig. 5.38.

5.8.1 System description

The schematic diagram of the integration of the reverse osmosis plant with nuclear reactor for electricity and freshwater production is shown in Fig. 5.39. This system has numerous components, namely a recuperator, reactor core, precooler, steam generator, small modular reactor (GT-MHR type), pumps, heat exchangers (HEXs), compressor, gas and steam turbines, mixing chamber, filter, condenser, Pelton turbine, throttle valves, three-way valve, and RO module. At state 1, compressed and hot helium leaves the nuclear reactor and is utilized to drive the gas turbine where electricity is generated by the use of an electric generator. At state 2, helium at reduced pressure and temperature leaves the gas turbine and is cooled in the recuperator. At state 3, helium leaves the recuperator and enters the steam generator where it is used to generate steam. The steam leaves the steam generator at state 11 and

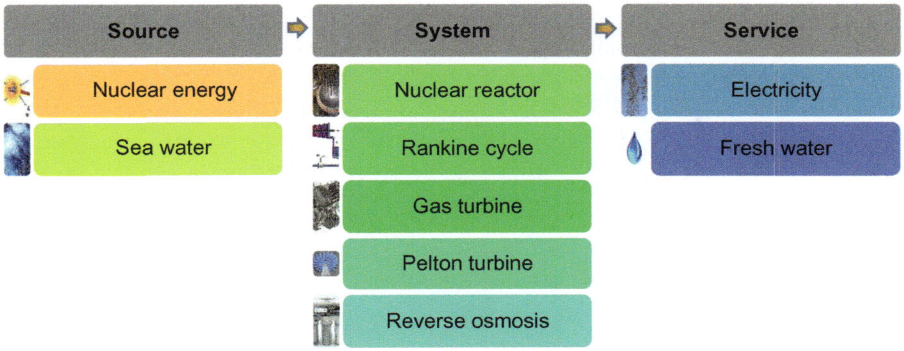

FIG. 5.38 3S approach for case study 4.

get expanded in the steam turbine to generate electrical power. At state 12, the steam leaves the steam turbine and is condensed in the condenser. The condensate leaves the condenser at state 13 and enters the pump 1 where it is pumped to state 14. The helium that leaves the steam generator is cooled in HEX1 by the use of Duratherm-T oil. The cooled helium at state 5 is further cooled in the precooler before entering the compressor (state 6). The helium gets compressed in the compressor at state 7 and gets heated in the recuperator before entering the reactor core (state 8). The hot Duratherm-T oil is used to preheat the seawater RO feed in HEX2. The heated seawater that comes out from the HEX2 (state 20) goes to pump 2 where it gets pressurized and enters the filter at state 21. The pressurized seawater leaves the filter at state 22, and splits into two streams, the stream at state 31 and the stream at state 23. The stream at state 23 enters the chemical pre-treatment plant where it is chemically treated. The treated stream leaves the chemical pre-treatment plant at state 24 and gets pressurized through pump 3 to state 25. The stream at state 25 enters the RO module and produces two outputs: fresh water (state 26) and brine (state 29). The brine (state 29) leaves the RO module at high pressure, and energy is recovered from it using the Pelton turbine. The stream at state 31 gets throttled by throttle valve 1. This throttle stream (state 32) goes to the mixing chamber where it gets mixed with another stream of water (state 26). The mixed streams leave the mixing chambers as one stream (state 27) and enter the throttle valve 2 where the pressure gets reduced (state 28). The stream at 28 is the stream of fresh water produced by the desalination plant considered in this case study.

5.8.2 Assumptions

In order to perform thermodynamic analysis of this integrated system, there is a need to make conceptually correct assumptions. The following main reasonable assumptions are considered in this case study:

- The systems work under steady state.
- The changes in kinetic and potential energies and exergies are negligible.
- The turbines, throttle valves, and pumps work adiabatically.
- A 95% effectiveness of recuperator and precooler is sufficient.
- The pressure losses in pipelines and heat exchangers are negligible.

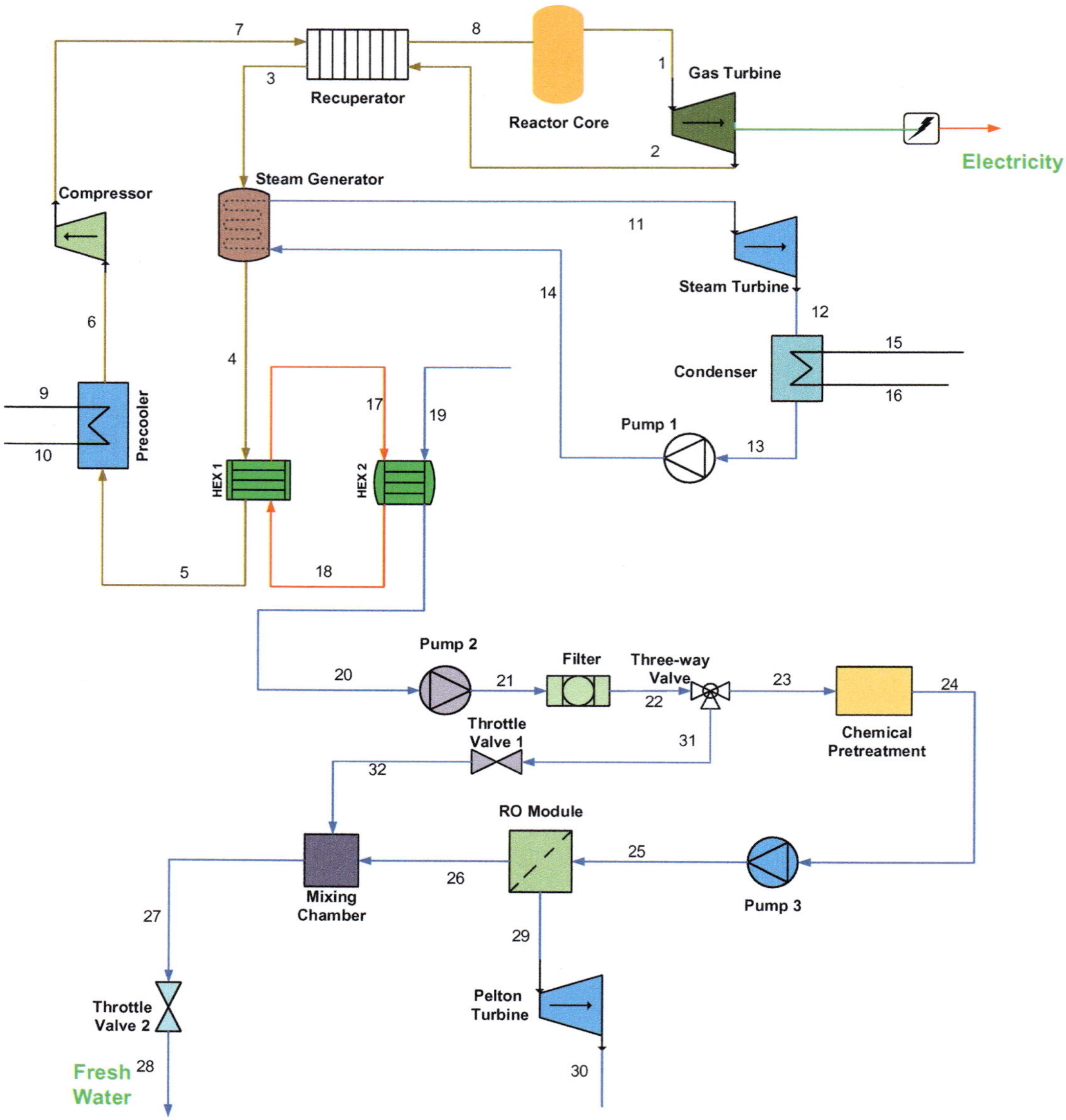

FIG. 5.39 An RO-plant coupled with GT-MHR for electricity and fresh water. *Modified from Khalid F, Dincer I, Rosen MA. Analysis and assessment of a gas turbine-modular helium reactor for nuclear desalination. J Nucl Eng Radiat Sci 2016;2:31014. doi:10.1115/1.4032508*

5.8.3 Analysis

To evaluate the thermodynamic performance of the RO process, electrical power generation without steam turbine and electrical power generation with steam turbine processes are considered in this case study. The exergy efficiency of the RO unit is defined as follows:

$$\eta_{ex,ro} = \frac{\dot{m}_{28}ex_{28} + \dot{m}_{30}ex_{30} - \dot{m}_{19}ex_{19}}{\dot{W}_{Pump3} + \dot{W}_{Pump2} - \dot{W}_{Pelton}} \quad (5.18)$$

where $\dot{W}_{Pump3} + \dot{W}_{Pump2}$ are the electrical power consumed by the pumps and $\dot{W}_{Pelton}$ is the electrical power generated by the Pelton turbine.

$$\eta_{ex,elec1} = \frac{\dot{W}_{e1}}{\dot{E}x^{Q_{rc}}} \quad (5.19)$$

where $\dot{W}_{e1}$ is the net electrical power generated by the system and $\dot{Q}_{rc}$ is the heat rate required by the reactor core.

The net electrical power is evaluated as

$$\dot{W}_{e1} = \left[\left(\eta_m \dot{W}_{t,net1} - \dot{W}_{cons,net1}\right)\eta_g\right](1 - \varepsilon_{ut})\eta_{switch} \quad (5.20)$$

where η_m, η_g and η_{switch} are the efficiencies of motor, generator, and switch, respectively. $\dot{W}_{t,net1}$ and $\dot{W}_{consumed,net1}$ are the net work rate generated by the turbine and net work rate consumed by the integrated system, respectively.

The exergy efficiency of the electrical process including steam turbine work is written as

$$\eta_{ex,elec} = \frac{\dot{W}_{e}}{\dot{E}x^{Q_{rc}}} \quad (5.21)$$

The integrated system presented in this case study produces two useful outputs, namely fresh water and electrical power. Thus, exergy efficiency of the integrated system in this case study is written as

$$\eta_{ex,ov} = \frac{\dot{m}_{28}ex_{28} + \dot{m}_{30}ex_{30} - \dot{m}_{19}ex_{19} + \dot{W}_{e}}{\dot{E}x^{Q_{rc}}} \quad (5.22)$$

5.8.4 Results and discussion

Khalid et al. [5] have identified all the state points with all their relevant thermodynamic data such as P (pressure in kPa), T (temperature in °C), $\dot{m}$ (mass flow rate in kg/s), ex (specific exergy in kJ/kg), and h (specific enthalpy in kJ/kg. Table 5.11 lists the value of some of the useful parameters employed in the present case study. By determining the rates of exergy destruction of various units in the integrated energy systems, the ratios of exergy destruction of some major components of the integrated system are tabulated in Table 5.12. It is to be noted from the table that the reactor core contributes the most to the exergy destruction (nearly 65% of the total exergy destruction of the system). The exergy destruction rate in the recuperator is also high, i.e., 10% of the total exergy destruction rate. Thus, one should make efforts to reduce these exergy destruction rates with the condition that these efforts need to be economical as well.

As the pressure ratio of the gas turbine cycle changes, the value of the various processes' exergy efficiencies changes except for the reverse osmosis process (see Fig. 5.40). The change

TABLE 5.11 Values of useful parameters used in the present case study.

Parameter	Value
Inlet temperature of the coolant for GT-MHR (°C)	528.6
Outlet temperature of the coolant for GT-MHR (°C)	850
Mass flow rate of the coolant for GT-MHR (kg/s)	401
Inlet temperature of RO feedwater for GT-MHR (°C)	45
Amount of fresh water produced by GT-MHR (kg/s)	1234
Pressure ratio in the GT-MHR	2.5

Source: *Based on data provided from Khalid F, Dincer I, Rosen MA. Analysis and assessment of a gas turbine-modular helium reactor for nuclear desalination. J Nucl Eng Radiat Sci 2016;2:31014. doi:10.1115/1.4032508.*

TABLE 5.12 Ratios of exergy destruction of some major components of the integrated system.

Component	The ratio of exergy destruction
Steam generator	0.037
Compressor	0.044
Pump 3	0.021
Steam turbine	0.022
Recuperator	0.10
Precooler	0.013
Reactor core	0.65
HEX1	0.016
HEX2	0.05

Source: *Approximate values based on the data provided by Khalid F, Dincer I, Rosen MA. Analysis and assessment of a gas turbine-modular helium reactor for nuclear desalination. J Nucl Eng Radiat Sci 2016;2:31014. doi:10.1115/1.4032508.*

in exergy efficiency of the overall integrated system and electrical processes can be explained by the fact that, after the optimal value of pressure ratio, the rate at which compressor work increases is more than the rate at which the turbine produces electric power, leading to the overall decrease in electricity generated by the integrated system and electrical processes.

The gas turbine inlet temperature effects on the exergy efficiencies of different processes are shown in Fig. 5.41. As the inlet temperature rises from 700 °C to 900 °C, the exergy efficiency of the electrical process excluding steam turbine work changes from 24.2% to 31.3%. However, no change is observed in the efficiency of the reverse osmosis process.

The variation of exergy efficiencies of different processes defined for integrated energy with inlet temperature of RO feedwater is illustrated in Fig. 5.42. It is depicted in the figures that, as the inlet temperature of the feed increases, all the exergy efficiency values increase.

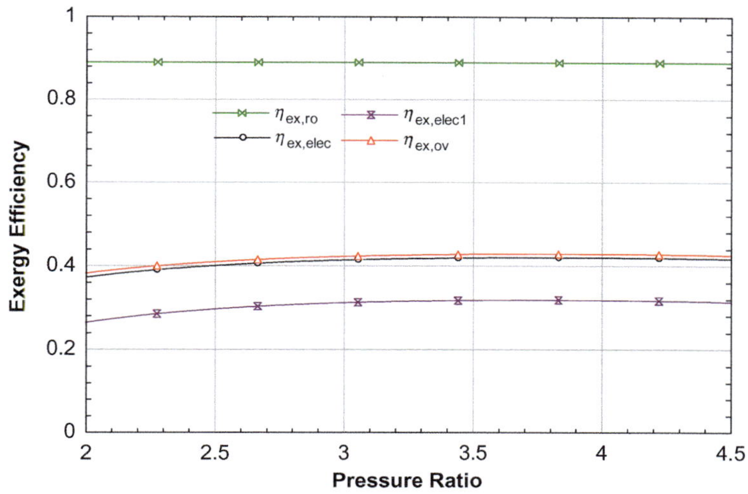

FIG. 5.40 Pressure ratio effects on various exergy efficiencies.

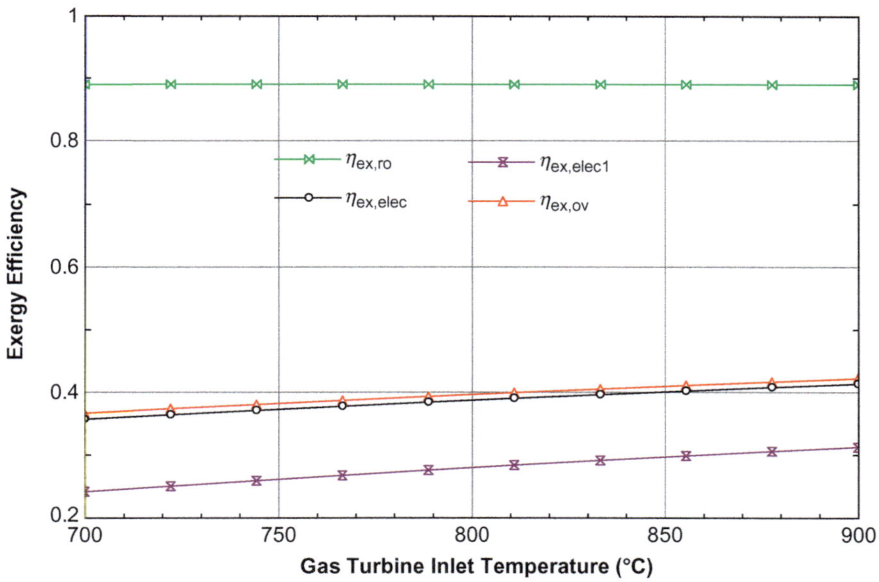

FIG. 5.41 Gas turbine inlet temperature effects on various exergy efficiencies.

This can be explained by the fact that the seawater viscosity decreases with the rise in feed temperature, permitting the membrane to conduct more ions. Thus, the results presented in this figure suggest that preheating of RO feedwater is beneficial for the integrated system and reverse osmosis process as well.

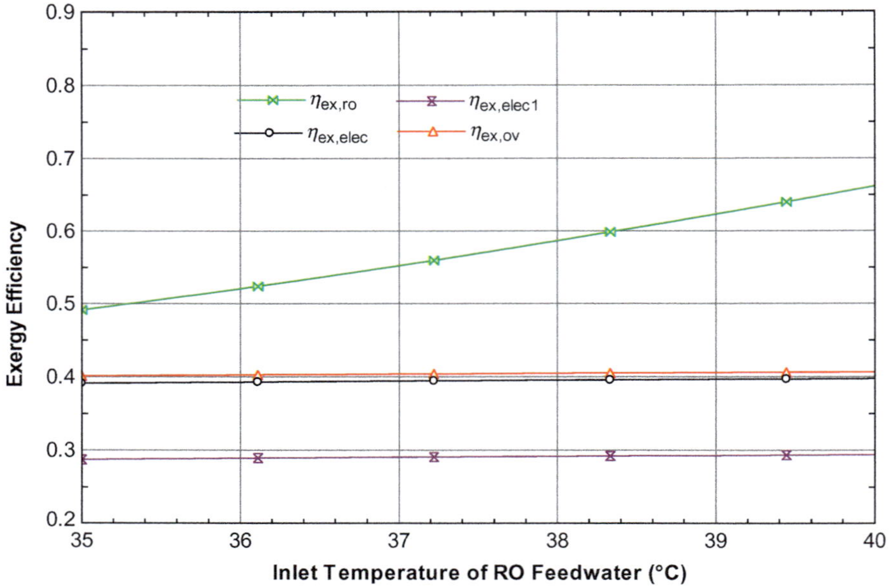

FIG. 5.42 Effect of the inlet temperature of RO feedwater on exergy efficiencies of various processes of the integrated energy system.

As the recovery ratio varies from 40% to 60%, the exergy efficiency of the electrical process without steam work decreases from 29.9% to 29.6% while the exergy efficiency of the integrated energy system changes very slightly from 40.9% to 41.0%, as shown in Fig. 5.43. The exergy efficiency of the reverse osmosis process changes significantly. This can be explained by the fact that the Pelton turbine work increases with the rise in recovery ratio.

5.8.5 Closing remarks

This case study reports thermodynamic analysis of a high-temperature nuclear reactor for nuclear desalination and electricity generation. The reverse osmosis (RO) based desalination method is considered in the case study. To capture the waste heat, a steam turbine is used, and it is found that around 10% exergy efficiency improvement can be achieved by capturing waste heat. Additionally, the feed of the RO plant is preheated, and to recover the energy from the pressurized brine, a Pelton turbine is used. The results presented in this case study will help researchers consider coupling of a nuclear power plant with an RO plant to produce fresh water and electricity.

5.9 Case study 5: Supercritical water cooled nuclear reactor for co-production of compressed hydrogen and electricity

This case study reports thermodynamic analysis of a supercritical water-cooled reactor (SCWR) for hydrogen and electricity production, as the required inputs and conversion

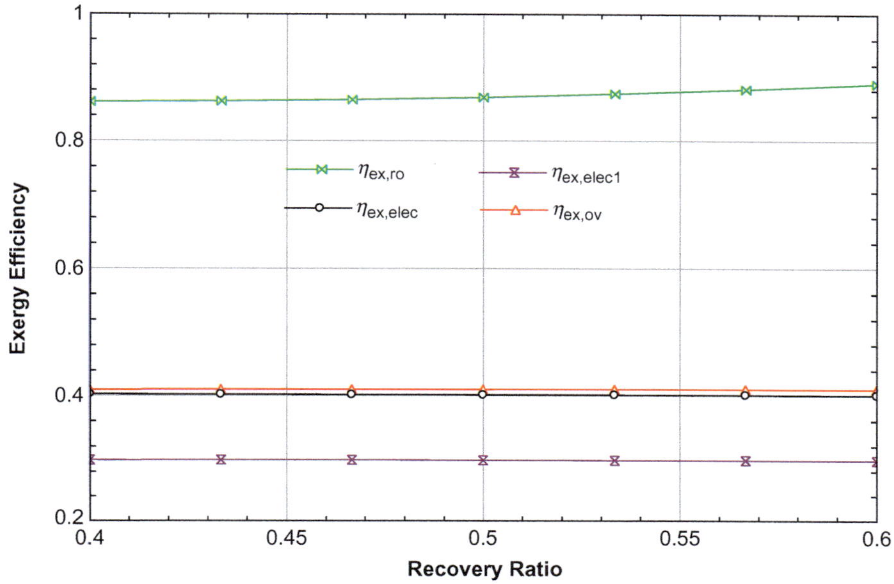

FIG. 5.43 Variation of exergy efficiencies of different processes for the integrated system with varying recovery ratio.

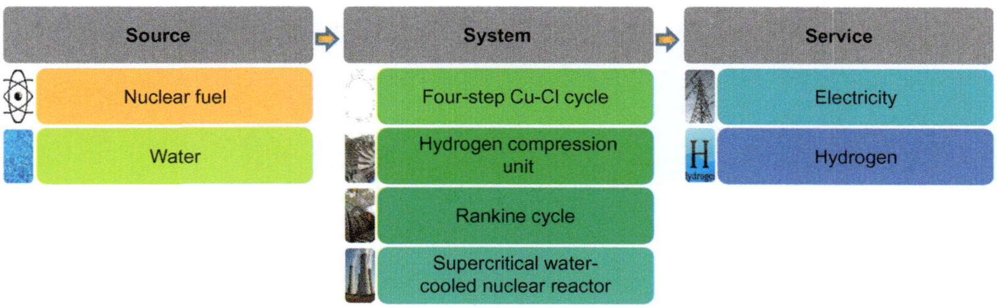

FIG. 5.44 3S approach for case study 5.

systems are shown in Fig. 5.44. A four-step thermochemical cycle, namely a Cu—Cl four step cycle, is considered in the case study for hydrogen production. This case study is based on the investigation conducted by Al-Zareer et al. [6]. The salient feature of this case study is that hydrogen produced by this system is in a compressed state, which eventually makes it (hydrogen) more physically viable to store or transport. Additionally, a detailed interaction of the Cu—Cl cycle with SCWR is also presented showing how heat/electricity is provided to each step of the Cu—Cl cycle.

5.9.1 System description

The schematic diagram of the integration of the four step Cu—Cl thermochemical cycle with supercritical water cooled nuclear reactor for electricity and hydrogen production is shown in Fig. 5.45. The integrated system has several components: a supercritical water nuclear reactor, set of steam turbines, heat exchangers, mixers, pumps, vapor-liquid separator, dryer, hydrolysis reactor, electrolyzer, a set of compressors, condenser, and oxygen production reactor. The supercritical steam leaves the nuclear reactor and passes through the oxygen production reactor and leaves, so that it can provide the heat at the desired temperature. After leaving the oxygen production reactor, the supercritical steam enters the hydrolysis reactor where it provides the required heat at the temperature. The supercritical steam leaves the hydrolysis reactor and enters the steam generator unit, where it is utilized to provide heat to the pressurized steam cycle. The supercritical water leaves the steam generator unit and gets pumped through the supercritical pump and then returns to the nuclear reactor. The Cu—Cl cycle has four reactors: hydrolysis, oxygen production, dryer and electrolyzer. The hydrogen gets produced in the electrolyzer and then is compressed to the storage pump with the use of multistage

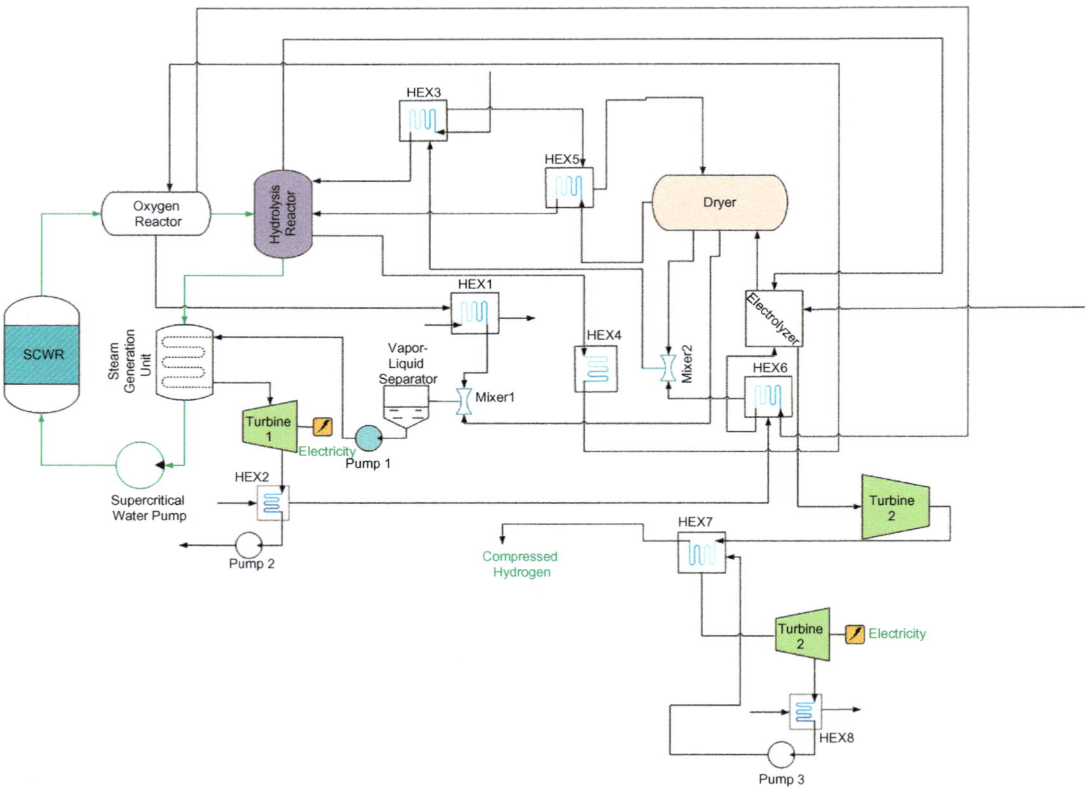

FIG. 5.45 Schematic diagram of integrated system having the four step Cu—Cl thermochemical cycle with supercritical water-cooled nuclear reactor for electricity and hydrogen production.

compression. To cool down the compressors, a water circuit is used to produce electrical power via steam turbines. All the reactants or products in the thermochemical cycle are reused except water to produce hydrogen and oxygen.

5.9.2 Assumptions

In order to perform thermodynamic analysis of this integrated system, there is a need to make conceptually correct assumptions. The following main reasonable assumptions are considered in this case study:

- The systems work under steady state conditions.
- No startup period is taken.
- The changes in kinetic and potential energies and exergies are negligible.
- The turbines, throttle valves, and pumps work adiabatically.
- A 95% efficiency of the electrical generator is used.
- The pressure losses in pipeline and heat exchanger pressures are negligible.

5.9.3 Analysis

The performance of the Cu—Cl cycle and overall system is measured by defining energy and exergy efficiencies.

For the Cu—Cl cycle, the following expressions of exergy and energy efficiencies are used, respectively:

$$\eta_{ex,cycle} = \frac{\dot{m}_{H_2} ex_{H_2}}{\dot{W}_{elect} + \dot{Ex}^{Q}_{required}} \tag{5.23}$$

where the thermal exergy associated with net heat required by the cycle is $\dot{Ex}^{Q}_{required}$. Here, $\dot{W}_{elect}$ denotes the electrical power consumed by the electrolyzer. The energy efficiency of the cycle is defined as

$$\eta_{en,cycle} = \frac{\dot{m}_{H_2} LHV_{H_2}}{\dot{W}_{elect} + \dot{Q}_{required}} \tag{5.24}$$

Likewise, for the overall system performance the following energy and exergy efficiencies expressions are utilized, respectively:

$$\eta_{en,ov} = \frac{\dot{m}_{H_2} LHV_{H_2} + \dot{W}_{net,electric}}{\dot{Q}_{core}} \tag{5.25}$$

where mass flow rate of compressed hydrogen produced is $\dot{m}_{H_2}$, and the lower heating value of the hydrogen is LHV. Here, $\dot{Q}_{core}$ denotes the heat required by the reactor core, and $\dot{W}_{net,electric}$ denotes the net electrical work rate of the overall system.

$$\eta_{ex,ov} = \frac{\dot{m}_{H_2} ex_{H_2} + \dot{W}_{net,electric}}{\dot{Ex}^{Q}_{core}} \tag{5.26}$$

5.9.4 Results and discussion

Al-Zareer et al. [6] have identified all the state points with all their relevant thermodynamic data, such as P (pressure in kPa), T (temperature in °C), $\dot{m}$ (mass flow rate in kg/s), ex (specific exergy in kJ/kg), and h (specific enthalpy in kJ/kg). Table 5.13 lists the values of some of the useful parameters calculated in the present case study. The results show that the integrated system can achieve an energy efficiency of 31.6% and an exergy efficiency of 56.2% at the given state points. The system is able to deliver a net electric output of 553 MW with a daily production of 175 t of hydrogen. The Cu—Cl cycle considered in this case study can achieve an exergy efficiency of 60.2% and energy efficiency of 40.1% (see Table 5.13). The exergy destruction of various subsystems in the integrated energy systems are calculated and tabulated in Table 5.14. It is to be noted from the table that the Rankine cycle supporting system contributes the most to the exergy destruction (nearly 74% of the total exergy destruction of the system). The exergy destruction in the four step Cu—Cl cycle is also high, i.e., 24% of the total exergy destruction. Thus, one should make efforts to reduce these exergy destructions with the condition that these efforts need to be economical as well.

The thermal energy and exergy associated with the different components of the Cu—Cl cycle are plotted in Fig. 5.46. The thermal energy associated with the oxygen production reactor is approximately 436 kJ per mole of hydrogen produced while thermal exergy associated with it is 274 kJ per mole of hydrogen produced. From the figure, it is to be noted that thermal energy associated with the drying process is the maximum while its thermal exergy is not. The reason for this trend is due to the fact that the drying is the low temperature process. Thus, exergy

TABLE 5.13 Calculated values of different parameters used in the thermodynamic study.

Parameter	Value
Energy efficiency of the four step Cu—Cl cycle (%)	40.1
Exergy efficiency of the four step Cu—Cl cycle (%)	60.2
Hydrogen produced daily (tonnes)	175
Energy efficiency of the overall system (%)	31.6
Exergy efficiency of the overall system (%)	56.2
Pressure ratio in the gas turbine cycle	2.5
Net electric power produced by the system (MW)	553
Rate of exergy destruction in oxygen production reactor (kJ/mol H_2)	85.2
Net electric power required by the four step Cu—Cl cycle (kJ/mol H_2)	55.0
Pressure of compressed hydrogen (bar)	700
Rate of exergy destruction in overall system (MW)	630

Source: *Modified from Al-Zareer M, Dincer I, Rosen MA. Development and assessment of a novel integrated nuclear plant for electricity and hydrogen production. Energ Conver Manage 2017;134:221–34.*

TABLE 5.14 The rates of exergy destruction of major subsystems in the integrated system.

Component	Exergy destruction[a] (kJ/mol H_2)
Rankine cycle supporting system	465
Four step Cu—Cl cycle	156
Hydrogen compression system	9
Other	3

[a] *Approximate values, modified from Al-Zareer M, Dincer I, Rosen MA. Development and assessment of a novel integrated nuclear plant for electricity and hydrogen production. Energ Conver Manage 2017;134:221–34.*

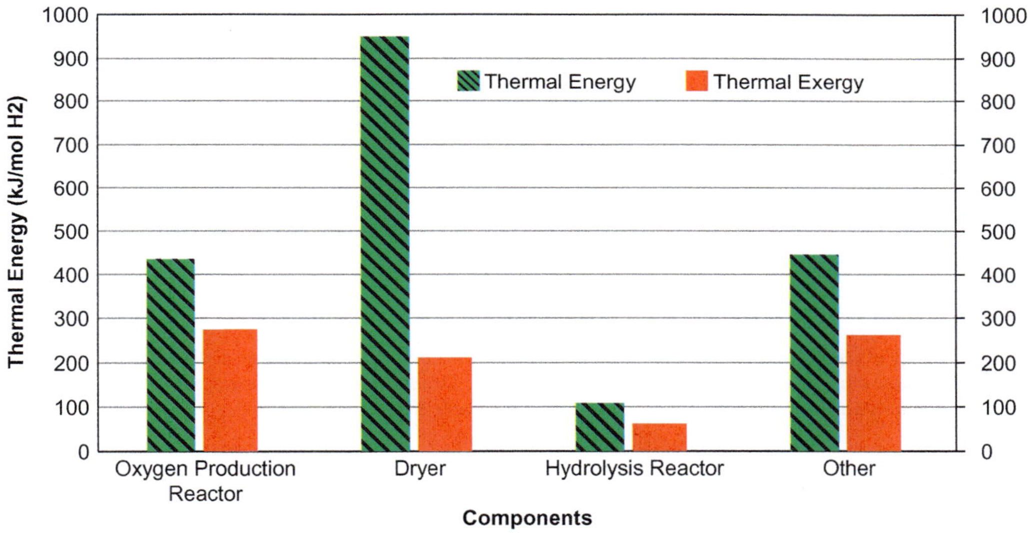

FIG. 5.46 Thermal energy and exergy of the various components associated with the four step Cu—Cl cycle. *Data from Al-Zareer M, Dincer I, Rosen MA. Development and assessment of a novel integrated nuclear plant for electricity and hydrogen production. Energ Conver Manage 2017;134:221–34*

associated with it is much less compared to thermal energy. These results clearly show the advantage of exergy analysis over conventional energy analysis and suggest that one should use low grade energy for the drying purpose. On the contrary, the exergy associated with the oxygen production reactor is the maximum in spite of low thermal energy compared to other components. This trend can be explained by the fact that the temperature requirement in the oxygen production is the highest among all the components, thus it has higher thermal exergy.

The exergy efficiencies and exergy destructions of various components of the integrated system are plotted in Fig. 5.47. The exergy destruction in the electrolyzer is given in kJ per mole of hydrogen produced with an exergy efficiency of 70%. This result suggests that the

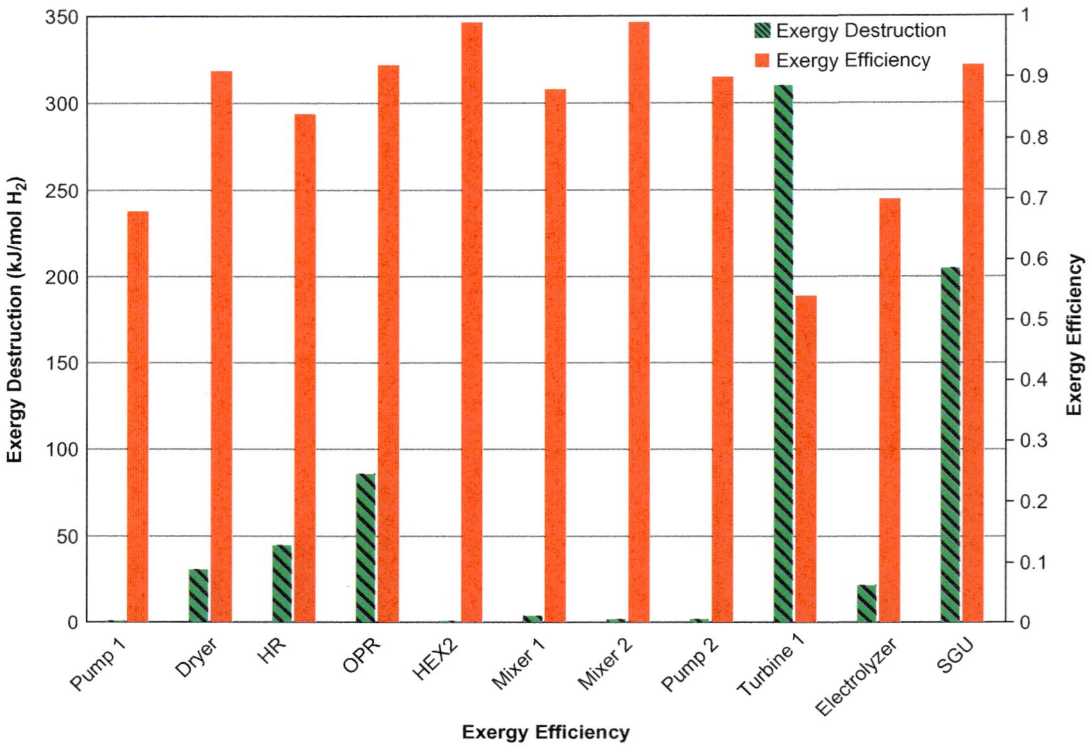

FIG. 5.47 Exergy efficiency and exergy destruction of various components of the integrated system. *Approximated values based on the data provided by Al-Zareer M, Dincer I, Rosen MA. Development and assessment of a novel integrated nuclear plant for electricity and hydrogen production. Energ Conver Manage 2017;134:221–34*

exergy destruction in the electrolyzer is significant. Turbine 1 has the highest exergy destruction (304 kJ per mole of hydrogen produced) among all the main components of the integrated system. This is because it has the maximum interaction of energy among all components.

5.9.5 Closing remarks

In the present case study, the energy and exergy assessment of a nuclear energy-based energy system, with an SCWR for co-production of hydrogen and electricity, is reported. The results show that hydrogen on a large scale can be produced with the use of the present system, i.e., roughly around 2.02 kg/s. The presented system can achieve an overall exergy efficiency of 56.2% with an overall energy efficiency value of 31.6%. The results presented in this case study will help the researchers consider coupling of thermochemical cycles with SCWR for cogeneration or extending to trigeneration/multigeneration depending upon the need.

5.10 Conclusions

In this chapter, a general overview of nuclear energy systems is presented along with several nuclear energy integrated case studies. Nuclear energy is currently a significant energy source worldwide. Due to the nuclear fission reaction, these sources have huge amounts of radiative energy release, which can later be converted into heat and power via conventional power cycles. The nuclear fission reaction yields fundamentally neutron radiation that is converted into high-temperature thermal energy. In order to remove the heat generated from the reaction, and to attain pressurized superheated steam, a steam generation unit is usually employed.

Nuclear fuel is considered as the highest energy density fuel as compared to many other fuels such as hydrogen, gasoline, diesel, etc. Nuclear power plants receive uranium as feedstock nuclear fuel, which is extracted from uranite mineral resources. Presently, most nuclear power plants utilize uranium as the feedstock, although there are a few reactors operating with thorium. There are several sub-processes for preparation of nuclear fuel starting from mining and fuel enrichment until fuel recycling, materials recovery, and disposal. The nuclear waste includes several radioactive isotopes and transuranic elements, which have a very long half-life. Hence, a careful waste management is required for nuclear power plants.

Nuclear reactors have significantly advanced over time. The models of the first generation are a boiling water reactor (BWR), gas-cooled reactor, liquid metal fast breeder reactor (LMFBR), and pressurized heavy water reactor (PHWR). There are now Generation IV nuclear reactors, such as a gas-cooled fast reactor (GFR) and the sodium-cooled fast reactor (SFR) which are being advanced.

Over time, utilization of nuclear waste heat has evolved as a promising approach for various applications. One of the possible utilization pathways is through nuclear thermochemical cycles (e.g., CuCl, MgCl), in which there is a closed sequence of chemical reactions (e.g., hydrolysis, copper oxychloride decomposition) driven thermochemically. These are commonly used for water splitting. When the cycle contains an electrochemical reaction, it is called a hybrid thermochemical cycle. The main advantage of nuclear power plants is the capability of converting this waste heat for generating a high-pressure and high-temperature vapor, which can be used to produce heating, cooling, or fresh water. Desalination plants require huge amounts of thermal or electrical energy. This is commonly provided from fossil fuels, including coal, oil, and natural gas. In order to reduce the environmental footprint of desalination, the process can be integrated into nuclear power plants. Furthermore, nuclear energy can be utilized for water splitting to yield hydrogen through numerous techniques such as nuclear water electrolysis, nuclear thermochemical cycles, nuclear high-temperature steam electrolysis, etc. We conclude that there is an important potential of nuclear power integration in various sectors, including synthetic fuel production, ammonia and fertilizer production, steam-assisted gravity drainage (SAGD), coal gasification, aluminum production, desalination, hybridization with renewables, district heating and cooling, distributed energy generation, among others.

Nomenclature

$\dot{Ex}$ exergy rate (kW)
ex specific exergy (kJ/kg)
h specific enthalpy (kJ/kg)

$\dot{m}$ mass flow rate (kg/s)
P pressure (kPa)
$\dot{Q}$ heat transfer rate (kW)
Q heat (kJ)
s specific entropy (kJ/kg K)
T temperature (°C)
$\dot{W}$ work rate (kW)

Acronyms

BWR boiling water reactor
CANDU Canadian deuterium
EES engineering equation solver
GCR gas-cooled reactor
GFR gas-cooled fast reactor
GT gas turbine
HEX heat exchanger
HF hydrofluoric acid
HR hydrolysis reactor
LFR lead-cooled fast reactor
LHV lower heating value
LMFBR liquid metal fast breeder reactor
MHR modular helium reactor
MSR molten salt reactor
OPR oxygen production unit
PWHR pressurized heavy water reactor
PWR pressurized water reactor
RO reverse osmosis
SAGD steam-assisted gravity drainage
SCWR supercritical-water-cooled reactor
SFR sodium-cooled fast reactor
SGU steam generation unit
UC uranium carbide
VHTR very-high-temperature reactor

Greek letters

η efficiency
ε effectiveness

Subscripts and superscripts

1, 2, ... state numbers
c compressor
cog cogeneration
con condenser
cons consumed
d destruction
e electrical
elec electric power
elect electrolyzer
en energy
ex exergy

g	generator
gen	generation
gt	gas turbine
hex	heat exchanger
htf	heat transfer fluid
m	mechanical or motor
ov	overall
r	reactor
RC	Rankine cycle
rc	reaction or reactor core
ro	reverse osmosis
ut	utilization

References

[1] Dincer I, Zamfirescu C. Chapter 6—Nuclear power generation. Adv. Power Gener. Syst. Boston: Elsevier; 2014. p. 311–68. https://doi.org/10.1016/B978-0-12-383860-5.00006-7.

[2] The Generation IV International Forum (GIF). Gener IV Int Forum 2013. https://www.gen-4.org [accessed 14 December 2018].

[3] Khalid F, Bicer Y. Energy and exergy analyses of a hybrid small modular reactor and wind turbine system for trigeneration. Energy Sci Eng 2019;00:1–15. https://doi.org/10.1002/ese3.327.

[4] Khalid F, Dincer I, Rosen MA. Comparative assessment of CANDU 6 and sodium-cooled fast reactors for nuclear desalination. Desalination 2016;379:182–92.

[5] Khalid F, Dincer I, Rosen MA. Analysis and assessment of a gas turbine-modular helium reactor for nuclear desalination. J Nucl Eng Radiat Sci 2016;2:31014. https://doi.org/10.1115/1.4032508.

[6] Al-Zareer M, Dincer I, Rosen MA. Development and assessment of a novel integrated nuclear plant for electricity and hydrogen production. Energ Conver Manage 2017;134:221–34.

CHAPTER

6

Integration of renewable energy systems for multigeneration

6.1 Introduction

Multigeneration using renewable energy resources has attracted significant interest in recent years due to its several advantages and benefits compared to conventional energy sources. For instance, multigeneration systems can offer better energy security, enable decentralized energy systems and reduce environmental impacts. Renewable energy sources are usually accessible at any location in the world, depending on the type. In fact, they are more scattered around the world than conventional fossil fuels. For example, in some regions, there are dominant wind resources on land or tides on the sea, whereas other regions can take advantage of geothermal energy or biomass from agricultural areas. Solar radiation is even more diversified compared to other types of renewable energy sources, as it falls on the earth's surface everywhere, with varying intensities.

One of the driving forces behind the wide implementation of renewable energy sources is the gradually declining cost. The realization of renewable energy-based multigeneration systems is influenced by economics based on the desire for the payback period to be reasonably short. Therefore, the main target becomes to effectively design and optimize integrated energy systems having high energy conversion efficiencies with an acceptably low cost. In fact, this is a challenging trade-off problem for designers due to the fact that if efficiency is higher, the system cost mostly tends to be greater. Since the renewable sources are available as long as the world exists, mankind seeks for efficient ways of harvesting these sources. The term *harvesting* is a convenient expression because we obtain useful products from the sources. However, harvesting these renewable energy sources and transforming them into useful products is a challenging task for several major reasons, namely their fluctuating and intermittent characteristics and relatively low resource intensity (larger required land area). Though renewable energy is freely available as a source, the conversion infrastructure required for harvesting can be expensive as a result of low resource availability per unit of area and relatively lower energy conversion efficiencies.

On the other hand, some specific types of renewable energy resources have considerably higher capacity factor with minimal fluctuations throughout the year, e.g., geothermal and ocean thermal energy. Geothermal energy is a continuous type of renewable, so energy storage is not as essential as solar and wind energy when the energy is utilized on-site or near the plant. This enables a geothermal system to be directly connected to the grid to generate base-load power. Similarly, an ocean thermal energy conversion (OTEC) plant working in a sunbelt region can function steadily since its heat source and heat sink temperature are almost constant throughout the year. Nevertheless, OTEC requires energy storage for transport of energy to the land. Therefore, energy conversion to a transportable type of media is a significant area of research. Recently, synthetic fuels such as hydrogen, ammonia and methanol have become an alternative solution for these purposes.

The classification of fundamental renewable energy systems is illustrated in Fig. 6.1.

In this section, we introduce mostly established renewable energy systems and provide fundamental details regarding their operation principles and types.

6.2 Solar energy systems

Solar energy is the main source of most types of renewables, since it is the key cause of wind, hydro, biomass, and ocean energy cycles. Solar energy causes wind streams through air temperature differences. Solar energy drives photosynthesis, which is the fundamental step in biomass formation. Solar energy evaporates water in the lakes, seas, and rivers and completes the rain cycle, which eventually forms hydro energy. The tidal currents and waves are formed by the temperature gradient caused by solar energy, as well as the Sun's position with respect to Earth. Hence, solar energy is the key source and it is better to be converted into useful forms in the first place by engineered devices and systems.

Solar radiation, solar spectrum, and intensity of irradiation are significant parameters defining the performance of solar energy systems. As shown in Fig. 6.2, concentrated solar collectors can generate high-temperature heat, which can be used for various purposes, e.g., process heat, thermal storage, and power generation. Photovoltaics can generate direct electricity using photons carried by sunlight. In addition, solar thermal collectors are utilized for recovering the thermal energy from sunlight radiation. Most of the solar thermal collectors,

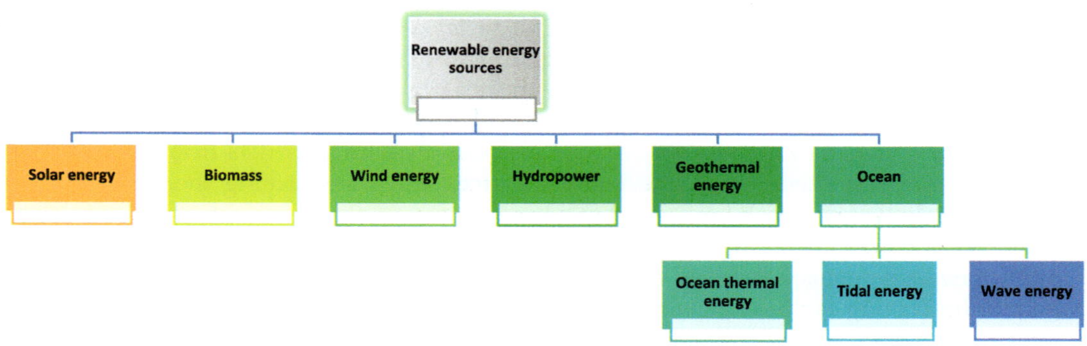

FIG. 6.1 Classification of fundamental renewable energy sources.

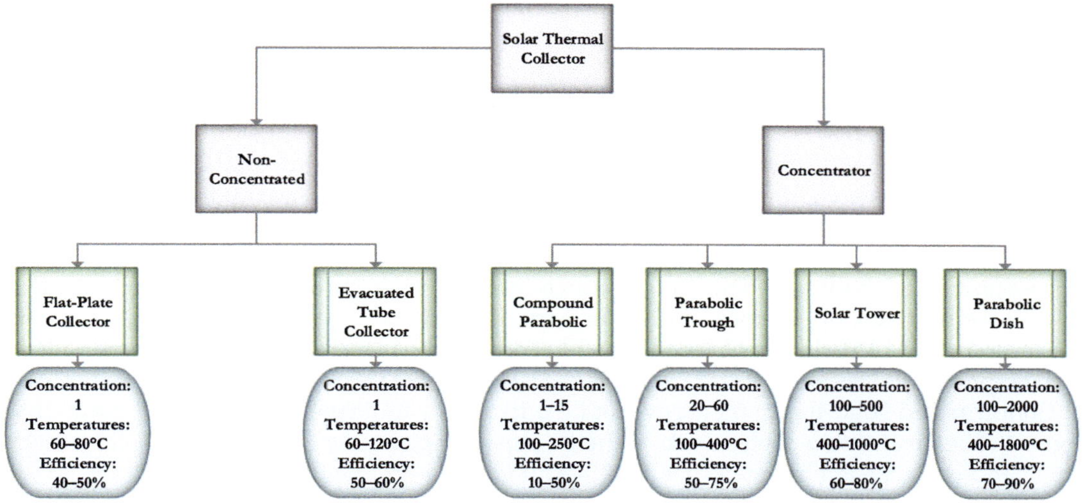

FIG. 6.2 Classification of solar thermal energy systems.

which are mainly used for thermodynamic cycles for power generation, are solar towers with heliostat field, solar through power plants, and solar dish heat engines, as classified in Fig. 6.2. Low/medium temperature solar collectors, which are flat-plate and evacuated tube collectors, are mainly used for hot water production. However, they can also be used for other purposes such as absorption cooling, drying, etc. In addition to concentrated solar thermal, electricity can also be produced by concentrated photovoltaic modules.

The benefits of concentrated photovoltaic (CPV) systems can be listed as follows:

- Multi-junction tandem cells can be used, which can exceed the theoretical Shockley-Queisser limit compared to non-concentrated single junction cells.
- Lower cell area requirement is achieved by concentrating the sunlight.
- Cogeneration can be applied for production of by-products as heat (CPV/T), which increases the overall system efficiency and reduces the costs.

Since solar energy has intermittent characteristics, proper energy storage methods are essential. Although there are diverse storage techniques, three types of energy storage methods are mostly used for solar energy applications: (i) batteries for electricity storage, (ii) thermal storage for photo-thermally produced heat, and (iii) chemical media for conversion of solar energy into chemical products such as hydrogen, ammonia, etc. There are several disadvantages of each method. For instance, batteries have low specific energy densities requiring large areas and proper cooling; thermal storage utilizes low-quality energy and requires an additional step for conversion to electricity; chemical storage also requires an additional step for electricity generation and can have some safety issues. On the other hand, chemical storage has a significant advantage, which is being transportable.

Some power generation systems use Rankine cycles using organic fluids such as isobutene, toluene, etc. They are called organic Rankine cycles (ORCs), and are more suitable for low-grade thermal energy sources such as solar collectors and geothermal sources. Supercritical CO_2 cycles or transcritical cycles can also be considered in this category. Pure ammonia,

ammonia-water, and other refrigerants can also be employed in organic Rankine engines having low capacity, which can utilize low-cost refrigeration compressors of scroll or screw type in reverse, namely as expanders. This specification raises the marketability of low-power solar driven generators. Another heat engine possibility is the Stirling engine that works at very high pressures (in the range of 200 bar) and high temperatures (in the range of 700 °C to 800 °C), working with helium or hydrogen as discussed in Chapter 4. Hydrogen is a flammable gas, which can cause safety issues if not properly managed. In these engines, hydrogen and helium can leak simply due to having a very small molecular structure, which brings maintenance issues. However, Stirling systems are very compact and can attain high engine efficiencies corresponding to about 40%. This can yield an overall electricity production efficiency of about 20–23% for 10 h/day operation and installed capacity of 10 to 25 kW [1]. Nevertheless, the long warm-up period of Stirling engines is considered one of the major disadvantages, especially in solar applications.

Concentrated solar thermal energy can be utilized for multiple products as illustrated in Fig. 6.3. The air Brayton cycle is one of the promising options for concentrated solar energy applications because the system has low thermal inertia and hence it can adjust to solar radiation variations. In the Brayton cycle, the constant pressure heat addition process can be realized by using specially designed solar reactors to replace the combustion chambers.

There are also thermochemical processes (e.g., Mg-Cl, Cu-Cl cycles) that can utilize concentrated solar energy to drive the internal steps, as shown in Fig. 6.4. In addition, photochemical hybrid cycles utilize direct sunlight to drive some of the reactions inside the cycles [2, 3]. A theoretical method for photochemical disproportionation of CuCl was also suggested by Zamfirescu et al. [4] as a potential method to reduce the power consumption and develop the performance of hybrid copper-chlorine cycles. A four-step thermo-photochemical process was also considered by Perret [5]. The system contained a photolysis reaction (a photochemical reaction of hydrolysis) which forms hydrogen at 353 K, and a thermal decomposition reaction with oxygen evolution at 1143 K.

Gasification, cracking and reforming are three of the most commonly employed techniques for hydrocarbon conversions. We are in an era of attempting to greenize environmentally polluting processes. Therefore, implementing solar energy in these hydrocarbon conversion processes is an important solution to reduce carbon footprints. Concentrated solar energy can be used to drive all of the processes by using specially designed reactors and systems. Fig. 6.4

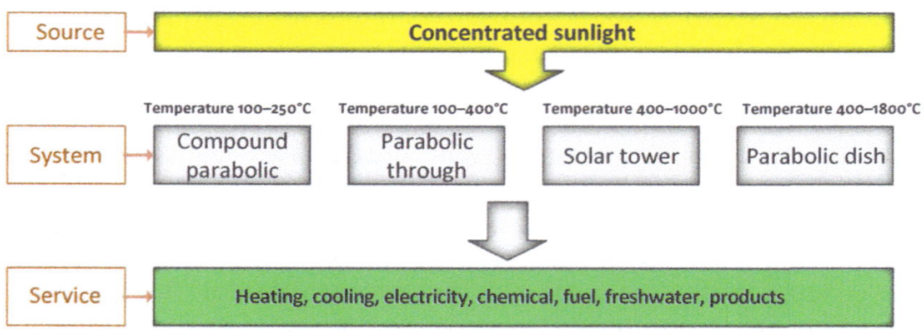

FIG. 6.3 Concentrated solar light-based energy conversion systems.

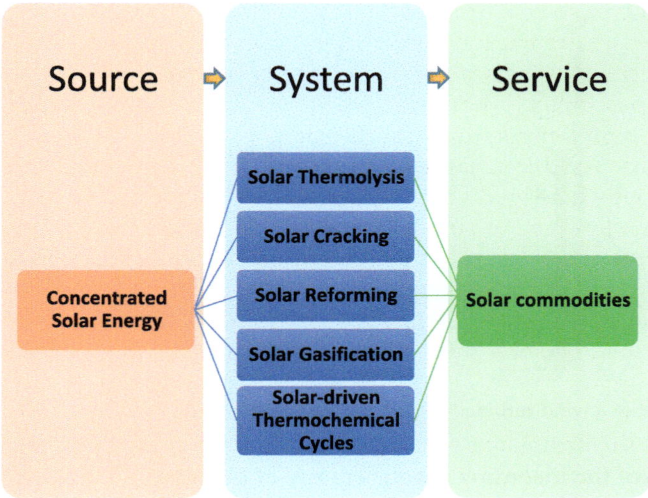

FIG. 6.4 Solar thermochemical energy conversion routes for production of useful solar commodities.

shows that concentrated solar energy can be employed to produce many useful products (heating, cooling, chemicals, fresh water, etc.) by conversion of various feedstock (water, coal, natural gas, oil, etc.).

6.3 Wind energy systems

Wind energy is a noteworthy renewable energy source that has shown rapid growth in recent years. The levelized cost of electricity from wind power plants has decreased considerably, competing with other conventional sources. Due to the wide availability of wind streams around the globe, wind power plants have been constructed in many regions of the world. The fundamental source of winds is actually solar energy due to temperature gradients. Winds carry energy in the form of mechanical energy. The available kinetic energy in the wind is converted into other useful forms, mainly electricity, by appropriate machines called wind turbines. Similar to solar energy, one of the main issues of wind energy conversion systems is the intermittent and fluctuating nature of the wind throughout the year and day. The wind turbines need to utilize the wind whenever available by adapting to the changes in its direction and intensity. In order to overcome the intermittency issue, the mechanical energy obtained from wind needs to be stored in different energy forms. There are numerous energy storage techniques that can be integrated to wind energy applications, such as kinetic (flywheels), electrochemical (batteries), chemical (hydrogen), thermomechanical (compressed air), etc. However, the widely utilized option is local storage in batteries equipped with grid connected systems. The energy conversion efficiencies of wind turbines can reach up to 35–50% with the existing technology [6]. The theoretical limit of wind energy conversion was studied by Albert Betz with considerations of conservation of mass and momentum of the air stream flowing over an actuator disk [7] as shown in Fig. 6.5. The theoretical

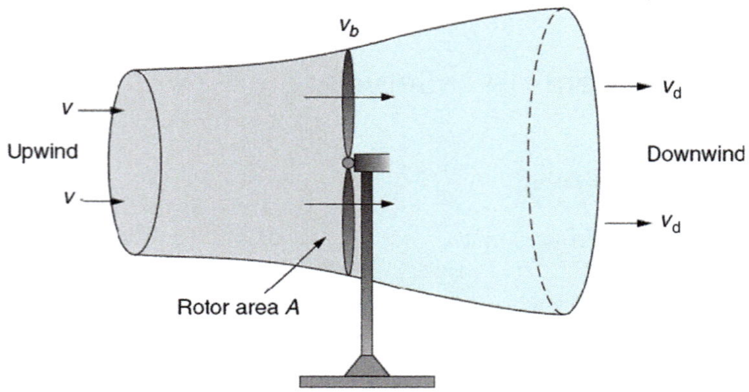

FIG. 6.5 Wind streams at the inlet and outlet of a wind turbine.

limit (Betz limit) corresponds to 59.3% of the incoming kinetic energy of the wind. In modern wind turbines, the captured wind energy is transformed into rotational energy by a wind turbine and then to electrical work via generators. However, direct usage of the mechanical energy is also possible, as it has historically been used for water pumping, grain milling, wood cutting, etc.

Two main groups of wind turbine design exist, namely the horizontal axis wind turbine (HAWT) and the vertical axis wind turbines (VAWT) as depicted in Fig. 6.6. The horizontal axis wind turbines have emerged as the leading technology compared to vertical axis ones. They have the rotor axis horizontal and in line with the wind direction. The common structures have the generator and the blade fixed on a rotating assembly at the top of a wind tower. The mechanical system also includes a gearbox that multiplies the rotation of the blades. Compared to the vertical axis, the horizontal axis wind turbines can attain higher

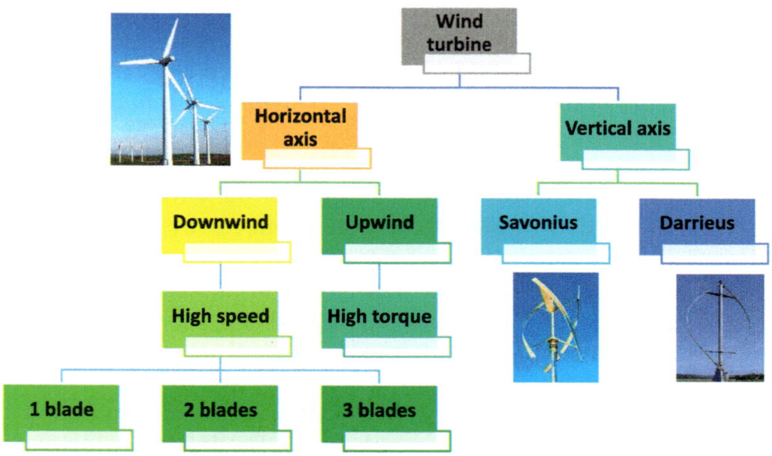

FIG. 6.6 Main types of wind turbines.

efficiencies. The vertical axis ones operate well in fluctuating winds if the wind direction remains quasi-constant.

One important feature of the VAWT is that it does not have to be pointed in the direction of the wind. Therefore VAWT installation is suitable where wind direction variations are more frequent [6]. An additional feature of the VAWT is that it permits location of the generator and gearbox on the lower side, which can reduce the maintenance costs and increase the tower support. The HAWT structure can vary based on the number of blades. Commonly, wind turbines have two or three blades. The HAWT uses airfoils to produce lift under the wind action and rotate the propeller. In addition to the wind turbine itself, wind power plants involve numerous other essential subsystems, such as yaw control, a gearbox, voltage-frequency and rotation controllers, voltage-raising and -lowering transformers, protection systems for over current, over speed, overvoltage, and atmospheric outbreaks, and other anomalous forms of operation. All these subsystems are structured by a robust mechanical assembly.

Recent conceptual studies (e.g. [8, 9]) show that wind turbines can also be used to produce thermal energy via a couple of techniques, namely (i) friction between solid materials, (ii) agitation of fluid, and (iii) electromagnetic induction. Similar to concentrated solar tower systems, wind-to-heat generation systems convert an energy source into thermal energy. The source of energy is kinetic energy of the blowing wind in this case, rather than solar irradiation in the solar towers. The mechanical energy is utilized to generate electromagnetic induction, which eventually heats the fluid flowing inside the electromagnetic field. Rotating energy is converted to thermal energy by using electromagnetic induction. The heat energy produced at the top of the wind tower is transferred to the bottom utility by the circulation of suitable heat transfer fluid. The heated heat transfer fluid can be directly utilized in a conventional power generation cycle (e.g., Rankine cycle) via heat exchangers or it can be stored in the thermal energy storage tanks, where it can be later on converted into electricity according to the demand.

The induction heating can be incorporated into the wind turbine nacelles for heating the heat transfer fluid, which is recycled in the thermal energy storage tanks. In this way, the system can operate continuously by storing the wind energy in thermal energy form using thermal energy storage tanks and molten salt. The main advantages and disadvantages of electromagnetic induction for wind thermal energy systems (WTES) are tabulated in Table 6.1.

TABLE 6.1 Advantages and disadvantages associated with electromagnetic induction for WTES

Advantages	Disadvantages
Suitable for large-scale applications	Complicated three phase windings
Low maintenance	The weight and cost of this generator are heavier and higher respectively than the simple heat generator
Production of high temperature for thermal energy storage	Challenging estimation of the resulted energy cost due to new concept

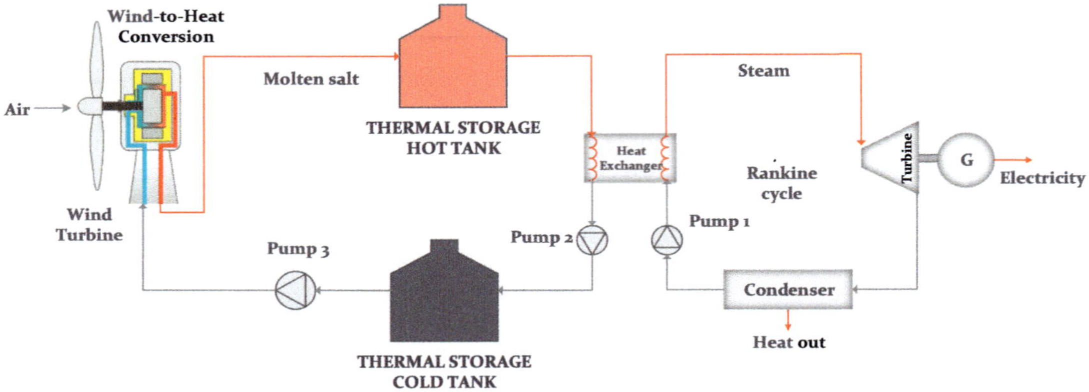

FIG. 6.7 Wind-to-thermal energy conversion system integrated to molten salt thermal storage and Rankine cycle for more stable power generation. *(Modified from Karasu H, Dincer I. Analysis and efficiency assessment of direct conversion of wind energy into heat using electromagnetic induction and thermal energy storage. J Energy Resour Technol 2018;140:071201. https://doi.org/10.1115/1.4039023).*

A concept study was initially conducted by Okazaki et al. [8] and a thermodynamic analysis of a molten salt integrated wind energy system was performed by Karasu and Dincer [9]. As shown in the system layout in Fig. 6.7, the molten salt is circulated through highly specialized piping in the heat generator of the wind turbine during the day and held in storage tanks at night. The tanks store the salt at atmospheric pressure. Use of molten salt for both heat transfer and thermal energy storage minimizes the number of storage tanks and salt volume needed. Molten salt can be stored at about 560 °C until electricity is needed for day or night. On demand, the molten salt is dispatched from the hot tank through a heat exchanger to create super-heated steam, which then drives a conventional steam turbine in the Rankine cycle.

A sample wind thermal energy storage (WTES) system consists of wind heat generators using electromagnetic induction, which are installed on top of wind turbine towers, molten salt circulation and storage cycle, steam transfer fluid, steam generation system, a Rankine steam turbine/generator cycle, and a thermal storage backup system as illustrated in Fig. 6.7.

6.4 Hydro energy systems

Hydro energy is available in many forms: (i) potential energy from high heads of water retained in dams, (ii) kinetic energy from current flow in rivers and tidal barrages, and (iii) kinetic energy from the movement of waves on relatively static water masses.

Similar to wind energy, the fundamental source of hydropower is also solar irradiation as shown in Fig. 6.8. The solar irradiation causes evaporation over the water bodies and forms clouds. Later on, the water falls as rain and forms water reservoirs on land, which are used to convert mechanical energy into electricity by hydropower plants.

The classification of hydropower plants is mainly conducted based on the capacity and size as shown in Fig. 6.9.

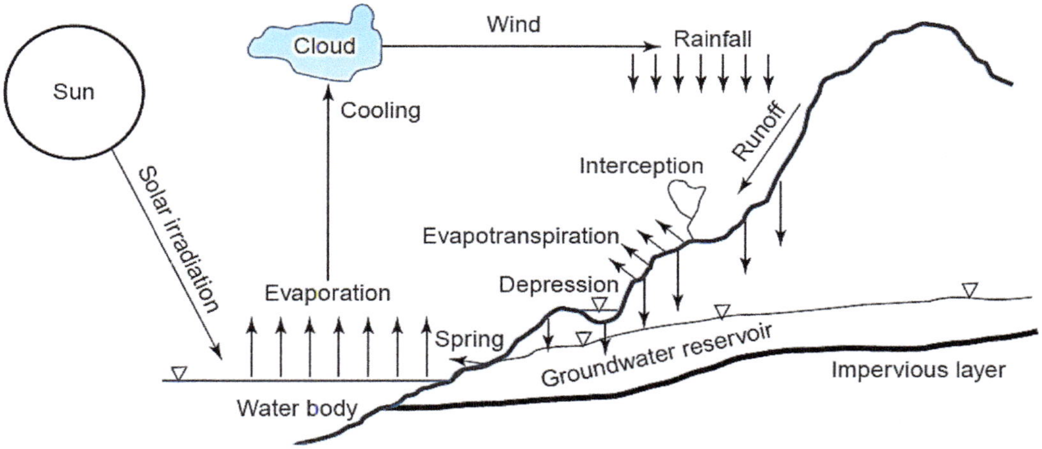

FIG. 6.8 Hydrological cycle on earth. *(Adapted from Sen Z. Practical and applied hydrogeology. Oxford, UK: Elsevier; 2015. https://doi.org/10.1016/C2013-0-14020-2).*

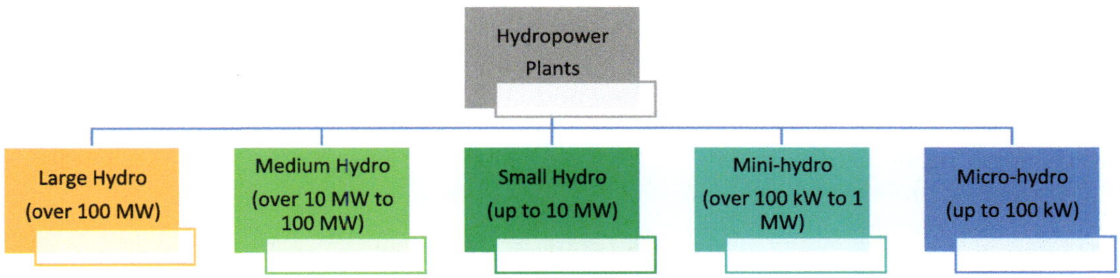

FIG. 6.9 Classification of hydropower plants based on capacities.

Large hydropower and medium hydropower plants usually feed into a large electricity grid with very high efficiencies. Mini-hydropower plants can either be installed stand-alone or more often feeding into the grid, whereas micro-hydropower plants usually provide power for a small community or rural industry in remote areas away from the grid. Due to environmental concerns of large-scale hydropower plants, small and micro-hydropower systems receive increasing attention.

As shown in Fig. 6.10, the main conversion mechanism is the use of water turbines. The generator is linked to the turbine through a shaft, and the rotation of the generator is driven by the turbine rotation. In the generator, there is an electromagnetic field converting the mechanical energy to electrical energy. Hydroelectric energy is the most established and significant renewable and clean energy source, especially in the subtropical climate belt of the world. It is one of the most reliable, technically exploitable and environmentally friendly renewable energy alternatives [10].

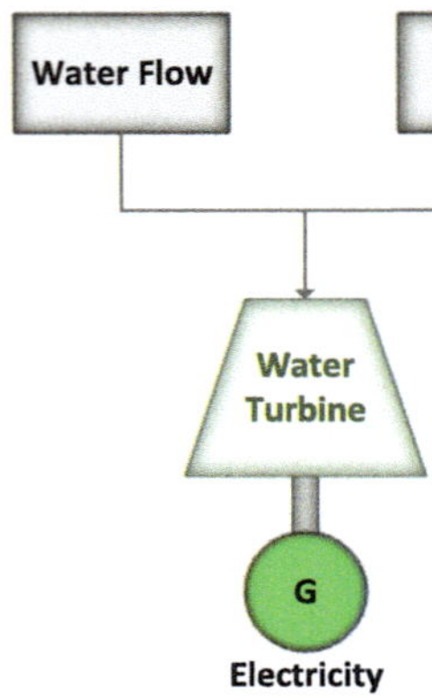

FIG. 6.10 Power production pathways from hydro energy.

Since hydroelectric generators can be initiated or turned off almost immediately, hydropower is more responsive than most of the other energy types to provide peak demands. Water can be stored overnight in a reservoir and kept until needed during the day, and then released to turbines for power generation to help supply the peak-load demand. The electric energy generation is sustainable as long as the rainfall and snow occasions are dynamic in the area. Nevertheless, current global warming and climate change impacts cause rainfall and snow irregularities, such as increase or decrease of snow/rain levels, throughout the year based on the location [10].

The chief environmental influence of reservoirs is the alteration of the natural ecological life, particularly at the downstream areas. In the design of a dam, the possible maximum rainfall and its resulting possible maximum flood circumstances need to be taken into account for dam stability works, flood dangers, and for the spillway design, which allows unnecessary flood water to bypass the dam reservoir [10].

There are commonly four main types of hydropower facilities:

- *Impoundment*: This type uses a dam to store river water in a reservoir. The most commonly used hydropower plant structures have an impoundment facility regularly behind a dam to store drainage basin river flow in a reservoir. Water is led through a pipe system to a turbine to rotate it, and eventually activates a generator for electricity production.
- *Diversion*: Also called a run-of-river facility, a diversion is another type of hydropower facility that does not have any impoundment. It diverts a part of a river through a canal or penstock in which water turbines are situated for electricity generation.
- *Pumped storage*: This type stores energy by pumping water uphill to a reservoir at a higher elevation from a second reservoir at a lower elevation. It works like a battery by storing the electricity produced by water turbines as discussed in Chapter 3. It converts the electrical work to potential energy by pumping water uphill; then, when there is low energy demand, this system stores energy and during high electrical demand periods, the water is released back to the lower reservoir and rotates a turbine for electricity generation.

TABLE 6.2 Main types of hydro turbines utilized in hydro energy applications

Pelton turbine		The usage of Pelton turbines is most suitable when there are high heads with low discharges, which is possible in a wide range of circumstances with heads that vary from 15 m up to almost 2000 m. High-pressure heads deliver fast water jet impingements in the blades yielding very high turbine rotational speed.
Francis turbine		The Francis turbine works in a fully submerged position with water flow entering in a radial direction toward the axis and exiting in the axis direction. These type of turbines are appropriate for rather low water heads of 500 m or less, and they are the most frequently used high power turbines. Large-scale turbines are capable of over 500 MW power delivery from a water head 100 m with efficiencies of up to 95%.
Propeller and Kaplan turbines	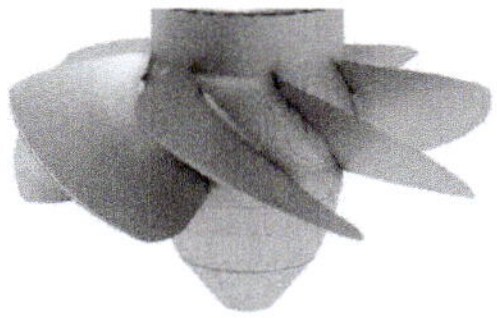	The propeller turbine is a reaction type turbine working in full submergence. It is similar in its geometrical form to a ship propeller, which is the most appropriate design for low head water sources with a high flow rate such as those in slow running rivers.

- *Other types*: Tidal, wave, heat, and current flows for electricity production through numerous technical system designs can be placed in this category. Tidal and wave energy are commonly categorized under ocean energy as well. Hence, there will be more explanation in the next sections.

The type of hydro turbines selected for a hydro facility depends on the following criteria:

- Height of standing water or "head"
- Volume flow, or volume of water
- Efficiency of turbine, and cost

There are mainly two types of hydro turbines, which are impulse turbines (Pelton turbine) and reaction turbines (Francis turbine, propeller turbine, Kaplan turbine); details and schematics are shown in Table 6.2.

6.5 Geothermal energy systems

Geothermal energy is a form of thermal energy, which is present in some specific areas of the world at different temperature levels in the range of about 35 to 350 °C. Above 250 °C is extremely rare, whereas most geothermal locations can provide temperature levels up to 250 °C [6]. Geothermal heat is utilized either for process heating or is transformed into electricity via suitable heat engines. A large portion of heat-consuming processes can be delivered by geothermal energy, including space and water heating, industrial processes, and several processes in agriculture and the food industry. As a source of heat, geothermal energy conversion is also subject to the Carnot limit. Hence, it is vital to evaluate the range of the Carnot factor for geothermal reservoirs. Moreover, it is beneficial to analyze applicable irreversibilities specific to geothermal energy conversion.

Heat from the underground is conveyed to the surface in three possible ways that are categorized by the type of geothermal field: (i) hot water, (ii) wet steam, and (iii) dry steam. Therefore, depending on the fluid type, appropriate conversion systems are employed. Hot water fields comprise reservoirs of water with temperatures in the range of 60 to 100 °C, and are most proper for space heating and agricultural uses. For hot water fields to be practically viable, they need to contain a huge amount of water with a temperature of minimum 60 °C and lie within 2000 m of the surface [6]. Wet steam fields consist of water under pressure and typically measure in the range of 100 °C. When the water is transported to the surface, some of the water flashes into steam and the steam can run the turbines to produce electrical work.

Dry steam fields are geologically similar to wet steam areas, except that superheated steam is removed from the aquifer. Dry steam fields are relatively rare. Since superheated water explosively converts into steam when exposed to the atmosphere, this type of geothermal source is much safer and usually more cost-effective for electricity generation, which is much more easily transported.

Table 6.3 shows the maximum conversion efficiencies from low-, medium-, and high-temperature geothermal sources when sensible heat is extracted from the geothermal fluid. The values are calculated based on the Carnot efficiency definition. In some cases, the geothermal fluid is steam. One can assume in such cases that steam is condensed through an isothermal process. This process is followed by an isentropic expansion to ambient conditions. Although this is an idealization, it is useful to identify the conversion efficiency for this case to find out the upper limit.

Throughout the historical advancements, several thermodynamic cycles have been industrialized for geothermal power generation. The collection of the cycle depends on the type of geothermal fluid, its flow rate and the temperature level. The geothermal fluid can be dry steam, low-pressure brine, or high-pressure brine as classified in Fig. 6.11. The geothermal basin can or cannot require reinjection of the fluid after its use in the power plant. The power plant design must be such that it removes as much exergy as possible from the geothermal fluid.

TABLE 6.3 Thermodynamic limits of geothermal energy conversion

Geothermal source	Temperature	Sensible heat exchange	Latent heat exchange
Low temperature	150 °C	17%	29%
Medium temperature	220 °C	24%	39%
High temperature	500 °C	44%	61%

Source: *Dincer I, Zamfirescu C. Chapter 7—Renewable-energy-based Power generating systems. Adv. Power Gener. Syst., Boston: Elsevier; 2014, p. 369–453. https://doi.org/10.1016/B978-0-12-383860-5.00007-9.*

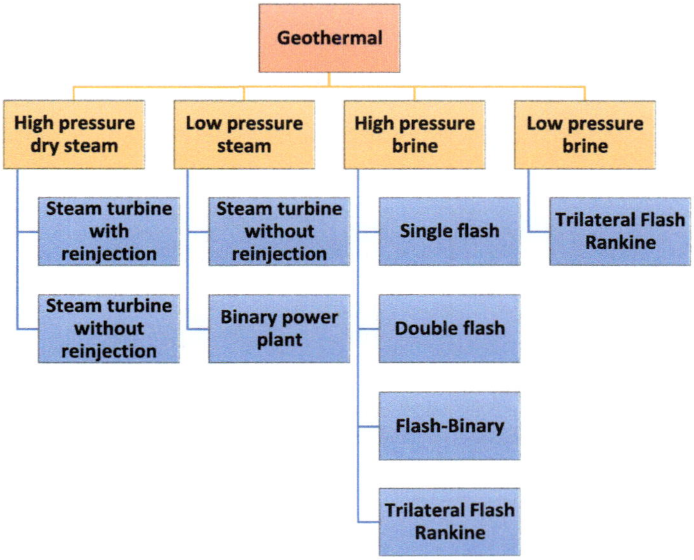

FIG. 6.11 Classification of geothermal energy conversion based on fluid types.

If the geothermal heat is accessible in the form of dry steam, the energy conversion into work can be attained with steam turbines. The steam after expansion can be condensed and reinjected into the geothermal well, or it can be just sent into the atmosphere. Some geothermal basins reject steam at a high or lower pressure, which can be expanded in a turbine for power generation. Based on the kind of geothermal well, reinjection of the geothermal fluid is or is not needed. If a geothermal plant yields dry steam, a turbine can be directly used to generate power, as pointed out in Fig. 6.12. For the cases when the steam pressure is low, or there is no need to recycle the geothermal fluid, then the steam can be released to the atmosphere after expansion down to atmospheric pressure, as shown in Fig. 6.12. Or it can be used for cogeneration purposes. Nonetheless, if the steam pressure and temperature are high enough, the quantity of produced power permits driving a recirculation pump and still producing reasonable yield from the turbine. Fig. 6.12 shows that the steam can be expanded in a vacuum, condensed, and the generated water pressurized in a pumping station and reinjected. It is noted that the reinjection pressure must be very high, in the range of minimum a couple hundred bars [6]. The reinjection is required in several cases to keep the geothermal source at steady generation. If the geothermal basin has a hot rocky layer, then the water injected through one well is boiled and extracted as steam through another well.

The geothermal sources yielding low-pressure dry steam, which requires reinjection, can be combined with organic Rankine cycle generators as shown in Fig. 6.13. These cycles are known as binary cycles, as occasionally they function with binary mixtures of refrigerants such as ammonia-water mixture. The dry steam extracted from the geothermal well is condensed and subcooled before water reinjection. The heat discharged by the geothermal brine is then conveyed to a working fluid. The working fluid is boiled and expanded as saturated toluene vapor. After expansion, it tracks internal heat recovery in the heat exchanger, followed by condensation, pressurization of the saturated liquid, preheating and boiling [6].

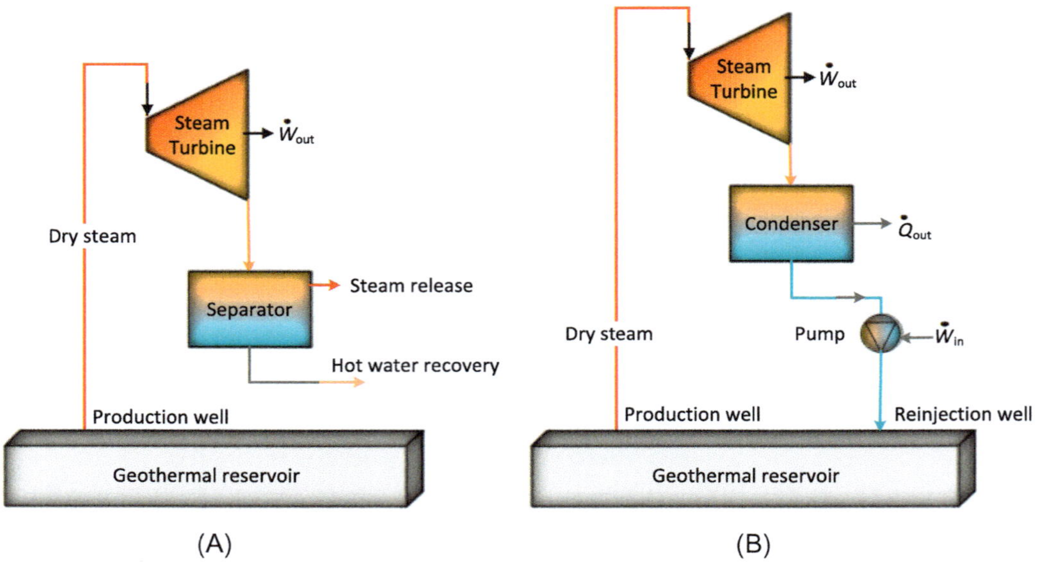

FIG. 6.12 Schematic diagram of dry steam geothermal power plants with direct expansion: (A) without reinjection, (B) with reinjection.

FIG. 6.13 Schematic diagram of a geothermal binary cycle power plant driven by dry steam.

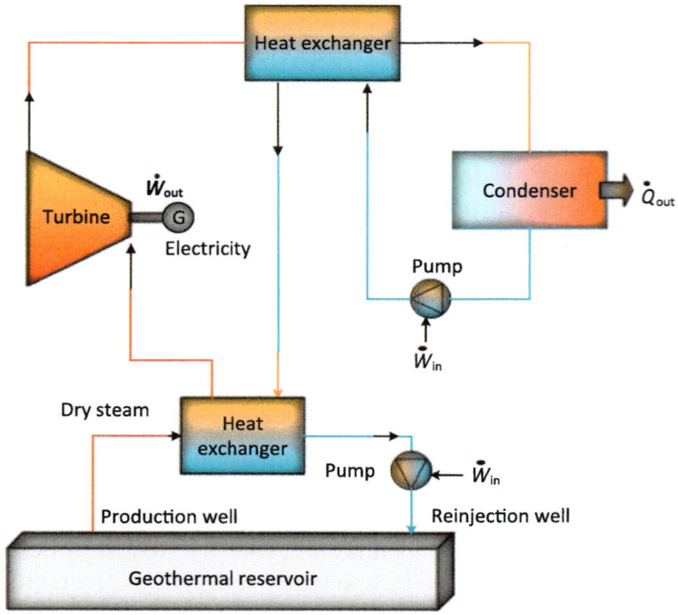

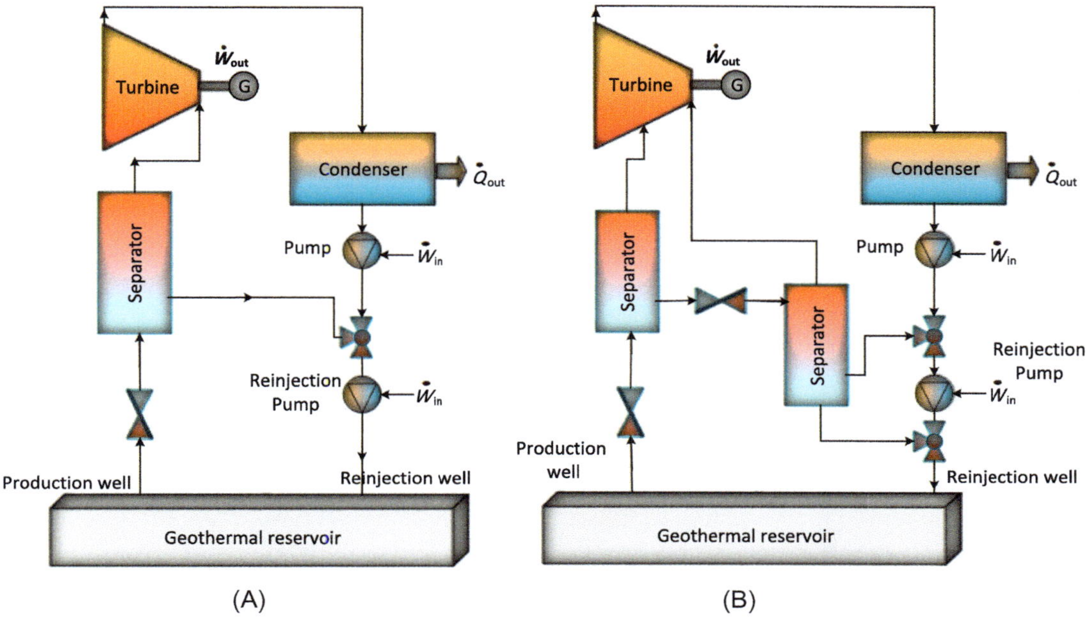

FIG. 6.14 Schematic diagram of steam flash geothermal power plants (A) single flash, (B) double flash.

Fig. 6.14 shows the single and double flash type geothermal plants. In a single flash, there exists one separator, which splits the steam and water. The split water is merged with condensed water (after the turbine and condenser) and reinjected. In the double flash arrangement, the remaining hot fluid from the first flashing stage is flashed again to make steam with a lower pressure than the primary steam. It produces 15–25% more power, but is more complex and hence more costly.

In binary geothermal plants, high-temperature fluids from a well flow through heat exchangers in which a secondary loop contains a low boiling point fluid such as butane, as illustrated in Fig. 6.15. They are a combination of flash and binary plants. It helps these types of plants generate electricity at much lower temperatures, as low as 55 to 60 °C, but there are low conversion efficiencies due to Carnot limitation. The system shown in Fig. 6.15 is a combination of flash and binary types. Binary cycles have lower efficiencies than other types of geothermal plants.

6.6 Ocean energy systems

There are mainly five different types of energy extraction methods from oceans, namely tides, waves, ocean currents, thermal energy, and osmotic (salinity gradient). When tides hit the shore, they can be imprisoned in reservoirs behind dams. Then when the tide drops, the water behind the dam can be sent, similar to hydropower plants. Tidal energy is an alternative with potential for cleaner power generation. A classification of ocean energy for energy conversion applications is illustrated in Fig. 6.16.

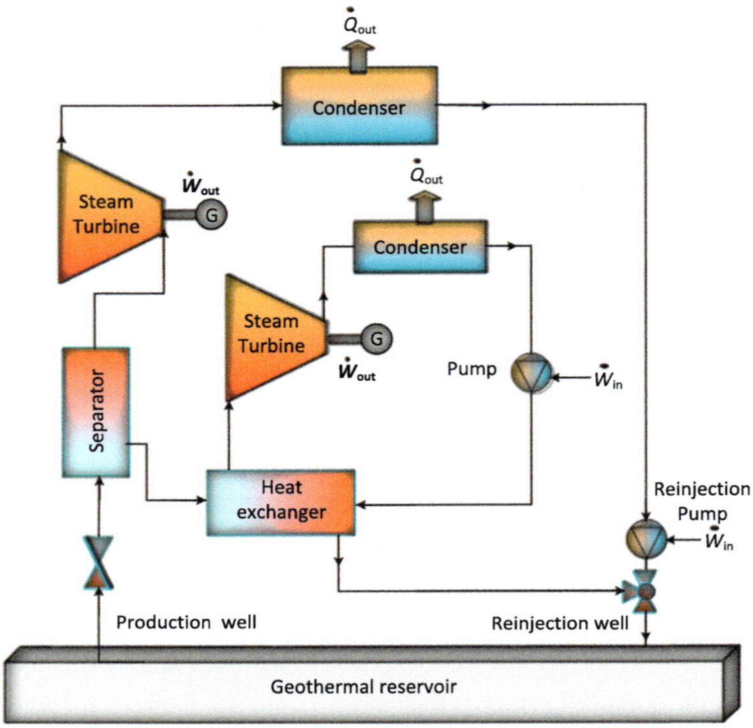

FIG. 6.15 Schematic diagram of a flash-binary geothermal power plant.

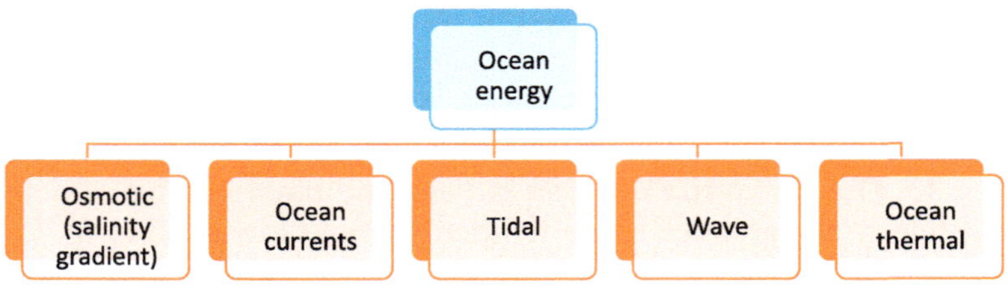

FIG. 6.16 Ocean energy conversion pathways.

The kinetic energy due to the moon's gravitational pull on the oceans during the Earth's rotation produces a durable tidal effect and forms the potential to use the kinetic energy of waves hitting continental shores for energy conversion. Wave energy can be collected with floating bodies that perform elliptic movement under the action of waves.

The thermal cycle of tropical solar energy distresses the oceans during the Earth's rotation and produces kinetic energy that is used directly to rotate immersed turbine generators. The temperature gradient of the ocean at deep levels yields a temperature difference, which can be enough to extract thermal energy at low efficiencies. These type of systems are called ocean thermal energy conversion (OTEC) for which a system layout is shown in

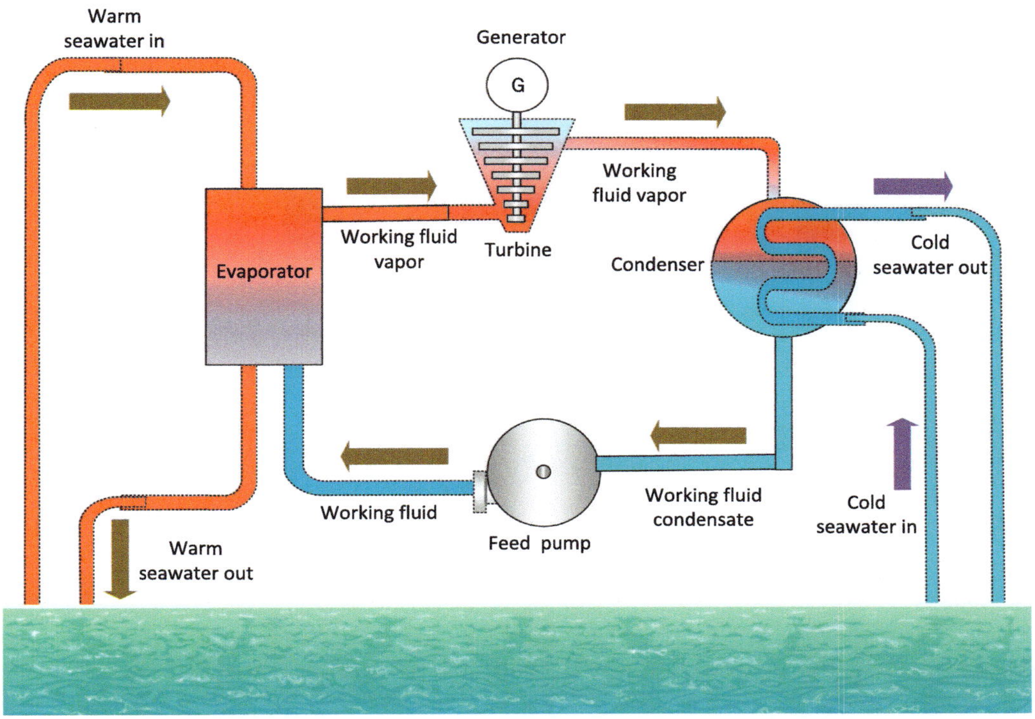

FIG. 6.17 Layout of a closed-type ocean thermal energy conversion (OTEC) system.

Fig. 6.17 of a closed-type. For conversion systems, ammonia-water Rankine and Kalina cycles have been proposed for OTEC applications. Using depth seawater temperature as a sink at about 4 °C, and the ambient temperature at 25–35 °C as a heat source, an energy conversion efficiency of about 5% can be achieved [6]. A sample OTEC system layout is exemplified in Fig. 6.17. The system employs a Rankine cycle using ammonia as the working fluid. The system can be built on a floating platform or on a ship. It utilizes the surface water as the source. Typically the water at the ocean surface is at a higher temperature than in the deep ocean. The surface water is circulated with pumps through a heat exchanger that performs as a boiler for ammonia fluid. Water from the deep ocean at 4 °C is pumped to the surface and used as a heat sink in an ammonia condenser. There are also open type OTEC plants as depicted in Fig. 6.18. In this open type, surface water from the ocean initially enters a flash evaporator and then the extracted vapor enters a turbine. The exhaust of the turbine is condensed using cold ocean water. Furthermore, the condensed water can be useful if collected using a suitable technique.

Tides also have power and a tidal power facility is based on the daily fluctuations of water level. The probability of energy generation is very high and it is also possible to produce power during high demand periods if there are convenient reservoirs. The most common types do not have a reservoir and produce power from water kinetic energy or undammed sources such as undershot waterwheels. It is similarly possible to capture

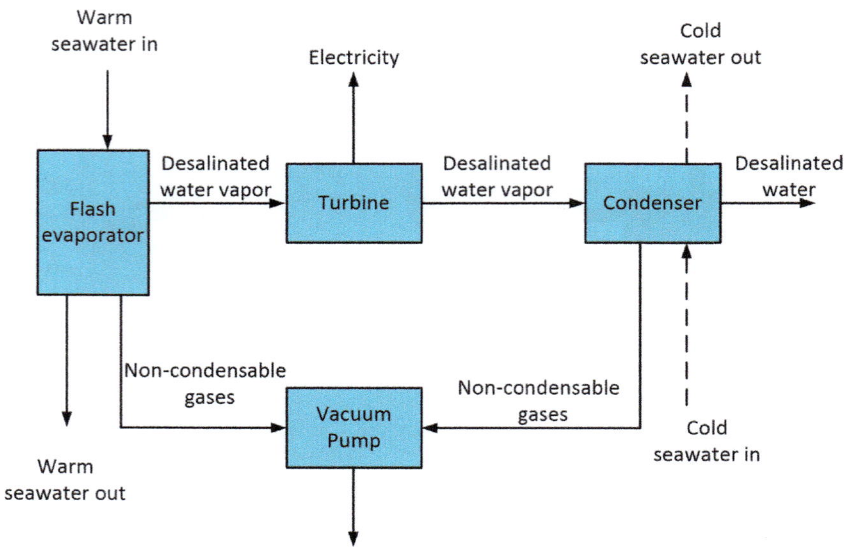

FIG. 6.18 Layout of an open-type ocean thermal energy conversion (OTEC) plant.

hydropower from the tides by employing bidirectional turbines in the path of the tidal water flow in bays, and particularly in river inlets [10]. It is essential to have a large tidal range with a barrier across the bay or inlet to funnel the water via the turbines as the tide comes in and out.

Surface wave motion is limited and very challenging to capture, but numerous original systems have been proposed for power generation. Very small installations generate electricity commercially and most have been hindered by practical difficulties. Some of the wave energy deployment methods are: oscillating float system, oscillating paddle system, oscillating snake system, oscillating water column, pressure transducer system, wave capture systems, overtopping wave system, etc.

A tidal barrage system can have three types of operation, based on the phase at which it produces power. When ebb generation is applied, the basin is filled during the flood tide. In night operation, additional water can be pumped into the basin, as energy storage during off-peak hours. If the tide ebbs low enough, water is released over turbine systems to produce electricity (ebb tide is the period between high tide and low tide.). In flood generation, the dam gates are closed such that the water level upsurges on the ocean side until it extends the maximum. Then the water flows over the turbine systems and fills the basin while producing energy. In two-way production, electricity is produced in both the flood and ebb phases of the tide [6]. The world's largest tidal power production facility is located in France at La Rance, having an installed capacity of 240 MW, operating with 24 reversible turbines (each has 10 MW) and a hydrostatic head of 5 m [11].

Ocean water currents are produced by tides, the Earth spinning, the thermal cycle of tropical solar energy, and superficial winds. The energy carried by water currents can be extracted through current turbines immersed in water. Principally, water current turbines have a working mechanism similar to wind turbines, with the difference that the density and the viscosity of water are 100 and 1000 times higher than that of air, respectively. Hence, there are a couple of differences between the operation of water current turbines and wind turbines. For instance, water current turbines are manufactured with horizontal axis or vertical axis arrangements similar to wind turbines. The thrust produced by water current turbines is much greater compared to wind turbines, so they need to be made of stronger material.

Osmotic energy is a form of energy derived from different salt concentrations in seawater and freshwater. This type of energy resource is typically located in large river inlets and fjords. The system employs a semi-permeable membrane for equalizing the salt concentrations, therefore raising the pressure in the seawater section. An osmotic power plant produces power from salinity gradients by guiding freshwater and saltwater into separate chambers, divided by a membrane. The salt molecules pull the freshwater through the membrane, creating a pressure on the salt water side that can drive a turbine. Salinity gradients can likewise provide power by reversed electrodialysis. This method is currently being researched for advancements (e.g. [12–14]) that are mainly devoted to enhancing the membrane characteristics, since the power intensity obtained from the membranes is in the range of a few watts per square meter.

6.7 Biomass energy systems

Biomass is also categorized under renewable energy sources because its inherent energy is also sourced from the sun and it can regrow given sufficient time. Any type of fossils belonging to living species is a form of biomass. In fact, biomass has been a primary energy source on the Earth for thousands of years. Biomass comes from the photosynthesis aspect of solar energy distribution and comprises all plant life, all succeeding types in the food chain, and ultimately all organic waste. Biomass resources have diverse origins, ranging from wood and crop forms to waste forms. The main feature of biomass is its chemical composition, in such forms as sugar, starch, cellulose, hemicellulose, lignin, resins, and tannins.

Photosynthesis can be described by a simplified chemical reaction as follows:

$$6\,H_2O + 6\,CO_2 + \text{energy} \rightarrow C_6H_{12}O_6 + 6\,O_2$$

where $C_6H_{12}O_6$ is glucose, which is subsequently transformed into other sugars, cellulose, lignin, etc. A typical photosynthesis process may have an average efficiency of 5%, as illustrated in Fig. 6.19.

Biomass energy can be described as the energy extracted from biomass for conversion into useful forms such as heat, electricity, and transportation fuel applications. The most common and easiest method of biomass conversion is by combusting and producing heat. The high-temperature heat can be further transformed into electricity through heat engines. The fundamental biomass conversion pathways are exemplified in Fig. 6.20. Gasification converts the biomass into a gaseous fuel such as syngas, whereas liquefaction produces a liquid fuel. Another route of biomass energy conversion is biochemical. Fermentation and photosynthesis are classified under biochemical energy conversion [6]. In this case, the fermentation process (anaerobic or aerobic digestion) can yield biogas, alcohol, and hydrogen. Furthermore, the

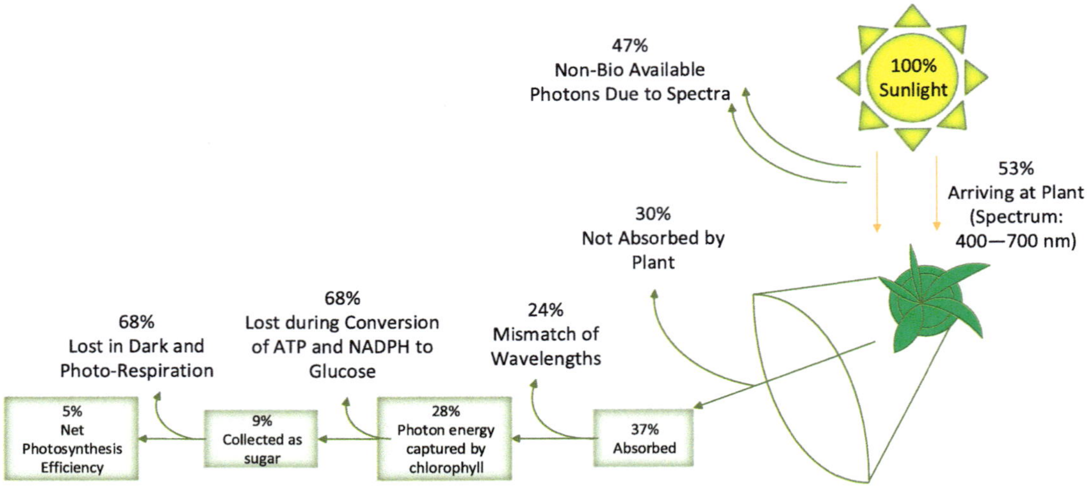

FIG. 6.19 Schematic illustration of approximate net photosynthetic efficiency. *(Data from Kui LICHLIU. Analysis of photosynthesis efficiency of maize hybrids with different yield in the later growth stage. Acta Agron Sin 2002;3:17; Zhu X-G, Long SP, Ort DR. What is the maximum efficiency with which photosynthesis can convert solar energy into biomass? Curr Opin Biotechnol 2008;19:153–9. doi:10.1016/J.COPBIO.2008.02.004; Chow WS, Melis A, Anderson JM. Adjustments of photosystem stoichiometry in chloroplasts improve the quantum efficiency of photosynthesis. Proc Natl Acad Sci U S A 1990;87:7502–6. doi:10.1073/PNAS.87.19.7502; Sinclair TR, Horie T. Leaf nitrogen, photosynthesis, and crop radiation use efficiency: a review. Crop Sci 1989;29:90. doi:10.2135/cropsci1989.0011183X002900010023x; Photosynthesis. Energy efficiency of photosynthesis | Britannica.com 2018. https://www.britannica.com/science/photosynthesis/Energy-efficiency-of-photosynthesis (Accessed 23 November 2018)).*

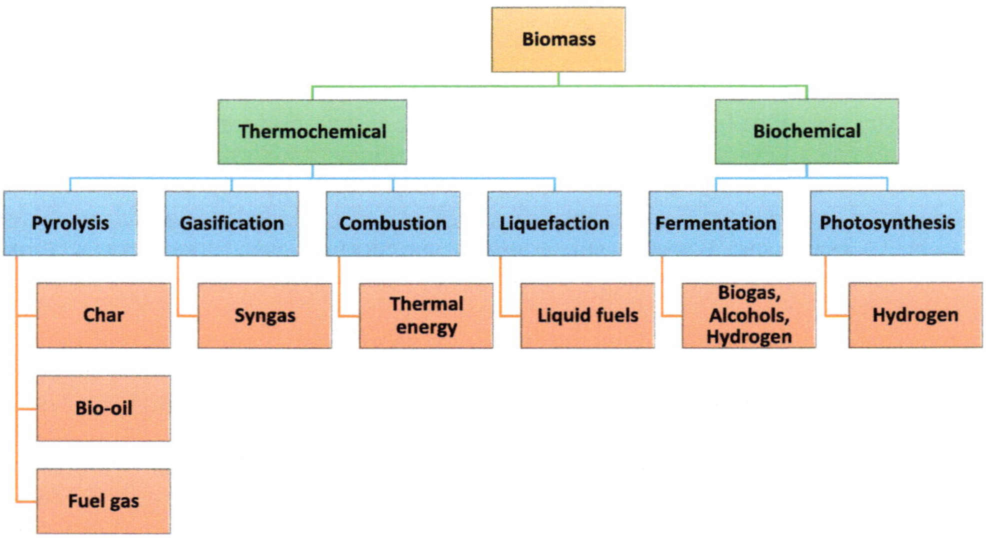

FIG. 6.20 Biomass energy conversion pathways.

photosynthesis process conducted by some phototrophic organisms can ultimately generate hydrogen.

Biomass can be transformed to a gaseous fuel by two methods, namely the thermochemical and biochemical conversion options. In the thermochemical conversion method, bio-syngas is attained, whereas in the biochemical conversion path, either biogas or hydrogen can be generated.

Some specific features of biofuels can be listed as follows:

- Primary biofuels are used in an unprocessed form, primarily for heating, cooking or electricity production, e.g., fuelwood, wood chips and pellets, organic materials.
- Secondary biofuels are products of processing of biomass, e.g., liquid biofuels such as ethanol and biodiesel that can be used in vehicles and industrial processes.
- Bioenergy is predominantly used domestically (80%), and less industrially (18%).
- Liquid biofuels for transport are growing but still represent only 1% of total transport fuel and less than 0.5% of total global energy consumption

Biodiesel is ethyl or methyl ester that is produced through a process that combines organically derived oils with ethanol or methanol in the presence of a catalyst. Common sources of biodiesel include new and used vegetable oils, animal fats, and recycled restaurant greases. The higher heating value of biodiesel is about 40,700 kJ/kg, which is about 9% less than that of petroleum diesel (44,800 kJ/kg). The most common biodiesel mixture used in the United States is B20, which is 20% biodiesel and 80% conventional diesel. Biodiesel is also used as a single fuel in compression ignition engines, called B100. Whereas B20 can be used in a diesel engine without any modification, B100 usually requires the use of special material for parts such as hoses and gaskets. Due to lower heating values, B100 provides less power from the engine. Some maintenance issues can also arise and some manufacturers may not cover biodiesel use in their warranties. B100 could increase nitrogen oxides emissions while significantly reducing hydrocarbon, sulfur and carbon monoxide emissions.

Numerous options exist for power generation using biomass, with direct combustion of biomass for driving a steam generator being one of the most common. Biomass combustion facilities can also be combined with ORCs, which are categorized by decent turbine efficiency at low installed capacities. The main features of gasification, combustion and pyrolysis are explained in Table 6.4.

When biomass is transformed to biogas, it can be used as a fuel for internal combustion engines to generate electrical work. It can be utilized for stove heating, water heaters, torches, motors, and other apparatus. It is also possible to combine biogas production amenities with micro-power plants comprising gas turbines cascaded with a reciprocating internal combustion engine. For such micro-power plants, alcohol and gasoline motors can use methane without affecting their operating integrity. This alteration can be performed by installing a biogas cylinder in place of conventional fuel. For gas flow regulation, a reducer can be located close to the motor [6].

Combined heat and power generation by means of biomass gasification techniques integrated to gas-fired engines or gas turbines can attain considerably greater electrical efficiencies, between 22% and 37%, than those of biomass combustion options with steam generation and a steam turbine, at 15–18% [6]. If the generated gas is utilized in fuel cells for power

TABLE 6.4 Biomass conversion through combustion, gasification and pyrolysis

Process type	Basic process diagram	Details
Combustion	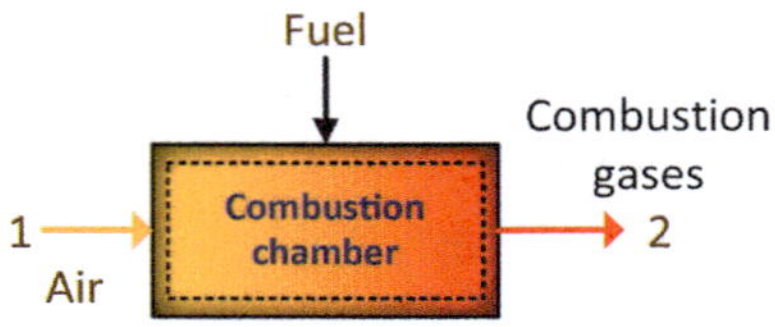	It is a burning process, including the sequence of exothermic chemical reactions between a fuel and an oxidant accompanied by the production of heat and conversion of chemical species.
Gasification	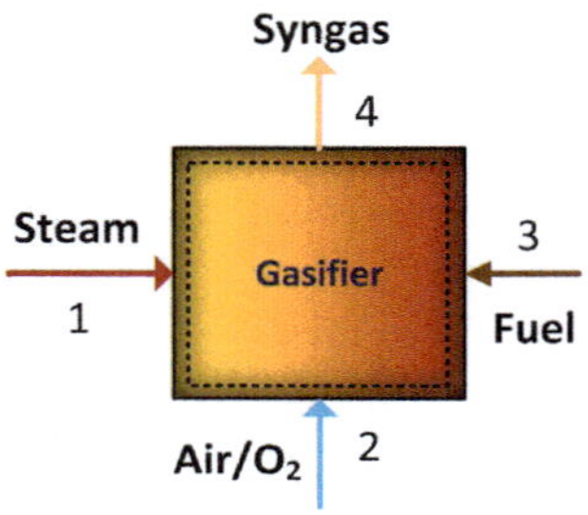	It is a process that converts organic or fossil based carbonaceous materials into CO, H_2, CO_2 and CH_4 as well as other gases. It is achieved by reacting the material at high temperatures (greater than 600–700 °C), without combustion, with a controlled amount of oxygen and/or steam. The resulting gas mixture is called syngas (from synthesis gas or synthetic gas) or producer gas and is itself a fuel.
Pyrolysis	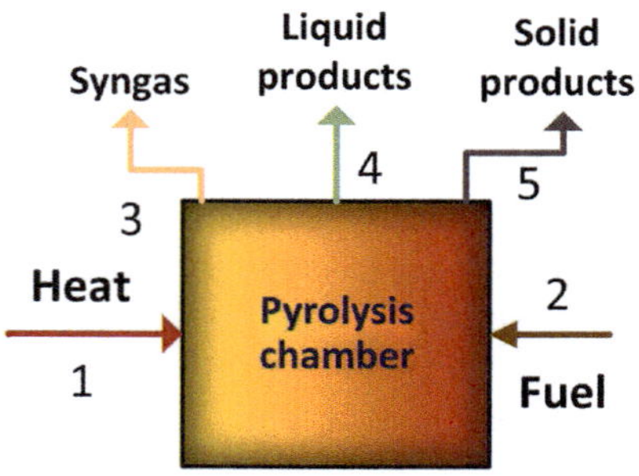	It is a thermochemical decomposition of organic material at elevated temperatures in the absence of oxygen, occurring typically at atmospheric pressure and operating temperatures between 400 and 800 °C.

generation, an even higher overall electrical efficiency can be attained, in the range of 25–60%, which can even be enhanced by recovering the excess heat in a fuel cell.

Because of the enriched electrical efficiency of energy conversion via gasification, the potential reduction in CO_2 is larger than with combustion. The creation of NOx molecules can also be mostly eliminated, and the elimination of contaminants is easier for numerous materials. The NOx advantage, however, may be partly lost if the gas is subsequently used in gas-fired engines or gas turbines. Considerably smaller releases of NOx, CO, and hydrocarbons can be anticipated if the generated gas is used in fuel cells rather than gas-fired engines or gas turbines.

6.8 Brainstorming for system integration

Multigeneration energy systems can contribute significantly to achieving a sustainable energy era. The useful outputs can range from hot water, potable and fresh water, to space heating/cooling, chemical fuels and electricity. Integrating renewable energy resources is one of the most attractive options. The efficiency levels of multigeneration systems are often greater than those of trigeneration and cogeneration systems with less environmental impacts.

There are numerous renewable energy systems that can be used for system integration; some of them are listed as follows:

- Photovoltaics (PV)
- Photovoltaics/thermal (PV/T)
- Concentrated photovoltaics (CPV)
- Concentrated photovoltaics/thermal (CPV/T)
- Concentrated heliostat solar tower
- Parabolic trough solar collector
- Parabolic dish solar collector
- Wind turbines for power and heat generation
- Biomass gasification/combustion/pyrolysis
- Ocean thermal energy conversion
- Wave energy and tidal energy
- Geothermal power generation cycles
- Geothermal district heating
- Hydropower

The following paragraphs discuss several ideas for renewable-based integrated energy systems, as shown in Fig. 6.21 and Fig. 6.22. Numerous case studies based on a variety of concepts are presented in the forthcoming sections.

- *Thermal energy storage*: Storing thermal energy to overcome the intermittent nature of renewable sources is a promising technique. Especially for thermally driven processes (e.g., Rankine cycle, Brayton cycle), it can eliminate an additional energy conversion step and directly run the bottoming cycle. Biomass combustion, solar energy, and wind thermal energy systems (WTESs) are among the most suitable options for this type of system integration.
- *Cooling cycles (e.g., absorption)*: Cooling is one of the fundamental requirements of many industries. In particular, absorption cooling systems can be integrated into any low-temperature heat source (above 80 °C) from renewables. Flat plate and evacuated tube type solar collectors are among the suitable options for system integration. Other types of concentrated solar systems such as concentrated photovoltaics/thermal can also be integrated into absorption chillers. Concentrated solar power (CSP) systems typically have higher temperatures, hence they are more suitable for integration after the power cycles (e.g., Rankine cycle condenser). In addition to absorption coolers, vapor-compression type refrigeration systems can always be integrated into all types of renewables once electricity is produced.
- *Space heating/hot water*: Any low-grade (below 70 °C) thermal energy from renewable sources can be used for space heating and hot water requirements of the community. Flat

FIG. 6.21 Brainstorming on system integration for renewable energy systems.

Wind energy	Wind thermal energy system (WTES) integrated to Rankine cycle, Brayton cycle, process heat
Concentrated solar energy	Integrated to thermal energy storage for cooling, desalination, electricity
Geothermal energy + Solar energy	Power cycles and concentrated solar plant for heating, hot water, cooling, chemical fuel and electricity
Concentrated solar energy	Thermochemical energy storage (reversible reactions) for process heat, cooling, desalination, chemical fuel
Wind energy + Solar energy	Electricity, heating, cooling, chemical fuel, energy storage

FIG. 6.22 A few ideas for system integration using renewable energy resources.

plate, evacuated tube solar collectors, low-grade geothermal plants and waste heat are among the most suitable options for system integration.

- *Rankine cycle*: The required steam to drive the Rankine cycle can be sourced from biomass combustion, concentrated solar energy, wind thermal energy system (WTES), or a high-grade geothermal system. When the temperature levels are below 200 °C–250 °C, organic Rankine cycles can also be employed to achieve power generation from medium temperature sources.
- *Brayton cycle*: The Brayton cycle requires high-temperature heat addition, which can be sourced from concentrated solar energy, biomass combustion, wind thermal, etc. Hence, the heat addition in the heat exchanger or combustion chamber can be supplied from renewables.
- *Synthetic fuel production*: Most of the synthetic fuel production processes require a range of medium/high temperature as well as electrical work. For instance, synthetic methanol production can take place at 220 °C and needs compressed hydrogen and carbon dioxide. The compressor power and thermal energy for the reaction can be sourced from renewables.
- *Ammonia and urea production*: Conventional ammonia production requires compressed hydrogen and nitrogen in a Haber-Bosch reactor. Nitrogen is commonly produced from air separation. Renewable electricity can be used in these air separation plants as well as for the compressor work inside the Haber-Bosch cycle. On the other hand, electrochemical ammonia synthesis routes do not generally require high pressures. They rather necessitate electricity and/or thermal energy. Hence, they are more suitable for integration with renewables. This topic will be further discussed in Chapter 7.
- *CO_2 capture and utilization*: CO_2 is one of the main causes of global warming. It can be captured from the air or industrial facilities. It is a feedstock for many other chemicals such as methanol, ethanol, urea, etc. Among renewables, biomass has the highest potential for system integration with CO_2 utilization due to the fact that combustion and gasification yield CO_2.
- *Process heating*: Several industries require medium-high temperature heat to drive their processes. Natural gas or coal combustion is the common source of this heat. Concentrated solar, biomass, geothermal energy, or wind thermal can provide the necessary high-temperature thermal energy. The renewables have the advantage of small/medium scale and modularity. Hence, they can be installed near the industrial plants to reduce the thermal energy transfer distance.
- *Desalination*: Desalination plants necessitate either thermal and/or electrical energy depending on the type selected (reverse osmosis, multi-effect distillation, etc.). The pump work in reverse osmosis can be sourced from renewables. In addition, thermal energy-producing renewables can drive the desalination process. Concentrated solar, biomass, geothermal, and wind thermal are among the most suitable options.
- *Thermochemical cycles*: Thermochemical cycles require thermal energy, whereas hybrid thermochemical cycles require both heat and electricity. Hence, they can be appropriate for system integration with renewables.
- *Fuel cells*: The fuel input for fuel cells is diverse (hydrogen, methane, ammonia, etc.). As previously explained, synthetic fuels such as methanol and ethanol can also feed fuel cells. These fuels can be produced via renewables and can act as a storage medium and then be utilized in fuel cells on demand. In addition, biogas or syngas can also be fed to fuel cells for cleaner operation.

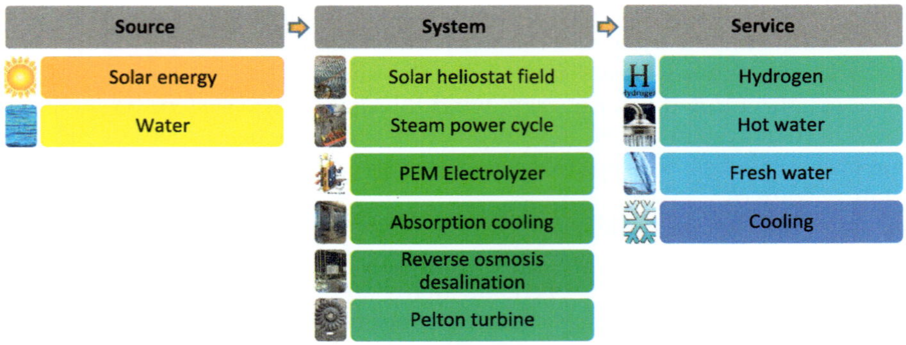

FIG. 6.23 3S approach for case study 1.

6.9 Case study 1: Solar heliostat-based multigeneration system

This case study integrates a heliostat solar system with a steam turbine as well as with a seawater reverse osmosis desalination unit and absorption cooling system. This case study is adapted from El-Emam and Dincer [15]. The desalination unit works with energy recovered by a Pelton turbine. The useful outputs from the system are cooling, heating, fresh water and hydrogen through electrolysis as shown in Fig. 6.23. It is planned to cover the demand of 4 MW electricity with the production of 1.25 kg/h of hydrogen and 90 kg/s of fresh water. The sources, systems and services (3S) in this case study are illustrated in Fig. 6.23.

6.9.1 System description

The layout of the integrated system is presented in Fig. 6.24. The concentrated solar energy is captured by molten salt in the solar tower. A steam power cycle with reheat option is combined with the solar system for electricity production. High-temperature steam is produced for the steam cycle by the solar heat exchanger. A portion of the steam exiting the high-pressure turbine is utilized for delivering the required heat to the water electrolysis unit for hydrogen generation. The water entering the electrolysis unit is heated up to 80 °C, the operational temperature of the polymer electrolyte membrane (PEM) electrolysis. The thermal energy can likewise be delivered to the electrolyzer when necessary, based on the operational parameters. After departing the high-pressure turbine, the other portion of the expanded steam is led to the generator of the absorption cooling system, after delivering thermal energy for generating domestic hot water. Both steam streams are condensed and pumped back to the solar steam generation unit after the electrolyzer heat exchanger and the absorption cycle generator. A reverse osmosis (RO) desalination unit is combined for producing fresh water from seawater. The desalination unit is combined with a Pelton turbine for energy recovery. The auxiliary power needed for the pumps of the RO desalination unit, the electrolyzer, and the absorption chiller is delivered from the electricity produced from the steam turbine. As previously stated, the system is planned to primarily deliver 4 MW of net electricity, attain 90 kg/s of fresh water and generate 1.25 kg/h hydrogen.

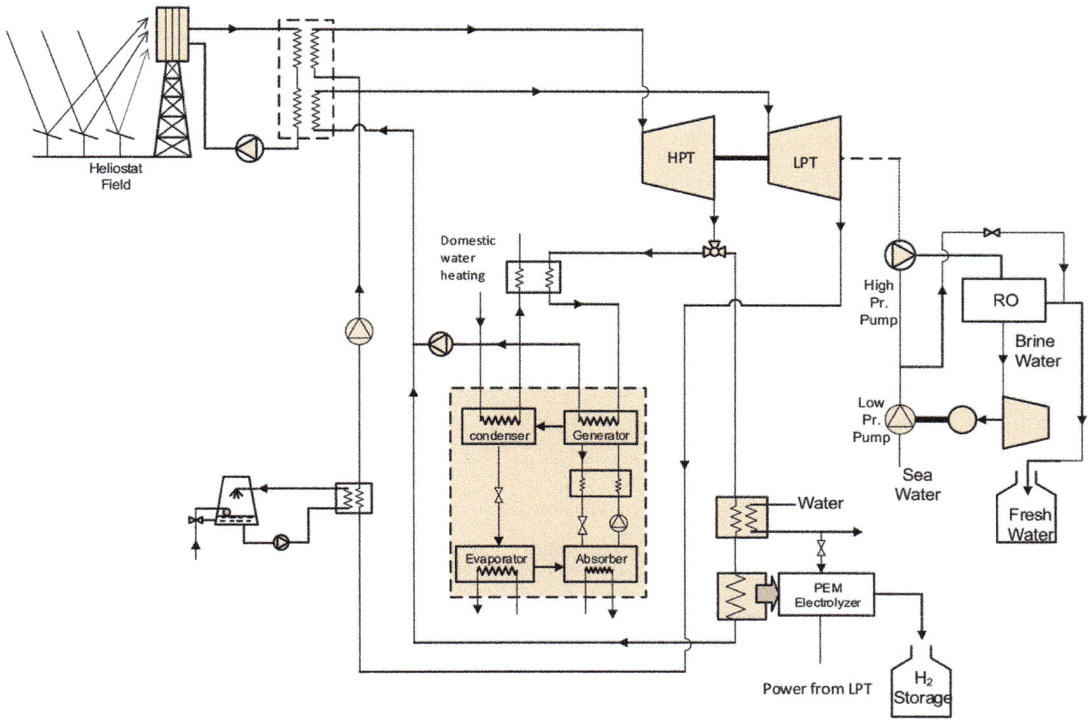

FIG. 6.24 The schematic diagram of the integrated energy system in case study 1. *(Adapted from El-Emam RS, Dincer I. Development and assessment of a novel solar heliostat-based multigeneration system. Int J Hydrogen Energy 2018;43:2610–20).*

6.9.2 Assumptions

In order to perform thermodynamic analysis of this multigeneration system, conceptually correct assumptions are necessary. Here, the main assumptions undertaken in this case study are listed:

- The system works at steady state with constant solar irradiation.
- The changes in kinetic and potential energy and exergy terms are neglected.
- The pressure and heat losses in the pipelines are neglected.

6.9.3 Analysis

The mass, energy and exergy stream balance equations are written for all system components. Seawater RO systems are analyzed evaluating three core courses: source intake, pretreatment and the RO separation process. The features of the seawater are considered to be those for a diluted solution, which can be estimated as a perfect solution. This is interpreted such that the influence of different particles constituting the solution is insignificant. The main solar field values are given in Table 6.5.

TABLE 6.5 Solar field parameters used in the analysis

Parameter	Value
Heliostat width and height	10 m
Mirror reflectance	0.9
Minimum heliostat distance to tower ratio	0.75
Maximum heliostat distance to tower ratio	7.5
Mirror washing water	0.7 l/m^2
Annual washes	60
Receiver diameter	8 m
Coating emittance	0.88
Receiver tube diameter	40 mm
Receiver tube thinness	1.25 mm
Minimum receiver height to diameter ratio	0.8
Maximum receiver height to diameter ratio	1.6
Maximum tower height	250 m

Source: *El-Emam RS, Dincer I. Development and assessment of a novel solar heliostat-based multigeneration system. Int J Hydrogen Energy 2018;43:2610–20.*

The overall energy and exergy efficiencies of the system can be written as follows:

$$\eta_{en_{overall}} = \frac{\dot{W} + \dot{Q}_{cooling} + \dot{Q}_{heating} + \dot{m}_{H_2} h_{H_2}}{\dot{Q}_{sun}} \tag{6.1}$$

$$\eta_{ex_{overall}} = \frac{\dot{W} + \dot{Q}_{cooling}\left(\frac{T_o}{T_c} - 1\right) + \dot{Q}_{heating}\left(1 - \frac{T_o}{T_c}\right) + \dot{E}x_{H_2}}{\dot{Q}_{sun}\left(1 - \frac{T_o}{T_{sun}}\right)} \tag{6.2}$$

6.9.4 Results and discussion

The operation structure of the solar heliostat plant strongly hinges on the incident solar irradiation. Therefore, the solar irradiation level is varied from 200 to 1000 W/m^2, in order to calculate its effects on the solar heliostat system and the overall system efficiency. Fig. 6.25 displays the changes of the receiver energy and exergy efficiencies. By increasing solar irradiation, the efficiency of the receiver rises. The energy efficiency improved by 14%, rising to 89%, and the exergy efficiency rises to 51%. The highest losses taking place in the receiver are a function of the receiver surface temperature, which marginally changes within the parametric study limits as shown in Table 6.6, where the temperature differs within 50 °C. This can be interpreted such that with higher solar irradiation, the heat losses from the receiver upsurge marginally. The energy and exergy efficiencies of the overall solar heliostat system and the overall integrated system are illustrated in Fig. 6.26 and Fig. 6.27,

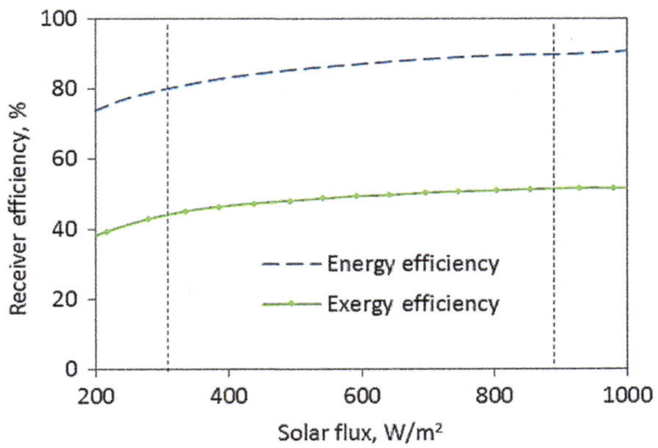

FIG. 6.25 The effects of solar irradiation on the energy and exergy efficiencies of the receiver. *(Adapted from El-Emam RS, Dincer I. Development and assessment of a novel solar heliostat-based multigeneration system. Int J Hydrogen Energy 2018;43:2610–20).*

TABLE 6.6 The changes of receiver surface temperature by increasing solar irradiation

Solar irradiation (W/m²)	Receiver surface temperature (°C)
200	510
400	515
600	523
800	538
1000	545

Source: *El-Emam RS, Dincer I. Development and assessment of a novel solar heliostat-based multigeneration system. Int J Hydrogen Energy 2018;43:2610–20.*

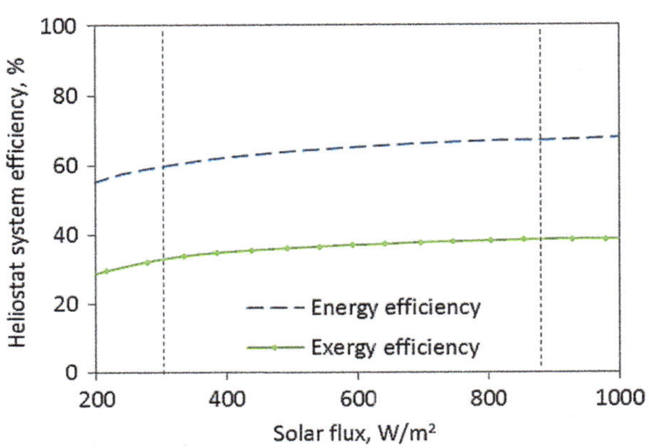

FIG. 6.26 The effects of solar irradiation on the energy and exergy efficiencies of the solar heliostat system. *(Adapted from El-Emam RS, Dincer I. Development and assessment of a novel solar heliostat-based multigeneration system. Int J Hydrogen Energy 2018;43:2610–20).*

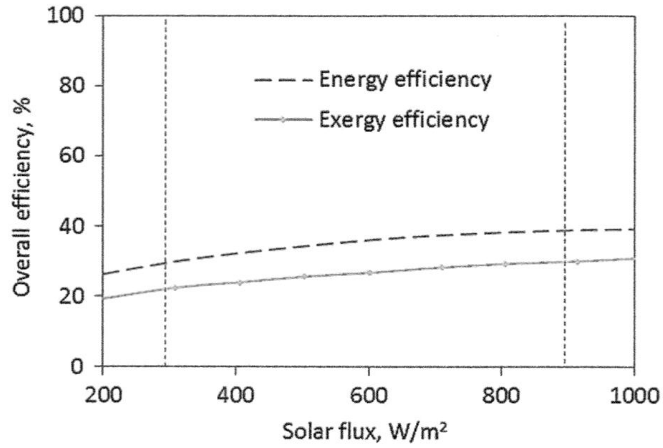

FIG. 6.27 The effects of the solar irradiation on the overall energy and exergy efficiencies. *(Adapted from El-Emam RS, Dincer I. Development and assessment of a novel solar heliostat-based multigeneration system. Int J Hydrogen Energy 2018;43:2610–20).*

respectively. The efficiency rises visibly through the low solar irradiation levels, whereas after 400 W/m^2, the improvement in the efficiencies is less.

The solar heliostat field and solar tower are optimized for this system by System Advisor Model (SAM) software [16, 17], based on the weather data of the city of Aswan in Egypt. All weather circumstances and the parameters required for the analysis are given in Table 6.7. The results of the optimized heliostat system are shown in Fig. 6.28 and Fig. 6.29 as well as in Table 6.8.

6.9.5 Closing remarks

In this case study, a solar heliostat-based integrated system is designed to produce electricity, cooling, fresh water, hydrogen and domestic hot water and analyzed using energy and exergy approaches. The effects of the solar irradiation levels on the energy and exergy efficiencies of the receiver are investigated. The overall system energy and exergy efficiencies are influenced by the receiver efficiency. The integrated system functions to provide an

TABLE 6.7 The specifications and parameters of the selected location

Parameter	Value
Position	Latitude 23.97 Longitude 32.78
Elevation	194 m
Annual direct normal irradiance	2433.5 kWh/m^2
Annual dry bulb temperature	26.2 °C
Average wind speed	5.1 m/s

Source: *El-Emam RS, Dincer I. Development and assessment of a novel solar heliostat-based multigeneration system. Int J Hydrogen Energy 2018;43:2610–20.*

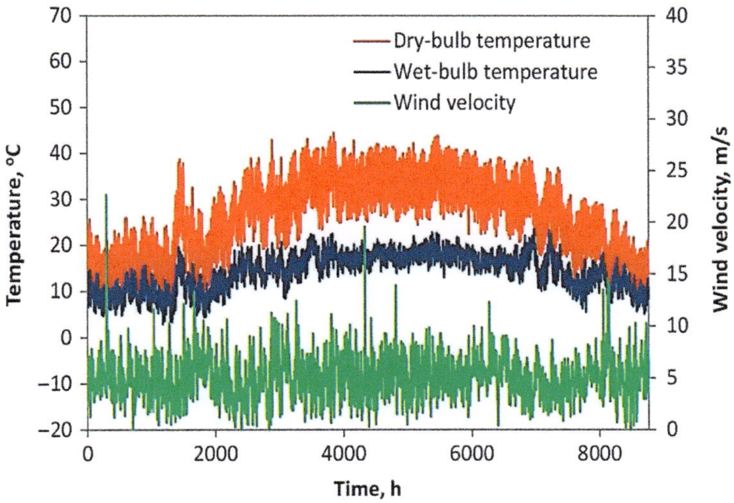

FIG. 6.28 Yearly distribution of dry and wet bulb temperatures and the wind velocity for the selected location. *(Adapted from El-Emam RS, Dincer I. Development and assessment of a novel solar heliostat-based multigeneration system. Int J Hydrogen Energy 2018;43:2610–20).*

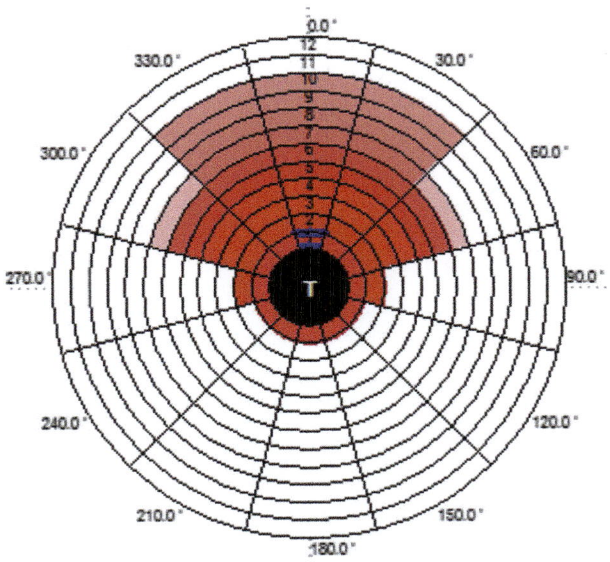

FIG. 6.29 The radial distribution of solar heliostat field with regard to solar tower located in the center. *(Adapted from El-Emam RS, Dincer I. Development and assessment of a novel solar heliostat-based multigeneration system. Int J Hydrogen Energy 2018;43:2610–20).*

TABLE 6.8 The optimization results of the heliostat field

Parameter	Value
Number of heliostats	396
Total reflective area	38,412 m^2
Maximum distance from the tower	403.7 m
Radial step size for layout	35.6 m
Receiver height	6.4 m
Tower height	63.3 m

Source: *El-Emam RS, Dincer I. Development and assessment of a novel solar heliostat-based multigeneration system. Int J Hydrogen Energy 2018;43:2610–20.*

electrical work rate of 4 MW, 1.25 kg/h of hydrogen and 90 kg/s of fresh water. The solar heliostat field is optimized based on the system operational factors. The optimized number of heliostats for the designated site is calculated as 396 and the tower height as 63.3 m.

6.10 Case study 2: Solar parabolic troughs integrated to thermoelectric generator for multigeneration

This case study presents comparative energy and exergy analyses of a solar energy-based integrated multigeneration system with a specific arrangement of thermoelectric generators. The system integrates the output of thermoelectric generators with the parabolic trough solar collectors (PTSCs). This case study is adapted from Islam et al. [18]. The sources, systems and services (3S) in this case study are illustrated in Fig. 6.30.

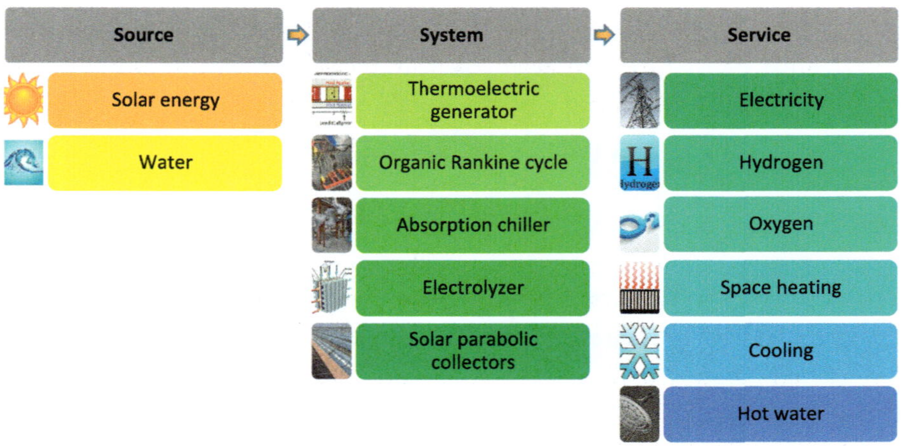

FIG. 6.30 3S approach for case study 2.

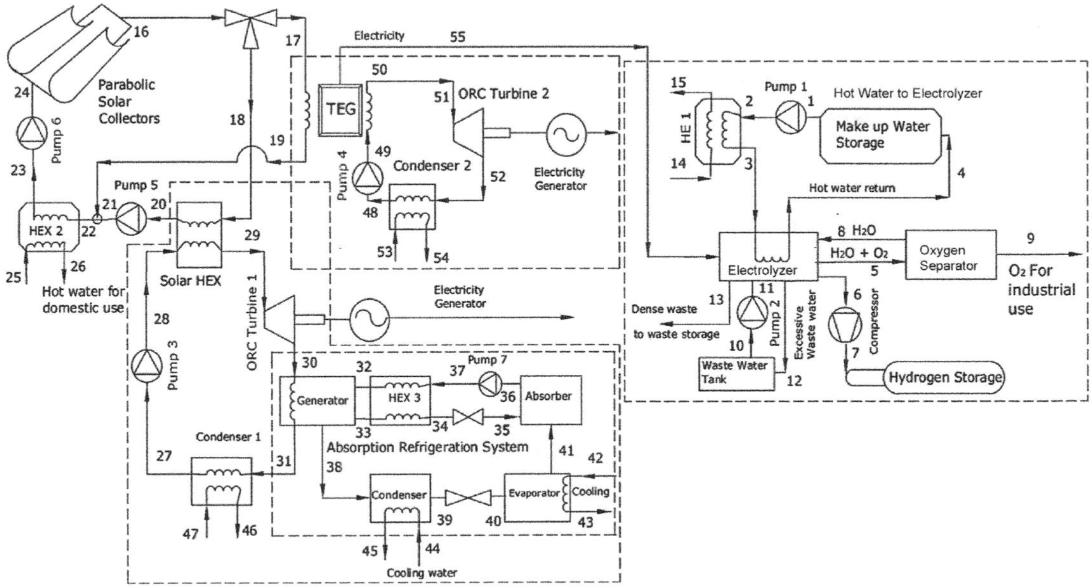

FIG. 6.31 The system layout of case study 2. *(Adapted from Islam S, Dincer I, Yilbas BS. Development, analysis and assessment of solar energy-based multigeneration system with thermoelectric generator. Energ Conver Manage 2018;156:746–56).*

6.10.1 System description

The system layout is given in Fig. 6.31. This system generates electricity, hydrogen, space heating, cooling and hot water as useful outputs. The temperature of the heat transfer fluid (HTF) is elevated by PTSCs. Therminol VP-1 is chosen as the HTF since it has great thermal stability and low viscosity. Furthermore, its performance is constant in the range of 12 °C–400 °C [19].

Thermoelectric generators are combined within the multigeneration system as presented in Fig. 6.31. A portion of the heat transfer fluid (Therminol VP-1) is removed to divert through the high-temperature part of the thermoelectric generator. The ORC working fluid, which is R113, is recycled through the low-temperature connection to keep the waste heat and produce extra electricity.

6.10.2 Assumptions

In order to perform thermodynamic analysis of this multigeneration system, there is a need to make conceptually correct assumptions. The fundamental assumptions considered in this study are listed as follows:

- The reference temperature and pressure values are taken as 298 K and 101.325 kPa, respectively.
- Steady state operating conditions are assumed.
- The kinetic and potential energy changes are neglected.

- R113 and water are treated as actual fluids, but air is treated as an ideal gas.
- The pumps and turbines run adiabatically.
- Isobutane and R113 are selected as working fluids for ORC 1 and ORC 2, respectively.
- There is no change in chemical compositions within the processes, so no chemical energy changes occur.
- The isentropic efficiencies of all pumps and turbines are assumed to be 85%.

6.10.3 Analysis

The main operating parameters of the electrolyzer are listed in Table 6.9.

The energy efficiency of thermoelectric generators is dependent on the figure of merit, which is signified by ZT. The low- and high-temperature levels through the thermoelectric generators are denoted by T_h and T_c, respectively. The energy efficiency of thermoelectric generators is written as follows:

$$\eta_{TEG} = \frac{\Delta T\left(\sqrt{1+ZT}-1\right)}{T_h\left(\sqrt{1+ZT}+\frac{T_c}{T_h}\right)} \tag{6.3}$$

The value of the figure of merit ZT is selected as 1.

The energy efficiency of ORC 1 and ORC 2 are defined as follows, respectively:

$$\eta_{en_{ORC1}} = \frac{\dot{W}_{net,ORC1}}{\dot{m}_{29}h_{29} - \dot{m}_{30}h_{30}} \tag{6.4}$$

$$\eta_{en_{ORC2}} = \frac{\dot{W}_{net,ORC2}}{\dot{m}_{51}h_{51} - \dot{m}_{52}h_{52}} \tag{6.5}$$

The net work production within the system is calculated as follows:

$$\dot{W}_{net} = \dot{W}_{ORC1} + \dot{W}_{TEG} + \dot{W}_{ORC2} - \dot{W}_{P5} \tag{6.6}$$

TABLE 6.9 Specifications of a typical electrolyzer used for H_2 production

Electricity requirements	22 kW
Operating temperature	77 °C
Operating pressure	709.275 kPa
Yearly hydrogen production	10,772.64 m^3
Capacity	450 L/h
Area of electrodes	0.25 m^2

Source: *Islam S, Dincer I, Yilbas BS. Development, analysis and assessment of solar energy-based multigeneration system with thermoelectric generator. Energ Conver Manage 2018;156:746–56.*

The overall energy and exergy efficiencies of the complete system are respectively defined as follows:

$$\eta_{en_{overall}} = \frac{\dot{W}_{net} + \dot{Q}_{cond} + \dot{Q}_{evp} + \dot{Q}_{HE1} + \dot{Q}_{HE2} + \dot{Q}_{HE3} + \dot{m}_{H_2}h_{H_2} + \dot{W}_{TEG}}{\dot{Q}_{sun}} \tag{6.7}$$

$$\eta_{ex_{overall}} = \frac{\dot{W}_{net} + \dot{Q}_{cond}\left(1-\frac{T_O}{T_C}\right) + \dot{Q}_{evp}\left(\frac{T_O}{T_{evp}}-1\right) + \dot{Q}_{HE1}\left(1-\frac{T_O}{T_{HE1}}\right) + \dot{Q}_{HE2}\left(1-\frac{T_O}{T_{HE2}}\right) + \dot{Q}_{HE3}\left(1-\frac{T_O}{T_{HE3}}\right) + \dot{m}_{H_2}ex_{H_2} + \dot{W}_{TEG}}{\dot{Q}_{sun}\left(1-\frac{T_O}{T_{sun}}\right)} \tag{6.8}$$

6.10.4 Results and discussion

The main exergy destructions occurring in the multigeneration system are shown in Fig. 6.32. The values of major exergy destruction ratios showing the shares within the system are shown in Fig. 6.33, depicting the percentage contribution of irreversibilities in the main subsystems. It is obvious that the share of ORC 1 in exergy destruction is highest followed by the solar heat exchanger. The main output parameters as well as efficiencies of the subsystems and overall system are listed in Table 6.10.

The energy and exergy efficiencies of ORC 1, ORC 2 turbines as well as the overall system are presented in Fig. 6.34.

Fig. 6.35 depicts the effects of changing the rate of mass flow on system overall efficiencies as well as subcomponents. The rate of mass flow is changed from 35 to 40 kg/s through the solar heat exchanger. The energy efficiency of ORC 1 increases from 32.6% to 39.3%, due to the fact that increased rate of mass flow of the through heat exchanger increases the enthalpy of isobutane because of its constant flow rate between states 29 and state 30. Likewise, the exergy efficiency of this turbine 1 increases from 66.2% to 69.7% due to the increased difference in the temperature between isobutane and surroundings. The energy and exergy efficiencies of

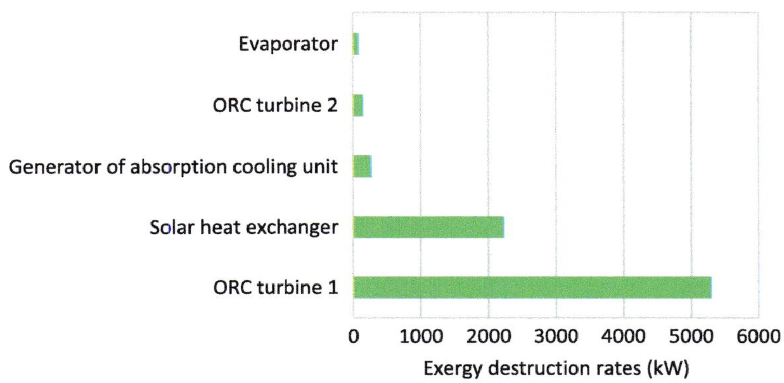

FIG. 6.32 The exergy destruction rates within the multigeneration system.

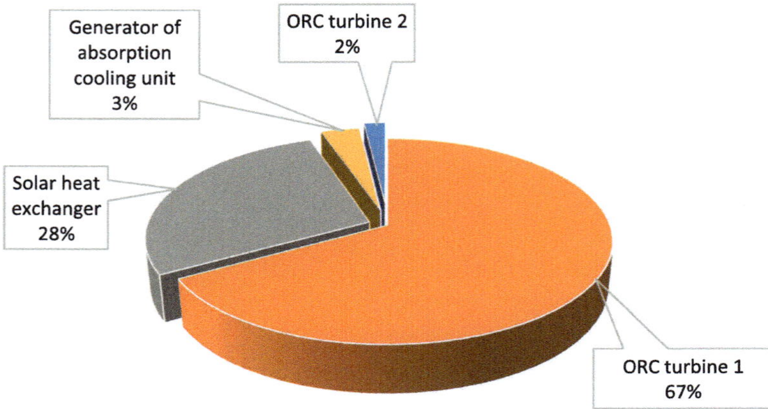

FIG. 6.33 The exergy destruction ratios within the multigeneration system. *(Data from Islam S, Dincer I, Yilbas BS. Development, analysis and assessment of solar energy-based multigeneration system with thermoelectric generator. Energ Conver Manage 2018;156:746–56).*

TABLE 6.10 The main output parameters as well as efficiencies of the subsystems and overall system

Component	Value
$\dot{W}_{\text{NetORC1}}$	3420 kW
$\dot{W}_{\text{Net}_{\text{ORC2}}}$	199.6 kW
$\dot{W}_{\text{TEG}}$	91.26 kW
$\dot{W}_{\text{Max TEG}}$	0.058 kW
Number of TEG required	1582
$\eta_{\text{en}\eta\text{ORC1}}$	28.8%
η_{exORC1}	65.7%
$\eta_{\text{en}\eta\text{ORC2}}$	34.6%
η_{exORC2}	51.2%
$\eta_{\text{en}_{\text{Overall}}}$	57.3%
$\eta_{\text{ex}_{\text{Overall}}}$	49.7%

ORC 2 also increase from 33.2% and 50.7% to 40.95% and 55.96%, respectively. This is due to the fact that the temperature of isobutane extracted increases with the increased mass flow of Therminol VP-1, which in turn increases the inlet enthalpy of R113, the working fluid of this ORC 2. The energy and exergy efficiency values of the overall system are improved from 59.6% to 79.9% and 51.7% to 56.4%, respectively. This significant rise in both efficiencies is because the waste heat of the solar cycle is used to run subunits: thermoelectric generators, electrolyzer, absorption chiller, heat exchanger 2 and production of heat at condenser 1 of ORC 1.

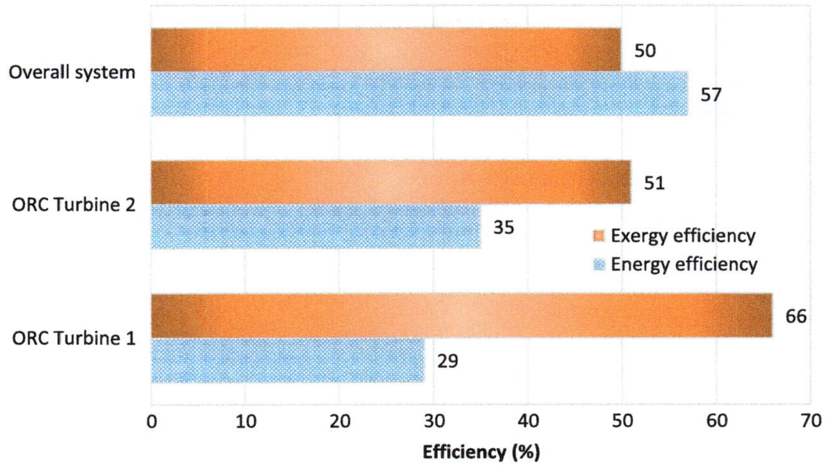

FIG. 6.34 The energy and exergy efficiencies of ORC turbines and overall system. *(Data from Islam S, Dincer I, Yilbas BS. Development, analysis and assessment of solar energy-based multigeneration system with thermoelectric generator. Energ Conver Manage 2018;156:746–56).*

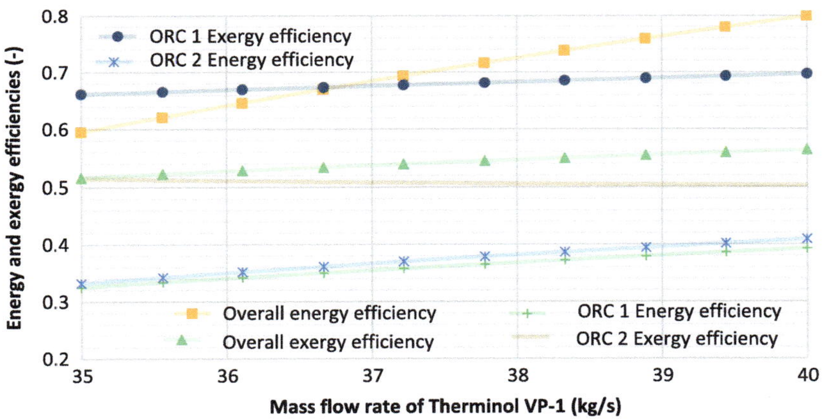

FIG. 6.35 The effects of changing mass flow rate of HTF on the energy and exergy efficiencies when isobutane is used. *(Data from Islam S, Dincer I, Yilbas BS. Development, analysis and assessment of solar energy-based multigeneration system with thermoelectric generator. Energ Conver Manage 2018;156:746–56).*

6.10.5 Closing remarks

This case study performed energy and exergy analyses of an integrated energy system including thermoelectric generators. The operational parameters are changed to examine the performance and find the best working conditions. The integration of thermoelectric generators and hydrogen production electrolyzer shows convincing results. The net amount of work done by the thermoelectric generators rises with the rise of temperature and rate of mass flow of the fluid flowing over its high-temperature connections. This case study demonstrates the possibility of integrating thermoelectric generators into solar energy applications as a waste heat recovery unit.

6.11 Case study 3: Combined solar and geothermal energy-based integrated system for multigeneration

In this case study, a solar and geothermal energy-based integrated energy system is analyzed for multigeneration purposes. The system comprises two organic Rankine cycles, two thermal energy storage systems, an absorption cooling system, a heat pump for space heating and a drying system. This case study is adapted from Islam and Dincer [20]. The sources, systems and services (3S) of case study 3 are illustrated in Fig. 6.36.

6.11.1 System description

The integrated system contains a geothermal-solar cycle, organic Rankine cycles (running on isobutane), absorption cooling unit (lithium bromide), water-based thermal energy storage, a heat pump (R410a) and a dryer to produce dry air for crop drying. The energy and the exergy efficiencies for the single generation, cogeneration, trigeneration and multigeneration systems are comparatively calculated. Solar and geothermal energy systems are combined to obtain a multigeneration system as the system layout is presented in Fig. 6.37. This system has five useful outputs from integrating solar and geothermal sources. The HTF is heated by parabolic trough collectors, and Therminol VP-1 is chosen as the heat transfer fluid owing to its high thermal stability and low viscosity.

6.11.2 Assumptions

In order to perform thermodynamic analysis of this multigeneration system, there is a need to make conceptually correct assumptions. The following assumptions are made for the analysis of the integrated system:

- The reference temperature and pressure are taken as 298 K and 101.325 kPa, respectively.
- Steady state operating conditions are assumed.

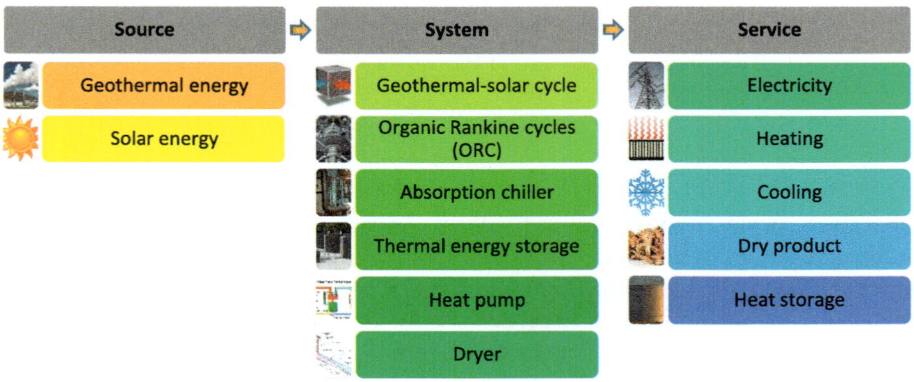

FIG. 6.36 3S approach for case study 3.

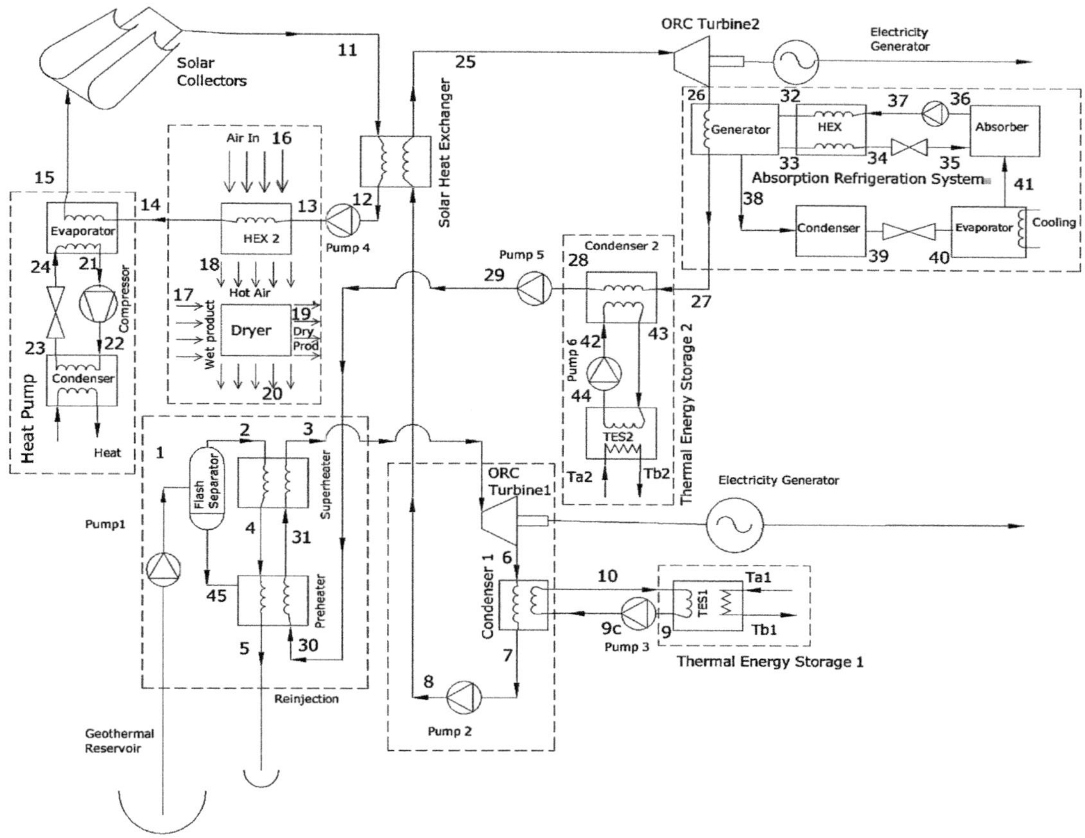

FIG. 6.37 The system layout of the integrated system for case study 3. *(Adapted from Islam S, Dincer I. Development, analysis and performance assessment of a combined solar and geothermal energy-based integrated system for multigeneration. Sol Energy 2017;147:328–43. https://doi.org/10.1016/J.SOLENER.2017.02.048).*

- The kinetic and potential energy terms are neglected.
- R410a and water are treated as actual fluid, whereas air is treated as an ideal gas.
- The pumps and turbines are adiabatic.
- The working fluid of the ORC cycle is isobutane.
- The condenser outlet is saturated liquid and the evaporator exit is saturated in the absorption cooling unit.
- The discharging losses in the thermal energy storage system are taken as 10%.
- The pressure losses in all pipelines except geothermal piping and heat exchangers are neglected.
- There is no change in chemical composition during the processes; therefore chemical energy changes are neglected.
- The velocity of geothermal fluid flow is taken as 0.82 m/s.
- The typical thermal efficiency of parabolic trough collectors is taken as 75% [21].

6.11.3 Analysis

In this section, the power outputs of single to multigeneration outputs as well as the respective efficiencies are determined. The net work production in single to multigeneration systems is written as follows:

$$\dot{W}_{net_{SINGLEGEN}} = \dot{W}_{net,T1} + \dot{W}_{net,T2} - \dot{W}_{P1} - \dot{W}_{P1a} - \dot{W}_{P4} \tag{6.9}$$

$$\dot{W}_{net_{COGEN}} = \dot{W}_{net,T1} + \dot{W}_{net,T2} - \dot{W}_{P1} - \dot{W}_{P1a} - \dot{W}_{P4} - \dot{W}_{P\,Chiller} \tag{6.10}$$

$$\dot{W}_{net_{TRIGEN}} = \dot{W}_{net,T1} + \dot{W}_{net,T2} - \dot{W}_{P1} - \dot{W}_{P1a} - \dot{W}_{P4} - \dot{W}_{P\,Chiller} - \dot{W}_{C} \tag{6.11}$$

$$\dot{W}_{net_{MULTIGEN}} = \dot{W}_{net,T1} + \dot{W}_{net,T2} - \dot{W}_{P1} - \dot{W}_{P1a} - \dot{W}_{P3} - \dot{W}_{P4} - \dot{W}_{P6} - \dot{W}_{C} \tag{6.12}$$

The energy and exergy efficiencies of single to multigeneration systems are written as follows:

$$\eta_{en_{SINGLEGEN}} = \frac{\dot{W}_{net_{SINGLEGEN}}}{\dot{m}_1 h_1 + \dot{Q}_{sun}} \tag{6.13}$$

$$\eta_{ex_{SINGLEGEN}} = \frac{\dot{W}_{net_{SINGLEGEN}}}{\dot{m}_1 ex_1 + \dot{Q}_{sun}\left(1 - \frac{T_o}{T_{sun}}\right)} \tag{6.14}$$

$$\eta_{en_{COGEN}} = \frac{\dot{W}_{net_{COGEN}} + \dot{Q}_{cooling}}{\dot{m}_1 h_1 + \dot{Q}_{sun}} \tag{6.15}$$

$$\eta_{ex_{COGEN}} = \frac{\dot{W}_{net_{COGEN}} + \dot{Q}_{cooling}\left(\frac{T_o}{T_{evp}} - 1\right)}{\dot{m}_1 ex_1 + \dot{Q}_{sun}\left(1 - \frac{T_o}{T_{sun}}\right)} \tag{6.16}$$

$$\eta_{en_{TRIGEN}} = \frac{\dot{W}_{net_{TRIGEN}} + \dot{Q}_{cooling} + \dot{Q}_{HP}}{\dot{m}_1 h_1 + \dot{Q}_{sun}} \tag{6.17}$$

$$\eta_{ex_{TRIGEN}} = \frac{\dot{W}_{net_{TRIGEN}} + \dot{Q}_{cooling}\left(\frac{T_o}{T_{evp}} - 1\right) + \dot{Q}_{HP}\left(1 - \frac{T_o}{T_s}\right)}{\dot{m}_1 ex_1 + \dot{Q}_{sun}\left(1 - \frac{T_o}{T_{sun}}\right)} \tag{6.18}$$

$$\eta_{en_{MULTIGEN}} = \frac{\dot{W}_{net_{MULTIGEN}} + \dot{Q}_{cooling} + \dot{Q}_{HP} + \dot{Q}_{TES} + \dot{Q}_{Dryer}}{\dot{m}_1 h_1 + \dot{Q}_{sun}} \tag{6.19}$$

$$\eta_{ex_{MULTIGEN}} = \frac{\dot{W}_{net_{MULTIGEN}} + \dot{Q}_{cooling}\left(\frac{T_o}{T_{evp}} - 1\right) + \dot{Q}_{HP}\left(1 - \frac{T_o}{T_s}\right) + \dot{Q}_{TES}\left(1 - \frac{T_o}{T_s}\right) + \dot{Q}_{Dryer}\left(1 - \frac{T_o}{T_s}\right)}{\dot{m}_1 ex_1 + \dot{Q}_{sun}\left(1 - \frac{T_o}{T_{sun}}\right)} \tag{6.20}$$

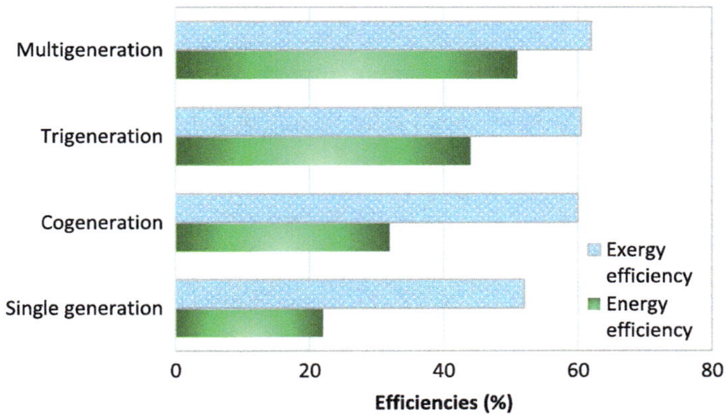

FIG. 6.38 The obtained energy and exergy efficiencies of single generation, cogeneration, trigeneration and multigeneration systems. *(Data from Islam S, Dincer I. Development, analysis and performance assessment of a combined solar and geothermal energy-based integrated system for multigeneration. Sol Energy 2017;147: 328–43. https://doi.org/10.1016/J.SOLENER.2017.02.048).*

6.11.4 Results and discussion

The energy and exergy efficiencies of the overall multigeneration system are calculated as 51% and 62%, respectively. The maximum work potential of the input energy is realized by the combination of absorption cooler and TES with the exit of turbines. The heat pump (R410a) and drying system are combined to preserve the waste heat of Therminol VP-1. Fig. 6.38 comparatively shows the energy and exergy efficiencies for a single generation, cogeneration, trigeneration as well as multigeneration. The exergy efficiencies for the single generation, cogeneration and tri-generation systems are calculated as 54%, 60% and 60.4%, respectively.

The rise in the energy efficiency of the multigeneration is due to the role of the high COP value of the heat pump. In addition, the high output from turbine 2 is due to the high temperature obtained by the working fluid isobutane from Therminol VP-1 in the solar heat exchanger. Pump 4 needs the minimum quantity of work input because of the minor pressure difference between inlet and exhaust. The phase of the heat transfer fluid is kept the same in all heat exchangers. The output and efficiencies of each subunit and hence the multigeneration system are listed in Table 6.11 and shown in Fig. 6.39.

Fig. 6.39 displays the energy and exergy efficiencies of turbines, TES and dryer. The low values of the turbine energy efficiencies are due to the high temperature of the working fluid exiting the turbines, where it are used for cooling and thermal energy storage. On the other hand, the turbine exergy efficiencies are noticeably high due to less exergy destruction.

Table 6.12 tabulates the exergy destructions of the major components of the integrated system. The exergy destruction ratios of each subsystem are shown in Fig. 6.40. The largest exergy destruction occurs in the solar cycle followed by the geothermal. After these, the largest exergy destruction is observed in the heat pump system due to the high exergy losses in the compressor. The absorption refrigeration cycle has the fourth largest exergy destruction due to the greater temperature difference through the generator of the absorption cooling unit and ORC working fluid isobutene. It is possible to balance the quantity of exergy destruction occurring in the chiller by operating the generator over a lower temperature source and substituting the absorber with an effective one. The compressor of the heat pump is fifth in exergy destruction, due to the high temperature variance between ambient and working fluid R410a.

TABLE 6.11 The main output parameters of the major components and the overall system

Component	Values
Turbine 1 in organic Rankine cycle	4110 kW
Turbine 2 in organic Rankine cycle	5094 kW
Absorption cooling system energetic COP	1.5
Absorption cooling system exergetic COP	0.7
Heat pump energetic COP	3.3
Absorption cooling system exergetic COP	0.5
Compressor power input rate	2884 kW
Loss of pumping power because of friction	898 kW
Pump 1 power input rate	198 kW
Pump 2 power input rate	1072 kW
Solar normal irradiance	850 W/m^2

Source:*Islam S, Dincer I. Development, analysis and performance assessment of a combined solar and geothermal energy-based integrated system for multigeneration. Sol Energy 2017;147:328–43. https://doi.org/10.1016/J.SOLENER.2017.02.048.*

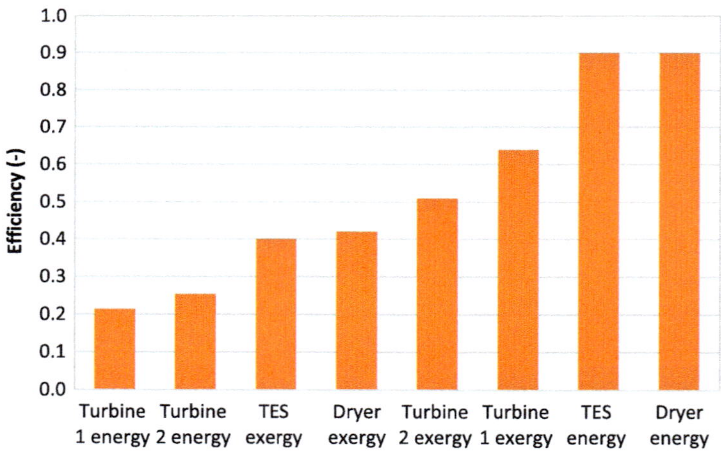

FIG. 6.39 The subsystem and component efficiencies. *(Data from Islam S, Dincer I. Development, analysis and performance assessment of a combined solar and geothermal energy-based integrated system for multigeneration. Sol Energy 2017;147:328–43. https://doi.org/10.1016/J.SOLENER.2017.02.048).*

The effects of changing the geothermal fluid temperature (from 440 K to 460 K) on the energy and exergy efficiencies of all configurations are shown in Fig. 6.41. The energy and exergy efficiencies of all configurations are directly affected by the temperature of brine; therefore, both efficiencies are improved when the brine temperature rises. Both the energy and exergy efficiencies of the single generation system increase from 22% to 24% and 54% to 59.3%, respectively. This rise in the efficiencies of the multigeneration system is less compared to the scenario in which the heat transfer fluid temperature is increased. The cause of

TABLE 6.12 The main exergy destruction rates of some components and subsystems of the multigeneration system (approximate values)

Component/subsystem	Exergy destruction rate (kW)
Solar collector	10,050
Geothermal cycle	5100
Heat pump	1050
Absorption cooling unit	1000
Compressor	980
Heat pump	810
TES 2	400
TES 1	110
Heat exchanger 2	14
Dryer	12

Source:*Islam S, Dincer I. Development, analysis and performance assessment of a combined solar and geothermal energy-based integrated system for multigeneration. Sol Energy 2017;147:328–43. https://doi.org/10.1016/J.SOLENER.2017.02.048.*

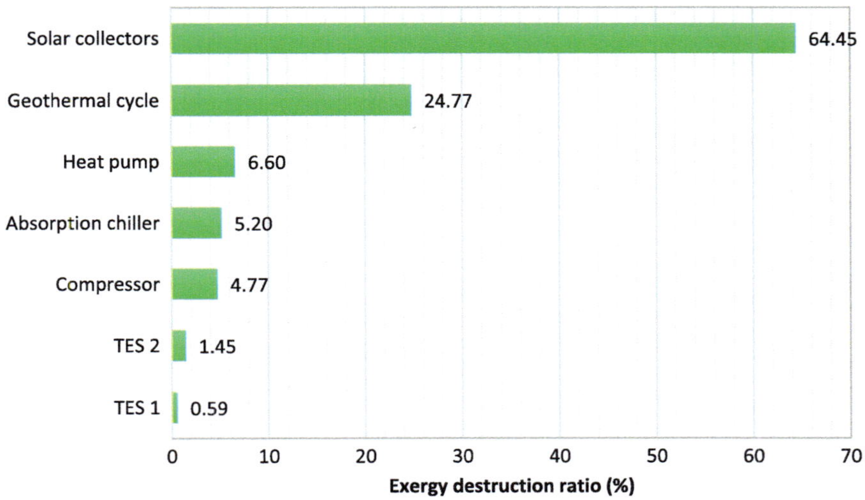

FIG. 6.40 The exergy destruction ratios of some major components and subsystems of the multigeneration system. *(Data from Islam S, Dincer I. Development, analysis and performance assessment of a combined solar and geothermal energy-based integrated system for multigeneration. Sol Energy 2017;147:328–43. https://doi.org/10.1016/J.SOLENER.2017.02.048).*

this is that only TES 1 is combined with the geothermal cycle whereas all other subsystems such as the absorption cooling unit, TES 2, dryer and heat pump are integrated with the solar cycle. The energy and exergy efficiencies of ORC 2 rise with the upsurge in PTSC efficiencies while the efficiencies of ORC 1 are constant.

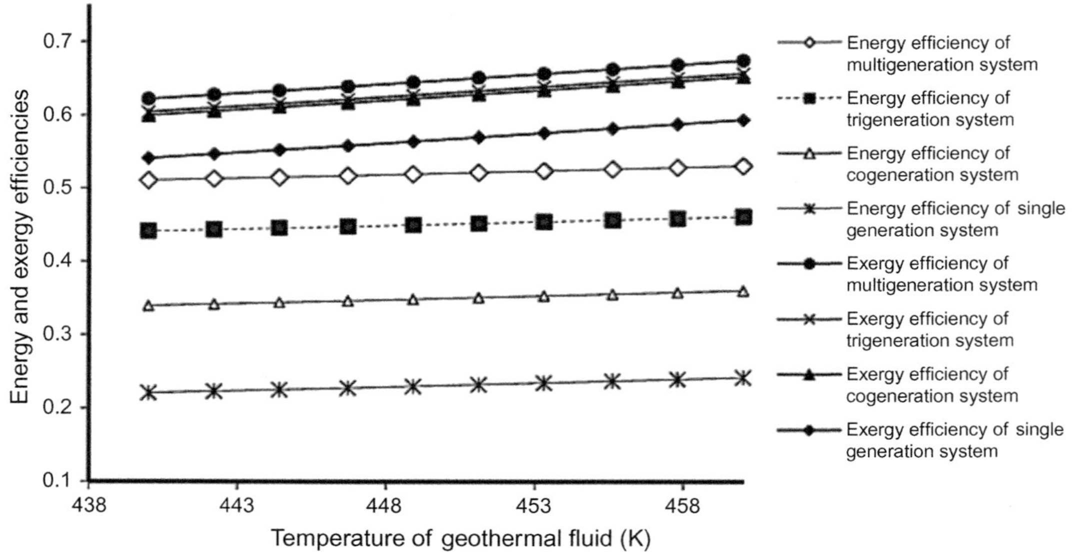

FIG. 6.41 The effects of changing the geothermal fluid temperature on the energy and exergy efficiencies. *(Adapted from Islam S, Dincer I. Development, analysis and performance assessment of a combined solar and geothermal energy-based integrated system for multigeneration. Sol Energy 2017;147:328–43. https://doi.org/10.1016/J.SOLENER.2017.02.048).*

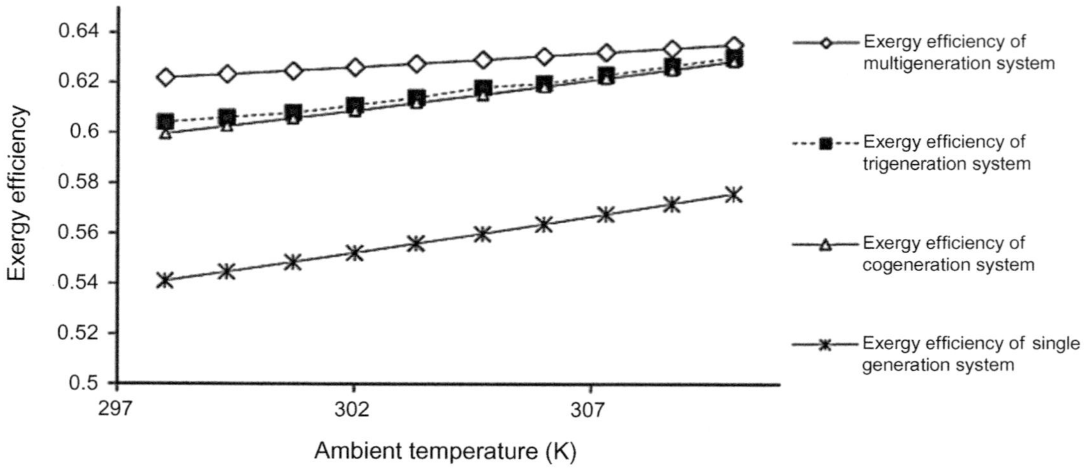

FIG. 6.42 The effects of changing reference temperatures on the exergy efficiencies of all configurations. *(Adapted from Islam S, Dincer I. Development, analysis and performance assessment of a combined solar and geothermal energy-based integrated system for multigeneration. Sol Energy 2017;147:328–43. https://doi.org/10.1016/J.SOLENER.2017.02.048).*

The efficiency of the multigeneration system can change based on the reference temperature. The impacts of deviation in the ambient temperature (from 298 K to 310 K) on the exergy efficiencies of all configurations are shown in Fig. 6.42. The exergy efficiencies of the single generation system and multigeneration system rise from 54% to 57.6% and 62% to 63.5%, respectively. This is caused by the reduction in the temperature difference between the system and ambient.

The overall results show that the energy efficiencies of turbine 1 and turbine 2 are 21.3% and 25.3%, respectively; the exergy efficiencies are 63.8% and 51%, respectively. Both efficiencies of the thermal energy storage system are found to be 90% and 40%, respectively. The dryer is found to have energy and exergy efficiencies of 90% and 41.7%, respectively. The coefficients of performance (COPs) of the R410a operated heat pump in terms of energy and exergy are 3.3 and 0.5, respectively, while the energetic and exergetic coefficients of performance (COPs) of the lithium bromide/water-based absorption cooler are 1.5 and 0.7, respectively.

6.11.5 Closing remarks

This case study compares the performances of a single generation to multigeneration systems combining solar and geothermal sources. The energy and exergy efficiencies of the single generation system are calculated as 22% and 54%, respectively. In contrast, the multigeneration system has quite higher efficiencies, corresponding to about 51% and 62% for energy and exergy efficiency, respectively. The parabolic trough collectors bear the largest quantity of exergy destruction after the geothermal system. The high exergy efficiency of multigeneration is realized due to the utilization of low-grade heat exiting from the condensers of turbine 1 and turbine 2. The energy and exergy efficiencies of all subsystems and the multigeneration system are improved noticeably with the upsurge in the efficiency of the solar collectors.

6.12 Case study 4: Integrated concentrated solar energy system for hydrogen, cooling, fresh water, domestic hot water and electricity production

In this case study, a concentrated solar energy system is integrated with an organic Rankine cycle, absorption cooling system, desalination unit and an electrolyzer for production of multiple useful outputs, namely hydrogen, cooling, fresh water, and domestic hot water as well as electricity, as shown in Fig. 6.43. This case study is adapted from El-Emam and Dincer [22].

6.12.1 System description

The integrated system contains concentrated solar thermal energy using parabolic trough technology with a solar thermal energy storage system. The layout of the multigeneration system for case study 4 is shown in Fig. 6.44. A two-tank type energy storage system is employed for delivering thermal energy to an organic Rankine cycle. The main purpose of storage tank integration is to achieve continuous operation. It is achieved by providing solar heat during solar radiation accessibility for 12 h counting the storage time, and the heat is delivered from the molten salts in the tanks to the heat transfer fluid of the solar trough system during the rest of the day when the solar irradiation is inaccessible. The type of working fluid in the solar trough system is Therminol-66. It is a common thermal oil, having an operational temperature

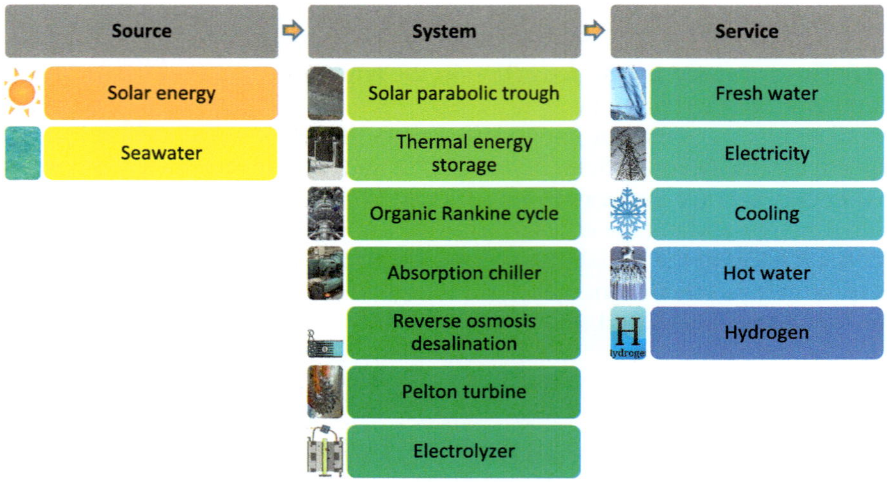

FIG. 6.43 3S approach for case study 4.

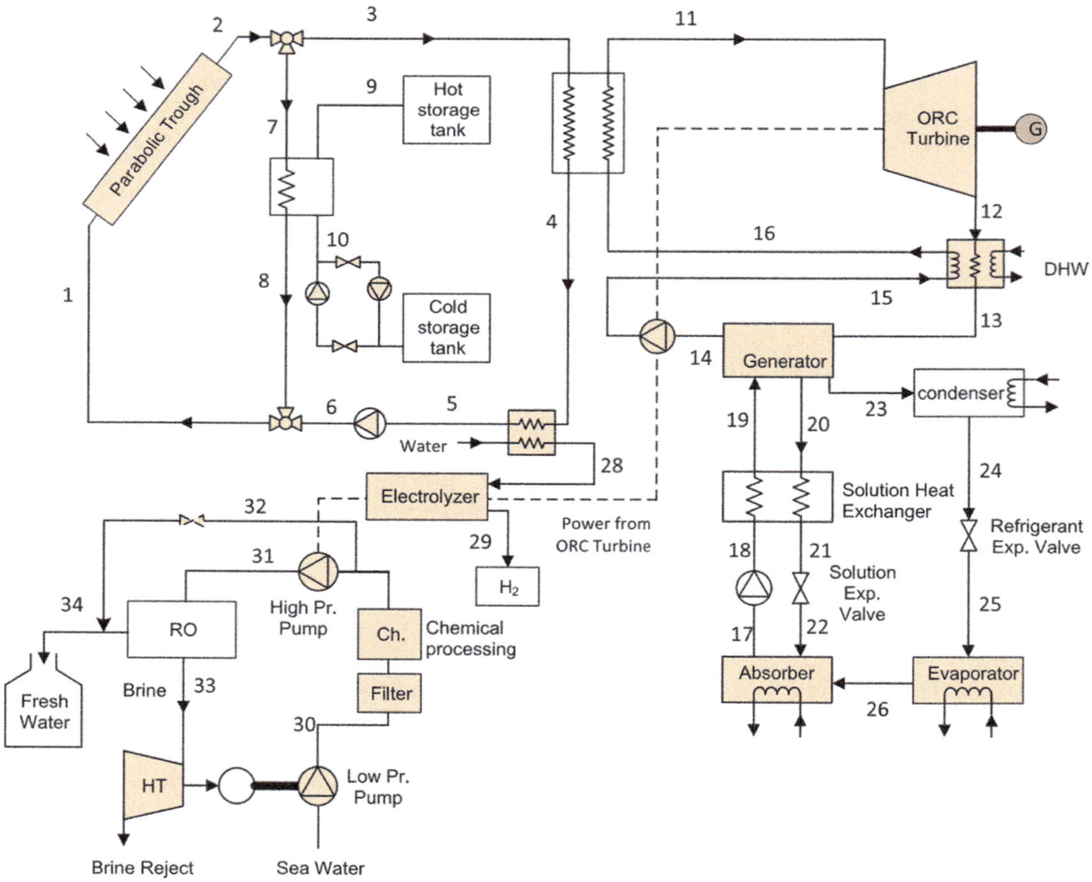

FIG. 6.44 The system layout of the integrated multigeneration system for case study 4. *(Adapted from El-Emam RS, Dincer I. Investigation and assessment of a novel solar-driven integrated energy system. Energ Conver Manage 2018;158:246–55).*

ranging from 0 °C to 345 °C [23]. Given the operational temperature range, this type of oil could likewise be used as the working fluid for the storage tanks [24]. However, the commercial nitrate salt Hitec XL is utilized as the heat transfer fluid for the storage system in this case study [25]. This salt has a moderately low freezing point temperature in the range of 120 °C, which avoids the risk related to freezing [26]. Additionally, molten salts are low-cost, non-flammable, and non-toxic materials, which increases their implementation [27]. Hitec XL is a ternary salt containing 48 wt% $Ca(NO_3)_2$, 45 wt% KNO_3 and 7 wt% $NaNO_3$ [28]. At the operating temperature range of the storage system, Hitec XL is of double the density of Therminol-66 and has greater thermal conductivity. The choice of the working fluid for the solar energy-driven organic Rankine cycle hinges on several conditions, among which the maximum temperature of the cycle is the most significant.

The parabolic trough concentrator combined to the two-tank molten salt storage system delivers thermal energy to the organic Rankine cycle. The system power output is 200–500 kW. The PEM electrolyzer, integrated with the system, can generate hydrogen. It is also heated by the solar heat transfer fluid after it exits the main ORC heat exchanger. The ORC turbine delivers the essential electricity to the electrolyzer as well as to the RO desalination unit (for fresh water at a production rate of 40 kg/s). The available heat of organic fluid exiting the turbine is used to obtain domestic hot water at about 50 °C, and then to the absorption cooling unit with evaporator thermal energy of 500–800 kW.

6.12.2 Assumptions

In order to perform thermodynamic analysis of this multigeneration system, there is a need to make conceptually correct assumptions. Here, the main assumptions undertaken in this case study are listed as follows:

- The changes in kinetic and potential energies are neglected.
- The pressure losses over the main system components are neglected.
- The reference temperature and pressure are taken as 25 °C and 100 kPa, respectively.

6.12.3 Analysis

The overall energy and exergy efficiencies of the system are calculated based on the useful outputs of the systems. This includes the useful work output, and energy and exergy flows related to the generated hydrogen, fresh water produced, cooling, and domestic hot water streams as follows:

$$\eta_{en_{overall}} = \frac{\dot{W} + \dot{Q}_{cooling} + \dot{m}_{H_2} LHV_{H_2} + \dot{m}_{DHW} h_{DHW} + \dot{m}_{freshwater} h_{freshwater}}{\dot{Q}_{sun}} \tag{6.21}$$

$$\eta_{ex_{overall}} = \frac{\dot{W} + \dot{Q}_{cooling}\left(\frac{T_o}{T_c} - 1\right) + \dot{m}_{H_2} ex_{H_2} + \dot{m}_{DHW} ex_{DHW} + \dot{m}_{freshwater} ex_{freshwater}}{\dot{Q}_{sun}\left(1 - \frac{T_0}{T_{sun}}\right)} \tag{6.22}$$

TABLE 6.13 The purchase cost values of the main system components and other financial parameters in the multigeneration system

Component	Value
Heat transfer fluid cost for solar parabolic trough	3$/kg
Cost of concentrators per unit area	320 $/m^2
Specific cost of storage	22 $/kWh$_{th}$
Turbine purchase cost in organic Rankine cycle	4750 ($\dot{W}_T$ [kW])$^{3/4}$ $
Evaporator purchase cost in organic Rankine cycle	309.14 (A_{ev} [m^2])$^{0.85}$ $
Pump purchase cost in organic Rankine cycle	200 ($\dot{W}_P$ [kW])$^{0.65}$ $
Domestic hot water heater purchase cost in organic Rankine cycle	516.62 (A_h [m^2])$^{0.6}$ $
Specific capital cost of PEM electrolyzer	1000 $/kW
Purchase equipment cost of absorption cooling system	1154 ($\dot{Q}_{Evap}$[kW]) $^{0.66}$ $
Specific capital cost of reverse osmosis desalination unit	1700 $/(m^3/day)
Interest rate, i	6%
Maintenance factor, ∅	1.09
Equipment life time, n	15 years
Yearly operational hours, N	7000 h

Source: *El-Emam RS, Dincer I. Investigation and assessment of a novel solar-driven integrated energy system. Energ Conver Manage 2018;158:246–55.*

Here, $\dot{m}_{H_2}LHV_{H_2}$, $\left(\dot{m}h\right)_{DHW}$, and $\left(\dot{m}h\right)_{\text{fresh water}}$ are the energy related to the flow stream of the generated hydrogen, domestic hot water and fresh water from the desalination system, respectively. The exergy related to these streams is used in the exergy efficiency considering the mass flow rate and the specific exergy of each stream. $\dot{W}$ signifies the net work production in both equations. The purchase cost values of the main system components in the multigeneration system are listed in Table 6.13, which are utilized in exergoeconomic analysis and optimization study.

6.12.4 Results and discussion

Fig. 6.45 depicts the changes of the solar trough parameters such as area and HTF exit temperature based on the ORC turbine inlet temperature. The other factors are preserved in the reference case. Increasing the turbine inlet pressure yields an upsurge in the required solar concentrator area, causing a greater cost. The rate of the HTF to the organic fluid mass ratio is controlled by the inlet temperature of the ORC pump. The values shown on the lines denote the resultant ORC turbine pressure value of the operating point. In this case, the HTF exit temperature refers to state 4 in Fig. 6.44.

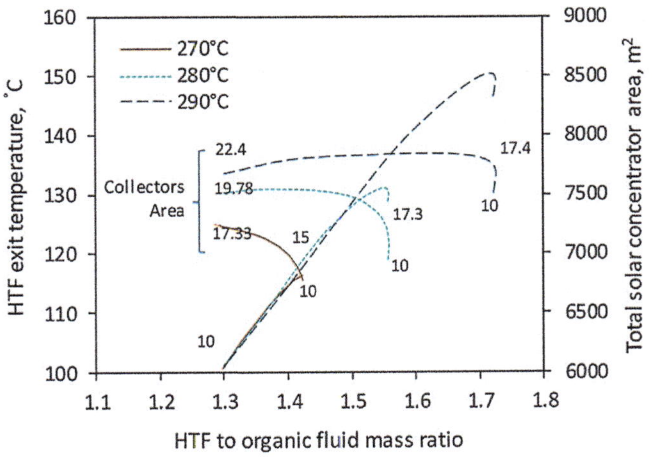

FIG. 6.45 The effects of HTF mass ratio on the solar parabolic trough area and HTF exit temperature. *(Adapted from El-Emam RS, Dincer I. Investigation and assessment of a novel solar-driven integrated energy system. Energ Conver Manage 2018;158:246–55).*

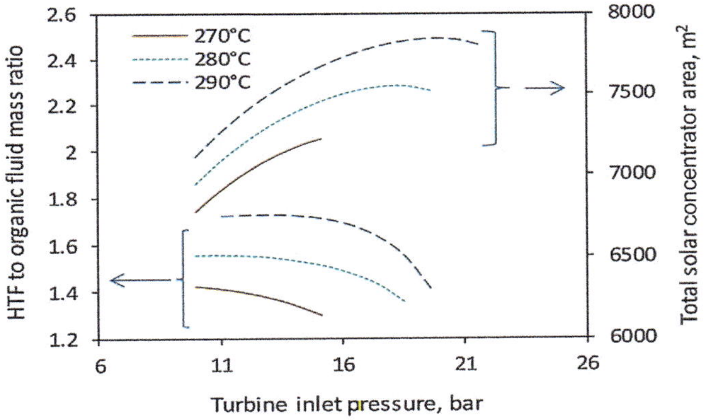

FIG. 6.46 The effects of changing turbine inlet pressures on the total solar concentrator area and the HTF mass ratio at different values. *(Adapted from El-Emam RS, Dincer I. Investigation and assessment of a novel solar-driven integrated energy system. Energ Conver Manage 2018;158:246–55).*

In Fig. 6.46, the effects of changing the turbine inlet pressure on the total solar concentrator area are illustrated. The essential HTF mass flow rate as a ratio of the ORC fluid mass flow rate is also depicted in the figure. Increasing the turbine pressure directly causes a decrease in the quantity of thermal energy needed to be fed to the ORC, taking a constant inlet temperature. The total solar trough area is related to the turbine work with upsurges with greater turbine pressure. Increasing the superheating temperature of the organic fluid yields an upsurge of the HTF flow rate and the total concentrator area.

Fig. 6.47 displays the impacts of the inlet pressure of the turbine on the total cost rate and the absorption cooling system. The lower cost rate rises with the pressure parameters, since the purchase cost of the equipment would increase; the cost rate begins to drop when a smaller concentrator area is necessary and this is related to the reduction in the net power of the system after certain pressure values. The higher the superheating temperature, the higher the pressure value at which the cost rate begins to decline.

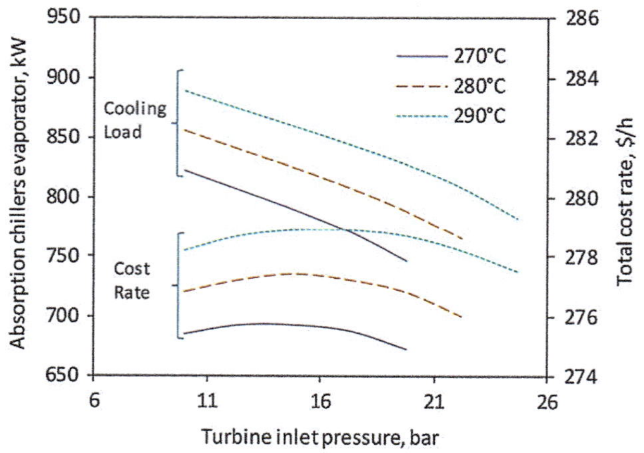

FIG. 6.47 The effects of ORC turbine inlet pressure on the cost rate and the absorption cooling system. *(Adapted from El-Emam RS, Dincer I. Investigation and assessment of a novel solar-driven integrated energy system. Energ Conver Manage 2018;158:246–55).*

6.12.5 Closing remarks

This case study integrates concentrated solar collectors with various energy systems such as desalination, absorption cooling and electrolyzer. The results show that increasing the turbine inlet pressure causes an upsurge in both energy and exergy efficiency parameters. The exergy efficiency declines after reaching a certain peak point. The pressure value at which the maximum exergy is projected rises at higher turbine inlet temperature. The maximum exergy efficiency values of 29.7%, 30% and 30.3% are attained with corresponding cost rates of 275.7 $/h, 277.2 $/h and 278.9 $/h for the ORC turbine temperatures of 270 °C, 280 °C and 290 °C, respectively. The optimization study determines some optimized points satisfying the operation at maximum efficiency with minimum cost rate, in regard to the set of decision variables deliberated for the analysis.

6.13 Case study 5: Optimization of flashing pressure for geothermal power plants using single to quadruple flashing steps

Geothermal power plants are becoming increasingly important, requiring optimization of the use of the geothermal resource. Flash steam power plants represent a significant portion of worldwide geothermal capacity. In this case study, a detailed analysis of the optimum flashing pressures required for single, double, triple, and quadruple flash systems with reinjection is presented. The source is geothermal energy that is used for power generation as illustrated in Fig. 6.48. A series of equations allowing the determination of optimum flashing pressure for the entire operation range of geothermal power plants has been determined. The efficiencies are assessed from both energy and exergy points of view. The exergy analysis shows a clear improvement in efficiency as the number of flashing steps increases. This shows the contribution of multi-staging for better efficiency as discussed in Chapter 3. This case study is taken from a course project [29] submitted by Ron R. Roberts for Advanced Energy Systems course in UOIT taught by Dr. Ibrahim Dincer.

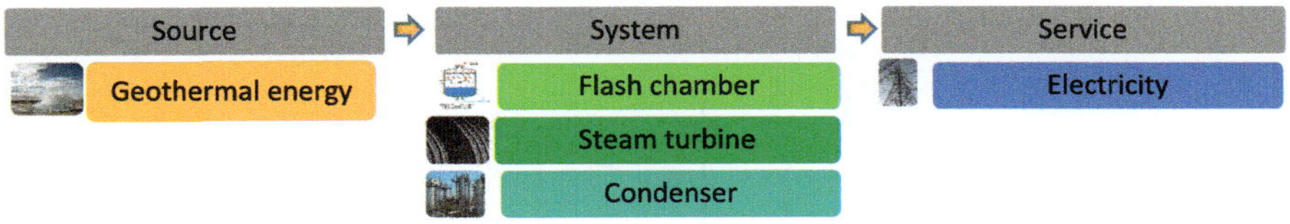

FIG. 6.48 3S approach for case study 5.

6.13.1 System description

For single flash systems, optimizing the parameters to maximize turbine power can be accomplished easily by plotting turbine power vs. flashing pressure for a range of pressures between the condenser pressure and the supply pressure. For double, triple and quadruple flash systems, the problem becomes increasingly difficult, as an optimum flashing pressure must be found at each stage. Each of these methods establishes an optimum point for some of the possible range of supply pressures or temperatures, recognizing that, as the number of flashing steps is increased, each new system requires optimization of the new first stage only. The other stages remain the same as the previous system. For example, once the parameters for the optimum flash pressure are known for a single flash system, only optimization of the first stage flash pressure in a double flash system is required. The optimum pressure of the second stage in the double flash system is the same as that of a single flash system with the first stage flash pressure as its supply pressure. Similarly, once the parameters for the optimum flash pressure of a double flash system are known, a triple flash system can be evaluated to determine the optimum first stage flash pressure, treating the second and third stages as a double flash system with the first stage flash pressure as its supply pressure. The equation to determine optimum pressure can be treated in one of two ways: first, a direct equation that determines flash pressure as a function of supply pressure, and, second, an indirect equation that determines the pressure drop as a percentage of the supply pressure. In this method, the flash pressure is then determined as the supply pressure minus the pressure drop. A schematic diagram of the single flash geothermal system is shown in Fig. 6.49.

A schematic diagram of the double flash geothermal system is shown in Fig. 6.50.

A schematic diagram of the triple flash system components is shown in Fig. 6.51.

A schematic diagram of the quadruple flash system components is shown in Fig. 6.52.

6.13.2 Assumptions

In order to perform thermodynamic analysis of this multigeneration system, there is a need to make conceptually correct assumptions. Here, the main assumptions undertaken in this case study are listed as follows:

- The systems operate under steady state conditions.
- The kinetic and potential energy changes are negligible.
- The pressure and temperature drops through the piping components are negligible.
- The flashing process in the valves is isenthalpic.
- The separators operate adiabatically.

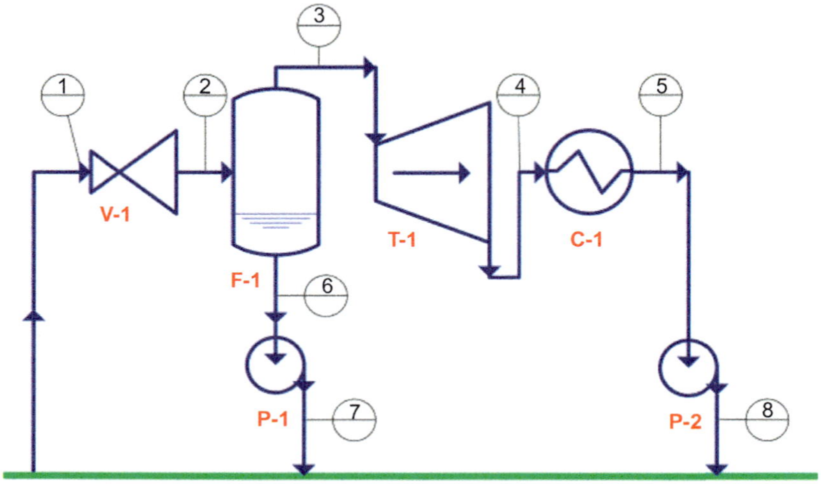

FIG. 6.49 Schematic diagram of a single flash system with reinjection.

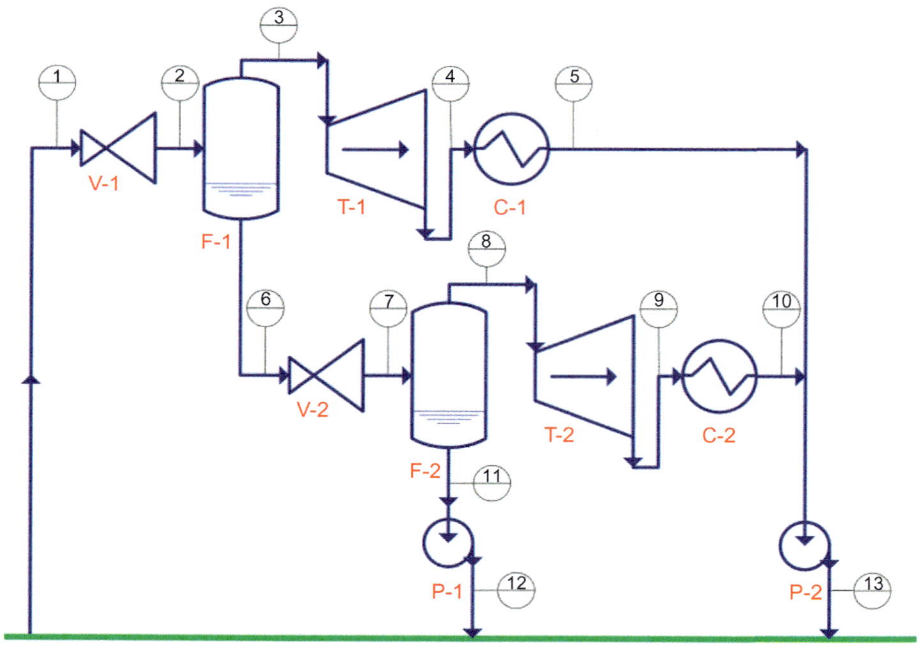

FIG. 6.50 Schematic diagram of a double flash geothermal system with reinjection.

- Turbines and pumps are assumed as isentropic for simplicity.
- The pressure drops through the condensers are negligible.
- The values for thermodynamic properties are determined using Engineering Equations Solver (EES) software.

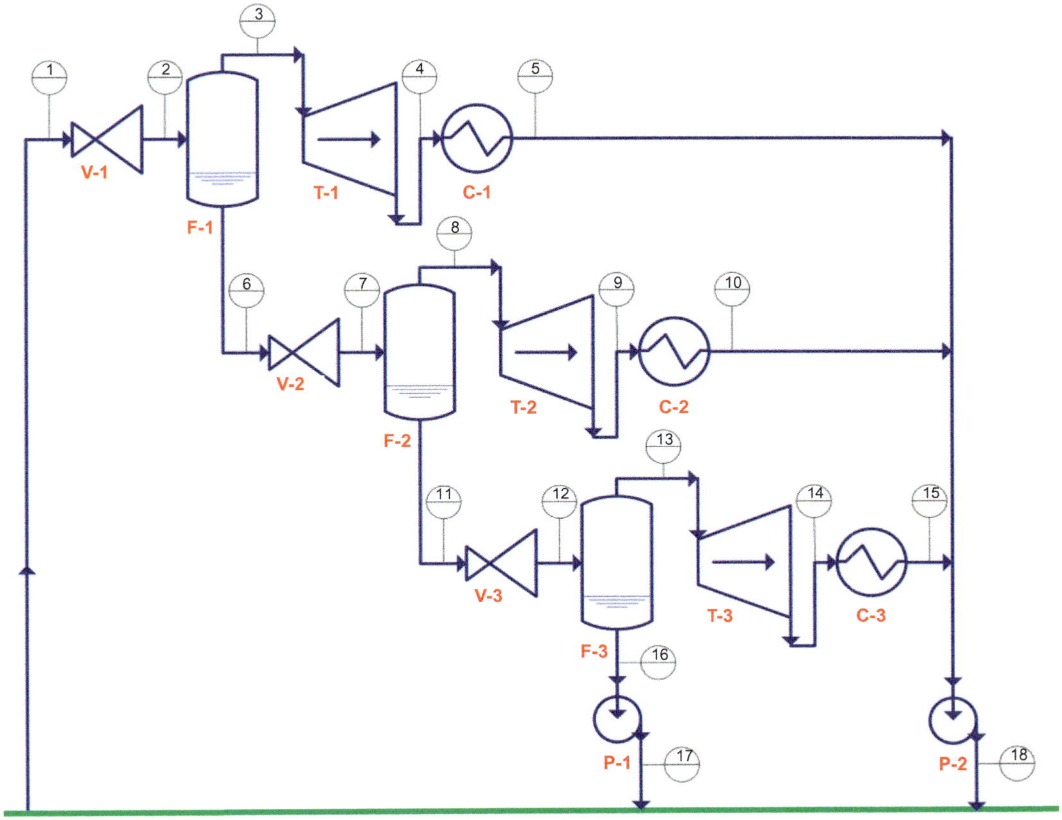

FIG. 6.51 Schematic diagram of a triple flash geothermal system with reinjection.

The quadruple flash systems are analyzed using the following assumptions:

- Source is water at saturation temperature (quality, $x = 0$).
- All condensers operate at saturation pressure at 40 °C (7.381 kPa).
- Reference conditions are temperature of 10 °C and pressure of 101.325 kPa.

6.13.3 Analysis

The difference between a geothermal power plant with reinjection and one without reinjection is the return of some of the energy/exergy to the ground for renewal. For a system without reinjection the energy output is the sum of the turbine energies and the input energy is the source energy. Hence, the energy efficiency can be written as follows:

$$\eta_{en} = \frac{\sum \dot{W}_T}{\dot{Q}_{source}} \tag{6.23}$$

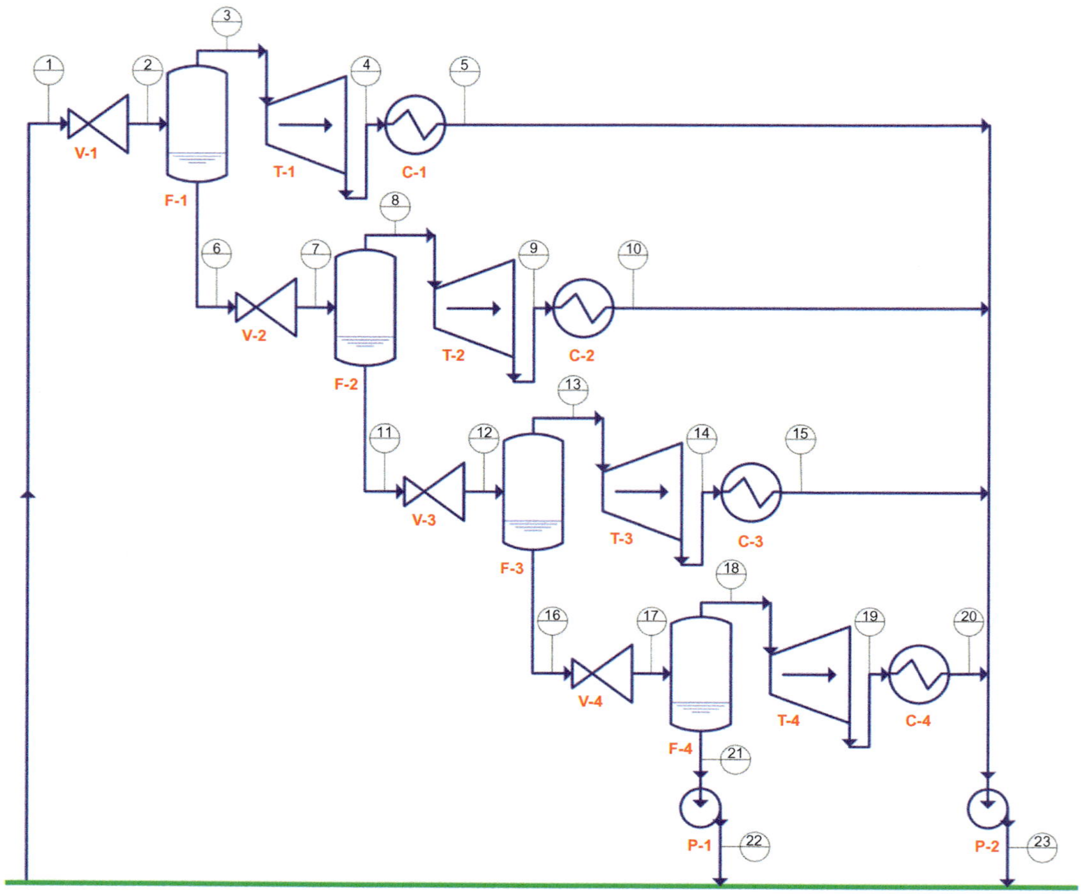

FIG. 6.52 Schematic diagram of a quadruple flash geothermal system with reinjection.

Similarly, using the exergy of the source, the exergy efficiency can be written as follows:

$$\eta_{ex} = \frac{\sum \dot{W}_T}{\dot{E}x^{Q_{source}}} \tag{6.24}$$

For a system with reinjection the sum of the reinjection energy must be subtracted from the source energy and the net work is used. The net work is the sum of turbine work minus the sum of the pump work required for reinjection. Hence, the energy efficiency can be written as

$$\eta_{en} = \frac{\sum \dot{W}_T - \sum \dot{W}_{pump}}{\dot{Q}_{source} - \sum \dot{Q}_{reinjection}} \tag{6.25}$$

Similarly, the sum of reinjection exergy can be subtracted from the source exergy yielding an exergy efficiency as

$$\eta_{ex} = \frac{\sum \dot{W}_T - \sum \dot{W}_{pump}}{\dot{E}x^{Q_{source}} - \sum \dot{E}x^{Q_{reinjection}}} \tag{6.26}$$

6.13.4 Results and discussion

For a given supply pressure the amount of energy that can be extracted using a single turbine varies with the inlet pressure. The turbine inlet pressure is equal to the flashing pressure. As flashing pressure is increased, the amount of energy that can be extracted rapidly rises. The amount of steam produced is rising fast enough to make up for the decrease in enthalpy of the steam. The energy then levels off, reaching a peak. The flashing pressure at the peak represents the optimum flashing pressure, which is also the optimum turbine inlet pressure. After the peak, the amount of energy that can be produced slowly decreases as the flashing pressure continues to rise. At this point, the increase in the amount of steam being produced can no longer make up for the continued decrease in enthalpy of the steam. A plot of various supply pressures results in a group of curves is shown in Fig. 6.53 for low supply pressures (from 100 kPa to 1000 kPa) and in Fig. 6.54 for high supply pressures (from 1000 kPa to 10,000 kPa).

Once the turbine inlet pressure has been determined, the maximum power can also be determined. The equations shown on each of these graphs are only valid for turbine inlet pressures given in kPa and for the ranges shown. The graphs for double, triple and quadruple flash systems are presented in a similar manner.

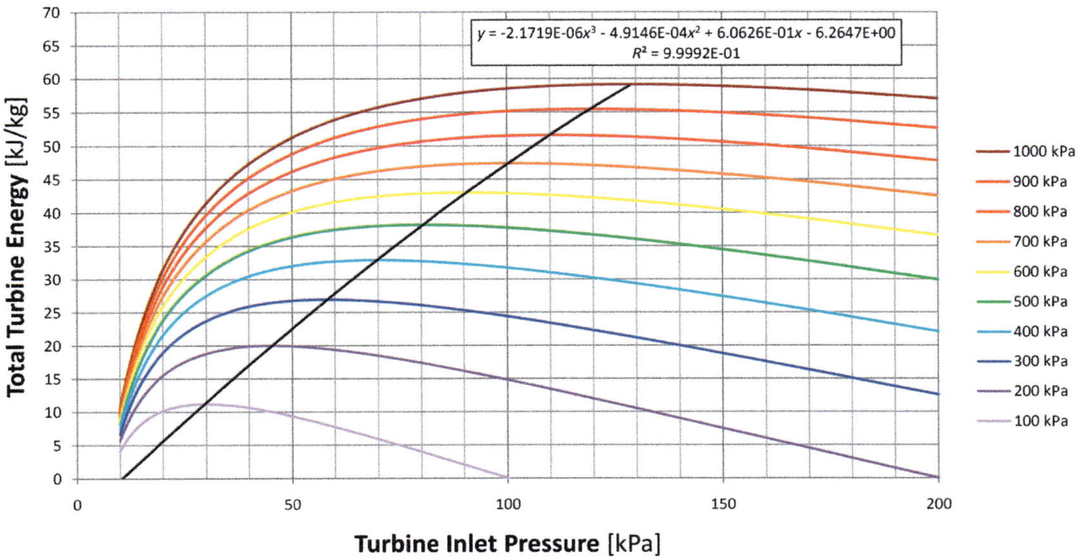

FIG. 6.53 The change of total turbine specific energy with respect to the turbine inlet pressure for low pressure supply to single flash system.

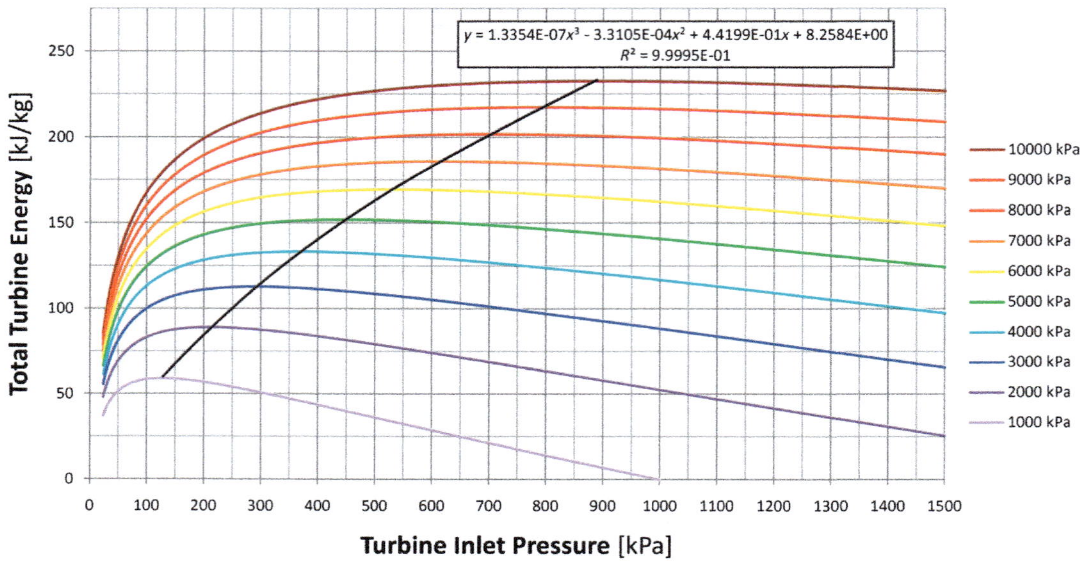

FIG. 6.54 The change of total turbine specific energy with respect to the turbine inlet pressure for high pressure supply to single flash system.

6.13.4.1 *Double flash system*

After the peak, the amount of energy that can be produced slowly decreases as the flashing pressure continues to rise, as shown in Figs. 6.55 and 6.56. At this point the increase in the amount of steam being produced can no longer make up for the continued decrease in enthalpy of the steam. A plot of various geothermal supply pressures is shown in Fig. 6.55 for low supply pressures (from 100 to 1000 kPa) and in Fig. 6.56 for high supply pressures (from 100 to 10,000 kPa).

6.13.4.2 *Triple flash system*

Figs. 6.57 and 6.58 show that after the peak, the amount of energy that can be produced slowly decreases as the flashing pressure continues to rise. A plot of various supply pressures results in a group of curves, as shown in Fig. 6.57 for low supply pressures (from 100 to 1000 kPa) and in Fig. 6.58 for high supply pressures (from 100 to 10,000 kPa).

6.13.4.3 *Quadruple flash system*

Similar to other systems, the amount of energy that can be produced slowly decreases as the flashing pressure continues to rise. Figs. 6.59 and 6.60 show the effects of various supply pressures on the turbine energy yield for low supply pressures and for high supply pressures, respectively.

6.13.4.4 *Efficiency comparisons*

A plot of energy efficiency with respect to the number of turbines shows only tiny improvements for a flash system with reinjection, as depicted in Fig. 6.61. Even when supply pressures

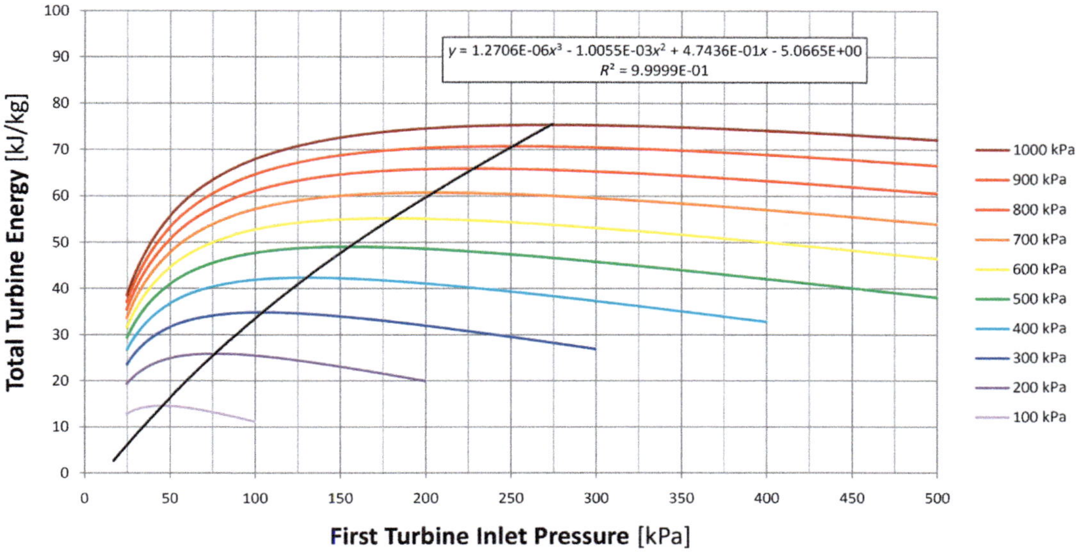

FIG. 6.55 The change of total turbine specific energy with respect to the turbine inlet pressure for low pressure supply to double flash system.

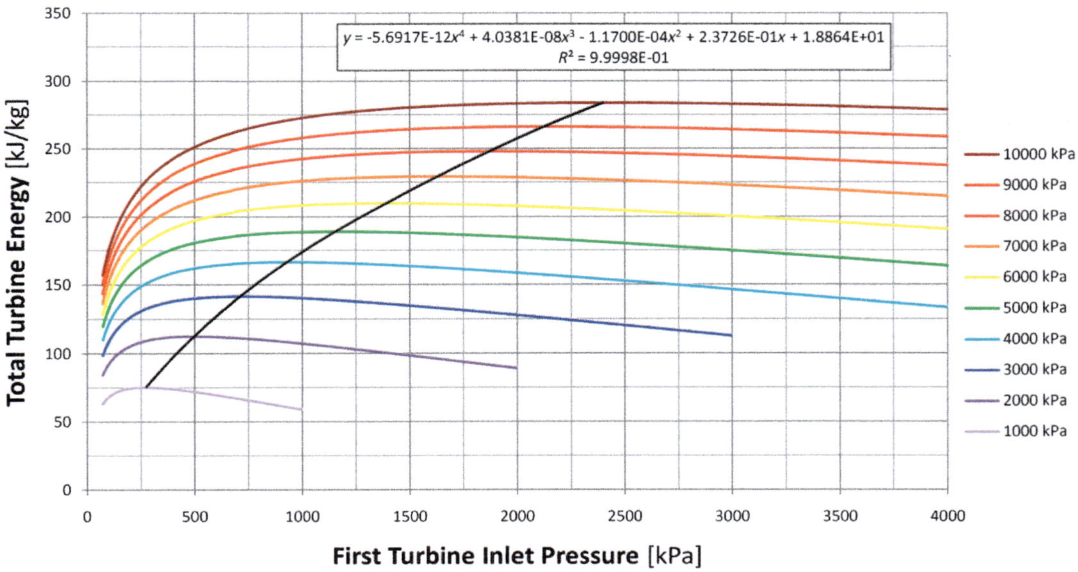

FIG. 6.56 The change of total turbine specific energy with respect to the turbine inlet pressure for high pressure supply to double flash system.

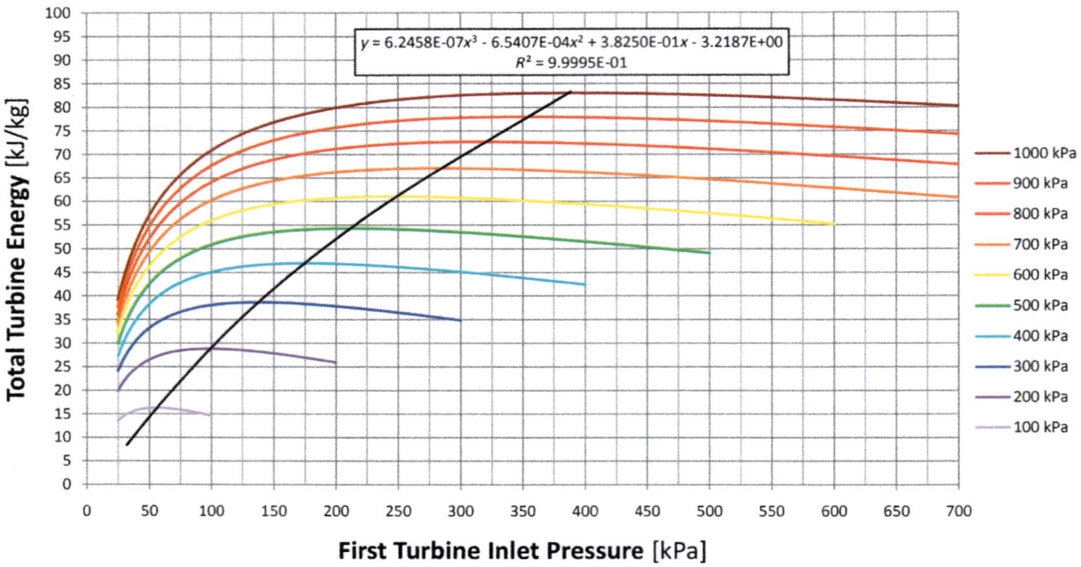

FIG. 6.57 The change of total turbine specific energy with respect to the turbine inlet pressure for low pressure supply to triple flash system.

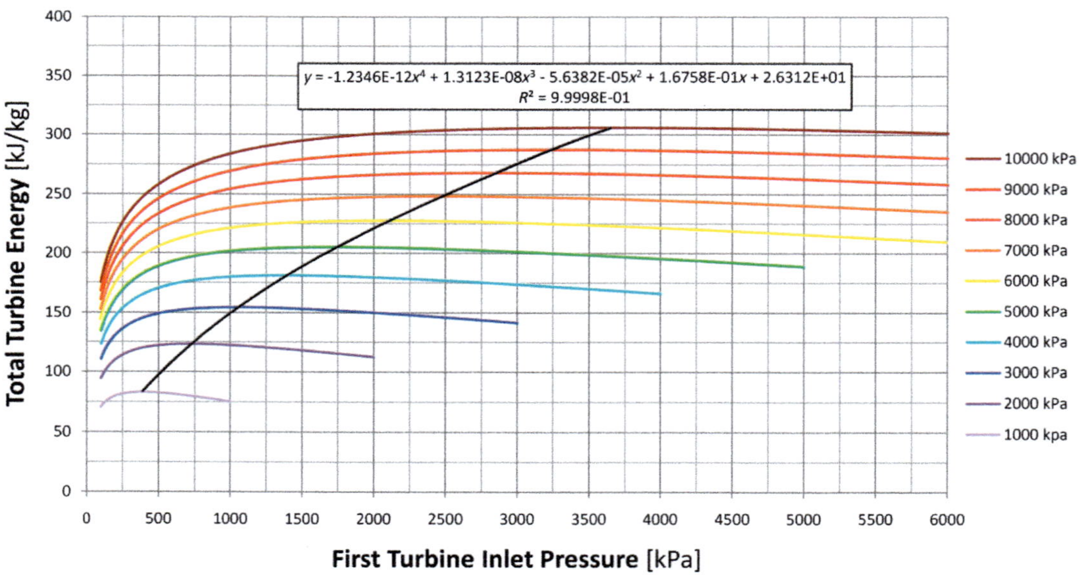

FIG. 6.58 The change of total turbine specific energy with respect to the turbine inlet pressure for high pressure supply to triple flash system.

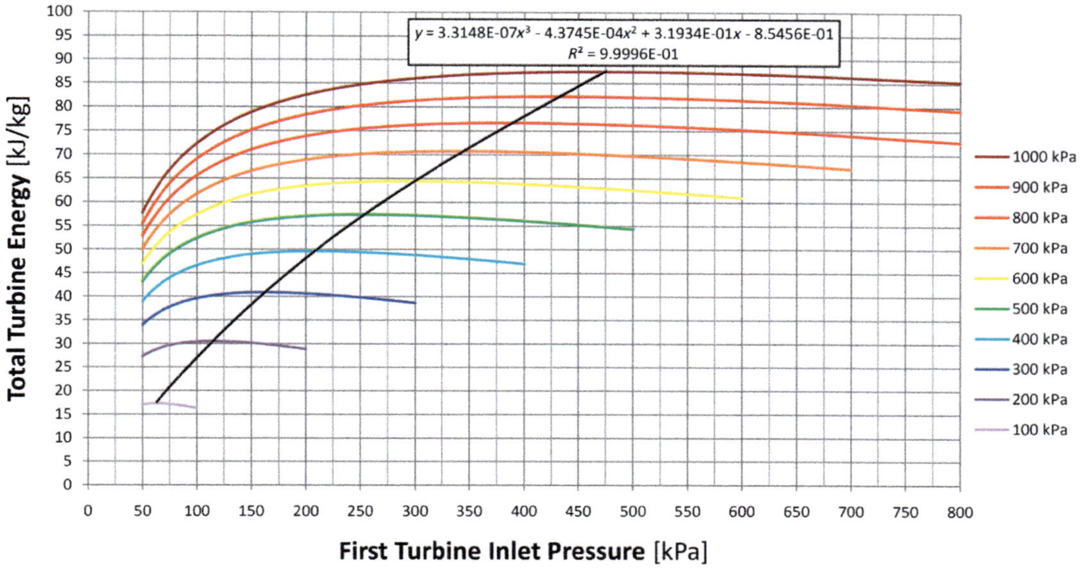

FIG. 6.59 The change of total turbine specific energy with respect to the turbine inlet pressure for low pressure supply to quadruple flash system.

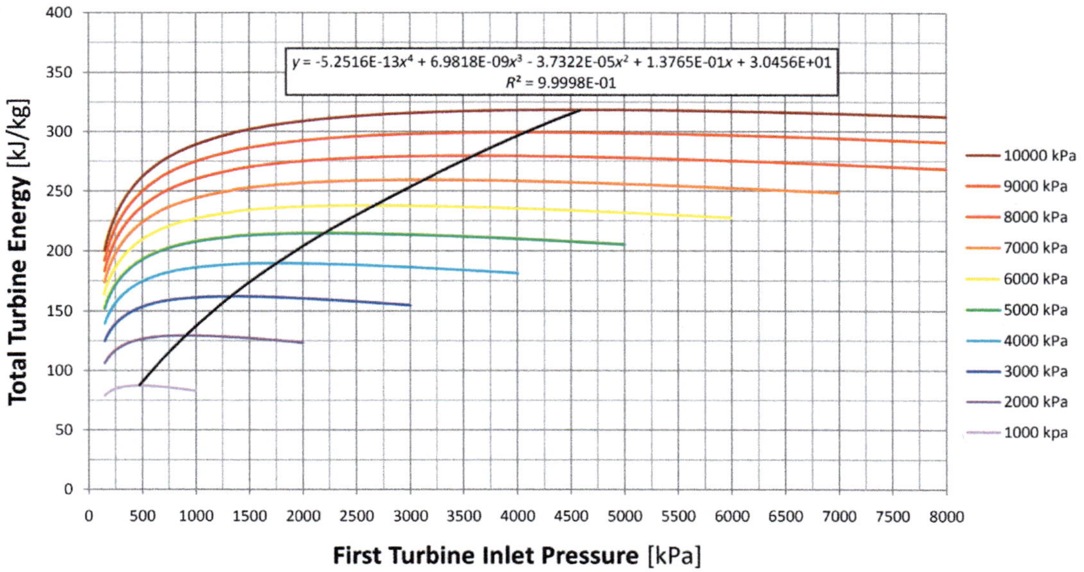

FIG. 6.60 The change of total turbine specific energy with respect to the turbine inlet pressure for high pressure supply to quadruple flash system.

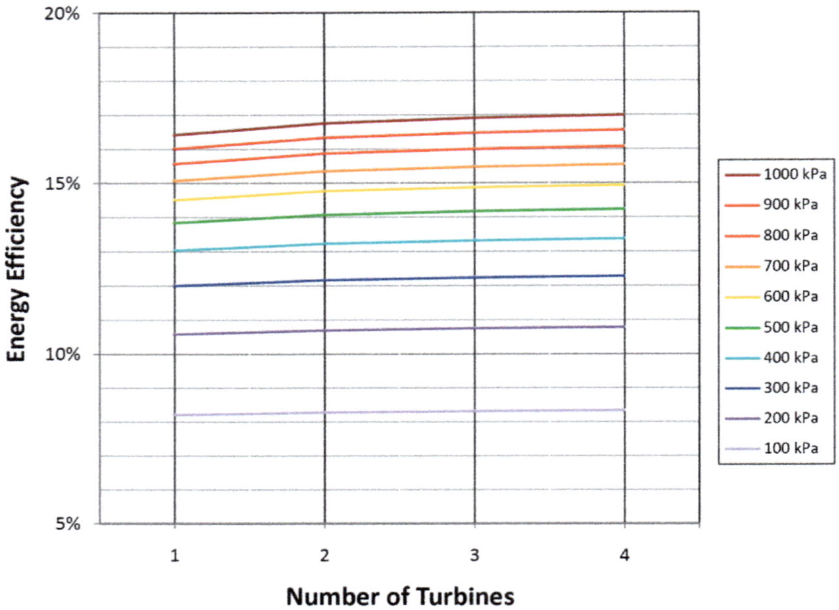

FIG. 6.61 The change of the energy efficiency with respect to the number of turbines for low pressure supply with reinjection.

are very high, less than a 1.5% improvement in overall efficiency is seen in Fig. 6.62. This would imply that there is not much point increasing the number of turbines from a first law of thermodynamics point of view.

The plots of exergy efficiencies with respect to the number of turbines are shown in Figs. 6.63 and 6.64, respectively, for low and high pressure inlets. The exergy efficiencies increase more considerably as compared to energy efficiencies. This shows the importance of exergy analysis. The exergy efficiencies shown are the maximum possible. Actual exergy efficiencies will be lower than this due to exergy destruction because of irreversible losses in each system component (since turbines and pumps are assumed to be isentropic). However, energy analysis alone cannot differentiate between good quality heat and poor quality heat.

The plots of specific work increase from single to double flash system reveal that the lower the source pressure the larger the increase is in relative work, as shown in Figs. 6.65 and 6.66. However, these values are all fairly close. As a good approximation, for supply pressures between 100 kPa and 1000 kPa, the increase is approximately 29%. Similarly, the change between a double and triple flash system is approximately 10% over the same range, decreasing to 6% when the number of flashes increases to four. It is projected that this continues to decrease until the improvement would only be 4% when increasing from a quadruple flash system to quintuple.

As the pressure increases, the relative improvement in net work decreases. For the pressure range from 1000 kPa to 10,000 kPa, the improvement from a single flash to a double flash system is approximately 25% (with a low of 23% for 10,000 kPa). Similarly the improvement

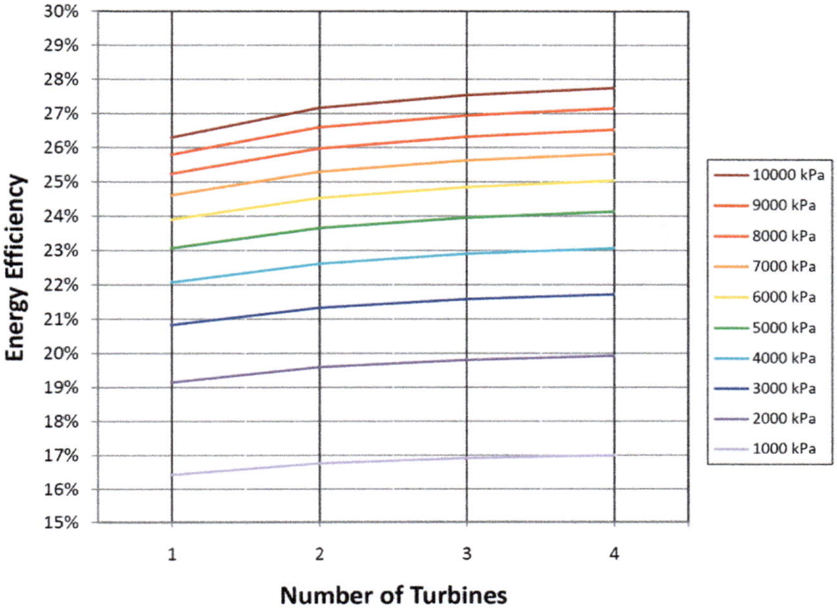

FIG. 6.62 The change of the energy efficiency with respect to the number of turbines for high pressure supply with reinjection.

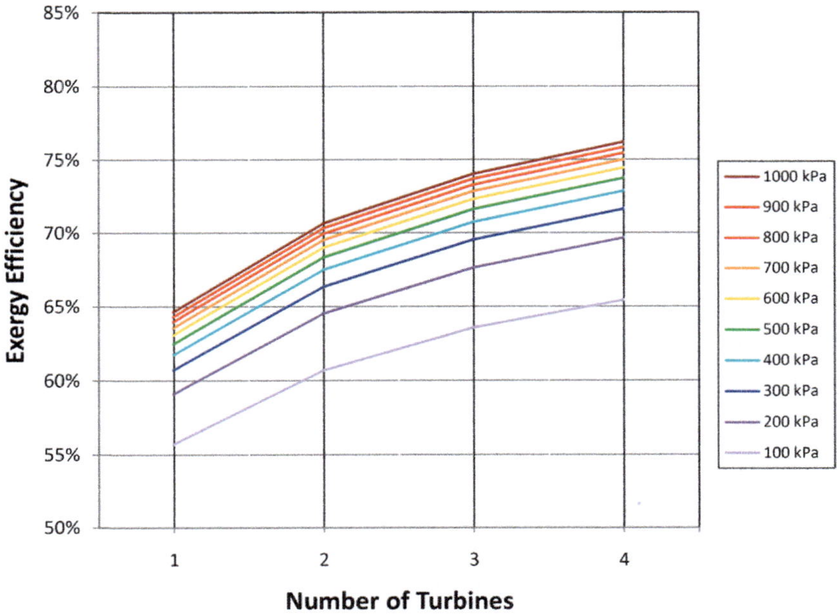

FIG. 6.63 The change of the exergy efficiency with respect to the number of turbines for low pressure supply with reinjection.

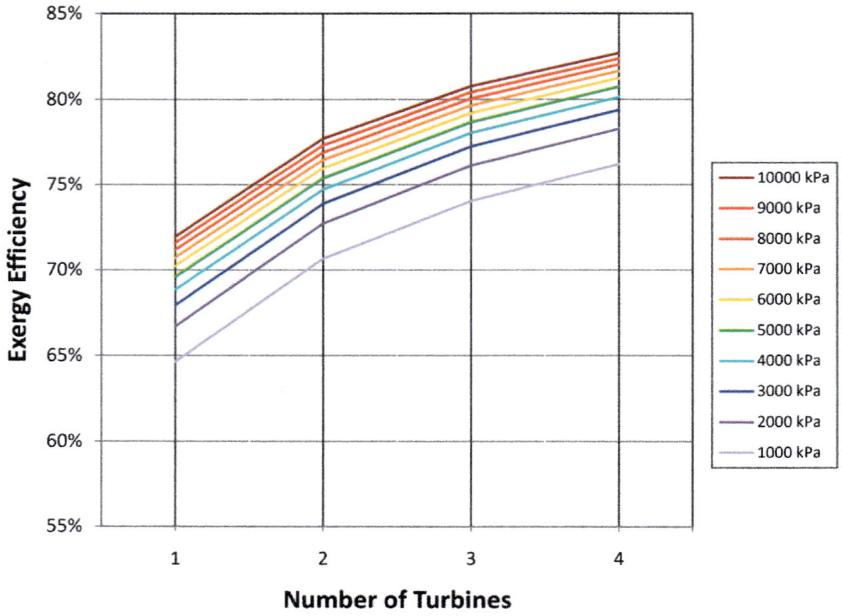

FIG. 6.64 The change of the exergy efficiency with respect to the number of turbines for high pressure supply with reinjection.

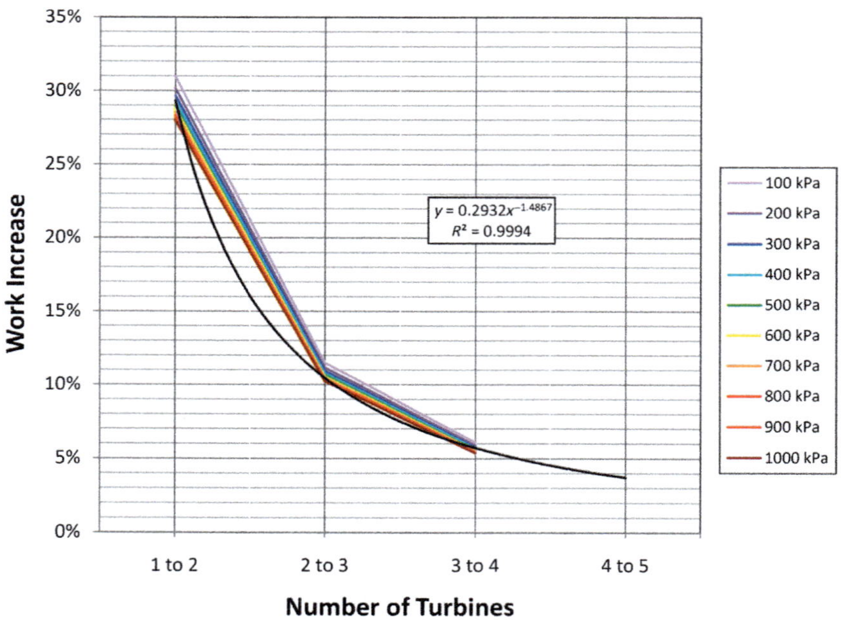

FIG. 6.65 The net work increase by changing number of flash stages for low pressure supply with reinjection.

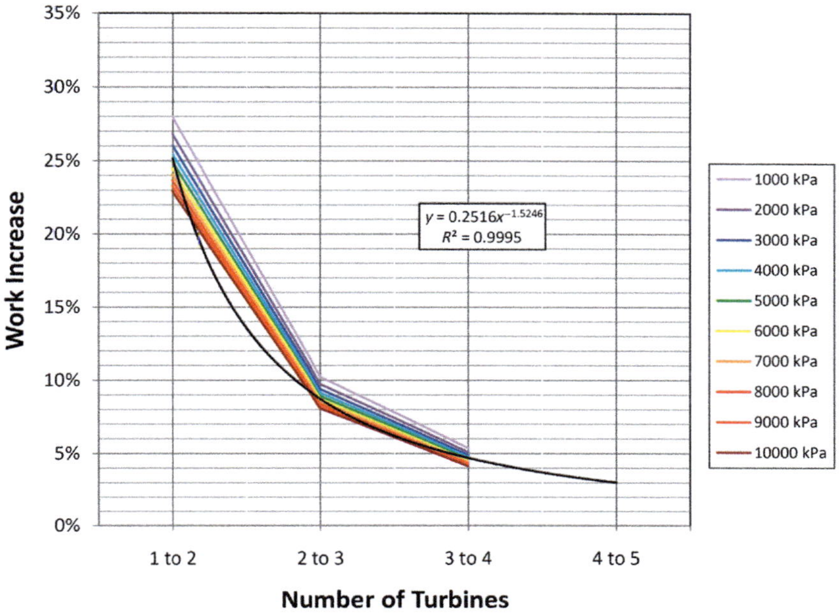

FIG. 6.66 The net work increase by changing number of flash stages for high pressure supply with reinjection.

from a double to triple flash system is only 9%. As the number of flashing steps increases, the improvement continues to decrease, with only a 5% improvement when switching to a quadruple flash system. These values are consistent with the double flash results seen in the literature.

6.13.5 Closing remarks

A detailed analysis is carried out for determination of the optimum flashing pressure(s) for saturated water supply pressures from 100 kPa to 10,000 kPa in geothermal power plants. This range is extended as low as 15 kPa for single flash units, 25 kPa for double flash units, 50 kPa for triple flash units, and 100 kPa for quadruple flash units. An evaluation of energy efficiency revealed few changes as the number of flashing steps is increased, even though the output power increases significantly each time. This is determined to be due to the inability of the energy analysis to differentiate between high and low quality energy. However, extension of the analysis to exergy efficiency clearly shows the improvement as the number of flashing steps is increased. This improvement becomes small in absolute terms as the number of flashing steps is further increased.

An analysis of relative increase in efficiency reveals that low supply pressure systems benefit more than high supply pressure systems. When increasing the number of flashing steps from one to two, the improvement in output power varies from 23% at 10,000 kPa supply to a high of 31% at 100 kPa supply. With an average improvement around 28% for single to double flash, the improvement from double to triple flash system is about 10%.

Further increment to a quadruple flash unit would see a minor increase, with an increase from 3% to 4% projected for a theoretical quadruple flash unit. The use of triple and quadruple flash units will require careful evaluation of the cost, given the small increase in efficiency observed.

6.14 Case study 6: Integrated solar and geothermal based multigeneration system for buildings

In this case study, thermodynamic analysis of a multigeneration system consisting of geothermal and solar energy sources is performed. The useful outputs obtained from this integrated system are electricity, cooling, heating, hydrogen as well as hot water for residential applications. The sources, systems and services in this multigeneration system are illustrated in Fig. 6.67. This case study is adapted from Khalid et al. [30].

6.14.1 System description

The integrated multigeneration system contains a parabolic trough solar collector, a geothermal well, Rankine cycles, and an absorption cooling unit according to the system layout shown in Fig. 6.68. The selected heat transfer fluid is Duratherm A oil, which enters the solar collector and exits after being heated. The HTF then arrives at heat exchanger 2, which also performs as a storage tank and is used in water heating, entering the storage tank at state 20 to deliver the energy needed by Rankine cycle 1. The HTF then arrives at heat exchanger 3, which is utilized to heat water at state 35 to state 23 for running high-pressure steam turbine 2 in Rankine cycle 2. The thermal energy is exchanged from the hot oil to the generator of the absorption cooling unit and the HTF exits the generator at state 13 and is sent back to the solar collector. Water arrives at high-pressure turbine 1 for production of electricity. A percentage of electricity produced by HPST 1 is utilized to generate hydrogen through water electrolysis, and which is then used as a feed gas for the fuel cell to generate additional

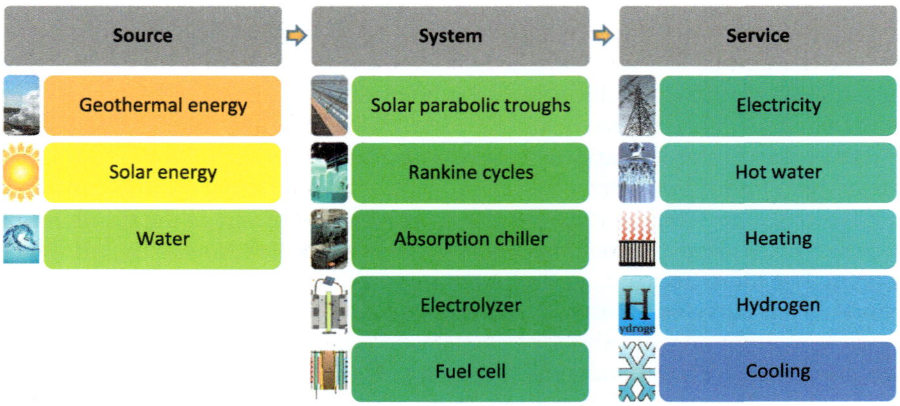

FIG. 6.67 3S approach for case study 6.

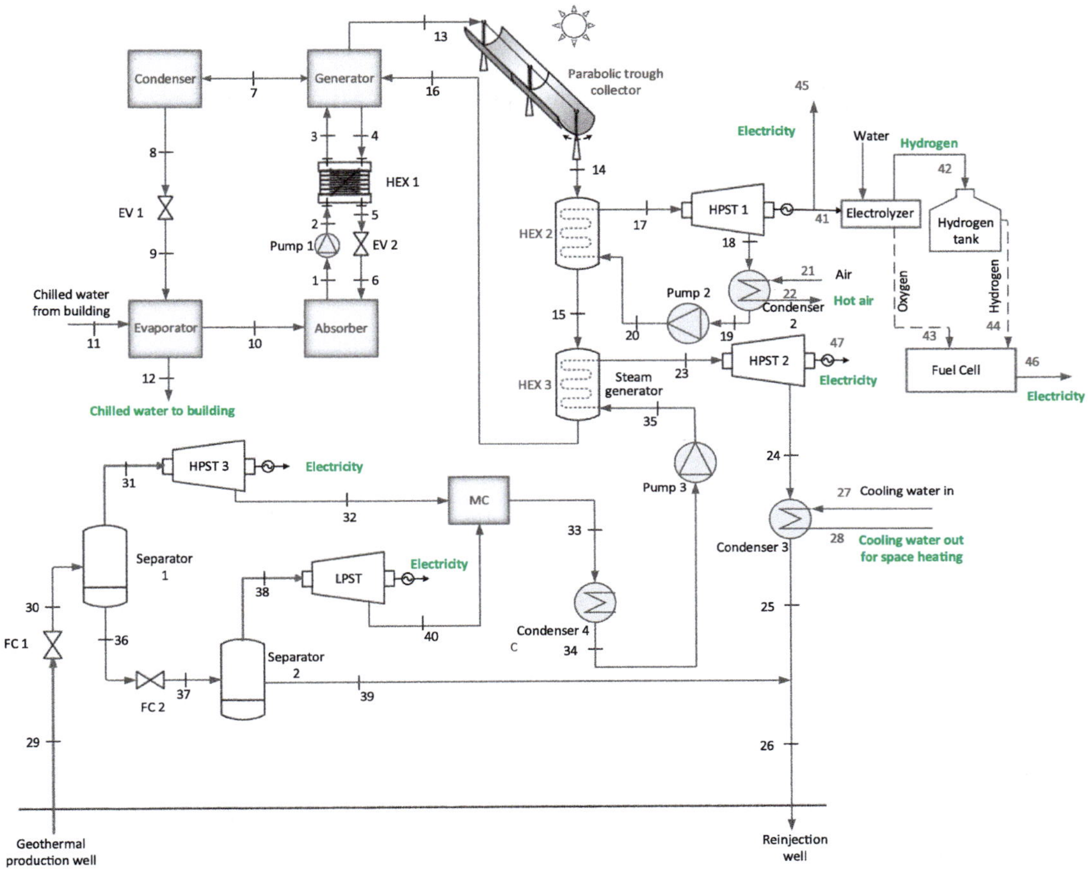

FIG. 6.68 The schematic diagram of case study 6. *(Adapted from Khalid F, Dincer I, Rosen MA. Techno-economic assessment of a solar-geothermal multigeneration system for buildings. Int J Hydrogen Energy 2017;42:21454–62).*

electricity. After exiting HPST 1, stream 18 is sent to condenser 2. Stream 19 enters a pump and is reverted to heat exchanger 2 for heat exchange by the oil. The stream leaving heat exchanger 3 is directed to HPST 2 to yield electrical work. After HPST 2, stream 24 is directed to condenser 3, providing hot water for the residential community. The stream exiting the condenser is sent to the reinjection well. The stream from the geothermal production well is directed to a flash chamber 1 which is later on distributed to separator 1 where it is separated into two flows. Stream 31 as vapor arrives at high-pressure steam turbine 3 (HPST 3) and then the mixing chamber. Stream 36 as liquid is additionally flashed in flash chamber 2 and then divided in separator 2 and the liquid portion is directly sent to the geothermal reinjection well. The vapor from the second separator is directed to a low pressure steam turbine, from which the exit stream is transported to the mixing chamber along with stream 32. The exit of the mixing chamber is condensed and sent to heat exchanger 3 for being heated and used in high-pressure turbine 2.

6.14.2 Assumptions

In order to perform thermodynamic analysis of this multigeneration system, there is a need to make conceptually correct assumptions. Here, the main assumptions undertaken in this case study are listed as follows:

- The changes in kinetic and potential energy terms are neglected.
- The pressure losses within the pipes are negligible.

6.14.3 Analysis

The energy efficiencies can generally be defined as the ratio of useful energy output to total energy input. The energy and exergy efficiencies of Rankine Cycle 1 are respectively written as

$$\eta_{en,rc1} = \frac{\dot{W}_{HPST1} - \dot{W}_{P1}}{\dot{m}_{14}h_{14} - \dot{m}_{15}h_{15}} \tag{6.27}$$

$$\eta_{ex,rc1} = \frac{\dot{W}_{HPST1} - \dot{W}_{P1}}{\dot{m}_{14}ex_{14} - \dot{m}_{15}ex_{15}} \tag{6.28}$$

The energetic and exergetic COP of the absorption cooling unit are respectively expressed as follows:

$$COP_{en,ac} = \frac{\dot{Q}_{eva}}{\dot{Q}_{generator}} \tag{6.29}$$

$$COP_{ex,ac} = \frac{\dot{Q}_{eva}\left(\dfrac{T_0}{T_{eva}} - 1\right)}{\dot{Q}_{generator}\left(1 - \dfrac{T_0}{T_s}\right)} \tag{6.30}$$

In this system, hydrogen acts as an energy storage medium. Therefore, it is not constantly fed to the fuel cell, but only when necessary during peak electrical demands. Therefore, in order to examine the performance of the overall system, two different energy and exergy efficiency definitions are used. In the first expression, the electricity generated by the fuel cell is excluded as written in the following formulation:

$$\eta_{en,ov1} = \frac{\left(\begin{array}{c}\dot{Q}_{ev} + \dot{m}_{28}h_{28} - \dot{m}_{27}h_{27} + \dot{m}_{22}h_{22} - \dot{m}_{21}h_{21} + \dot{W}_{HPST1} + \dot{W}_{HPST2} + \\ \dot{W}_{HPST3} + \dot{W}_{LPST} - \dot{W}_{P2} - \dot{W}_{P3} + \dot{m}_{H_2}LHV - \dot{W}_{electrolyzer}\end{array}\right)}{\dot{m}_{29}h_{29} - \dot{m}_{26}h_{26} + \dot{Q}_{sol}} \tag{6.31}$$

In the second expression, the electricity generated by the fuel cell is involved, and the energy efficiency of the overall system is written as follows:

$$\eta_{en,ov2} = \frac{\left(\begin{array}{c}\dot{Q}_{ev} + \dot{m}_{28}h_{28} - \dot{m}_{27}h_{27} + \dot{m}_{22}h_{22} - \dot{m}_{21}h_{21} + \dot{W}_{HPST1} + \dot{W}_{HPST2} + \\ \dot{W}_{HPST3} + \dot{W}_{LPST} - \dot{W}_{P2} - \dot{W}_{P3} + \dot{W}_{fuelcell} - \dot{W}_{electrolyzer}\end{array}\right)}{\dot{m}_{29}h_{29} - \dot{m}_{26}h_{26} + \dot{Q}_{sol}} \tag{6.32}$$

The exergy efficiencies can be written in a similar fashion for the overall system using two different expressions:

$$\eta_{ex,ov1} = \frac{\left(\begin{array}{c}\dot{Ex}^{Q}_{cooling} + \dot{m}_{28}ex_{28} - \dot{m}_{27}ex_{27} + \dot{m}_{22}ex_{22} - \dot{m}_{21}ex_{21} + \dot{W}_{HPST1} + \dot{W}_{HPST2} + \\ \dot{W}_{HPST3} + \dot{W}_{LPST} - \dot{W}_{P2} - \dot{W}_{P3} + \dot{m}_{H_2}ex_{H_2} - \dot{W}_{electrolyzer}\end{array}\right)}{\dot{m}_{29}ex_{29} - \dot{m}_{26}ex_{26} + \dot{Ex}^{Q}_{sol}} \quad (6.33)$$

$$\eta_{ex,ov2} = \frac{\left(\begin{array}{c}\dot{Ex}^{Q}_{cooling} + \dot{m}_{28}ex_{28} - \dot{m}_{27}ex_{27} + \dot{m}_{22}ex_{22} - \dot{m}_{21}ex_{21} + \dot{W}_{HPST1} + \dot{W}_{HPST2} + \\ \dot{W}_{HPST3} + \dot{W}_{LPST} - \dot{W}_{P2} - \dot{W}_{P3} + \dot{W}_{fuelcell} - \dot{W}_{electrolyzer}\end{array}\right)}{\dot{m}_{29}ex_{29} - \dot{m}_{26}ex_{26} + \dot{Ex}^{Q}_{sol}} \quad (6.34)$$

6.14.4 Results and discussion

The thermodynamics analysis results depict that the energy efficiencies of the overall system are 18.4% when the fuel cell unit is exempted and 17% when the fuel cell unit is accounted for. The fundamental output parameters are listed in Table 6.14. The energy efficiency of Rankine Cycle 1 is calculated as 20.2%. This integrated system has the ability to provide about 140 kW of cooling load.

The exergy destruction rates of most of the key modules of the system are shown in Fig. 6.69. The highest exergy destruction rate happens in the solar parabolic trough collector, which is followed by HEX2 and the electrolyzer. Therefore, this implies that for performance improvement of the overall system, a special care needs to be given to decrease the exergy destructions in these units, keeping in mind that such efforts are desired to be cost-effective.

The deviations of the energetic and exergetic COPs of the vapor absorption cycle with reference temperature are illustrated in Fig. 6.70. By increasing ambient temperature (from 15 °C

TABLE 6.14 Parameter values from modeling and energy and exergy analyses of the system

Parameter	Value
Output of low pressure steam turbine (kW)	64.0
Output of high pressure steam turbine (kW)	131.4
Cooling capacity (kW)	140
Rankine cycle 1 energy efficiency (%)	20.2
Rankine cycle 1 exergy efficiency (%)	49.8
Energy efficiency of the overall system excluding fuel cell (%)	17.4
Exergy efficiency of the overall system excluding fuel cell (%)	16.6
Energy efficiency of the overall system including fuel cell (%)	16.0
Exergy efficiency of the overall system including fuel cell (%)	14.6

Source: *Khalid F, Dincer I, Rosen MA. Techno-economic assessment of a solar-geothermal multigeneration system for buildings. Int J Hydrogen Energy 2017;42:21454–62.*

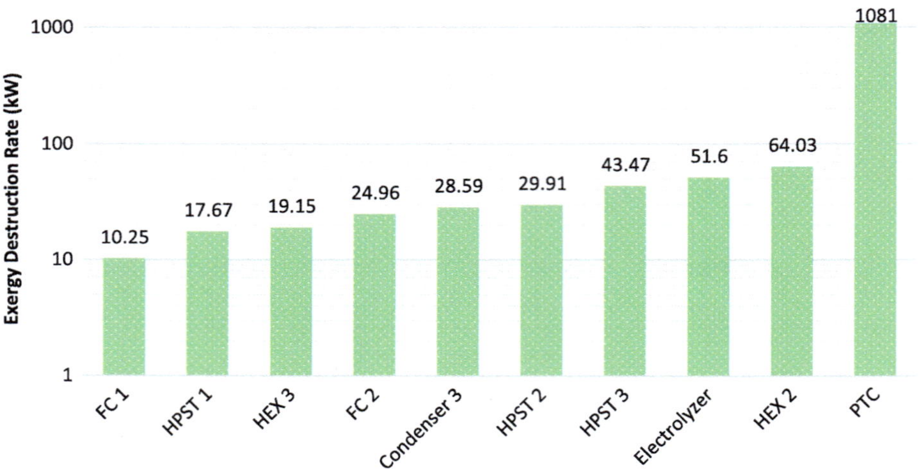

FIG. 6.69 Major exergy destruction rates of the subsystems.

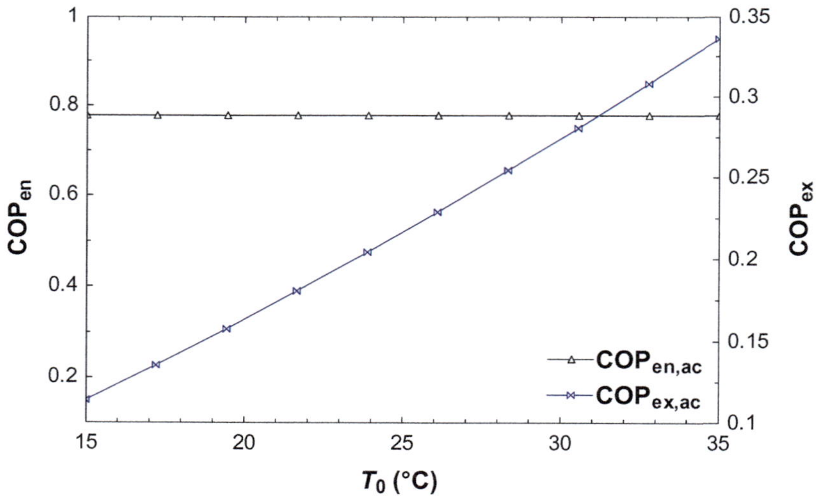

FIG. 6.70 The effects of changing reference temperature on the energetic and exergetic COPs of the absorption cooling unit.

to 35 °C), the exergetic COP rises from 0.11 to 0.33. This is attributed to the fact that when reference temperature upsurges, the exergetic output of the evaporator also rises, yielding an improved exergetic COP.

Fig. 6.71 shows the impacts of the reference temperature on the energy and exergy efficiencies of the overall system. When the reference temperature increases (from 15 °C to 35 °C), there is no alteration in the energy efficiencies although the exergy efficiency of the complete

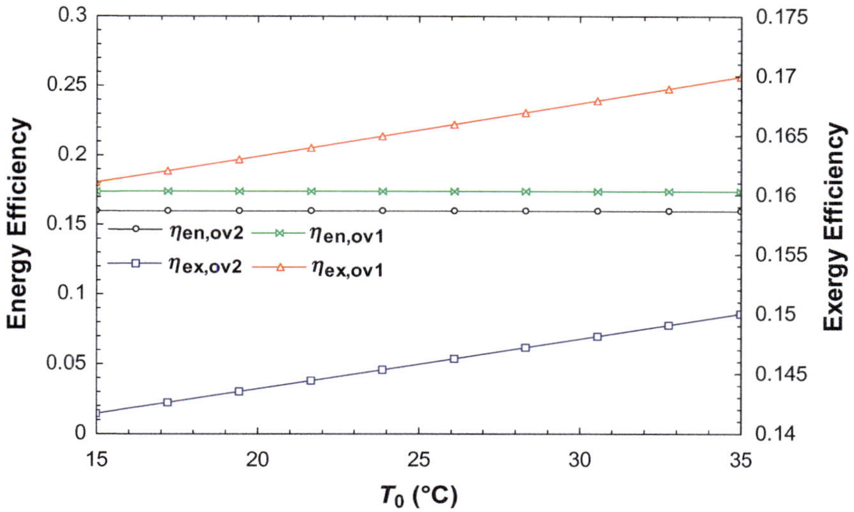

FIG. 6.71 The effects of reference temperature on the energy and exergy efficiencies of the complete system.

system (excluding fuel cell) rises from 16.2% to 17%, and the exergy efficiency of the complete system (counting fuel cell) rises from 14.2% to 15%. This is explained as follows: as the reference temperature rises, the exergetic input of the sun drops and the exergetic output of the absorption cooling unit rises, yielding the upsurge in system exergy efficiency.

The influences of HTF outlet temperature on the energy and exergy efficiencies of the complete system are illustrated in Fig. 6.72. When the HTF outlet temperature increases (from 250 °C to 300 °C), the exergy efficiency of the complete system (excluding fuel cell) rises from 15.1% to 21.4%, and the exergy efficiency of the complete system (including fuel cell) rises from 13.5% to 18.8%.

When the HTF outlet temperature rises, both energy and exergy efficiencies of the complete system rise. In addition, as shown in Fig. 6.73, when the hydrogen generation rate increases, both the energy and exergy efficiencies of the complete system drop.

6.14.5 Closing remarks

This case study analyzes an integrated multigeneration system consisting of solar and geothermal resources to yield multiple useful commodities, namely electricity, cooling, heating, hydrogen and hot water for a residential community. The performance of the integrated system is quantified using energy and exergy efficiencies. In addition, it is shown that hydrogen can act as an energy storage medium and be converted into electricity on demand via fuel cells. The hydrogen generation rate from the electrolyzer is about 2.7 kg/h. It is also calculated that the highest exergy destruction rate happens in the solar parabolic trough collector. This case study demonstrates that hydrogen storage can be a solution for uninterrupted energy supply for residential communities.

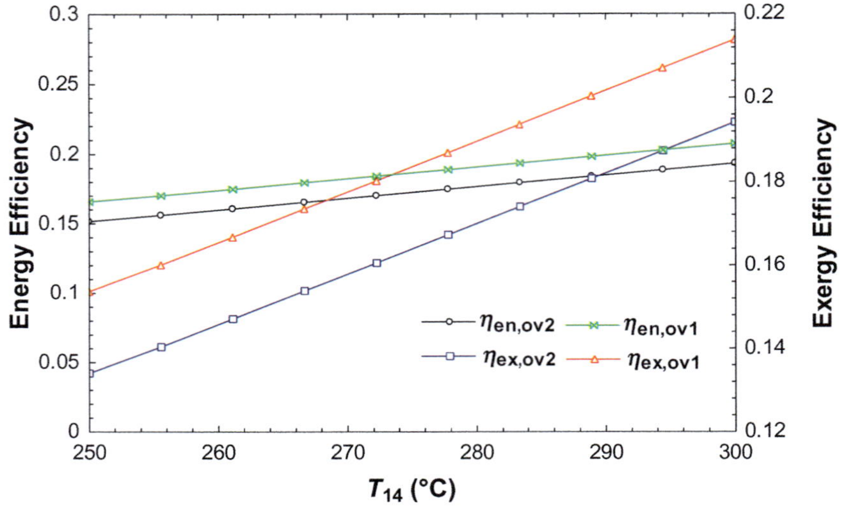

FIG. 6.72 The effects of HTF outlet temperature on the energy and exergy efficiencies of the complete system.

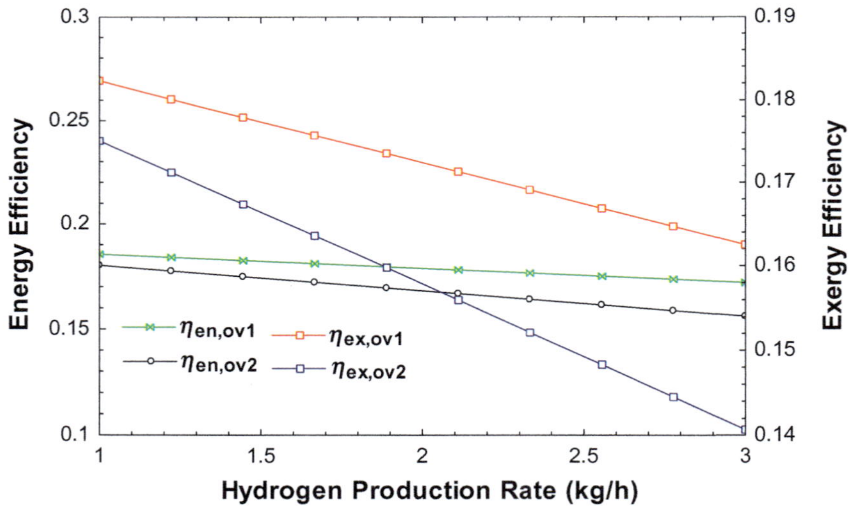

FIG. 6.73 The effects of hydrogen generation rate on the energy and exergy efficiencies of the complete system.

6.15 Case study 7: Photovoltaic/thermal (PV/T) and geothermal-based multigeneration system

In this case study, an integrated system using solar and geothermal energy types, for hydrogen generation, electricity production, cooling and heating, is proposed and analyzed. This case study is adapted from Bicer and Dincer [31, 32]. This integrated renewable energy

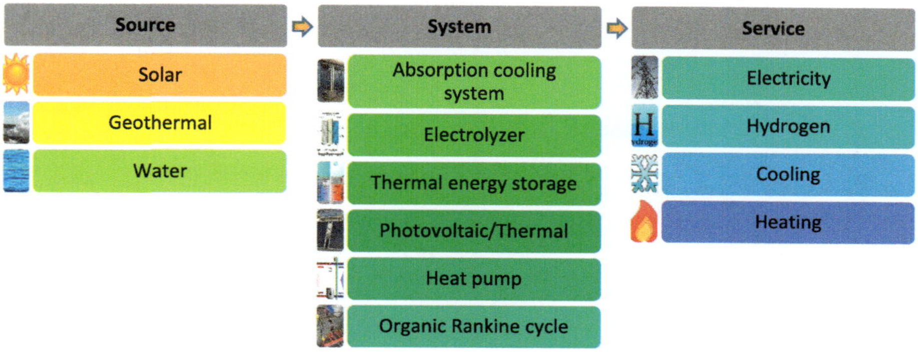

FIG. 6.74 3S approach for case study 7.

system comprises solar PV/T panels for heating, water heating and hydrogen production purposes and geothermal energy for electricity, cooling and hydrogen generation as well. The sources, systems and services for this integrated system are illustrated in Fig. 6.74.

6.15.1 System description

The designed integrated system consists of an organic Rankine cycle, a heat pump, an absorption cooling unit, a thermal energy storage and a hydrogen generation unit according to the system layout as shown in Fig. 6.75. Since geothermal energy is quite location specific, it is better suited for consumers located in a geothermal area. By integration of solar and geothermal subsystems, hourly variations of the solar irradiation on the solar collectors can be overcome. In this way, a more reliable multigeneration system can be achieved for satisfying the requirements of the community. The working fluid of the organic Rankine cycle is isobutene, and it is connected to a medium-high temperature geothermal well. The inlet temperature of Evaporator 1 is about 200 °C and the outlet temperature is about 150 °C, which is then used in Heat Exchanger 1 for preheating the electrolysis water. The inlet turbine temperature is about 145 °C and the steam exits the turbine at about 100 °C transferring its heat to the generator of the absorption cooling unit. The generated electricity powers the electrolyzer for hydrogen production. The condensers of the system are cooled by cold groundwater, which is delivered by the pump. The temperature of the groundwater is about 10 °C. For the PV based-hydrogen generation unit, the groundwater is heated through a duct system positioned under the PV/T panels. In this way, the electrolysis efficiency can be enhanced. The obtained cooling effect from Evaporator 2 of the absorption cooling system is utilized to cool the residential buildings. For heating purposes of the residential buildings, a heat pump (working fluid R134a) is used. The energy supply of the heat pump is stored heat in the thermal energy storage unit with a capacity of 15 kW at a temperature of 60 °C. The water for electrolysis is filtered and stored in the water tank. The obtained oxygen from electrolysis is also stored in an oxygen tank for later use. R134a exits Evaporator 3 with a temperature of about 9.5 °C. After the compression, it enters Condenser 3 at 75 °C and exits at 35 °C. The heat output of Condenser 3 is used for residential heating. The solar thermal system can reach up to 85 °C and the hot water is stored in the thermal energy storage unit to enable continuous operation of the heat pump.

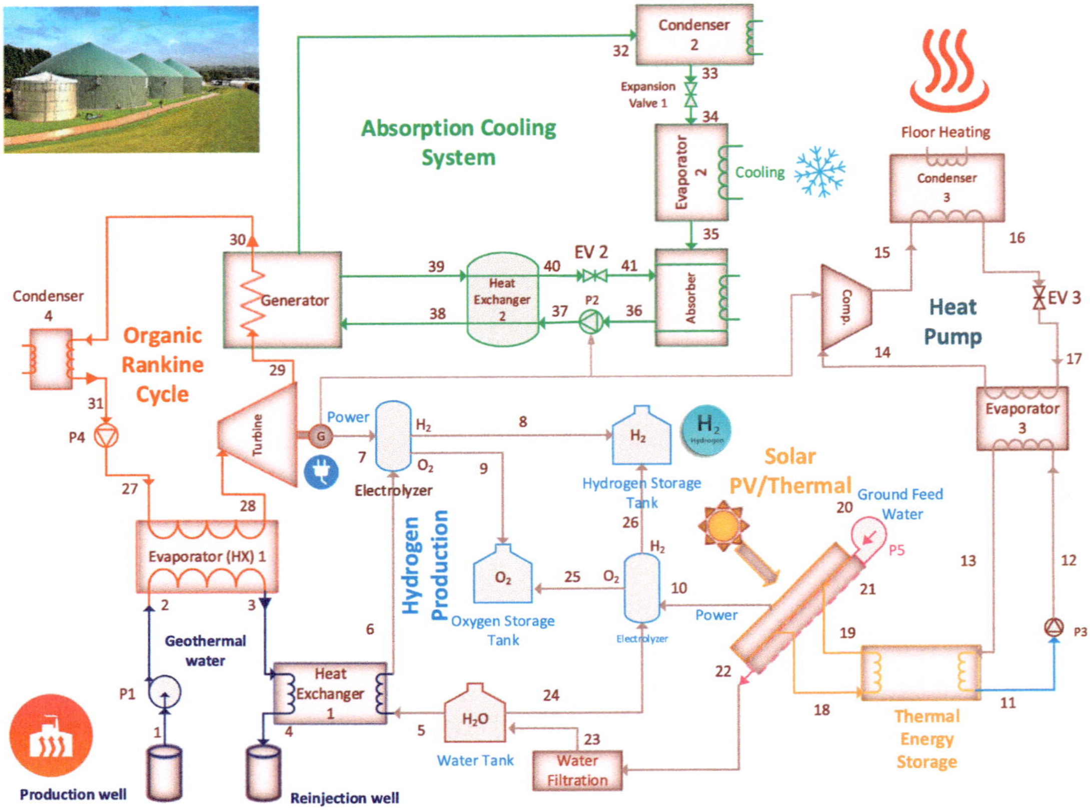

FIG. 6.75 The system layout of the integrated geothermal and solar energy system for multigeneration in case study 7. *(Modified from Bicer Y, Dincer I. Development of a new solar and geothermal based combined system for hydrogen production. Sol Energy 2016;127:269–84. doi:10.1016/j.solener.2016.01.031; Bicer Y, Dincer I. Analysis and performance evaluation of a renewable energy based multigeneration system. Energy 2016;94. https://doi.org/10.1016/j.energy.2015.10.142).*

As the system is suited to a residential area, heating, cooling and electricity are constantly necessary. By utilizing the geothermal energy source and thermal energy storage unit, uninterrupted power generation, heating, cooling and hot water can be provided. The generated electricity from the PV/T modules is utilized to power some auxiliary pumps.

6.15.2 Assumptions

In order to perform thermodynamic analysis of this multigeneration system, there is a need to make conceptually correct assumptions. Here, the main assumptions undertaken in this case study are listed as follows:

- The expansion valves, compressors, pumps and turbine are adiabatic.
- The air, hydrogen and oxygen are treated as ideal gases.

- The ambient has a temperature of 25 °C and a pressure of 100 kPa.
- The changes in kinetic and potential energy and exergy terms are ignored.
- The processes taking place are of steady state and steady flow.
- There is no chemical reaction taking place between the refrigerant and absorbent in the absorption cooling unit. Thus, chemical exergy is ignored and only physical exergy is taken into account in this unit.
- The working fluid of the organic Rankine cycle and the heat pump are isobutene and R134a, respectively.
- The working fluid of the absorption unit is a LiBr-water mixture.
- The solar PV/T back surface temperature is taken as 75 °C.
- The chemical reactants and products are at the reaction temperature and a pressure of 1 atm.

6.15.3 Analysis

The organic Rankine cycle exergy efficiency is the ratio of net work generated by turbine over total exergy input:

$$\eta_{ex,ORC} = \frac{\dot{W}_{out,T} - \dot{W}_{in,P4}}{(\dot{m}_2 ex_2 - \dot{m}_3 ex_3)} \tag{6.35}$$

The energetic COP of the absorption cooling unit are written as

$$COP_{en,AC} = \frac{\dot{Q}_{eva}}{(\dot{m}_{29} h_{29} - \dot{m}_{30} h_{30})} \tag{6.36}$$

where $\dot{Q}_{eva}$ is heat absorbed from the space by Evaporator 2.
The COP reversible can be found as follows:

$$COP_{reversible} = \left(\frac{T_{32} - T_0}{T_{32} + 273}\right)\left(\frac{10 + 273}{T_0 - 10}\right) \tag{6.37}$$

The exergetic COP of the absorption cooling unit can be written as the ratio of energetic COP over reversible COP:

$$COP_{ex,AC} = \frac{COP_{en,AC}}{COP_{reversible}} \tag{6.38}$$

In the heat pump system, the useful output is the heat output from Condenser 3. Thus, both energetic and exergetic COPs of the heat pump unit are written as follows:

$$COP_{en,HP} = \frac{\dot{Q}_{out,Cond3}}{\dot{W}_{in,C}} \tag{6.39}$$

$$COP_{ex,HP} = \frac{\dot{Q}_{out,Cond3}\left(1 - \frac{T_0}{T_s}\right)}{\dot{W}_{in,C}} \tag{6.40}$$

The overall system energy and exergy efficiencies are defined by considering all useful outputs and total inputs. Here, 50% of the electrical work output of ORC turbine is used in the geothermal-based electrolyzer. Similarly, 85% of the electricity from PV/T is used in the PV-based electrolyzer. Therefore, the useful work outputs are multiplied by 0.5 and 0.15, for turbine and PV, respectively.

$$\eta_{en,ov,sys}=\frac{\dot{W}_{out,T}\,0.5+\dot{W}_{PV}\,0.15-\sum\dot{W}_{Pumps}-\dot{W}_{in,C}+\dot{m}_{26}h_{26}+\dot{m}_{25}h_{25}+\dot{m}_{9}h_{9}+\dot{m}_{8}h_{8}+\dot{Q}_{out,Cond3}+\dot{Q}_{AC}}{\dot{m}_{1}h_{1}+\dot{Q}_{Solar}} \tag{6.41}$$

$$\eta_{ex,ov,sys}=\frac{\dot{W}_{out,T}\,0.5+\dot{W}_{PV}\,0.15-\sum\dot{W}_{Pumps}-\dot{W}_{in,C}+\dot{m}_{26}ex_{26}+\dot{m}_{25}ex_{25}+\dot{m}_{9}ex_{9}+\dot{m}_{8}ex_{8}+\dot{Ex}^{Q_{Cond3}}+\dot{Ex}^{Q_{AC}}}{\dot{m}_{1}ex_{1}+\dot{Ex}^{Q_{Solar}}} \tag{6.42}$$

6.15.4 Results and discussion

Solar PV/T panels can produce both hot water and electricity. A selected single PV/T panel has a nominal power of 235 W. The exergy destruction rates as well as energy and exergy efficiencies of most of the components are listed in Table 6.15. The largest exergy destruction rates are observed in Heat Exchanger 1, the electrolysis process (from geothermal source) and Evaporator 1. Because Heat Exchanger 1 and Evaporator 1 include high amounts

TABLE 6.15 The main results obtained for the multigeneration system components

Component	Exergy efficiency (%)	Exergy destruction rate (kW)	Work or heat transfer rate (kW)
Generator	93	9.943	–
Electrolysis Geothermal	32	961.4	1833
Compressor	55	0.922	2.042
Condenser 2	81	2.962	450.2
Condenser 3	12	0.368	8.357
Condenser 4	70	81.93	7806
Evaporator 1	58	1426	–
Evaporator 2	95	2.131	–
Evaporator 3	28	0.1955	–
Thermal Energy Storage	18	0.9767	1.062
Electrolysis PV	77	6.89	8.594
Turbine	90	193.1	1834
Solar PV/T	13	6.382	15.44 / 7.061

of geothermal water mass flow rate. The absorber has an exergy destruction rate of about 113.3 kW which is followed by Pump 4 with 105.4 kW. The exergy efficiencies of Generator and Evaporator 2 are 93% and 95%, respectively. In the organic Rankine cycle, the turbine produces about 1.8 MW electricity with an exergy efficiency of 90%. The turbine exergy destruction rate is about 193 kW.

Around 50% of the generated electricity is used in the electrolysis process for hydrogen production. In the geothermal-based electrolysis unit, about 18.1 kg/h hydrogen is stored in the tank, while in PV/T, 0.1185 kg/h hydrogen can be produced and stored, as the quantity of generated electricity is only around 7 kW.

The overall energy and exergy efficiencies of the renewable energy-based multigeneration system are calculated as 5.5% and 20.4%, respectively, under given conditions at 25 °C and 100 kPa. The low efficiencies are attributed to the low conversion efficiency of geothermal and solar power plants. The energetic COP of the absorption cooling unit is 0.58. The energetic COP of the heat pump system is 4.1. Furthermore, the energy and exergy efficiencies are calculated to be 16.9% and 50.3%, respectively, for the organic Rankine cycle.

6.15.4.1 *Effects of the reference temperature*

The reference temperature is important for the performance of multigeneration systems. The effects of differences in ambient temperature on some of the subsystems and overall efficiency of the multigeneration system are revealed in Fig. 6.76 and Fig. 6.77. When ambient temperature rises, the energetic COP of both the absorption cooling system and the heat pump unit are constant; however, the exergetic COP of the absorption cooling system rises from 0.1 to 0.75.

The exergy efficiency of the organic Rankine cycle rises to 60% when the ambient temperature increases from 15 °C to 50 °C, as shown in Fig. 6.77. The overall multigeneration system energy efficiency is constant with increasing ambient temperature; however, the exergy efficiency of the system rises to 27% at 50 °C.

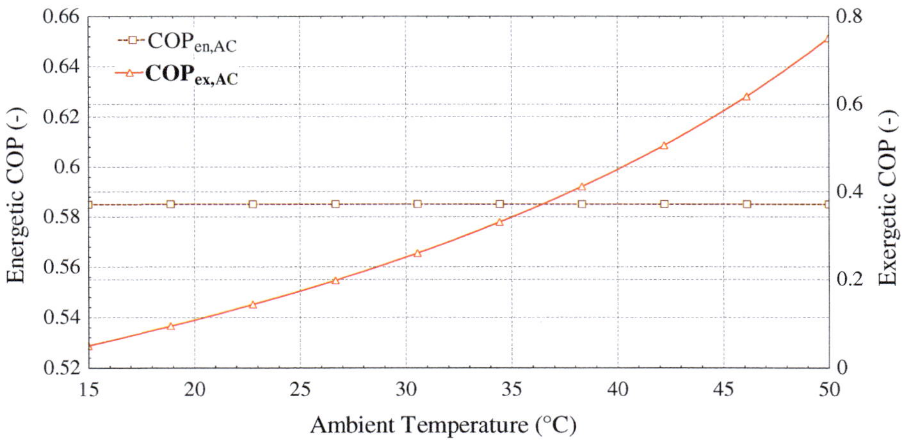

FIG. 6.76 The effects of ambient temperature on the COPs of absorption cooling system.

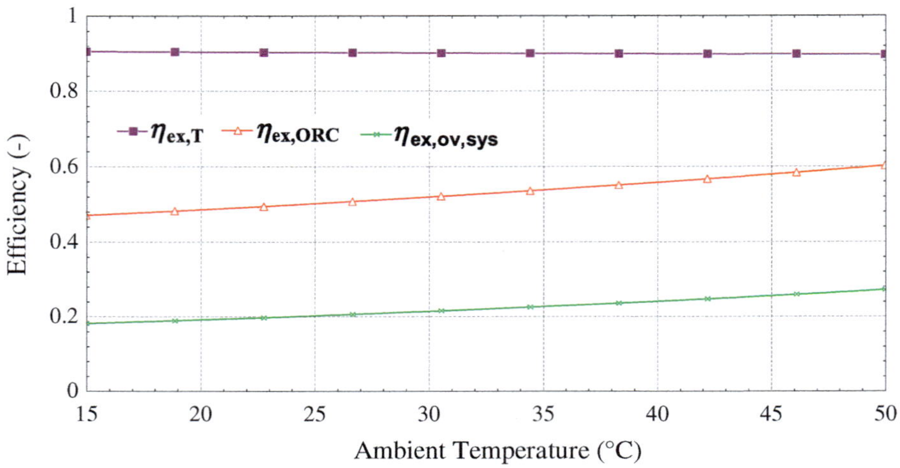

FIG. 6.77 The effects of ambient temperature on the turbine, organic Rankine cycle and overall system exergy efficiencies.

6.15.4.2 Effects of mass flow rates

The mass flow rates of the solar and geothermal systems are significant for the overall system efficiency. As depicted in Fig. 6.78, an escalation in the geothermal water mass flow rate reduces both the organic Rankine cycle and overall system efficiency. An increase in water mass flow rate in the solar energy system drops both energy and exergy efficiencies of the overall system as shown in Fig. 6.79. Since water mass flow rate in the PV/T unit is one of the two inputs of the overall system, the overall exergy efficiency slightly drops. Furthermore, increasing the isobutane mass flow rate in the organic Rankine cycle decreases the generator

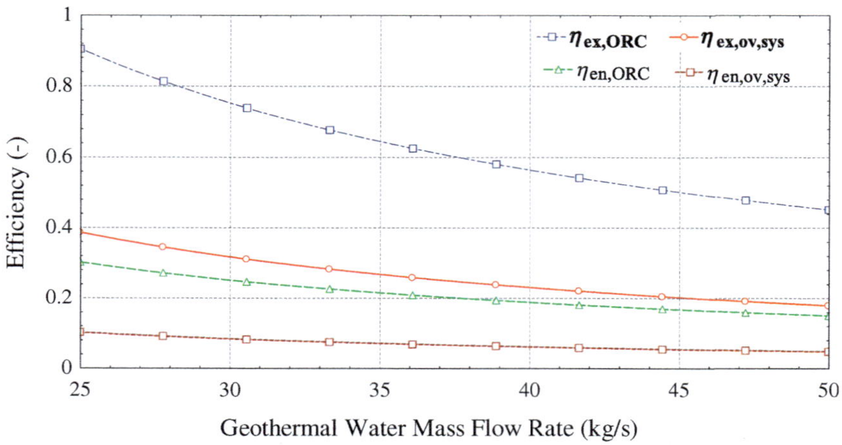

FIG. 6.78 The effects of geothermal water mass flow rate on the organic Rankine cycle and overall efficiencies.

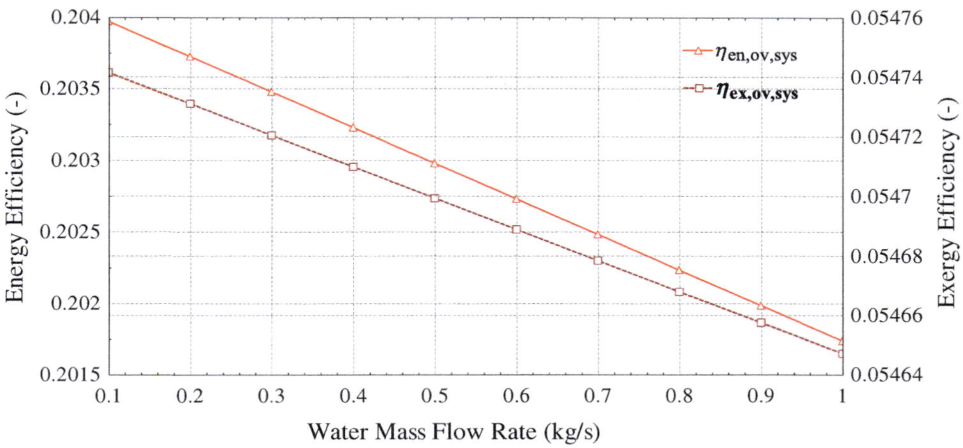

FIG. 6.79 The effects of water mass flow rate in the solar PV/T system on the overall efficiencies.

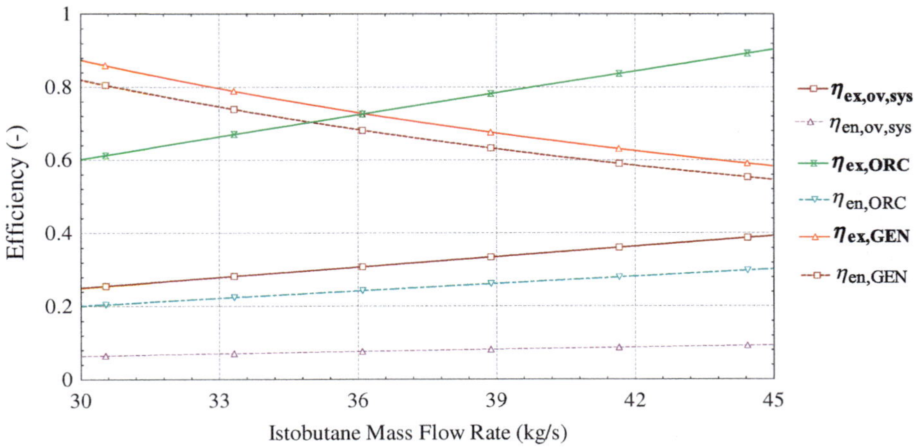

FIG. 6.80 The effects of working fluid isobutane mass flow rate on the generator, organic Rankine cycle and overall efficiencies.

efficiencies in Fig. 6.80. On the other hand, increasing isobutane mass flow rate in the organic Rankine cycle increases both the organic Rankine cycle and the overall efficiencies.

6.15.4.3 Effects of ambient temperature on exergy destruction rate and exergy efficiency

When the turbine exergy destruction rate increases from 185 kW to 210 kW by an increment of 35 °C ambient temperature as shown in Fig. 6.81, the exergy efficiency of the turbine slightly decreases. The solar PV/T system exergy efficiency drops to 12.5% from 13.5% with increasing temperature. Another high exergy destruction rate is observed in Evaporator 1.

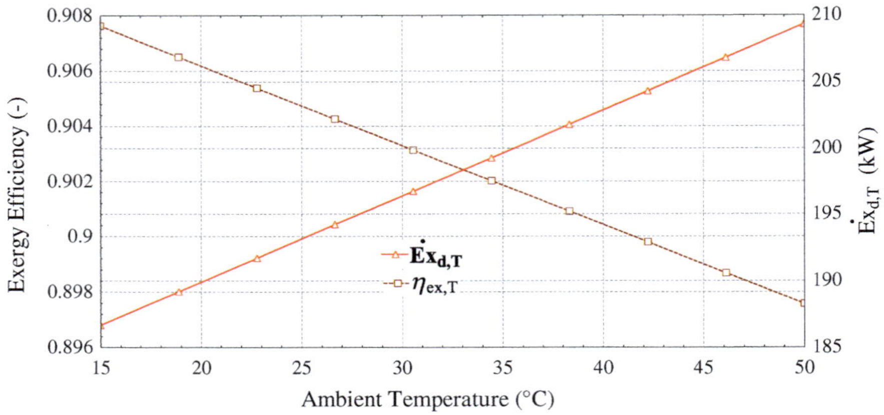

FIG. 6.81 The effects of ambient temperature on the turbine exergy destruction rate and exergy efficiency.

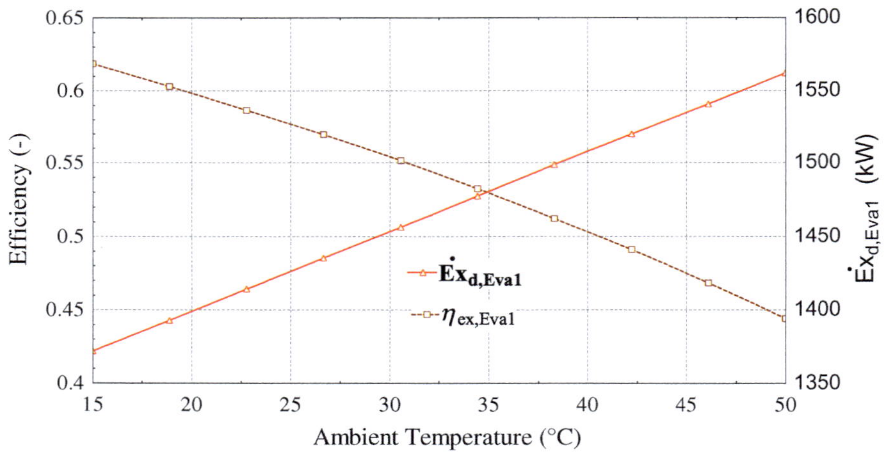

FIG. 6.82 The effects of ambient temperature on Evaporator 1 exergy destruction rate and exergy efficiency.

As depicted in Fig. 6.82, the exergy efficiency of Evaporator 1 drops to about 44% with increasing ambient temperature and exergy destruction rate.

6.15.4.4 Effects of working fluid

The type of working fluid used in organic Rankine cycle efficiency has a significant influence on system efficiencies. A parametric study is carried out by describing three different type of working fluids, namely isobutane, R123 and R245fa. Fig. 6.83 depicts the overall efficiencies depending on the ambient temperature in which isobutane has a steeper increase. Furthermore, as shown in Figs. 6.84 and 6.85, the highest exergy efficiency in the organic Rankine cycle is perceived when isobutane is utilized in the organic Rankine cycle. At 25 °C ambient temperature, the exergy efficiency is about 50% although it is around 22% and 13%

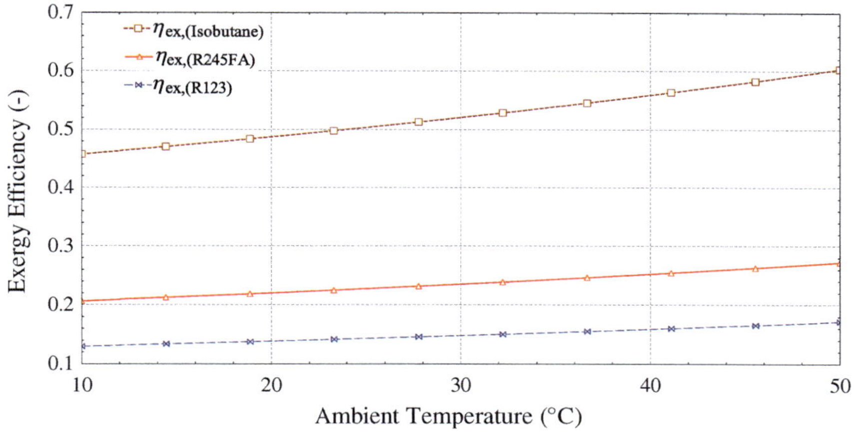

FIG. 6.83 The effects of ambient temperature on the exergy efficiency of the organic Rankine cycle with different working fluids.

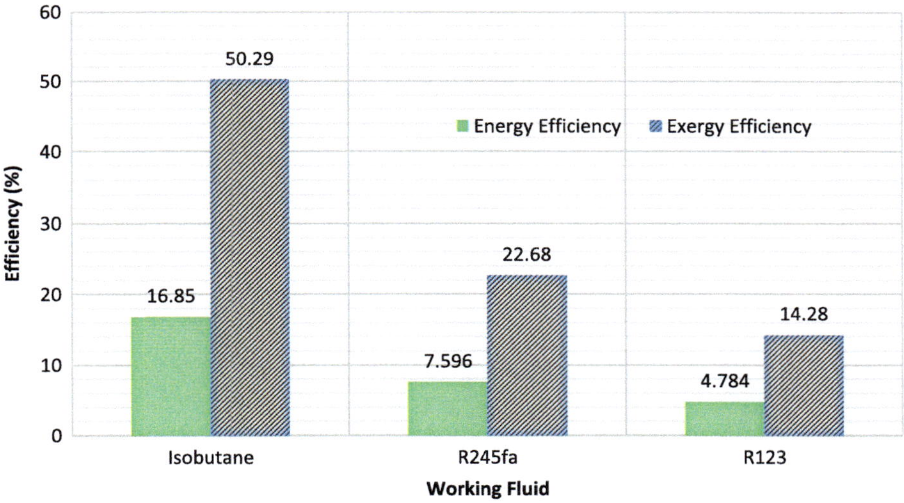

FIG. 6.84 Comparison of energy and exergy efficiency values of organic Rankine cycle under different types of working fluids at 25 °C and 100 kPa ambient conditions.

for R254fa and R123 respectively. This implies that the system can perform better when isobutane is the working fluid under given conditions.

When the geothermal water temperature upsurges to 240 °C, all working fluids perform better; however, the highest upsurge is observed in isobutene working fluid corresponding to about 60% organic Rankine cycle exergy efficiency at 240 °C. A comparison of efficiency values for three types of working fluids is depicted in Figs. 6.84 and 6.85 under the same values and ambient conditions.

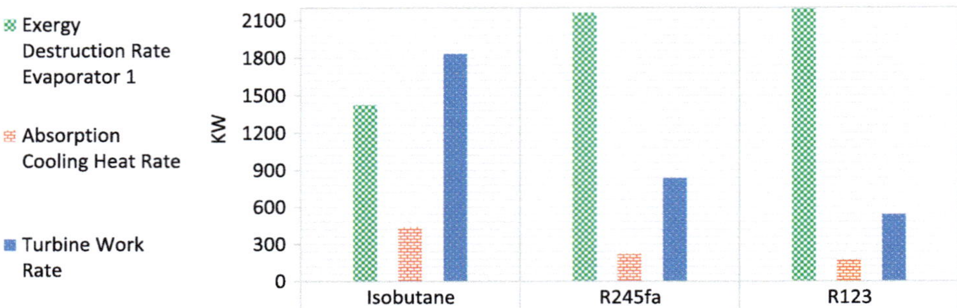

	Isobutane	R245fa	R123
Exergy Destruction Rate Evaporator 1	1426	2161	2192
Absorption Cooling Heat Rate	440.7	223.9	176.3
Turbine Work Rate	1834	837.1	541.4

FIG. 6.85 The effects of working fluid type on work/heat rates and exergy destruction rates of some components.

6.15.4.5 Effects of geothermal water temperature

Geothermal water temperature directly affects the total turbine work output and absorption cooling system. The produced work upsurges to 2.1 MW and the obtained cooling effect rises to 700 kW when geothermal water temperature is about 240 °C. The overall system efficiencies are dependent on geothermal water temperature as shown in Figs. 6.86 and 6.87. In the case of 220 °C geothermal water temperature, the energy efficiency of the multigeneration system can rise to 15% and the exergy efficiency rises to 65%. Therefore, the system can function more effectively on a high-temperature geothermal zone. The exergy results show that the organic Rankine cycle evaporator and Heat Exchanger 1 are the two chief components causing irreversibility.

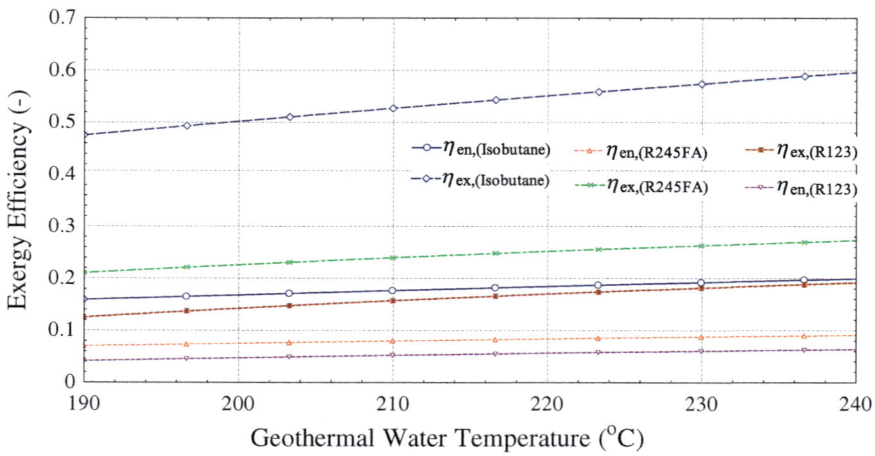

FIG. 6.86 The effects of geothermal water temperature on the energy and exergy efficiencies of the organic Rankine cycle with different working fluid types.

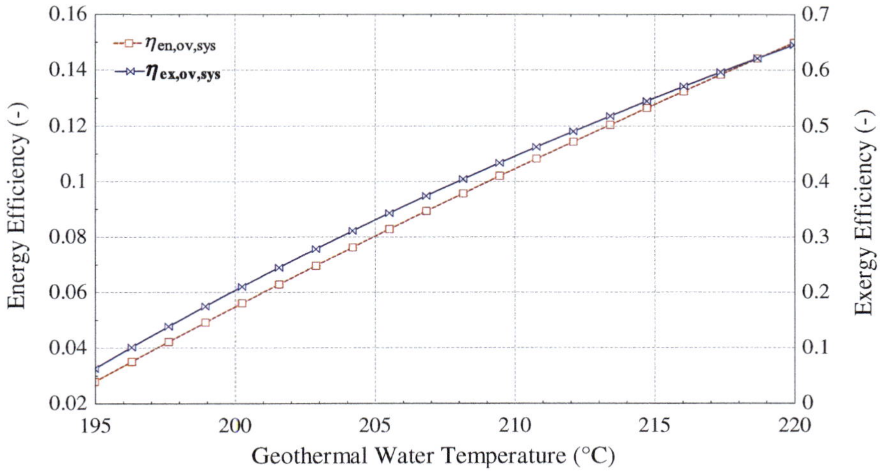

FIG. 6.87 The effects of geothermal water temperature on the overall energy and exergy efficiencies of the complete system.

6.15.5 Closing remarks

This case study presents a renewable energy-based multigeneration system for heating, cooling, hot water, power and hydrogen generation. The power generation is performed through a geothermal-based organic Rankine cycle as well as solar PV/T modules. By a duct system under the solar PV/T panels, the necessary water for the electrolysis process is preheated to raise the temperature of the water. The exergy efficiency of the overall system upsurges to 27% at the ambient temperature of 50 °C. Furthermore, the geothermal water temperature directly increases the overall system efficiency up to 60% when the geothermal water temperature is about 220 °C. An upsurge of 25 °C in geothermal water temperature from 195 °C to 225 °C enhances the energy efficiency of the multigeneration system up to 15%. The maximum energy and exergy efficiencies are found to be 15% and 65% respectively if the geothermal water temperature is about 220 °C. Under the given circumstances, isobutane is the best alternative as working fluid of for the organic Rankine cycle, yielding lower exergy destructions and increasing the turbine work output.

6.16 Case study 8: Concentrating photovoltaic/thermal (CPV/T) system using nucleate boiling heat transfer (NBHT) thermal management for multigeneration

In this case study, a multigeneration system utilizing concentrated photovoltaic/thermal (CPV/T) and hydrogen storage is considered and thermodynamically analyzed to meet electricity, hot and cold water, heating ventilation and cooling (HVAC) necessities of a residential community with the uninterrupted process as shown in Fig. 6.88. This case study is adapted from Khan et al. [33].

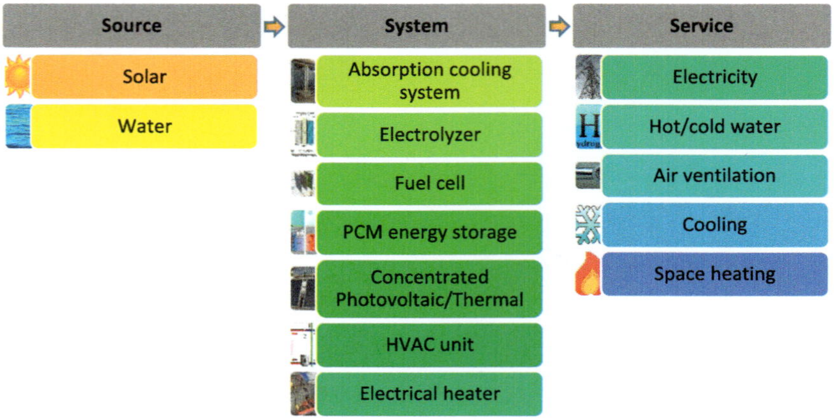

FIG. 6.88 3S approach for case study 8.

A portion of the produced power from CPV is utilized in the electrolyzer for hydrogen and oxygen production. The formed gases are stored and then utilized by proton exchange membrane (PEM) fuel cell to satisfy the system's electrical energy need during the nighttime and unfavorable circumstances in the daytime. The multigeneration system utilizes a specific type of thermal management, nucleate pool boiling heat transfer (NBHT), for CPV cooling. The thermal energy obtained from CPV/T is utilized in heating, hot water and cooling demands of the residential areas by using a lithium bromide absorption chiller. A humidity harvesting system is coupled at the exit of the absorption chiller, to transform humid air into water and the ventilation air demand of the building.

6.16.1 System description

The system is configured to capably extract the heat and power from the CPV system. The schematic diagram of the CPV/T multigeneration system is depicted in Fig. 6.89.

The removed thermal energy is then utilized for the multigeneration system to raise the overall efficiency of the system. Water and sunlight are the main inputs to the system while power, space heating and cooling, ventilation, and hot and cold water are the main outputs. The input parameters are listed in Table 6.16.

Initially, solar irradiation is concentrated on the photovoltaics modules by the CPV system with Fresnel lenses. After considering the losses at the lens and module levels, the main portion of the received energy to the PV modules is transformed into electrical and thermal energy. For the CPV, a full glass lens of 96% efficiency [37] and an inverted metamorphic multijunction (IMM) solar cell having an efficiency of 40.8% are considered [36]. The electrical output loss with the upsurge in temperature for the IMM solar cell is studied using a thermal coefficient of −0.052%/K, as calculated and reported by Steiner et al. [36]. A total of 10% of the losses have been considered for the CPV system, considering 6% for thermal losses and 4% for the glass lens [37]. The thermal energy produced by the CPV modules is then effectively extracted by the connected NBHT system.

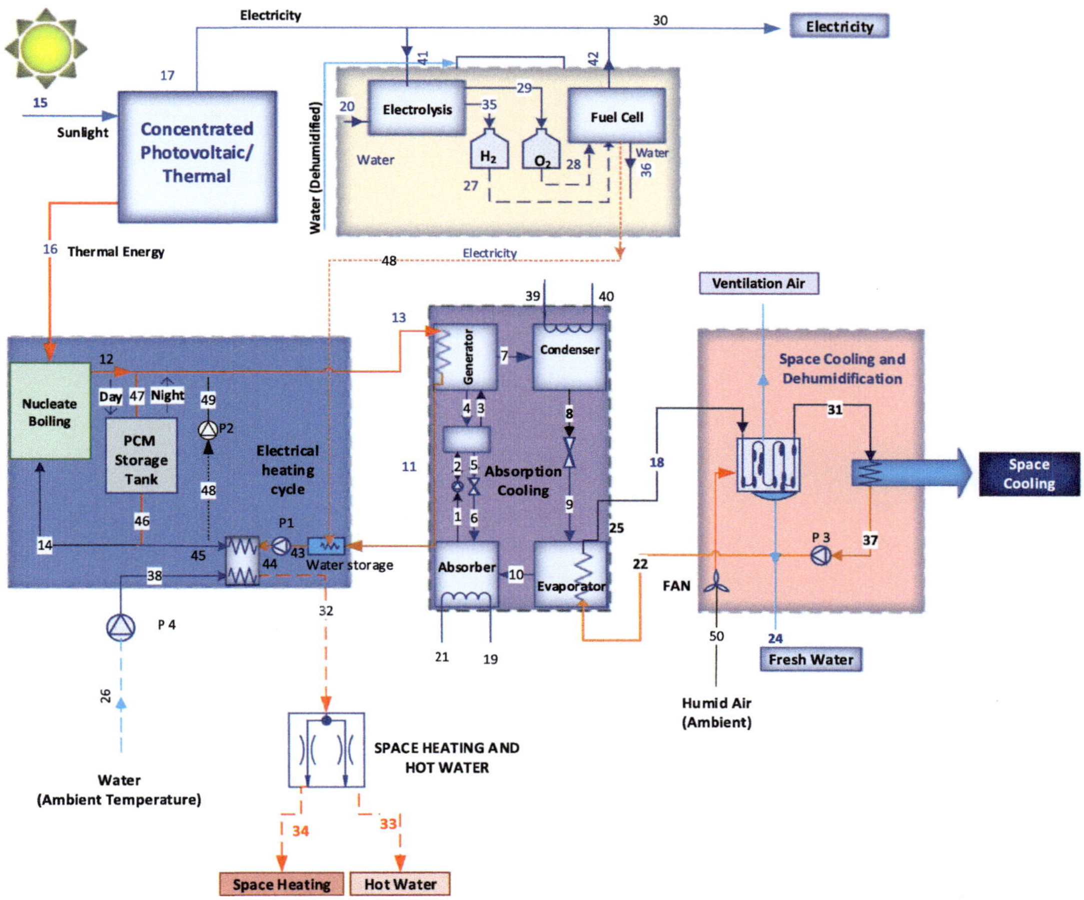

FIG. 6.89 The schematic diagram of the proposed integrated solar-based renewable energy system with thermal management. *(Modified from Khan SA, Bicer Y, Koç M. Design and analysis of a multigeneration system with concentrating photovoltaic thermal (CPV/T) and hydrogen storage. Int J Hydrogen Energy 2018. https://doi.org/10.1016/j.ijhydene.2018.12.047.*

In the second subsystem, the NBHT is employed as a thermal management method for the CPV cell. The performance of the NBHT method depends on the operational situations and chosen working fluid. The system is first analyzed for standard conditions of 101 kPa of pressure and the selected working fluid is water. At standard operating conditions, steam is generated at 373 K, which yields heat removal from the CPV modules. The operating circumstances are then changed to observe the effects on performance. Equivalent day and night times are considered as the base case, which can be altered based on the months and seasons. Thermal energy storage is also utilized during the night cycle. An electrical heating unit as well as a small water storage tank are also included to drive the subsystem via power from the fuel cell subsystem in the absence of solar irradiation due to weather circumstances.

TABLE 6.16 The input parameters for the designed system

Parameter	Explanation/value	Reference
METEOROLOGICAL DATA		
Ambient temperature	301 K	[34]
Ambient pressure	101 kPa	[34]
Relative humidity	0.5	[34]
Direct Normal Irradiance (I_{DNI})	0.850 kW/m^2	[35]
INSTALLED CPV SYSTEM		
Installed area	220 m^2 (Selected for the designed system)	–
Installed capacity	175.8 kW	–
Cell type	NREL, inverted metamorphic multijunction (IMM) solar cell	[36]
Cell efficiency ($\eta_{electrical,\ std}$)	40.8%	[36]
Thermal coefficient (α)	−0.052%/K	[36]
Optical efficiency	96% (Fresnel glass lens)	[37]
NBHT THERMAL MANAGEMENT SYSTEM		
Working fluid	Water	–
Saturation temperature (T_{NBHT})	373 K (Saturation temperature of water at the operating pressure of 101.3 kPa)	[38]
Latent heat	2257 kJ/kg (Latent heat of water at 101.3 kPa)	[38]
ABSORPTION COOLING SYSTEM		
Absorption chiller	LiBr-water chiller	[39]
Internal piping	Copper	–
Heat exchangers	Copper	–
PCM STORAGE SYSTEM		
PCM type	X95: Plus ICE PCM Solid-Solid (X) Range	[40]
Melting temperature	368 K	[40]
Latent heat of PCM	140 kJ/kg	[40]

Source: *Khan SA, Bicer Y, Koç M. Design and analysis of a multigeneration system with concentrating photovoltaic thermal (CPV/T) and hydrogen storage. Int J Hydrogen Energy 2018. https://doi.org/10.1016/j.ijhydene.2018.12.047*

In the third subsystem, a single effect absorption cooling unit is utilized with a lithium-bromide water mixture as a working fluid for the purpose of using the available heat. The system receives heat from the NBHT subsystem and generates cooling as the useful output. For the base scenario, water acts as the working fluid at 1 atm and the input stream is steam at 373.15 K. The stream exits the system as water at 373 K by transferring latent heat to the generator of the absorption cooling unit. Since the value of latent heat is superior compared to the sensible heat, there is not much effect from decreasing the output stream temperature to lower values. Nevertheless, the complete system is also analyzed by changing the NBHT temperature in order to observe the optimum operational temperature for this stream.

The output of the absorption cooling unit is sent to dehumidification, ventilation, and space cooling units. In the dehumidification unit, hot and humid air is distributed from outside using an electrical fan unit. The cold stream from the absorption cooling unit arrives at the dehumidifier at 278 K and exits at 288 K by taking heat from the humid air. This concept was adapted from Mohamed et al. [41] and Talaat et al. [42]. Because of the decrease in temperature, the air dehumidifies and yields condensation and cold air. The cold air is likely to be used as cooling and ventilation air to residential houses. The residual energy is utilized for providing a cooling effect into the buildings, using the circulation air from the building. In the space cooling unit, the temperature of the cold stream increases from 288 K to 296 K. The working fluid concludes the cycle and is sent back to the evaporator of the absorption cooling unit by the pump.

In the winter season, the heating and hot water demands of the residential houses are designed to be satisfied by heat generated through the CPV/T unit. The additional electricity from the CPV unit is utilized for production of hydrogen and oxygen through the electrolysis unit. For the designed conditions, about 40% of the power is utilized in the electrolysis unit during the daytime. The generated oxygen and hydrogen are stored and later on utilized to produce electricity via proton exchange membrane fuel cell (PEMFC). The regenerated power runs the system for the electrical energy requirements at nighttime and for electrical and thermal cycles in poor solar radiation circumstances such as nighttime and cloudy weather conditions.

6.16.2 Assumptions

In order to perform thermodynamic analysis of this multigeneration system, there is a need to make conceptually correct assumptions. Here, the main assumptions undertaken in this case study are listed as follows:

- The system operates at steady state.
- The changes in kinetic and potential terms are ignored.
- The heat as well as pressure losses through the pipes are insignificant and therefore the power needed for circulation pumps 1–4 is ignored.
- The electrolyzer and fuel cell work at 70% energetic efficiency [43, 44].
- The working fluid of the absorption cooling unit is LiBr-water.
- The saturated liquid situations are taken at state points 1, 4 and 8 and saturated vapor condition at state point 10 in Fig. 6.89.
- Any chemical reaction between absorbent and refrigerator is ignored along with any changes in the chemical exergy contents [45].

6.16.3 Analysis

The maximum ability of heat transfer of NBHT is its critical heat flux (CHF), calculated as follows [46, 47]:

$$q''_{CHF_{water}} = C_{crit} h_{fg} \rho_{v_{sat}} \left(\sigma g \frac{\left(\rho_{i_{sat}} - \rho_{v_{sat}}\right)}{\rho_{v_{sat}}^{2}} \right)^{0.25} \tag{6.43}$$

where q''_{CHF} (kW/m^2) is the critical heat flux (CHF) for water; C_{crit} for large surfaces is approximated as 0.16; the coefficient of the critical heat flux depends upon the working conditions of the phenomena, taken as 0.16; h_{fg} is enthalpy of vaporization of the working liquid; $\rho_{v_{sat}}$ is saturated vapor density; σ is surface tension, g is gravitational constant and $\rho_{i_{sat}}$ is saturated liquid density. This equation is applied to approximate the $q''_{CHF_{water}}$ value for the surfaces where the characteristic length of the surface is large relative to the bubble diameter.

The higher the CHF value, the higher is the ability to transfer the heat and thus the higher is the possibility of concentrating the light. Therefore, for each value of CHF, there is a conforming maximum concentration ratio (C.R$_{max}$), representing the maximum ratio of solar light that can be applied to the system. The C.R$_{max}$ for the considered system is calculated as follows:

$$\begin{aligned} C.R_{max} &= \frac{\text{Maximum ability of the heat transfer rate per unit area of the designed thermal management system}}{\text{Heat transfer rate per unit area generated in CPV cell}} \\ &= \frac{q''_{CHF_{water}}}{I_{DNI,t}\left(1 - \eta_{loss} - \eta_{electrical}\right)} \end{aligned} \tag{6.44}$$

where $I_{DNI,\,t}$ signifies the received solar light (kW/m^2), η_{loss} the proportion of received energy lost to the environment, and $\eta_{electrical}$ is the efficiency of received solar light transformed to electricity.

For each sub-system, the energy efficiency and exergy efficiency are written as per the following equations:

CPV:

$$\eta_{en,elect,CPV} = \frac{\dot{W}_{elec}}{\dot{Q}_{DNI,t}} \tag{6.45}$$

$$\eta_{ex,elect,CPV} = \frac{\dot{W}_{elec}}{\dot{Q}_{DNI,t}\left(1 - \left(\frac{T_o}{T_{sun}}\right)\right)} \tag{6.46}$$

where $\dot{W}_{elec}$ is the electrical energy production from the CPV unit, T_o and T_{sun} are the environment temperature and sun temperature, respectively.

The overall energy and exergy efficiencies of a CPV system taking into account electricity and heat can be written as follows:

$$\eta_{en,overall,CPV} = \frac{\dot{W}_{elec} + \dot{Q}_{CPV,thermal}}{\dot{Q}_{DNI,t}} \tag{6.47}$$

$$\eta_{ex,overall,CPV} = \frac{\dot{W}_{elec} + \dot{Q}_{CPV,thermal}\left(1 - \left(\frac{T_o}{T_{NBHT}}\right)\right)}{\dot{Q}_{DNI,t}\left(1 - \left(\frac{T_o}{T_{sun}}\right)\right)} \quad (6.48)$$

NBHT system:

$$\eta_{en,NBHT} = \frac{\dot{m}_{12}h_{12} - \dot{m}_{14}h_{14}}{\dot{Q}_{CPV,thermal}} \quad (6.49)$$

$$\eta_{ex,NBHT} = \frac{\dot{m}_{12}ex_{12} - \dot{m}_{14}ex_{14}}{\dot{Q}_{CPV,thermal}\left(1 - \left(\frac{T_0}{T_{cell}}\right)\right)} \quad (6.50)$$

where $\dot{Q}_{CPV,thermal}$ is heat transfer rate from the CPV cell to the NBHT system.

Absorption cooling system:

$$COP_{en,ABC} = \frac{\dot{m}_{22}h_{22} - \dot{m}_{18}h_{18}}{\dot{m}_{13}h_{13} - \dot{m}_{11}h_{11}} \quad (6.51)$$

$$COP_{ex,ABC} = \frac{\dot{m}_{22}ex_{22} - \dot{m}_{18}ex_{18}}{\dot{m}_{13}ex_{13} - \dot{m}_{11}ex_{11}} \quad (6.52)$$

Here, $COP_{en,\ ABC}$ and $COP_{ex,\ ABC}$ signify the energetic and exergetic coefficient of performance for the absorption cooling system, respectively.

Dehumidification:

$$\eta_{en,\ Dehumidification} = \frac{\dot{m}_{25}h_{25} - (\dot{m}_{24}h_{24} + \dot{m}_{23}h_{23})}{\dot{m}_{18}h_{18} - \dot{m}_{31}h_{31}} \quad (6.53)$$

$$\eta_{ex,\ Dehumidification} = \frac{\dot{m}_{25}ex_{25} - (\dot{m}_{24}ex_{24} + \dot{m}_{23}ex_{23})}{\dot{m}_{18}ex_{18} - \dot{m}_{31}ex_{31}} \quad (6.54)$$

The numerators of the equations denote the output hot-humid air stream from the environment whereas the denominator signifies the input cold stream of the absorption cooling unit.

Space cooling:

$COP_{en,\ Space\ cooling}$ is the energetic COP and $COP_{ex,\ Space\ cooling}$ is the exergetic COP of the space cooling unit, $\dot{Q}_{Spacecooling}$ signifies the cooling delivered, T_{37} is the temperature at the outlet of the cold steam from the space being cooled down:

$$COP_{en_{Space\ cooling}} = \frac{\dot{Q}_{Space\ cooling}}{\dot{m}_{37}h_{37} - \dot{m}_{31}h_{31}} \quad (6.55)$$

$$COP_{ex_{Space\ cooling}} = \frac{\dot{Ex}^{Q_{Space\ cooling}}}{\dot{m}_{37}ex_{37} - \dot{m}_{31}ex_{31}} \quad (6.56)$$

Overall system:

The overall energy and exergy efficiency of the entire integrated system can be expressed as follows:

$$\eta_{en_{overall}} = \frac{\dot{Q}_{stored} + \dot{Q}_{spaceHeating} + \dot{Q}_{spaceCooling} + \dot{Q}_{dehumidification,vent} + \dot{W}_{electricity} - \dot{W}_{in,Fan}}{\dot{Q}_{DNI,t}} \tag{6.57}$$

$$\eta_{ex_{overall}} = \frac{\dot{E}x^{Q_{stored}} + \dot{E}x^{Q_{space\ heating}} + \dot{E}x^{Q_{spaceCooling}} + \dot{E}x^{Q_{dehumidification,vent}} + \dot{W}_{electricity} - \dot{W}_{in,Fan}}{\dot{Q}_{DNI,t}\left(1 - \dfrac{T_0}{T_{sun}}\right)} \tag{6.58}$$

where $\dot{Q}_{DNI,t}$ signifies the total power input in terms of solar energy for which the units are kilowatts.

6.16.4 Results and discussion

For the standard designed conditions, the overall energy and exergy efficiency values of the considered integrated system are found as 68.63% and 34.74%, respectively. The electrical efficiency of CPV is found to be 36.64%. At standard conditions, the system can deliver about 29.46 kW of cooling, out of which 16.39 kW is utilized for the dehumidification unit to generate about 10.5 L/h of dehumidified water, and ventilation air of about 12.5 kW for space cooling. The system can deliver about 0.46 kg/h of hydrogen and 3.66 kg/h of oxygen gases. The exergy destruction of the major components/subsystems and their ratio is shown in Fig. 6.90. Furthermore, Fig. 6.91 denotes the exergy efficiency of the main modules of the system and the overall system.

6.16.4.1 Effects of ambient temperature and pressure

In order to examine the behavior of the integrated system, the ambient temperature is varied from 280 K to 330 K. This range is selected to represent a broader performance at

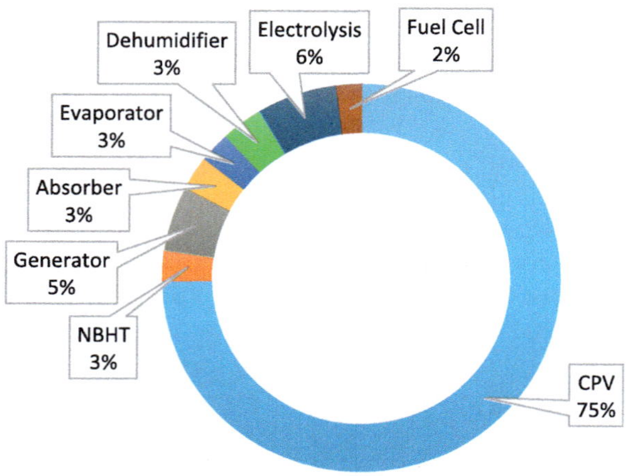

FIG. 6.90 Exergy destruction ratios of the major components within the integrated system.

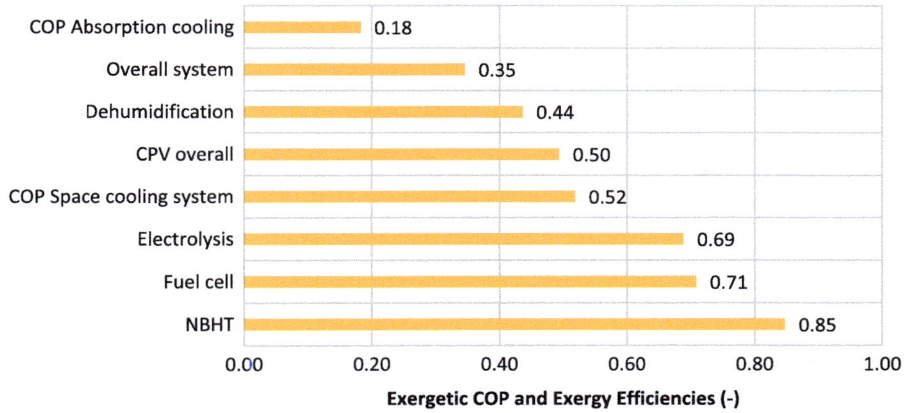

FIG. 6.91 Exergy efficiencies of the overall system and of major subsystems in the integrated system.

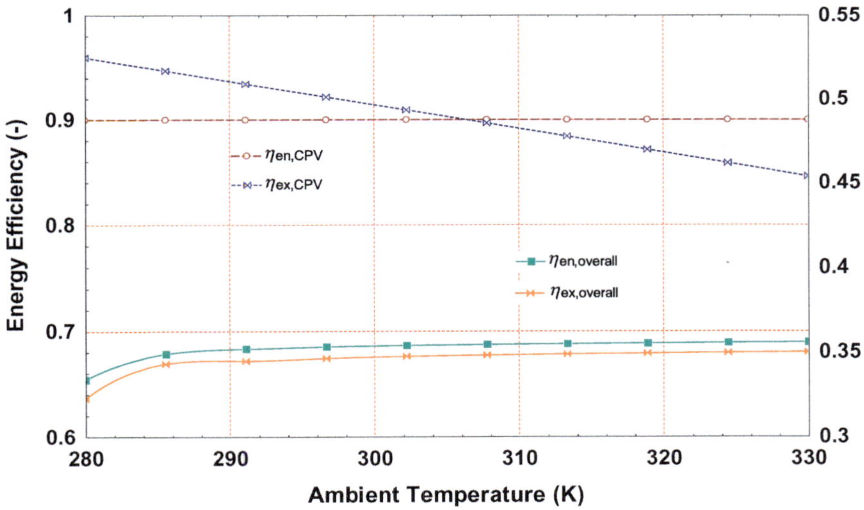

FIG. 6.92 Effects of ambient temperature on energy and exergy efficiency of CPV and overall system.

minimum and maximum temperatures of selected location, which is Doha, Qatar. The results imply a slight upsurge in overall energy and exergy efficiency of the integrated system, with an upsurge in reference temperature as shown in Fig. 6.92. Nevertheless, the exergy efficiency of the CPV unit drops from 52.5% to 45.4%, as depicted in Fig. 6.92. For the absorption cooling unit, the exergetic COP rises with an upsurge in ambient temperature, as depicted in Fig. 6.93. This situation is mostly due to larger temperature variance between condenser exit and the ambient environment. Although the energetic COP of the absorption cooling unit stays constant with an upsurge in ambient temperature.

The impacts of reference temperature on the exergy efficiency and exergy destruction rate of CPV are depicted in Fig. 6.94. The upsurge of ambient temperature has a negative impact

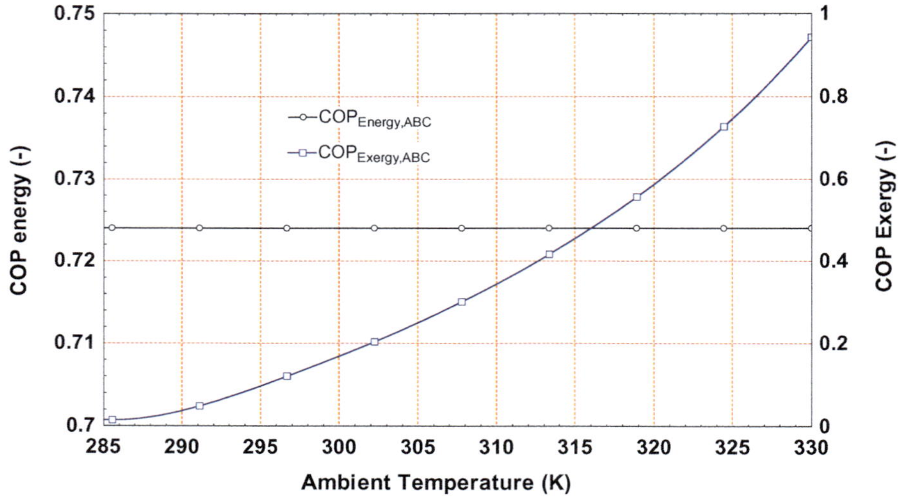

FIG. 6.93 Effects of ambient temperature on COP of absorption cooling system.

on the exergy efficiency of the CPV; this effect can be perceived from the exergy efficiency equation of CPV. The exergy destruction upsurges from 79.47 kW to 90.52 kW due to the upsurge in ambient temperature.

6.16.4.2 Effects of T_{NBHT}

While the performance of the CPV system is greatly reliant on operational temperature, a parametric study for boiling temperature, from 355 K up to 395 K, is executed to perceive

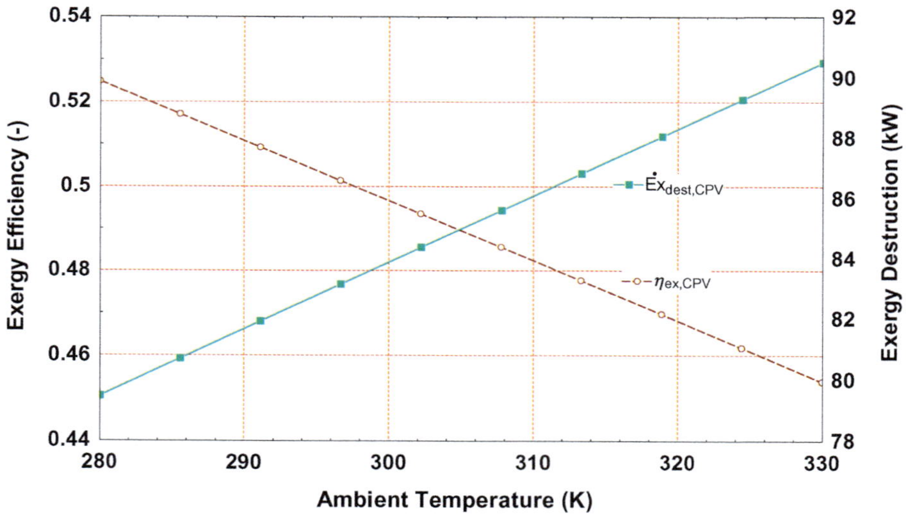

FIG. 6.94 The effects of ambient temperature on exergy efficiency and exergy destruction rate.

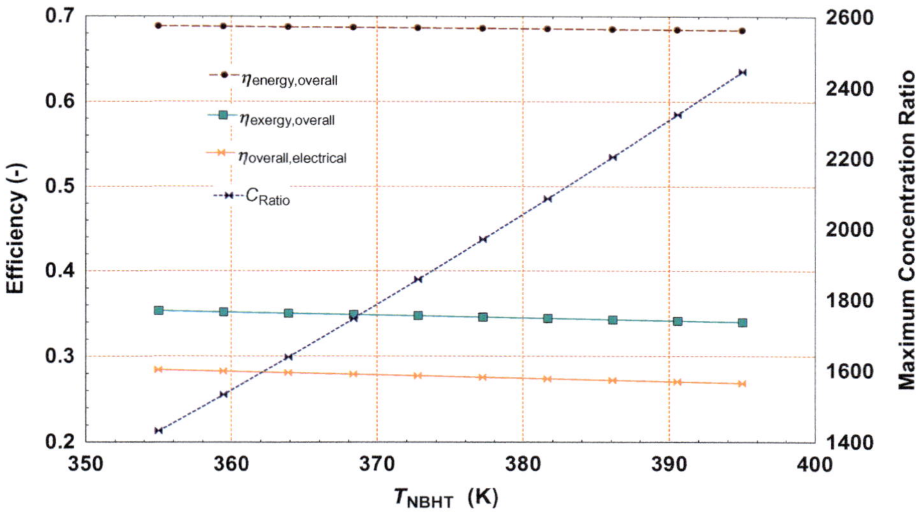

FIG. 6.95 The effects of boiling temperature of NBHT cooling technique on the overall system performance.

the efficiency of the CPV system, as depicted in Fig. 6.95. The lower limit of the parametric study is designated by taking the minimum temperature condition for the absorption cooling unit and the maximum limit by taking the reduction in the electrical efficiency of the CPV unit with an upsurge in the cell temperature. By raising the operational temperature such as the boiling temperature of the NBHT system, the cell temperature of the CPV rises. The upsurge in the cell temperature has a negative impact on electricity output and thus there is a reduction in electrical efficiency from 28.4% to 26.8%. The effects on the thermal efficiency with the upsurge in boiling temperature is insignificant. Nevertheless, there is a minor reduction on the overall energy and exergy efficiencies of the CPV/T unit. Another significant factor with an upsurges in boiling temperature is the $C.R_{max}$ of the system. The $C.R_{max}$ value increases from about 1400 to 2600. With the upsurge in boiling temperature, the critical heat flux (CHF) of the boiling phenomena rises. The critical heat flux is the maximum ability of the boiling system to transfer the heat. Therefore, with the upsurge in CHF, there is an upsurge in the ability of the system to endure a higher load of concentration of the received irradiation.

6.16.4.3 Effects of irradiation

As the DNI value changes with location and time, it is useful to perform a parametric study of the integrated system. An investigation is conducted of the DNI effect ranging from 0.5 kW/m^2 to 1 kW/m^2, taking the average low and high values for the chosen location. Fig. 6.96 displays a clear rise in the electricity and thermal energy output of the system. There is insignificant influence on the overall exergy and energy efficiencies of the integrated system. Since the efficiency of the CPV module is greatly reliant on the operational temperature an experimental thermal coefficient of −0.052%/K was selected for the specific CPV module. The operational temperature of the cell has been preserved nearly constant by the phase

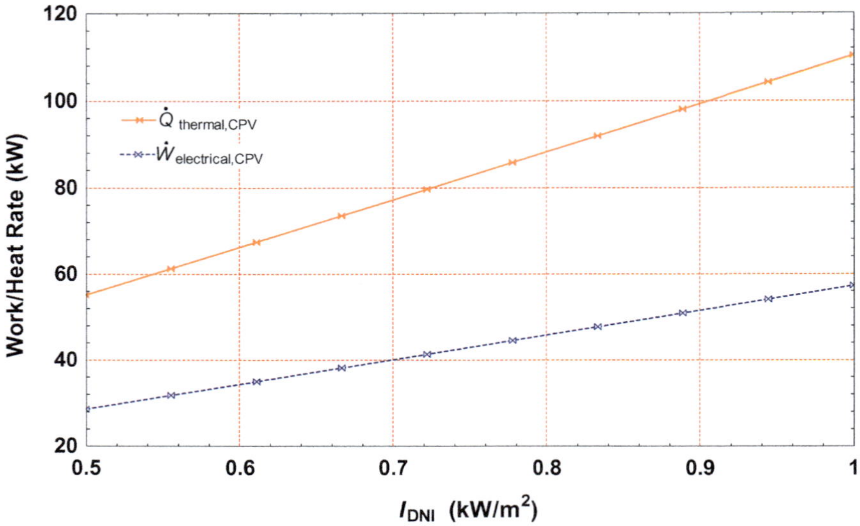

FIG. 6.96 Effects of solar irradiation level on overall thermal and electrical outputs of CPV.

change (boiling temperature) of the thermal management unit and is independent of the solar radiation level. Thus, there is a negligible change in operating or working temperature of the cell and hence in efficiency. Similarly, increase in irradiation level increases the amount incoming energy, which causes an increase of the resultant useful electrical and thermal energy. The $C.R_{max}$ decreases from 3165 to 1583 with the increase in DNI value from 0.5 kW/m^2 to 1 kW/m^2, because an increase in DNI results in an increase in the generated higher heat flux in the CPV. The impacts of DNI value on thermal energy outlet, i.e., cooling and ventilation energy, space cooling and space heating and dehumidified water output are depicted in Fig. 6.97. There is a clear upsurge in all thermal energy production because of the upsurge in received solar irradiation. A clear upsurge in the level of dehumidified water is observed from 4.73 L/h to about 13 L/h.

6.16.4.4 Effects of the percentage of cooling energy for water dehumidification

This parametric study displays the effects of an upsurge in the usage of generated cooling energy for dehumidification from 20% in summer to 100% in winter when there is no demand for residential cooling. For the standard design conditions, 55.6 kW of cooling can be achieved for dehumidification and ventilation demand. The proportion of this energy can be altered by varying the temperature of state point 31. As illustrated in Fig. 6.98, with a rise in the proportion of cooling for dehumidification, there is an evident decline in the space cooling. However, an upsurge in the energy content of the dehumidification process yields greater cooling energy through ventilation and dehumidification and a greater quantity of dehumidified water. The results imply that, for the designed conditions, the system can deliver dehumidified water up to about 20 L/h.

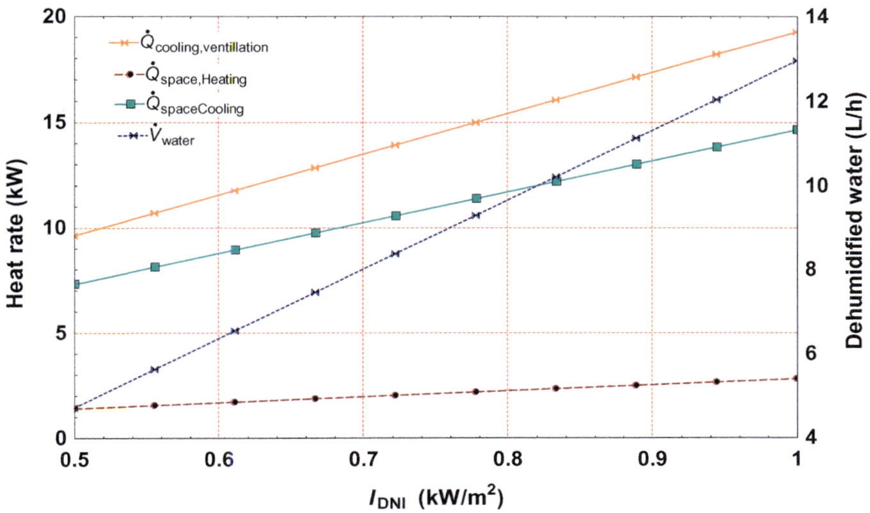

FIG. 6.97 The effects of the DNI values on the thermal and water outputs of the system.

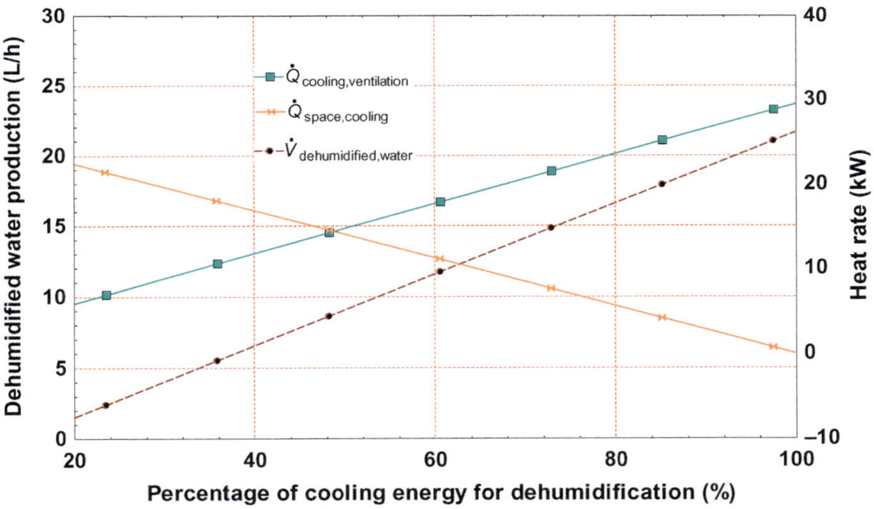

FIG. 6.98 The effects of using available cooling energy for dehumidification and ventilation.

6.16.4.5 Effects of installation capacity on system outputs

The considered installed power capacity of the CPV unit is about 187 kW under standard circumstances. Nonetheless, the system is scalable based on the demands of the residential area. For analyzing the outputs of the integrated system at higher or lower capacities, a parametric study is implemented as shown in Figs. 6.99 and 6.100. The essential installation area of CPV rises from 50 m^2 to 500 m^2. The generated amount of thermal, electrical, water and gases

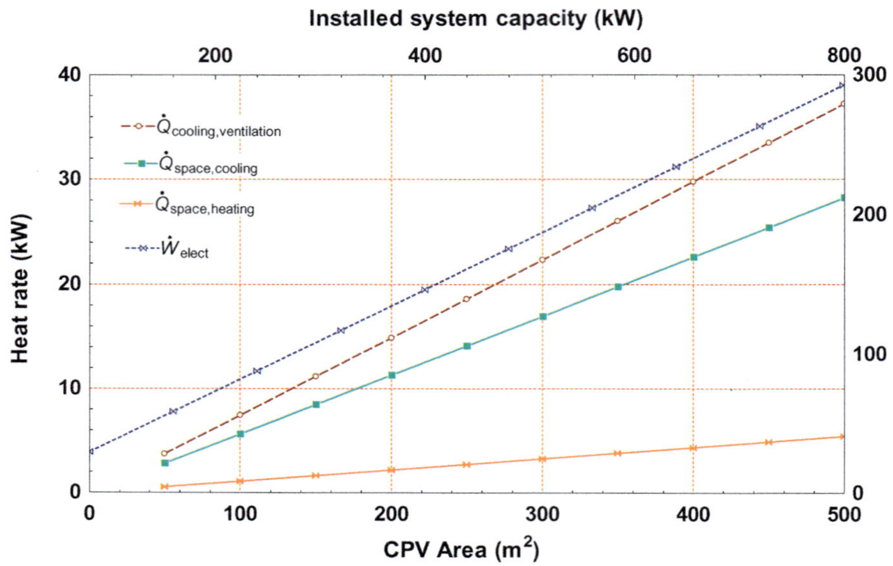

FIG. 6.99 Effects of the CPV area of the installed system on thermal and electrical outputs.

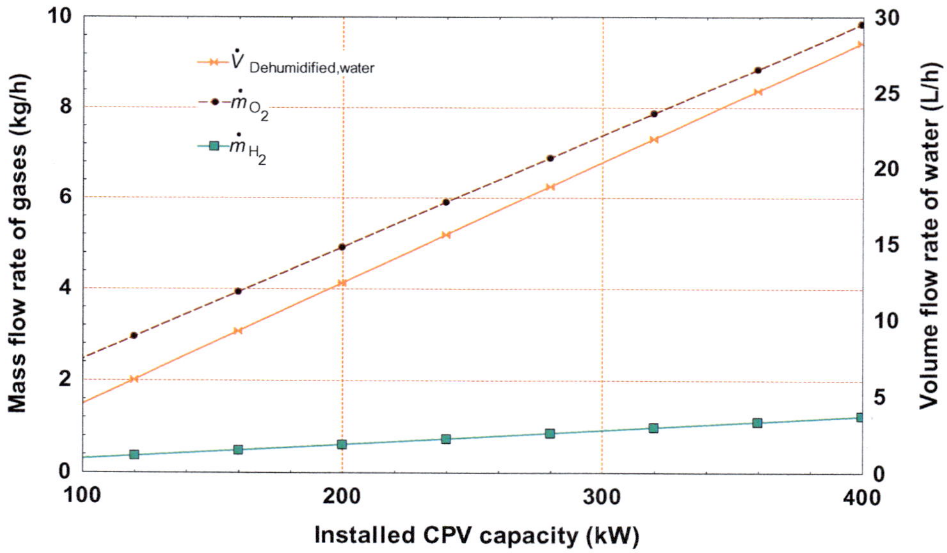

FIG. 6.100 Effects of the capacity of the installed system on produced gases and dehumidified water output.

outputs displays a direct connection with greater installed power. The parametric study confirms the successful implementation of this integrated system at higher or lower capacities. Nevertheless, the associated cost of the modules, subsystems and installation size because of an upsurge or reduction in capacity of the integrated system is not taken into account in this case study, but can be further studied in the future.

6.16.5 Closing remarks

In this case study, detailed thermodynamic analysis of an integrated CPV-based system is conducted. The storage system is considered for storing heat through PCMs. In addition, a reversible fuel cell system is employed, acting as a chemical energy storage unit. The reversible fuel cell unit is used for the needs of the residential building at nighttime and during poor solar irradiation levels. The conclusions of this case study can be summarized by the following points:

- At standard conditions, the electrical efficiency of the selected multi-junction CPV module is about 40.8%. However, by utilizing this hybrid multigeneration CPV/T system, the overall energy efficiency can be improved to about 68.6%. This implies that extending the useful outputs from the same source can significantly enhance the overall performances.
- The case study confirms that the suggested NBHT cooling method can be employed to attain high concentration ratios in CPVs, which can increase up to 1862 times at standard conditions, which are DNI of 870 W/m^2 and NBHT temperature of 373 K.
- The parametric study affirms the ability of the studied integrated system to function at various locations having diverse DNI values. Greater DNI values yield greater thermal and electrical productions; however, the ability to endure to maximum concentration ratio drops because of the high heat flux generation.
- The subsystem exergy efficiencies are calculated as 49.5% for CPV, 85% for NBHT, 34.7% for the overall integrated system, 43.8% for dehumidification and ventilation system, while 0.18 is the value of exergetic COP for absorption cooling and 0.52 is the value of exergetic COP for space cooling system.

6.17 Case study 9: Integrated concentrated solar and biomass energy system with phase change material (PCM) and ammonia synthesis

The integrated system in this case study provides five useful outputs: electricity, heating, cooling, ammonia synthesis and desalinated water, by employing phase change material (PCM) based energy storage for uninterrupted operation. The sources, systems, and services considered in this case are shown in Fig. 6.101. The results of this case study suggest that, with the use of the proposed multigeneration system, the energy, heating, cooling, and potable water for a remote location can be met in a more sustainable and efficient manner in addition to overseas energy transport using ammonia as an energy carrier. The overall system exergy and energy efficiencies are calculated as 18.9% and 28.0%, respectively. This case study is adapted from Shahid et al. [48].

6.17.1 System description

A schematic of the complete multigeneration energy system is depicted in Fig. 6.102 with the main system outputs including drinking water, electricity, ammonia, oxygen, and heating/cooling.

A parabolic dish solar collector is used as a heat source to heat up the PCM-based thermal energy storage (TES) system, which in turn is used as a boiler for the steam Rankine cycle (SRC). The TES is used to provide uninterrupted energy supply to the system during the

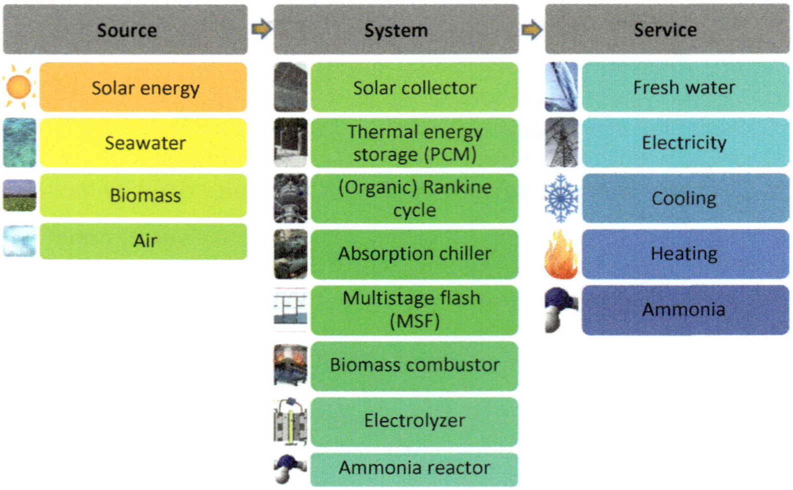

FIG. 6.101 3S approach for case study 9.

nighttime. The Rankine cycle is based on a two-stage turbine (high/low pressures) and the reheater between the two stages uses biomass-based steam. Part of the energy output from the Rankine cycle is used in the PEM electrolyzer for hydrogen production. The Rankine cycle is operated at 650 °C and 15.2 MPa, generating a gross combined power output of 30.6 MW. The net power available after internal consumption (PEM electrolyzer, compressors, and pump) is 20 MW.

As mentioned, the reheat requirement of the Rankine cycle is fulfilled using steam generated by a biomass combustor and the steam is subsequently used in the organic Rankine cycle as a heat source to the organic Rankine cycle boiler. Operated at 160 °C and 3.25 MPa, the organic Rankine cycle uses isobutane as the working fluid and provides power out of 1.5 MW.

The seawater is initially preheated using the brine reject leaving the multistage flashing (MSF) to a feed temperature of 45 °C. The feed is then fed to the internal reheaters of the MSF until it reaches the first effect at a temperature of 120 °C. The feed then travels through the multiple stages as the distillate is separated out and leaves the MSF at 46.5 °C and 9.6 kPa, where it is then pumped and discharged back into the sea. The designed output of the MSF is 2.3 million gallons per day.

Water at 80 °C is fed to the electrolyzer where it is split into hydrogen and oxygen via electrochemical reactions. The hydrogen produced is then supplied to the ammonia production system.

The fluid circuit for the absorption heating/cooling (AHT) flows between the desorber and absorber, where the absorber operates at a low pressure and the desorber at a higher pressure. This subsystem is used to attain cooling.

To facilitate the transport and storage of hydrogen, conversion of hydrogen to ammonia is a much more economically and technologically feasible option. For this reason, the output of the PEM electrolyzer is supplied to a Haber-Bosch ammonia synthesis plant, with a nitrogen supply obtained from an industrial plant. Hydrogen and nitrogen are supplied in a

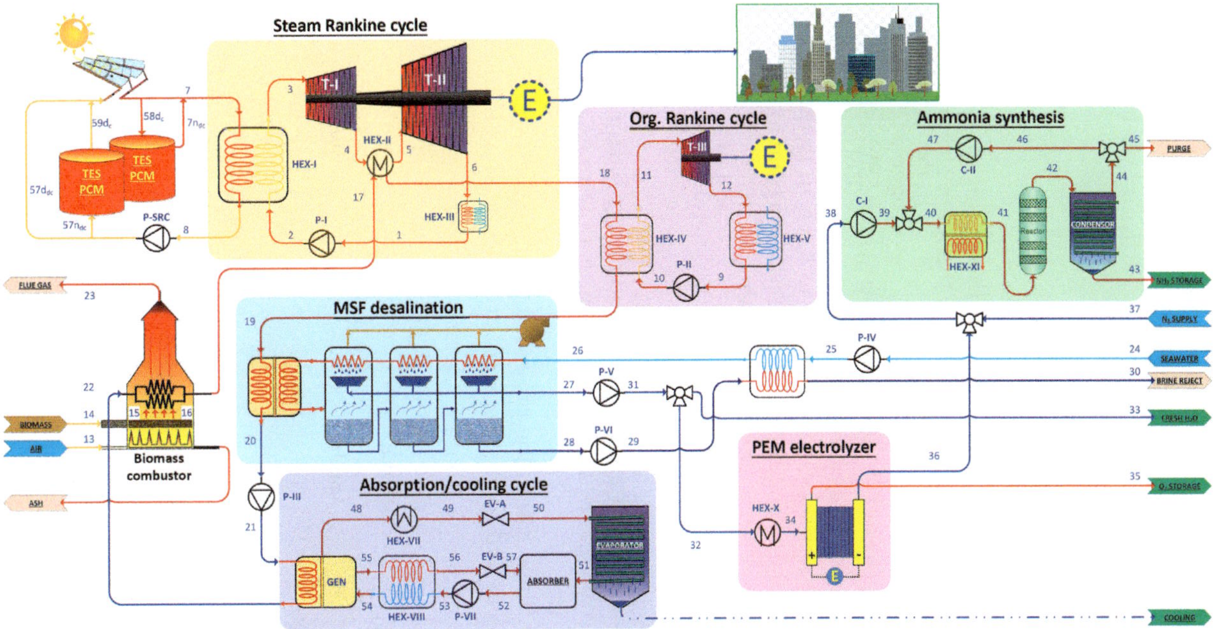

FIG. 6.102 Schematic diagram of case study 9.

stoichiometric ratio to the reactor at 450 °C and 150 MPa. A purge stream is extracted to limit the build-up of inert gas in the reactor and part of this is recycled back to the reactor feed to increase the overall conversion rate.

A biomass combustor is incorporated to supply steam for the desalination unit and subsequently absorption cooling/heating and product drying unit. The biomass is combusted in a high-temperature combustion chamber at a particular air-fuel ratio to provide steam at 500 °C and 0.2 MPa.

6.17.2 Assumptions

In order to perform thermodynamic analysis of this multigeneration system, there is a need to make conceptually correct assumptions. Here, the main assumptions undertaken in this case study are listed as follows:

- Reference state temperature (T_0) and pressures (P_0) are assumed as 298 K and 101.325 kPa.
- There are negligible heat losses from pipes.
- Pump, turbine, and compressors are assumed to have an isentropic efficiency of 85%.
- The kinetic and potential changes are considered negligible for the entire system.
- Negligible mass losses occur in the entire system.

6.17.3 Analysis

A detailed analysis of mass, energy, entropy and exergy balances is carried out for the entire system and each stream is evaluated for its mass flow rates, temperature, pressure, enthalpies, and exergies flows.

In order to quantitatively assess the proposed system's performance, the exergy efficiencies of the overall system and each subsystem are calculated as follows:
Steam Rankine cycle:

$$\eta_{exSRC} = \frac{\dot{W}_{T,I} + \dot{W}_{T,II} - \dot{W}_{P,I}}{\dot{Q}_{SC}.\left(1 - \frac{T_0}{T_{SC}}\right)} \tag{6.59}$$

Organic Rankine cycle:

$$\eta_{exORC} = \frac{\dot{W}_{T,III} - \dot{W}_{P,II}}{\dot{m}_{19}.(\dot{ex}_{19} - \dot{ex}_{18})} \tag{6.60}$$

Multistage flash:

$$\eta_{exMSF} = \frac{\dot{m}_{27}.\dot{ex}_{27}}{\dot{m}_{19}.(\dot{ex}_{19} - \dot{ex}_{20})} \tag{6.61}$$

Absorption cooling system:

$$COP_{ex} = \frac{\dot{m}_{47}.(\dot{ex}_{48} - \dot{ex}_{47})}{\dot{W}_{P-VII} + \dot{m}_{22}.(\dot{ex}_{21} - \dot{ex}_{22})} \tag{6.62}$$

PEM electrolyzer:

$$\eta_{\text{exPEM}} = \frac{\dot{m}_{36}.\dot{\text{ex}}_{36}}{\dot{W}_{\text{E,PEM}} + \dot{Q}_{\text{heat,H}_2\text{O}} \cdot \left(1 - \frac{T_0}{T_A}\right)} \tag{6.63}$$

Ammonia (NH_3) synthesis process:

$$\eta_{\text{exNH}_3} = \frac{\text{HHV}_{\text{NH}_3}\, \dot{m}_{\text{NH}_3} \left(1 - \frac{T_0}{T_{\text{Rxn}}}\right)}{\dot{W}_{\text{C,I}} + \dot{W}_{\text{C,II}} + \dot{Q}_{\text{HEX-XI}} \left(1 - \frac{T_0}{T_s}\right) + \dot{Q}_{\text{h,Rxn}} \left(1 - \frac{T_0}{T_{\text{Rxn}}}\right)} \tag{6.64}$$

Overall system:

$$\eta_{\text{exoverall}} = \frac{\dot{W}_{\text{T,I}} + \dot{W}_{\text{T,II}} + \dot{W}_{\text{T,III}} - \dot{W}_{\text{P, Net}} + \dot{m}_{27}\, \dot{\text{ex}}_{27} + \dot{Q}_{\text{e}} \left| 1 - \frac{T_0}{T_C} \right| + \dot{m}_{43}\, \dot{\text{ex}}_{43}}{\dot{Q}_{\text{SC}} \left(1 - \frac{T_0}{T_{\text{SC}}}\right) + \dot{Q}_{\text{biomass}} \left(1 - \frac{T_0}{T_S}\right) + \dot{Q}_{\text{HEX-XI}} \left(1 - \frac{T_0}{T_s}\right)} \tag{6.65}$$

6.17.4 Results and discussion

The energy efficiency and exergy destruction rates of the overall system and the subsystems are evaluated at the normal operating parameters as summarized in Figs. 6.103 and 6.104.

A parametric analysis of DNI for the overall system is evaluated as reported in Fig. 6.105. A DNI range of 300 to 1100 (W/m^2) is used to observe the resulting effect. The overall efficiency is stable with a slight dip in the efficiency levels at lower DNI values. Even though a higher DNI results in higher power output of the system, it must be noted that the overall

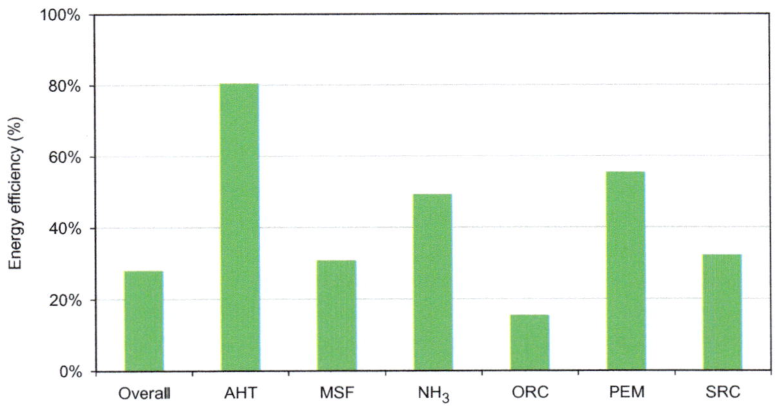

FIG. 6.103 Energy efficiencies of subsystems in case study 9.

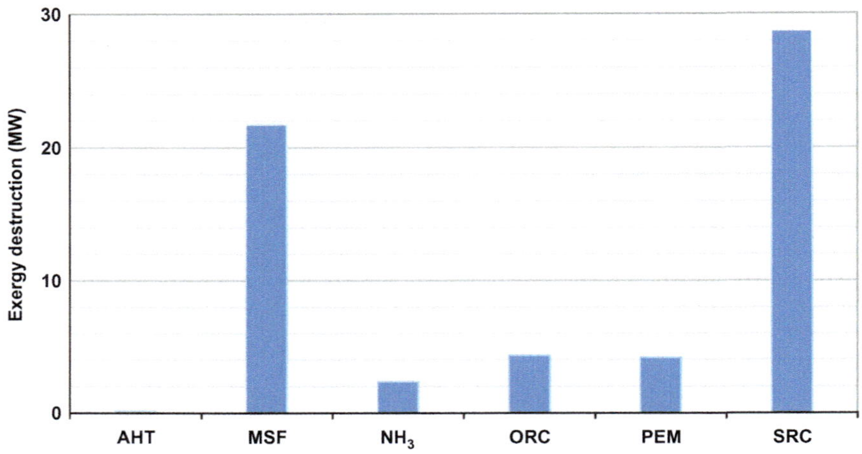

FIG. 6.104 Exergy destruction rates of subsystems at normal operation.

FIG. 6.105 Effects of DNI on overall energy and exergy efficiency.

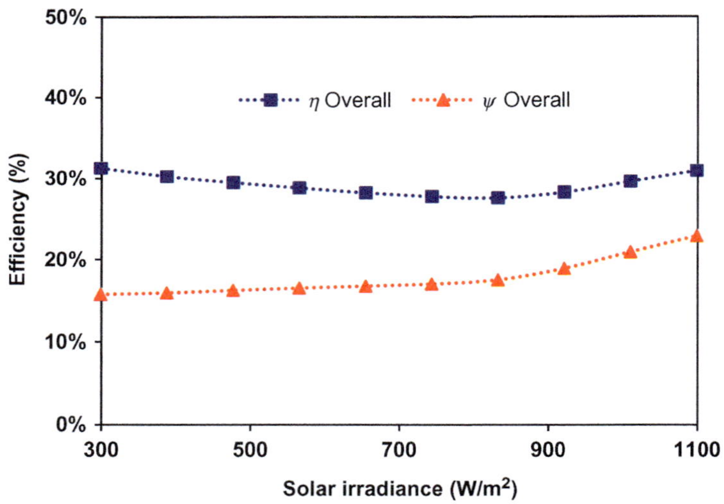

efficiency does not increase because other components of the system become over-stressed in terms of energy demand. In the detailed analysis of the DNI parametric study, it is observed that the reheater duty requirements also increase with the increasing DNI, as this results in a higher turbine input temperature. The reheater output temperature and quality are critical for the MSF desalination unit as they determine the final output and energy losses of the system. However, it should be noted that, despite the component losses, the overall efficiency of the system does not fall below 27.6%. The exergetic efficiency of the overall system, however, follows an increasing trend despite the slight dip in the energy efficiency.

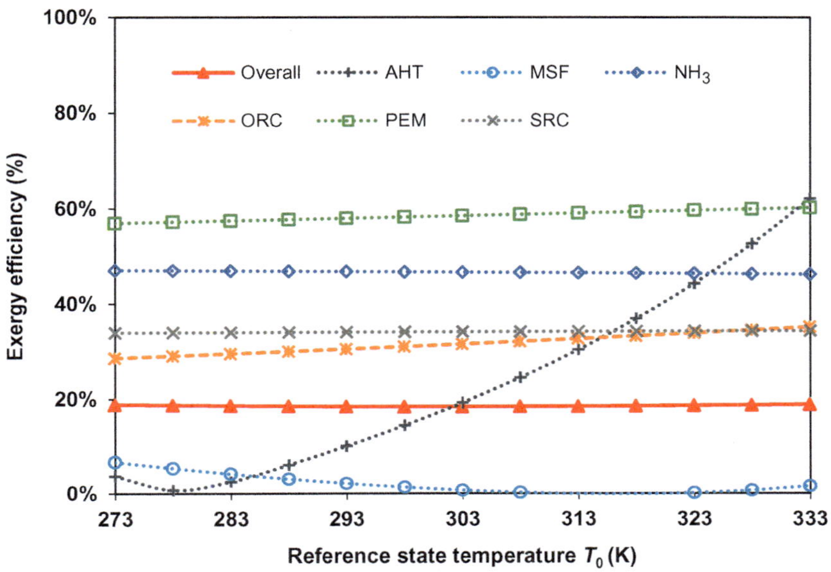

FIG. 6.106 Effects of ambient temperature on exergy efficiency.

For the purpose of the parametric study, the reference temperature (T_0) is varied from 273 to 333 K. The results of the parametric study are summarized in Fig. 6.106. As expected, the exergy efficiency of the overall system decreases slightly with an increasing reference temperature. However, it is very negligible. The same trend applies to MSF and the SRC subsystems. However, for the PEM, NH_3, and ORC, the opposite is true, as the exergy efficiency is observed to slightly increase.

The AHT subsystem, on the other hand, is observed to have very low initial exergy efficiency, which is associated with the irreversibilities of the processes. However, with an increasing reference temperature, it is observed to have higher overall exergy efficiency. The lowest point on the AHT exergy efficiency curve is the desired cooling temperature set at 6 °C, where the exergy flow is zero since the reference and desired state are at the same temperature.

The exergy destruction rates of each subsystem are also evaluated under the same parametric study. Fig. 6.107 shows a summary of the parametric study results. The highest exergy destruction rates are observed for the SRC and MSF subsystems, followed by PEM and ORC subsystems with almost the same exergy destruction rates. The NH_3 and SSAHT subsystems have the minimum exergy destruction rates. These destruction rates show a gradual increase as the reference temperature (T_0) increases. The massive destruction rates for the MSF and SRC are attributed to the irreversibility of the system. Moreover, the desired output of the MSF is pure water, which has a very low exergetic value in comparison to the high exergy potential of the concentrated brine discharge.

The effects are more pronouncedly observed in the SRC subsystem parameters, as depicted in Fig. 6.108. The net energy output from the two-stage turbine system in the SRC, i.e., W_{T_I} and

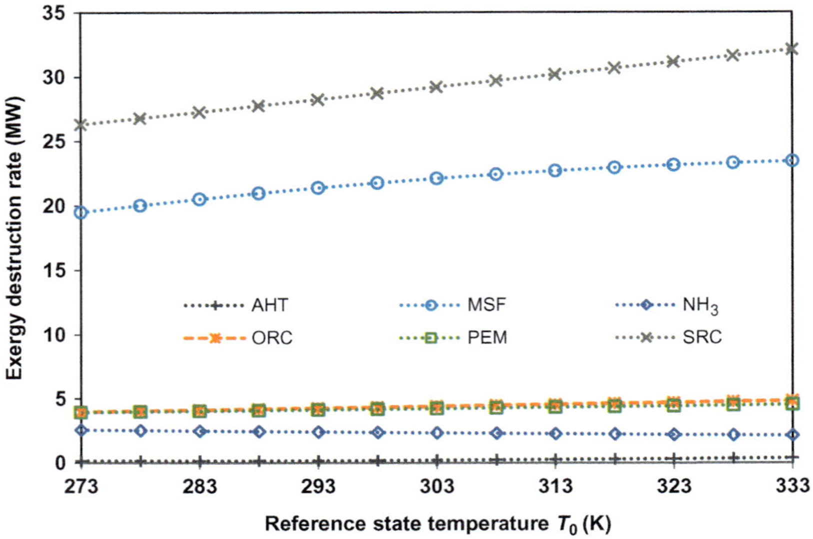

FIG. 6.107 The effects of ambient temperature on the exergy destruction rate.

FIG. 6.108 The variations with an inlet pressure of the electrical output from turbines I and II and the overall exergy destruction rate of the Rankine cycle subsystem.

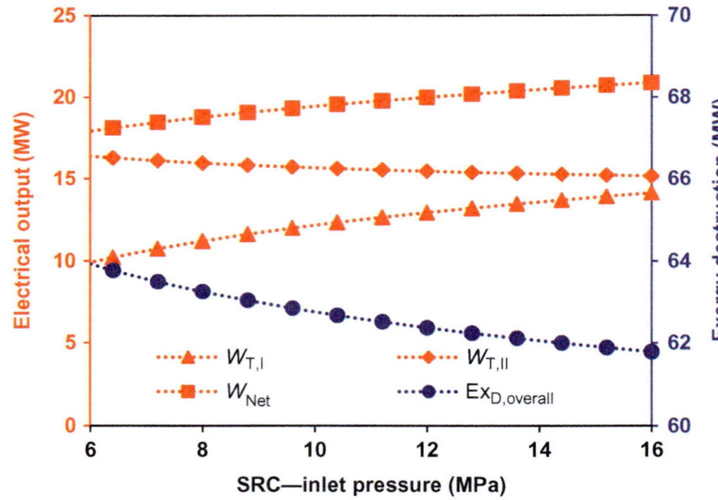

$W_{T_{II}}$, increases as the inlet pressure to the SRC increases. However, a closer look at the trend reveals that the first-stage turbine output drops as the second-stage turbine output rises at a faster rate. The output from the two turbines converges as the pressure nears the super-critical regime. The exergy destruction rate, however, drops by about 2 MW as the pressures increase, reflecting again on the benefits of higher exergetic values at high temperatures and their more effective conversion to useful outputs.

6.17.5 Closing remarks

In this case study, a multigeneration energy system based on renewable sources (biomass and solar energy) is developed and analyzed for energy and exergy flows. This proposed multigeneration energy system manages to utilize the resources more effectively in comparison to single generation systems. The efficiencies of the systems were evaluated on an energy and exergy basis and exergy destruction of each subsystem is also evaluated and observed, with the following concluding remarks:

- The overall exergy and energy efficiency of the proposed system stands at 18.9% and 28.0%, respectively.
- The steam Rankine cycle exergy and energy efficiencies are 33.9% and 32.2%, respectively.
- The organic Rankine cycle exergy and energy efficiencies are 28.6% and 15.5%, respectively.
- The PEM electrolyzer exergy and energy efficiencies are 57.0% and 55.4%, respectively.
- The MSF desalination unit exergy and energy efficiencies are 6.8% and 30.8%, respectively.
- The ammonia synthesis unit exergy and energy efficiencies are 47.1% and 49.2%, respectively.

6.18 Case Study 10: Integration of solar ponds with pressure retarded osmosis for desalination and photo-assisted chloralkali processes

In this case study, the integrated system uses the thermal energy storage characteristic of the solar pond to power the multieffect desalination process with a total production of approximately 10,000 m^3 of fresh water annually. Pressure retarded osmosis (PRO) along with a photo-assisted chloralkali reactor are employed to utilize the brine rejected from the desalination process. The inclusion of pressure retarded osmosis is proposed to take advantage of the salinity difference that occurs because of the preparation of the solar pond and desalination process. The quest for zero salt discharge in the proposed energy system configuration is undertaken through the introduction of a photo-assisted chloralkali reactor, which utilizes the power generated from the PRO, and a solar pond storage arrangement. This solar pond and desalination coupling system will provide independence from conventional energy sources, leading to a significant contribution to the reduction of greenhouse gas emissions caused by fossil-fuel driven desalination. Solar energy is received by a solar pond and photo-assisted chloralkali reactor to produce fresh water, electricity, hydrogen, chlorine and sodium hydroxide, as shown in Fig. 6.109. There are four major operating units: the solar pond, multieffect desalination, pressure retarded osmosis, and photo-assisted chloralkali reactor. This case study is adapted from Mohamed and Bicer [49].

6.18.1 System description

The integrated energy system is driven by a salinity gradient solar pond (SGSP) absorbing energy from the sun; a schematic diagram of the proposed system is shown in Fig. 6.110. The SGSP stores thermal energy, mainly in the lower convective zone (LCZ). Heat extraction from

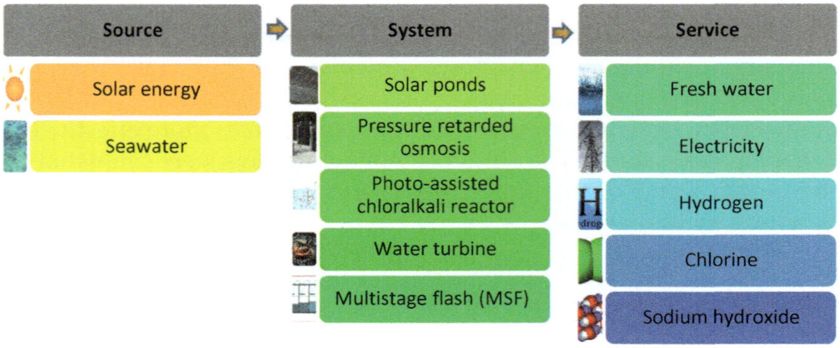

FIG. 6.109 3S approach for case study 10.

the SGSP is accomplished by employing external and internal heat exchangers. The heat gained by the LCZ is directly used to initiate the multieffect desalination (MED) unit by exchanging heat with seawater in the first effect (stage). In addition, a pipe heat exchanger is used to heat the inlet seawater to the desired top brine temperature. The MED system comprises six effects, where the pressure descends from the first to the last stage, and the vapor

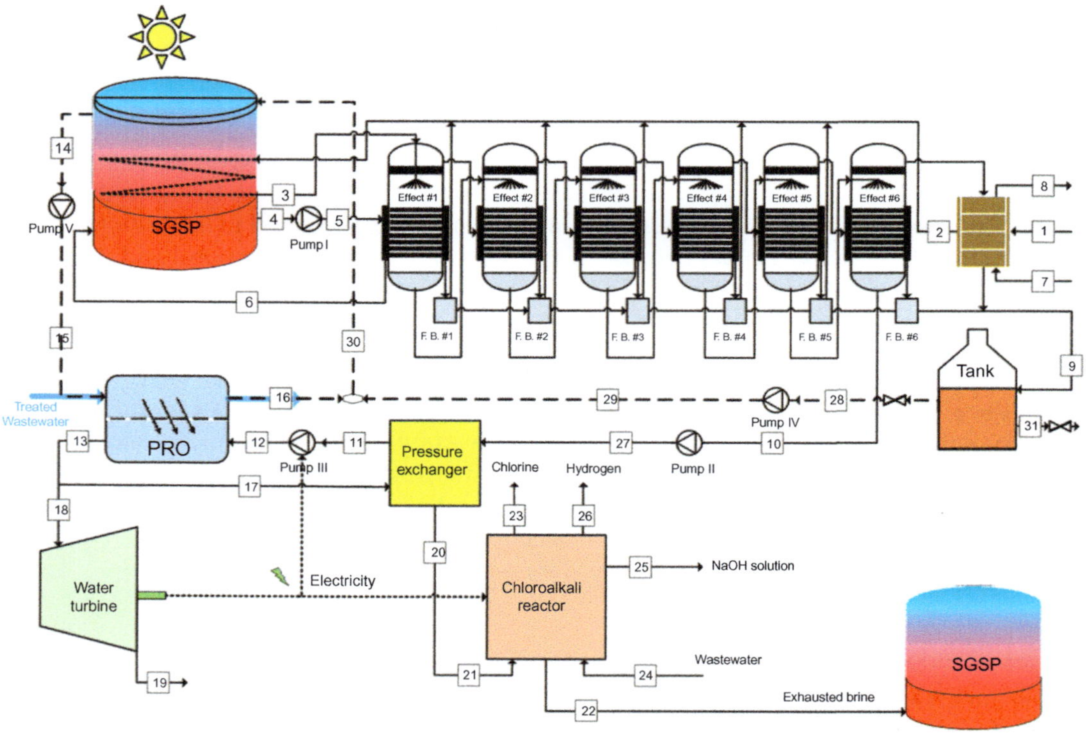

FIG. 6.110 Schematic diagram of the integrated system for case study 10.

generated from the previous step is used to run the next effect. The distillate is collected in the process, and the concentrated brine produced is used as the feed in the following stage. The combined distillate amount is received in a tank; the rejected brine is used to heat the feed to the MED utilizing a plate heat exchanger before it is sent to the LCZ of the SGSP. Fresh water is used to maintain the salinity gradient in the SGSP and compensate for part of the low salinity in the upper convective zone (UCZ). Usually, seawater is utilized to remove the excess heat from the MED and is discharged back to the sea. Instead, in this work, water is heated as a means of heat recovery from the condenser.

Following the heat supply to power the MED unit, the LCZ stream meets low salinity water pumped from the UCZ. Because of the salinity difference, power can be produced through the PRO system. The UCZ stream is sent back to the UCZ through a diffusor at a location lower than the freshwater compensation location. To optimize the pumping pressure of the LCZ stream, the flow is segmented into two parts, one that runs the turbine and the other exchanging hydraulic power to reduce the pumping power (pump III) required.

To achieve zero salt rejection to the environment, two integrated systems are implemented to accept the hypersaline water. The first idea involves building another solar pond with the primary goal of producing power through an organic Rankine cycle. The second involves the introduction of a photo-assisted chloralkali reactor that can utilize the power from the PRO unit to split salty water into hydrogen, chlorine, sodium hydroxide and salt. Treatment and purification processes are not included in this study.

6.18.2 Assumptions

In order to perform thermodynamic analysis of this multigeneration system, there is a need to make conceptually correct assumptions. Here, the main assumptions undertaken in this case study are listed as follows:

- The losses due to friction in the piping systems and fitting are assumed negligible.
- The system is in steady state.
- There are negligible changes in potential and kinetic energies.
- The pumps are assumed to have 90% isentropic efficiencies.

The specific assumptions for the analysis of the SGSP are:

- The temperature of LCZ must be 10 degrees higher than the top brine temperature.
- Reduction of solar radiation absorbed by the solar pond (θ')=0.85.
- The temperature of UCZ is equal to the ambient temperature.
- There are no heat losses from the sides.
- The ground temperature is equal to the ambient temperature.

The specific assumptions for the analysis of MED are:

- No vapor flashing occurs in the effects.
- Equal thermal loads in all effects.
- Constant heat transfer area between the effects.
- The formed vapors are free of salt.
- The heat losses to the surroundings are negligible.

6.18.3 Analysis

The energy and exergy efficiencies of the solar pond are calculated by the following equations, respectively:

$$\eta_{\text{enSolar Pond}} = \frac{\dot{m}_4(h_5 - h_6) + \dot{m}_1(h_3 - h_2)}{IA + \dot{W}_{\text{Pump I}}} \tag{6.66}$$

$$\eta_{\text{exSolar Pond}} = \frac{\dot{m}_4(\text{ex}_5 - \text{ex}_6) + \dot{m}_1(\text{ex}_3 - \text{ex}_2)}{IA\left(1 - \left(\frac{4}{3}\frac{T_o}{T_{\text{solar}}}\right) + \left(\frac{1}{3}\frac{T_o}{T_{\text{solar}}}\right)\right) + \dot{W}_{\text{Pump I}}} \tag{6.67}$$

where I is the irradiance and A is the area.

Two efficiencies are defined for the MED system. The first one considers the mass of produced fresh water as the only useful product. The second definition evaluates the efficiencies by considering the waste heat recovery in the condenser (cogeneration). In this respect, the exergetic efficiencies of the MED are calculated by the following equations:

$$\eta_{\text{exMED}} = \frac{\dot{m}_9\text{ex}_9}{\left(\dot{m}_4(h_5 - h_6) \times \left(1 - \frac{T_o}{T_4}\right)\right) + \left(\dot{m}_1(h_3 - h_2) \times \left(1 - \frac{T_o}{T_{\text{NCZ,source}}}\right)\right) + \dot{m}_1\text{ex}_1} \tag{6.68}$$

$$\eta_{\text{exMED,Cogen}} = \frac{\dot{m}_9\text{ex}_9 + \dot{m}_7(\text{ex}_8 - \text{ex}_7)}{\left(\dot{m}_4(h_5 - h_6) \times \left(1 - \frac{T_o}{T_4}\right)\right) + \left(\dot{m}_1(h_3 - h_2) \times \left(1 - \frac{T_o}{T_{\text{NCZ,source}}}\right)\right) + \dot{m}_1\text{ex}_1} \tag{6.69}$$

PRO efficiencies are commonly determined as the ratio of work produced over the maximum possible power as follows:

$$\eta_{\text{enPRO}} = \eta_{\text{exPRO}} = \frac{\dot{W}_{\text{out}}}{\dot{W}_{\text{ideal}}} \tag{6.70}$$

The efficiency of the PRO is set to 70% and the area of the membrane is calculated accordingly.

6.18.4 Results and discussion

The energetic and exegetic efficiencies of the solar pond can reach 13.5% and 1.6%, respectively. The irradiance value is averaged over the whole day; the efficiency values correspond to average values because the solar pond acts as a storage medium. It can be shown from the models that approximately 23% of the solar irradiation reaches the LCZ. One can directly calculate the efficiency using this number assuming that the total amount reaching the zone is used; however, the heat conduction from LCZ to NCZ (non-convective zone) and losses to the ground contribute to lower efficiency values.

MED produces 0.5 kg/s in these conditions, and 35.6% is utilized to compensate for the losses from the UCZ stream. In the calculation of MED performance, the amount of the pure water heated from the excess heat is attributed to the MED. The amount of heated water does increase the energetic efficiency from 22.4% to 67.6%. The energetic and exegetic efficiencies of the MED unit reach 67.6% and 5.8%, respectively.

The PRO unit is operated at a pressure of 3.7 MPa. The net power produced from the PRO system extended to 660 W by employing a 677 m^2 membrane. The chloralkali reactor uses the power generated from the PRO to produce 86.3 and 2.454 mg/s of chlorine and hydrogen gases with a 51% energy efficiency. Also, 0.37 kg/s of sodium hydroxide solution is produced in the cathode side of the reactor.

Concerning the contribution of the subsystems in the overall exergy destruction, the solar pond has one of the highest since it receives huge amounts of solar energy. Under the average annual conditions, the exergy destruction rates of the solar pond, MED, PRO and chloralkali are 1950 kW, 32.2 kW, 0.30 kW, and 0.12 kW, respectively. The overall energy efficiency is calculated to be about 12%, whereas the overall exergy efficiency is determined to be about 1%.

The performance of the proposed system is assessed by the monthly average efficiency, using irradiation data from a location in the Middle East (Kuwait City), as shown in Fig. 6.111. The maximum reported efficiencies are in June with a monthly average of 16.6% and 1.3% energy and exergy efficiencies, respectively.

The freshwater production from the MED system and overall is shown in Fig. 6.112. This case study demonstrates the impact of ambient air temperature on water production at constant sunlight irradiance of 600 W/m^2. One can observe the amount of water utilized in the PRO unit, which is reflected in the final amount in the figure. The production from the MED subsystem varies from 2.1 kg/s to 2.2 kg/s over the year. The average and maximum deviation between the produced MED amount and the final amount are approximately 6.7% and 7.4%, respectively.

Figs. 6.113 and 6.114 illustrate the impacts of irradiance on the PRO and chloralkali units. At 1000 W/m^2, the production of hydrogen and chlorine gases reaches 19.8 and 697.2 mg/s, respectively. Fig. 6.114 shows the net power production from the PRO unit, which exceeds 5 kW slightly at 1000 W/m^2. The required membrane area is around 5400 m^2 to deliver the intended power at an average efficiency of 70%.

The sizing of the solar pond is essential to optimize the performance of the proposed system. Fig. 6.115 shows the study of the depth of the NCZ on average efficiency of the solar pond.

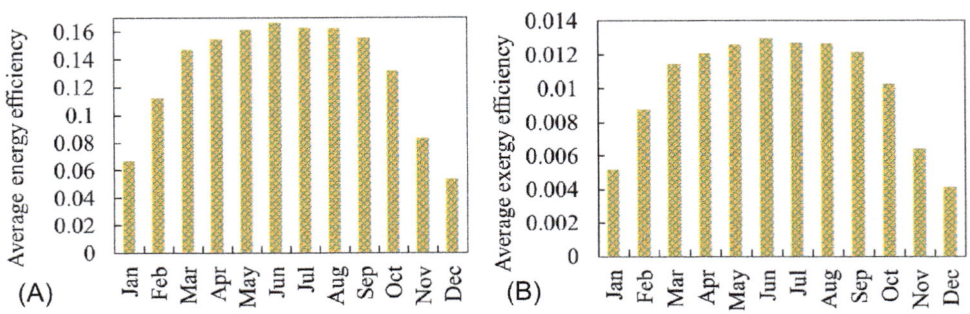

FIG. 6.111 Monthly average overall system (A) energy efficiencies (B) exergy efficiencies.

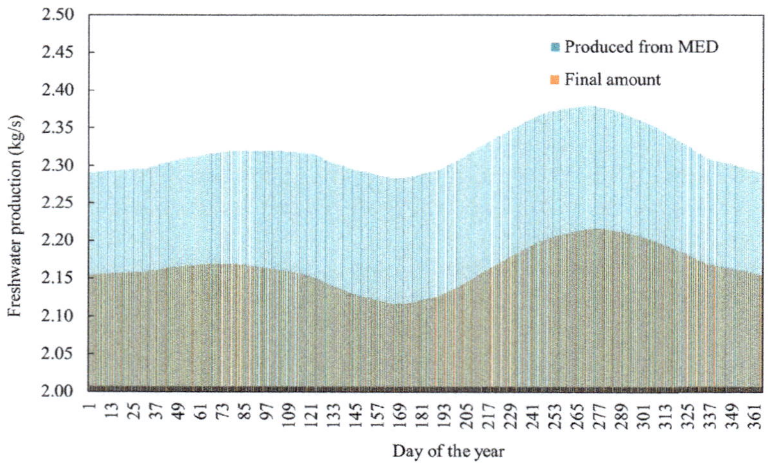

FIG 6.112 Freshwater production from the proposed system.

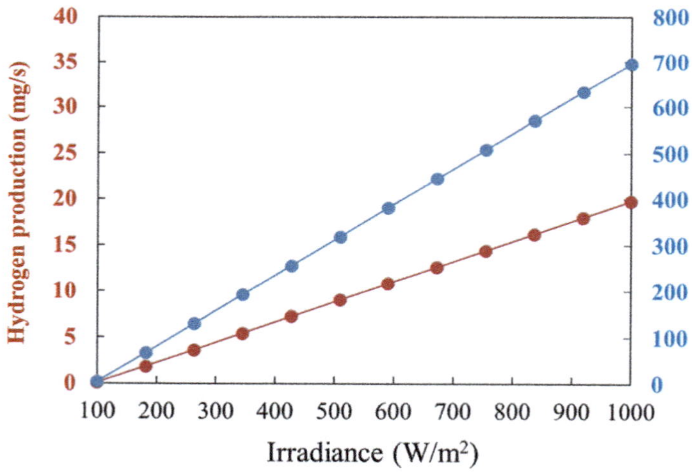

FIG. 6.113 Irradiance impact on the gaseous species production in the photo-assisted chloralkali process.

The efficiency increases as the NCZ increases because of the shielding effect (trapping heat in the LCZ). However, a deeper NCZ minimizes the amount of irradiation reaching the LCZ.

6.18.5 Closing remarks

This case study shows that the PRO system can recover some of the energy from the desalination brine. A photo-assisted chloralkali process is also incorporated to produce useful outputs such as hydrogen, chlorine and sodium hydroxide from the salty stream. Several

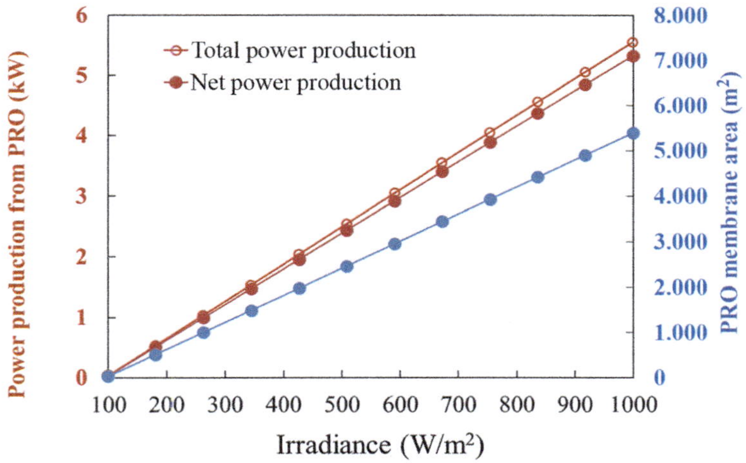

FIG. 6.114 Power production and required total membrane area in the PRO unit.

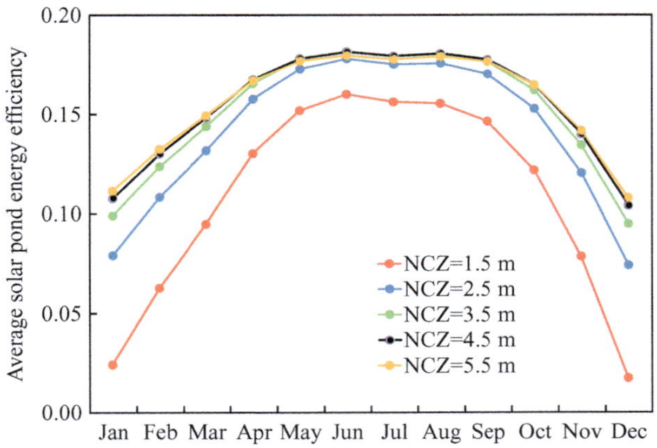

FIG. 6.115 Seasonal effect and NCZ depth impact on solar pond performance.

parametric studies are carried out to observe the impact of seasonal effects on the subsystems and overall performance. Increasing the length of the NCZ assists in trapping the heat in the LCZ while reducing the irradiation reaching the LCZ. Average energy and exergy efficiencies of 13% and 1.7% over the year are achieved by the proposed system.

6.19 Conclusions

Integrated renewable energy systems have numerous advantages and benefits compared to conventional energy systems, such as decentralized energy production, environmental impact reduction and better energy security. Renewable energy sources are typically available in most of the regions on Earth. Solar radiation is even more diversified compared to other types of renewable energy sources. Specific forms of renewable energy such as geothermal and ocean thermal energy have substantially higher capacity factors during the year. Renewable energy systems can be classified under the following categories: solar, wind, hydro, biomass, geothermal, and ocean. There are various energy conversion systems to transform these renewable sources into useful products.

Solar energy can be used as thermal or converted into electrical work. Conventional solar thermal systems are very suitable for residential application to provide electricity and heating as well as cooling and fresh water. Solar thermal systems include concentrated solar, which can drive several industrial processes requiring medium-high temperatures. Furthermore, concentrated photovoltaic systems achieve considerably higher solar-to-electricity efficiencies. The excess heat can also be utilized for other applications such as cooling, desalination, etc.

Wind energy is also broadly available on Earth. The available kinetic energy in the wind is converted into other useful forms, mainly electricity, by wind turbines. Two main groups of wind turbine designs exist, namely the horizontal axis wind turbine and the vertical axis wind turbine. Wind turbines can also be used to produce thermal energy via several techniques: (i) friction between solid materials, (ii) agitation of fluid, and (iii) electromagnetic induction. The mechanical energy is utilized to generate electromagnetic induction, which eventually heats the fluid flowing inside the electromagnetic field.

Hydro energy is available in many forms: (i) potential energy from high heads of water retained in dams, (ii) kinetic energy from current flow in rivers and tidal barrages, and (iii) kinetic energy from the movement of waves on relatively static water masses. Depending on the height and volume flow, suitable hydro turbines are utilized to convert available energy into electricity. Hydropower has the highest energy conversion efficiency among all renewables.

Geothermal energy is a form of thermal energy that exists in some specific areas of the world at different temperature levels. Geothermal heat is utilized either for process heating or is transformed into electricity via appropriate heat engines. A large portion of heat-consuming processes can be delivered by geothermal energy, including space and water heating, industrial processes, and several processes in agriculture and the food industry. The geothermal fluid can be in different forms such as hot water, wet steam or dry steam. The geothermal energy can be transformed via several cycles, including binary, flash, direct expansion and others.

There are mainly five different types of energy extraction from oceans: tides, waves, ocean currents, thermal energy, and osmotic (salinity gradient). Water current turbines have a working mechanism similar to wind turbines, with the difference that the density and the viscosity of water are 100 and 1000 times higher than that of air, respectively. The temperature gradient of the ocean at deep levels yields a temperature difference that can be enough to extract thermal energy at low efficiencies. This type of system is called ocean thermal energy conversion (OTEC). Wave energy can be collected with floating bodies that perform elliptic movement under the action of waves.

Biomass resources have a diverse origin ranging from wood and crop forms to food waste forms. The main feature of biomass is its chemical composition, in such forms as sugar, starch, cellulose, hemicellulose, lignin, resins, and tannins. Biomass energy can be converted into useful forms such as heat, electricity, and transportation fuel. The most common and easiest method of biomass conversion is by combusting and producing heat. Gasification converts the biomass into a gaseous fuel such as syngas, whereas liquefaction produces a liquid fuel.

Integration of these renewable energy resources can bring sustainable solutions and multiple products. Numerous energy systems are available for possible combinations, such as synthetic fuel production, absorption cooling, space heating, desalination, thermal energy storage, fuel cells and thermochemical cycles, among others.

In this chapter, a number of case studies are presented and analyzed thermodynamically. The case studies signify the potential integration options of renewable energy systems including energy storage. Numerous useful outputs are produced from selected integrated systems such as fresh water, electricity, cooling, heating, hydrogen, dried products, etc. Thence, it is evident that integrated renewable energy systems will play a significant role in achieving sustainable development.

Nomenclature

A	area (m^2)
CHF	critical heat flux (W/m^2)
COP	coefficient of performance
D	diameter of pipe (m)
DNI	direct normal irradiance (W/m^2)
ex	specific exergy (kJ/kg)
$\dot{Ex}$	exergy rate (kW)
h	specific enthalpy (kJ/kg)
H	total enthalpy of the flow (kJ)
L	length of pipe (m)
m	mass (kg)
$\dot{m}$	mass flow rate (kg/s)
MW	molar mass (kg/kmol)
P	pressure (kPa)
q''	heat flux (kW/m^2)
$\dot{Q}$	heat rate (kW)
Re	Reynolds number
s	specific entropy (kJ/kg K)
T	temperature (°C or K)
v	specific volume (m^3/kg)
V	velocity (m/s)
$\dot{W}$	work rate (kW)

Acronyms

AbC	absorption cooling
C.R	concentration ratio
$C.R_{max}$	maximum concentration ratio
CFD	computational fluid dynamics
COP	coefficient of performance
CPV	concentrated photovoltaics
CPV/T	concentrated photovoltaics/thermal
CSP	concentrated solar power
EES	Engineering Equation Solver
EGS	enhanced geothermal system

En energy
Ex exergy
FC fuel cell
HAWT horizontal axis wind turbine
HCPV highly concentrated photovoltaics
HEX heat exchanger
HHV higher heating value
HPST high pressure steam turbine
HTF heat transfer fluid
HVAC heating ventilation and cooling
HX heat exchanger
I irradiance
IMM inverted metamorphic multijunction
In input
LBAC lithium bromide absorption chiller
LHV lower heating value
MJSC multi-junction solar cells
NBHT nucleate pool boiling heat transfer
ORC organic Rankine cycle
OTEC ocean thermal energy conversion
Out output
PCM phase change material
PEM proton exchange membrane
PTC parabolic trough collector
PTSC parabolic trough solar collectors
PV photovoltaics
RO reverse osmosis
SAM System Advisor Model
SP state point
TEG thermoelectric generator
TES thermal energy storage
VAWT the vertical axis wind turbines
WTES wind thermal energy systems

Greek letters

$I_{DNI, t}$ total incoming irradiation power (kW)
ω humidity ratio (kg water/kg air)
f Darcy friction factor
ρ density (kg/m^3)
η efficiency
ψ Exergy efficiency

Subscripts and superscripts

0 ambient condition
1, 2, … i state points
a charging water inlet state
ABC absorption cooling system
abs absorber
AC absorption cooling
avg. average
C compressor

ch	chemical
cogen	cogeneration
cond	condenser
d	destruction
DHW	domestic hot water
DNI	direct normal irradiation
E	electrolyzer
elec	electrolysis
EV	expansion valve
Eva	evaporator
GEN	generator
geo	geothermal
HE	heat exchanger
HP	heat pump
HX	heat exchanger
i	state number
in	input
m	maximum
multigen	multigeneration
oc	open circuit
ORC	organic Rankine cycle
out	output
ov	overall
P	pump
ph	physical
PV	photovoltaic
RC	Rankine cycle
s	source or sink
sys	system
T	turbine
TEG	thermoelectric generator
TES	thermal energy storage
trigen	trigeneration

References

[1] Dincer I, Zamfirescu C. Advanced power generation systems. Boston, Elsevier; 2014.

[2] Kamiya N, Ohta T, Asano N. Hybridized hydrogen production system with Fe—I. Photochemical reaction in concentrated phosphoric acid. Int J Hydrogen Energy 1984;9:563–70. https://doi.org/10.1016/0360-3199(84)90235-0.

[3] Ohta T, Asakura S, Yamaguchi M, Kamiya N, Gotgh N, Otagawa T. Photochemical and thermoelectric utilization of solar energy in a hybrid water-splitting system. Int J Hydrogen Energy 1976;1:113–6. https://doi.org/10.1016/0360-3199(76)90063-X.

[4] Zamfirescu C, Naterer G, Dincer I. Solar light-based hydrogen production systems. In: Anwar S, editor. Encycl. Energy Eng. Technol. 1st ed. New York: Taylor and Francis; 2012.

[5] Perret R. Solar Thermochemical hydrogen production research (STCH). Sandia National Lab. (SNL-CA), Livermore, CA (United States); 2011.

[6] Dincer I, Zamfirescu C. Chapter 7—renewable-energy-based Power generating systems. Adv. Power Gener. Syst. Boston: Elsevier; 2014. p. 369–453. https://doi.org/10.1016/B978-0-12-383860-5.00007-9.

[7] Betz A. Introduction to the theory of flow machines. 1st ed. Karlsruhe: Pergamon Press; 1966.

[8] Okazaki T, Shirai Y, Nakamura T. Concept study of wind power utilizing direct thermal energy conversion and thermal energy storage. Renew Energy 2015;83:332–8. https://doi.org/10.1016/j.renene.2015.04.027.
[9] Karasu H, Dincer I. Analysis and efficiency assessment of direct conversion of wind energy into heat using electromagnetic induction and thermal energy storage. J Energy Resour Technol 2018;140:071201https://doi.org/10.1115/1.4039023.
[10] Şen Z. 4.12 Hydropower conversion. In: Dincer I, editor. Compr. Energy Syst. 1st ed. Cambridge, United States: Elsevier; 2018. p. 545–72. https://doi.org/10.1016/B978-0-12-809597-3.00416-8.
[11] EDF. Tidal Power | EDF France. La Rance Tidal Power Plant a World First 2018. https://www.edf.fr/en/the-edf-group/industrial-provider/renewable-energies/marine-energy/tidal-power (Accessed 23November 2018).
[12] Cheng ZL, Li X, Chung T-S. The forward osmosis-pressure retarded osmosis (FO-PRO) hybrid system: A new process to mitigate membrane fouling for sustainable osmotic power generation. J Memb Sci 2018;559:63–74. https://doi.org/10.1016/J.MEMSCI.2018.04.036.
[13] Long R, Lai X, Liu Z, Liu W. A continuous concentration gradient flow electrical energy storage system based on reverse osmosis and pressure retarded osmosis. Energy 2018;152:896–905. https://doi.org/10.1016/J.ENERGY.2018.04.022.
[14] Achilli A, Cath TY, Childress AE. Power generation with pressure retarded osmosis: An experimental and theoretical investigation. J Memb Sci 2009;343:42–52. https://doi.org/10.1016/J.MEMSCI.2009.07.006.
[15] El-Emam RS, Dincer I. Development and assessment of a novel solar heliostat-based multigeneration system. Int J Hydrogen Energy 2018;43:2610–20.
[16] System Advisor Model (SAM), National Renewable Energy Laboratory 2010. https://sam.nrel.gov/ (Accessed 25 January 2017).
[17] Blair N, Dobos A, Freeman J, Neises T, Wagner M, Ferguson T, et al. System advisor model, sam 2014.1. 14: General description. NREL rep no TP-6A20-61019. CO: Natl Renew Energy Lab Golden; 201413.
[18] Islam S, Dincer I, Yilbas BS. Development, analysis and assessment of solar energy-based multigeneration system with thermoelectric generator. Energ Conver Manage 2018;156:746–56. https://doi.org/10.1016/J.ENCONMAN.2017.09.039.
[19] Montes MJ, Abánades A, Martínez-Val JM, Valdés M. Solar multiple optimization for a solar-only thermal power plant, using oil as heat transfer fluid in the parabolic trough collectors. Sol Energy 2009;83:2165–76. https://doi.org/10.1016/J.SOLENER.2009.08.010.
[20] Islam S, Dincer I. Development, analysis and performance assessment of a combined solar and geothermal energy-based integrated system for multigeneration. Sol Energy 2017;147:328–43.
[21] Tzivanidis C, Bellos E, Korres D, Antonopoulos KA, Mitsopoulos G. Thermal and optical efficiency investigation of a parabolic trough collector. Case Stud Therm Eng 2015;6:226–37. https://doi.org/10.1016/J.CSITE.2015.10.005.
[22] El-Emam RS, Dincer I. Investigation and assessment of a novel solar-driven integrated energy system. Energ Conver Manage 2018;158:246–55.
[23] Harrison TD, Randall RA. Thermal storage experience at the MSSTF and plans for the future. Therm Energy Storage Fourth Annu Rev Meet NASA CP-2125, 653 Pages, Publ by NASA, Washington, DC, 1980, p125; 2125: 125.
[24] Al-Sulaiman FA, Dincer I, Hamdullahpur F. Exergy modeling of a new solar driven trigeneration system. Sol Energy 2011;85:2228–43. https://doi.org/10.1016/j.solener.2011.06.009.
[25] Kearney D, Kelly B, Herrmann U, Cable R, Pacheco J, Mahoney R, et al. Engineering aspects of a molten salt heat transfer fluid in a trough solar field. Energy 2004;29:861–70. https://doi.org/10.1016/S0360-5442(03)00191-9.
[26] Kearney D, Herrmann U, Nava P, Kelly B, Mahoney R, Pacheco J, et al. Assessment of a molten salt heat transfer fluid in a parabolic trough solar field. J Sol Energy Eng 2003;125:170. https://doi.org/10.1115/1.1565087.
[27] Reddy KS, Jawahar V, Sivakumar S, Mallick TK. Performance investigation of single-tank thermocline storage systems for CSP plants. Sol Energy 2017;144:740–9. https://doi.org/10.1016/J.SOLENER.2017.02.012.
[28] Blake DM, Moens L, Hale MJ, Price H, Kearney D, Herrmann U. New heat transfer and storage fluids for parabolic trough solar thermal electric plants, In: PACES International Symposium on concentrating Solar Power and Chemical Energy Technologies, Zurich, Switzerland; 2002 September 4–6.
[29] Roberts Ron R. Optimization of flashing pressure for geothermal Power plants using single through quadruple flashing steps, In: Oshawa, Ontario, University of Ontario Institute of Technology. Prep. ENGR 5100G Adv. Energy Syst; 2012. p. 26.

[30] Khalid F, Dincer I, Rosen MA. Techno-economic assessment of a solar-geothermal multigeneration system for buildings. Int J Hydrogen Energy 2017;42:21454–62.
[31] Bicer Y, Dincer I. Development of a new solar and geothermal based combined system for hydrogen production. Sol Energy 2016;127:269–84. https://doi.org/10.1016/j.solener.2016.01.031.
[32] Bicer Y, Dincer I. Analysis and performance evaluation of a renewable energy based multigeneration system. Energy 2016;94:https://doi.org/10.1016/j.energy.2015.10.142.
[33] Khan SA, Bicer Y, Koç M. Design and analysis of a multigeneration system with concentrating photovoltaic thermal (CPV/T) and hydrogen storage. In: Int J Hydrogen Energy. 2018https://doi.org/10.1016/j.ijhydene.2018.12.047.
[34] timeanddate.com. https://www.timeanddate.com/weather/qatar/doha/climate; 2018.
[35] Globalsolaratlas: World Bank Group. World Bank Group, 2018.
[36] Steiner MA, Geisz JF, Friedman DJ, Olavarria WJ, Duda A, Moriarty TE. Temperature-dependent measurements of an inverted metamorphic multijunction (IMM) solar cell, In: Conf Rec IEEE Photovolt Spec Conf; 2011https://doi.org/10.1109/PVSC.2011.6186461.
[37] Steiner M, Siefer G, Schmidt T, Wiesenfarth M, Dimroth F, Bett AW. 43% Direct sunlight conversion efficiency using 4J cells with achromatic full glass lens. AIP Conf Proc 2016;1766:080006https://doi.org/10.1063/1.4962104.
[38] Klein. EES: Engineering equation solver | F-chart software: Engineering Software 2018.
[39] Herold K, Radermacher R, Klein S. Absorption chillers and heat pumps. 2nd ed. New York, NY: CRC Press; 2016https://doi.org/10.1201/b19625.
[40] PlusICE. PlusICE Phase Change Materials Product limited. 2013.
[41] William GE, Mohamed MH, Fatouh M. Desiccant system for water production from humid air using solar energy. Energy 2015;90:1707–20. https://doi.org/10.1016/J.ENERGY.2015.06.125.
[42] Talaat MA, Awad MM, Zeidan EB, Hamed AM. Solar-powered portable apparatus for extracting water from air using desiccant solution. Renew Energy 2018;119:662–74. https://doi.org/10.1016/j.renene.2017.12.050.
[43] Bicer Y, Dincer I. Energy and exergy analyses of an integrated underground coal gasification with SOFC fuel cell system for multigeneration including hydrogen production. Int J Hydrogen Energy 2015;40:13323–37. https://doi.org/10.1016/j.ijhydene.2015.08.023.
[44] Larminie J, Dicks A, Larminie J, Dicks A. Proton exchange membrane fuel cells. In: Fuel Cell System Explained. John Wiley & Sons, Ltd; 2013. p. 67–119. https://doi.org/10.1002/9781118878330.ch4.
[45] Dincer I, Ratlamwala TAH. Integrated absorption refrigeration systems. Cham: Springer International Publishing; 2016https://doi.org/10.1007/978-3-319-33658-9.
[46] Nellis G, Klein SA. Heat transfer. Coulson Richardson's Chem. Eng. 7th ed. Elsevier; 20183–264. https://doi.org/10.1016/B978-0-08-102550-5.00001-8.
[47] Liang G, Mudawar I. Pool boiling critical heat flux (CHF)—Part 1: Review of mechanisms, models, and correlations. Int J Heat Mass Transf 2018;117:1352–67. https://doi.org/10.1016/j.ijheatmasstransfer.2017.09.134.
[48] Shahid UB, Bicer Y, Abdala A, Ahzi S. Thermodynamic assessment of an integrated renewable energy multigeneration system including ammonia as hydrogen carrier and phase change material energy storage. Energy Convers Manag 2019;198:111809. https://doi.org/10.1016/j.enconman.2019.111809.
[49] Mohamed AMO, Bicer Y. Integration of pressure retarded osmosis in the solar ponds for desalination and photo-assisted Chloralkali processes: energy and exergy analysis. Energy Convers Manag 2019;1(195):630–40.

Further reading

[50] Kui LICHLIU. Analysis of photosynthesis efficiency of maize hybrids with different yield in the later growth stage. Acta Agron Sin 2002;3:17.
[51] Zhu X-G, Long SP, Ort DR. What is the maximum efficiency with which photosynthesis can convert solar energy into biomass? Curr Opin Biotechnol 2008;19:153–9. https://doi.org/10.1016/J.COPBIO.2008.02.004.
[52] Chow WS, Melis A, Anderson JM. Adjustments of photosystem stoichiometry in chloroplasts improve the quantum efficiency of photosynthesis. Proc Natl Acad Sci U S A 1990;87:7502–6. https://doi.org/10.1073/PNAS.87.19.7502.
[53] Sinclair TR, Horie T. Leaf nitrogen, photosynthesis, and crop radiation use efficiency: a review. Crop Sci 1989;29:90. https://doi.org/10.2135/cropsci1989.0011183X002900010023x.

[54] Photosynthesis. Energy efficiency of photosynthesis | Britannica.com 2018. https://www.britannica.com/science/photosynthesis/Energy-efficiency-of-photosynthesis (Accessed 23 November 2018).
[55] Sen Z. Practical and applied hydrogeology. Oxford, UK: Elsevier; 2015https://doi.org/10.1016/C2013-0-14020-2.
[56] Islam S, Dincer I, Yilbas BS. Development, analysis and assessment of solar energy-based multigeneration system with thermoelectric generator. Energ Conver Manage 2018;156:746–56.
[57] Islam S, Dincer I. Development, analysis and performance assessment of a combined solar and geothermal energy-based integrated system for multigeneration. Sol Energy 2017;147:328–43. https://doi.org/10.1016/J.SOLENER.2017.02.048.

CHAPTER

7

Enhanced dimensions of integrated energy systems for environment and sustainability

7.1 Introduction

The relation between the environment and energy/exergy became more significant when environmental concerns started to be a main issue of energy systems. With the problem of climate change and other environmental matters, including ozone depletion and acid rain, there has been more intensive research on global environmental issues, with a specific focus on the connection between energy science and the environment. Since energy production, conversion and consumption impact the environment as well as living standards, the sustainability level of energy systems has become a major concern. In addition, increasing concerns about anthropogenic non-sustainable actions and their consequences on the environment have led to attempts to create systems for accomplishing sustainable development.

In this context, exergy analysis has emerged as a crucial assessment tool, because it is a true indicator of quality and can help increase the efficiency of energy conversion processes, as has been demonstrated in the previous chapters. Energy, exergy, and environment are directly related, which are critical aspects of sustainable development. Assessment tools based on exergy are more effective than pure energy-based methods for detailed analysis of integrated energy systems, because they can provide more in-depth, meaningful and realistic results. Nevertheless, environmental impacts of energy systems are not directly measured via exergy analysis. Hence, there is a need for additional tools to assess environmental impacts and other dimensions of energy systems.

In recent decades, an environmental impact assessment tool combining both exergy and environmental assessment was proposed, called exergoenvironmental analysis. A society seeking true sustainable development preferably should utilize only energy resources that have no or minimal environmental impact. Nonetheless, as all energy resources have some

Integrated Energy Systems for Multigeneration
https://doi.org/10.1016/B978-0-12-809943-8.00007-8

environmental impact, it is more realistic to propose that some of the issues concerning the limitations imposed on sustainable development by environmental emissions and their negative impacts can be partly overcome by increasing energy efficiency. Environmental concerns should primarily deal with the emissions of greenhouse gases and other gases into the atmosphere. Waste heat emissions to the environment are also serious concerns because waste heat can substantially raise the temperature of the environment, causing thermal air pollution. If it is not carefully controlled, local temperatures can increase and cannot be kept within safe and desirable levels. This thermal pollution can also disturb aquatic life and biological balances in lakes and rivers. These issues are in the scope of environmental assessment. Exergoenvironmental analysis considers the impact of all streams entering and exiting an overall system and each of its components. The environmental impact of the system components attained over their life cycles is applied to the component impact in the exergoenvironmental impact balance. Exergoenvironmental analysis is calculated using an indicator that defines the environmental impact per unit exergy. This indicator is the average environmental impact related to the production of a specific stream in the system per unit of exergy. Usually, the environmental impacts related to a stream are defined through the life cycle assessment (LCA) method.

In addition to environmental concerns, there are also social and economic aspects of sustainability. If ecosystems are irreversibly transformed by human actions, then succeeding human generations can be forced to live in an unsustainable environment. The heterogeneity of the environmental, social, and economic aspects of sustainability obviously needs to be taken into account. In order to have a better understanding of the economic aspects of energy systems, another potential tool can be used combining exergy and economics, which is called exergoeconomic analysis. This combination of exergy analysis with economic concepts considers the costs of the energy streams. Exergoeconomics can help reveal the cost of losses within a complex integrated energy system.

In summary, Fig. 7.1 clearly shows the progress toward comprehensive sustainability assessment of integrated energy systems, starting from exergy analysis and extending to exergo-enviro-sustainability assessment.

In this chapter, energy and exergy analyses of integrated energy systems are extended to environmental, economic and sustainability aspects and enhanced analysis of energy systems are introduced as illustrated in Fig. 7.1. Since exergy is a true measure of available energy and

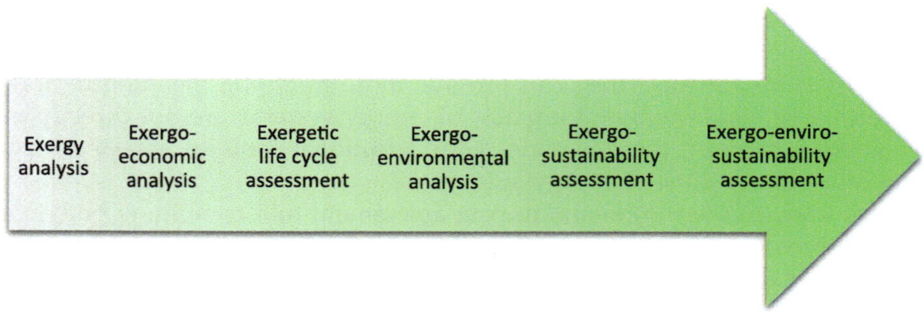

FIG. 7.1 Enhanced analysis of integrated energy systems.

achievable work, combining exergy analysis with economics and environment creates a beneficial and reasonable tool for investigating any energy system. Therefore, exergoeconomics is introduced in this chapter. Furthermore, environmental consequences of any energy system are very important decision parameters. Hence, a combination of exergy analysis and environmental impact assessment can make the overall assessment of any energy system more comprehensive. These analyses can further be extended to sustainability assessment by incorporating social aspects as well. This chapter also provides future directions in which several ideas for novel system integrations are discussed and presented.

7.2 Exergoeconomic analysis

Exergy analysis was associated with economic analysis of thermal energy systems by several pioneering researchers such as Bejan [1]. The main idea is to reduce the irreversibility within the system under economic constraints. In this way, an optimization of the integrated systems can be performed. The combination of exergy and economic analysis was later named exergoeconomics. The application of exergoeconomic analysis and an optimization approach to thermal systems was also detailed in Bejan [2, 3], Tsatsaronis [4], Tsatsaronis and Moran [5] and Szargut et al. [6]. A systematic approach for conducting exergoeconomic analysis based on fundamental balance equations (including mass, energy, entropy, exergy, and cost balances) was documented in Dincer and Rosen [7]. The origins of exergoeconomic analysis extend back to the 1980s. Tsatsaronis [4] defined exergy-economic cost accounting, exergy economic analysis, exergy-economic similarity number, and product/cost efficiency graphs as the chief methods for exergoeconomic analyses of thermal energy systems. In addition, Szargut et al. [6] presented further developments in exergoeconomic analysis. The concept of exergetic cost was proposed by Lozano and Valero [8] to account for several characteristics such as evaluation of alternative energy savings within the system, cost allocation, operation optimization, single- to multiobjective optimization, and local and partial optimization of subsystems. Exergoeconomic analyses of power plants were detailed by Kotas [9].

Humans and other living organisms desire sources with low entropy or high exergy. In fact, entropy and exergy are not conserved, unlike energy. A low-entropy source is transformed into a high-entropy waste by living organisms as a result of their activities. Similarly, high-exergy sources are destroyed by systems such as living organisms, Earth's natural cycles, etc. When the whole planet is considered as a boundary, the net exergy received by the planet is utilized to drive the Earth's water, wind, and other natural systems, as well as life on the planet; however, it is gradually destroyed due to irreversibility. Once exergy is destroyed, it cannot be reused. Therefore, we try to minimize the exergy destruction within systems. Exergy is associated with the reference environment, as it takes into account the temperature, pressure, composition, and species concentration. In other words, exergy embodies a portion of energy that is useful and beneficial to society, and thus it bears an economic value. In system design or analysis, suitable techniques should be used to define the prices and the costs associated with exergy content. This requires a specific type of formulation of costs based on exergy content. It has been proposed that, when analyzing a thermal system, it is rational to allocate costs in relation to outputs and accumulations of exergy. The cost terms of physical resources such as fuels and materials need to have a strict link with the exergy content of the resource in order to substitute resource-saving and effective technology. Once the cost of exergy is defined, additional models can be formed to create a costing system for other items of interest in an exergoeconomic analysis. The cost terms that are not directly related to energy such as material supply, environment remediation expenditure, incidental expenditures, and so on can be estimated using exergy content as a basis for cost accounting. The economic value of system outputs can be similarly distributed based on exergy.

7.2.1 EXCEM method

The EXCEM method is one of the models developed for exergoeconomic analysis. It is, in fact, a broadening of the usual exergy method. In addition to the four fundamental balance equations (mass, energy, entropy, and exergy), in this method, we add another balance equation, which is the cost balance equation. It is noted that in the EXCEM method the balance equations are expressed for non-conserved terms such as entropy, exergy, and costs. Therefore, it is important to consider the generation and consumption terms in order to attain a balance equation. This technique was initiated in the work by Rosen [10] and was also advanced in studies by Rosen and Dincer [11, 12] and Dincer and Rosen [7]. The main reasoning behind EXCEM analysis is that an understanding of the performance of a system necessitates an investigation of the flows of each of the amounts characterized by EXCEM into, out of, and at all points within a system, as is clearly depicted in Fig. 7.2.

In a cost balance equation for cost accumulation and cost outputs, a well-defined formulation does not exist. These types of costs are individually distributed, based on the system purpose and type and other financial factors. For instance, the costs can be allocated proportionally to the outputs and accumulations of a quantity (such as mass, energy, or exergy) or all non-waste outputs and accumulations of an amount.

The cost balance equation can be written in rate form or amount form, respectively, as follows:

$$\dot{C}_{in} + \dot{C}_{generation} - \dot{C}_{out} = \frac{dC}{dt} \tag{7.1}$$

$$C_{in} + C_{generation} - C_{out} = \Delta C \tag{7.2}$$

The cost input terms are usually determined by the information of the costs for each input and the rate of supply. For instance, the fuel rate delivered to any related system is commonly known. The specific cost of the fuel is known as well. In this case, finding the cost rate of fuel

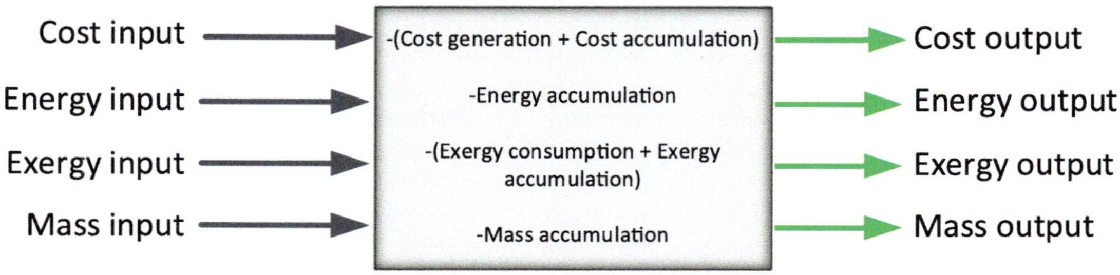

FIG. 7.2 EXCEM model illustrated as inputs and outputs.

input is simply attained by multiplication of the rate of fuel depletion and the specific cost of fuel. In the cost balance equation, the cost generation term represents the proper capital and other costs related to the construction/formation and maintenance of a system. The cost generation term can be expressed as

$$C_{generation} = C_{capital} + C_{O\&M} + C_{costcreation} \tag{7.3}$$

The cost generation includes the capital costs, operation and maintenance cost, and other cost creation for several reasons. The capital costs usually represent the most substantial factor of the total cost generation. The other cost generation terms can include interest, insurance, and so on. The cost generation rate is a differential cost balance signifying the total cost generation levelized over the operational lifetime of the system. Cost generation terms except for capital costs are commonly proportional to capital costs. Therefore, the tendencies defined are in qualitative agreement with those determined when the complete cost generation term is considered. The system evaluation in EXCEM can be conducted using parameters denoting the thermodynamic loss against the capital cost. Hence, the energetic loss ratio can be expressed as follows [13]:

$$R_{en} = \frac{E_{loss}}{C_{capital}} \tag{7.4}$$

where $E_{loss} = \sum E_{in} - \sum E_{product} - \triangle E_{accumulation}$ and $E_{product}$ denotes the energy recovered in the form of product, which is an output term; $\sum E_{out} = \sum (E_{product} + E_{loss})$.

The energetic loss ratio is expressed in general in terms of amounts, where E_{loss} denotes the total energy loss throughout the system lifetime and $C_{capital}$ denotes the current value of capital reduced for the whole lifetime. Similarly, the exergetic loss ratio can also be expressed as follows:

$$R_{ex} = \frac{Ex_{loss}}{C_{capital}} \tag{7.5}$$

where $Ex_{loss} = \sum Ex_{in} - \sum Ex_{product} - \sum Ex_{destruction} - \triangle Ex_{accumulation}$ signifies the exergy wasted into the ambient environment.

In the exergetic loss ratio, Ex_{loss} denotes the total exergy lost during the system lifetime. These equations can also be written in rate forms. In this case, the rate of lost exergy is divided by the rate of capital cost in which the reduced capital in present value is levelized for the whole lifetime. Though there can be few differences in the economic engineering analysis of the system, here a simple method is followed leading to the determination of the discounted capital in present worth, which does not consider loans, but only invested capital. In economic analysis, several financial parameters must be defined, as given in Table 7.1.

7.2.2 SPECO method

There is another approach for conducting exergoeconomic analysis, called the SPECO method, for Specific Energy Costing. This method was initially used to evaluate and improve numerous systems and components, such as energy production plants. This SPECO method has been detailed by several authors including Tsatsaronis and Moran [5], Lazzaretto and

TABLE 7.1 Numerous parameters generally used in economic analysis.

Parameter	Notation	Details
Tax on credit due to renewable energy	t_e	With energy use, a tax credit may apply
Tax on income	t_i	Income tax on money coming from output costs
Tax on property	t_p	Tax on property value
Tax on salvage	t_s	Tax on the salvage value of the system
Capital salvage factor	CSF	A portion CSF of invested capital is recovered at the end of lifetime
Levelized product price	LPP	Unit product sell price at present worth
System lifetime	n	Number of years in lifetime
Real discount rate	r	Rate of return for the discounted cash flow to determine value
Annual production	AP	Number of annual products
Annual consumption	AC	Annual consumption of consumables
Levelized cost of consumables	LCC	Summation of all inputs levelized for present worth
Operation and maintenance cost factor	OM_{CF}	A portion of either net income or invested capital
Invested capital	$C_{capital}$	Monetary value paid upfront to start the business

Tsatsaronis [14] and Bejan et al. [2]. Cost accounting with the SPECO approach utilizes cost balances and is commonly related to defining the real cost of goods or services, giving a reasonable basis for pricing them. Furthermore, the method provides a means for distributing and adjusting expenditures and defines useful assessment factors to support in making and assessing design and operating decisions.

For a system operating at steady state, there can be a number of incoming and outgoing material flows as well as heat and work relationships with the environment. As exergy is a more realistic measure of the value, and these effects and costs need to be only allocated to products of value, it is more accurate to utilize exergy as an origin for allocating costs in energy systems. Exergy costing delivers a sensible basis for allocating costs to the relationships that a thermal system has with the environment and to the causes of ineffectiveness within it. In the SPECO approach, the costs are defined based on the exergy content. In the SPECO analysis, a cost balance is typically expressed in a rate form, in which the cost rates are specified in dollars per hour and applied to the overall system operating at steady state.

In order to provide a better understanding, a general system having work, heat and mass interactions is considered here, as shown in Fig. 7.3. In this system, heat is applied and considered as an input, along with other types of inputs counting on invested capital. Heat is also output from the system, which can be attributed to heat losses or reaction heat. All streams, which are heat, work, and material, coming into and leaving the system are written by considering their exergy content.

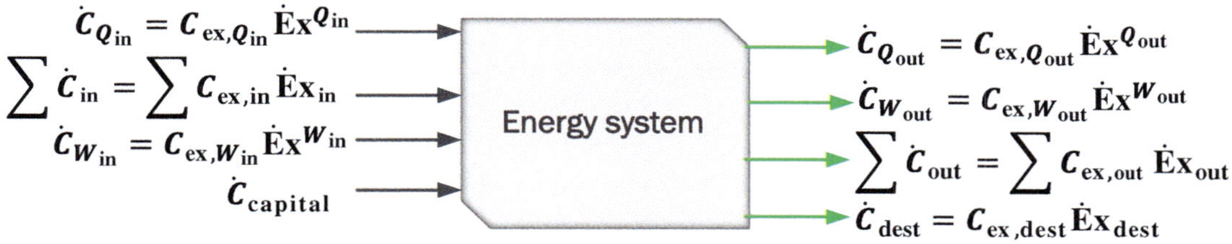

FIG. 7.3 Schematic illustration of exergetic cost balance equation.

There is an allocated $C_{ex,\,i}$ for each stream i. Fig. 7.3 illustrates a general energy system for the SPECO method application.

The streams are explained as follows:

$$\dot{C}_{Q_{in}} = C_{ex,Q_{in}} \dot{E}x^{Q_{in}} \tag{7.6}$$

signifies the rate of cost input due to the supplied heat to the system, where $C_{ex,\,Q_{in}}$ is a specific cost for exergetic heat and $\dot{E}x^{Q_{in}}$ is the thermal exergy rate delivered to the system.

$$\dot{C}_{W_{in}} = C_{ex,W_{in}} \dot{E}x^{W_{in}} \tag{7.7}$$

signifies the rate of cost input because of the provided work (mechanical, electrical, etc.) to the system, where $C_{ex,\,W_{in}}$ is a specific cost for work and $\dot{E}x^{W_{in}}$ is the exergy rate delivered to the system in work form.

$$\sum \dot{C}_{ex_{in}} = \sum C_{ex_{in}} \dot{E}x_{in} \tag{7.8}$$

signifies the cost rate in physical inputs; all types of inputs are considered (energy inputs, material inputs). For each input, the exergy rate is specified as $\dot{E}x_{in}$ and the specific cost of the particular type of exergy is denoted as $\dot{C}_{ex_{in}}$.

$$\dot{C}_{capital} \tag{7.9}$$

signifies the rate of capital cost, accounting for the operational lifetime of the system.

$$\sum \dot{C}_{ex_{out}} = \sum C_{ex_{out}} \dot{E}x_{out} \tag{7.10}$$

signifies the rate of cost in output stream (material, product, etc.), where $\dot{E}x_{out}$ is the exergy rate in product or material and $C_{ex_{out}}$ is the associated cost of exergy of the output stream (material, product, etc.).

$$\dot{C}_{Q_{out}} = C_{ex,Q_{out}} \dot{E}x^{Q_{out}} \tag{7.11}$$

signifies the rate of cost output due to the exiting heat (heat loss, reaction heat, etc.) from the system, where $C_{ex,\,Q_{out}}$ is a specific cost for exergetic heat output and $\dot{E}x^{Q_{out}}$ is the thermal exergy rate exiting the system.

$$\dot{C}_{W_{out}} = C_{ex,W_{out}} \dot{E}x^{W_{out}} \tag{7.12}$$

signifies the rate of cost output due to the obtained or produced work (mechanical, electrical, etc.) from the system, where $C_{ex,\ W_{out}}$ is a specific cost for obtained or produced work and $\dot{E}x^{W_{out}}$ is the exergy rate from the system in the form of work.

$$\dot{C}_{dest} = C_{ex_{dest}} \dot{E}x_{dest} \tag{7.13}$$

signifies the cost rate of exergy destruction that is associated with the distributed specific cost of exergy destruction $C_{ex_{dest}}$ and the rate of exergy destruction $\dot{E}x_{dest}$ occurring in the system.

The cost balance equation for the general system can be expressed as follows:

$$\dot{C}_{Q_{in}} + \dot{C}_{W_{in}} + \sum \dot{C}_{ex_{in}} + \dot{C}_{capital} = \dot{C}_{Q_{out}} + \dot{C}_{W_{out}} + \sum \dot{C}_{ex_{out}} + \dot{C}_{dest} \tag{7.14}$$

This cost balance equation expresses that the total costs of the outgoing exergy streams (including the cost in exergy destruction) are identical to the total expenditure to attain them, namely the cost of the entering exergy streams plus the capital and other costs. The capital investment rate can be written as follows:

$$\dot{C}_{capital} = \frac{C_{capital} CRF\ OM_{CF}}{n_{hours}} \tag{7.15}$$

where n_{hours} denotes the number of operating hours in a year. Note that if the other units are provided in seconds, then n_{hours} should be converted into seconds. The capital investment rate is expressed in dollars per hour (or dollars per second depending on the case) and it is calculated using the invested capital denoted in present value, the capital recovery factor (*CRF*), and operation and maintenance cost factor (OM_{CF}), which can be determined based on a specific percentage of the invested capital. A usual worth of about 1.06 can be taken for OM_{CF} for SPECO analysis, if not specified otherwise. It is important to check the units of each term since the exergy rates are mostly calculated in kJ/s.

The capital recovery factor (CRF) can be defined as

$$CRF = \frac{i(1+i)^n}{(1+i)^n - 1} \tag{7.16}$$

where i is the interest rate and n is the lifetime in years.

Summation of additional costs caused by exergy destruction, $\dot{C}_{dest}$ and final capital and operating cost rate $\dot{C}_{capital}$ gives a critical parameter called the total cost rate:

$$\dot{C}_{total} = \dot{C}_{dest} + \dot{C}_{capital} \tag{7.17}$$

Total cost rates of the system consist of the total investment cost and cost of exergy destruction.

The exergoeconomic factor, which is a measure of system effectiveness in terms of cost obtained through exergoeconomic analysis, is given as follows:

$$f = \frac{\dot{C}_{capital}}{\dot{C}_{dest} + \dot{C}_{capital}} \tag{7.18}$$

The exergoeconomic variables $\dot{C}_{capital}$ and $\dot{C}_{dest}$ imply the significance of a component in the system optimization, whereas the variable f, the exergoeconomic factor, is a relative measure of the component cost effectiveness.

The data to be used in the economic cost assessment must be converted in the same reference year that is the basis year for the calculations [13]. Based on a different time frame, an appropriate cost index can be used for normalization of the costs. The cost balance for each part of the system as well as for the complete system can be written. When cost balances are solved simultaneously, the assessment of the cost of exergy destructions is possible. Exergoeconomic factors are indicators of non-exergy related costs. The system evaluation may be performed with reference to the cost rate of exergy destruction divided by the cost rate for the capital or as the ratio of the variance between product and the fuel cost.

7.3 Exergoenvironmental analysis

Exergy analysis can also be associated with environmental impact analysis. When considering exergy in environmental impact analysis, the inventory analysis stage needs to be accounted more cautiously for mass and energy flows into and out of the system, and through complete phases of the life cycle. Then, the energy flows are associated with exergy flows. Ultimately, environmental impact indicators associated with exergy can be developed and studied, as reported by Dincer and Rosen [7].

When the environmental impacts are accounted for, the effects of all the system releases to the environment also need to be considered, as well as the contamination and resource reduction associated with the system building. Hence, the system boundary for the analysis is fixed at greater bounds than those of the real system. Indeed, the system construction process can be modeled in terms of required exergy for building and the environmental effect. Furthermore, the environmental impacts caused by system discharges can also be elaborated. Thus, the system is extended to two subunits: (i) upstream: the system construction process and (ii) downstream: waste environmental and decommissioning processes. For any system, the exergy stream in the type of fuels is stated as direct exergy. For the upstream and downstream units, the delivered exergy is denoted as the indirect exergy. Hence, the total exergy stream to be measured in a comprehensive exergoenvironmental analysis can be written as follows:

$$\dot{E}x_{supply} = \dot{E}x_{direct} + \dot{E}x_{indirect}^{upstream} + \dot{E}x_{indirect}^{downstream} \tag{7.19}$$

A similar equation can also be written for the environmental impact (*EI*), namely

$$EI_{total} = EI_{direct} + EI_{indirect} \tag{7.20}$$

where EI is environmental impact indicator; EI_{direct} signifies the downstream environmental impact, that is, the environmental impact initiated by the system because of the operation; and $EI_{indirect}$ signifies the environmental impact related to system construction. In most of the exergoenvironmental analyses, only the direct impacts are taken into account. However, when the exergoenvironmental analysis is extended to a more general exergetic life cycle analysis, all direct and indirect impacts are considered.

In most practical applications, the energy used to run any process is commonly electricity or a petrochemical fuel. The environmental impact and exergy efficiencies of these types of energy production are generally known or calculated through exergy analysis and life cycle assessment.

Therefore, when the rate of consumption of fuel is determined, indirect exergy associated with the supply of fuel and the environmental impact can also be identified. There are examples of the several environmental impact categories, for instance abiotic depletion, land use, climate change, stratospheric ozone depletion, eutrophication, human toxicity, acidification, photo-oxidant formation, waste heat generation, marine ecotoxicity and terrestrial ecotoxicity. The main purpose of the exergoenvironmental assessment is determining the effects of direct and/or indirect environmental impacts associated with the destruction of the process or system exergy. After determination of the impacts, they can be minimized by taking appropriate precautions and improvements. This can be performed by proposing alternative ways of efficiency enhancement. An additional method of minimizing the environmental impact is the lessening of the contaminating wastes.

The SPECO and EXCEM analyses, described previously, can possibly be broadened for exergoenvironmental analysis. Independent from the selected exergoenvironmental analysis method, the environmental impacts occurring within the entire system as well as its subcomponents should be initially defined. This step analyzes the pure ecological effects. Each exergy stream can be assigned a particular environmental effect factor [15].

A general exergetic environmental balance equation can be written considering all inputs and outputs in a system. For instance, if we consider a gas turbine (attached to a combustion chamber), the inlet and outlet of the gas turbine consist of gases. The exergy rates of the streams can be calculated through the exergy analysis. Based on the composition of the gases, total enthalpy, entropy and exergy values of inlet and outlet streams can be determined. We can also calculate the amount of work production and heat loss (if not adiabatic). Similarly, the environmental impacts of each flow, as well as component, need to be calculated on a rate basis. This can be accomplished using special techniques such as life cycle assessment. The degree of environmental impact to the amount of exergy ratio can be calculated for every stream of exergy (counting the exergy destruction as well). Consequently, a balance equation for the environmental impact of a general system (accounting for all possible interactions, e.g., heat, work and mass) can be written as follows:

$$\dot{E}I_{Q_{in}} + \dot{E}I_{W_{in}} + \sum \dot{E}I_{Min} + \dot{E}I_{sys} = \dot{E}I_{Q_{out}} + \dot{E}I_{W_{out}} + \sum \dot{E}I_{Mout} + \sum \dot{E}I_{dest,out} \tag{7.21}$$

where $\dot{E}I_{Q_{in}}$ is the environmental impact rate because of heating source required for a system, $\dot{E}I_{W_{in}}$ is the environmental impact rate because of work input required for a system, $\dot{E}I_{Min}$ is associated with any input stream of energy through mass, $\dot{E}I_{sys}$ denotes the environmental impact of a system itself (or its subsystems when the components are analyzed), $\dot{E}I_{Mout}$ is due to the release of mass into the environment of any material or product, and $\dot{E}I_{dest,out}$ denotes other outputs and exergy destruction.

When an interaction between any thermodynamic system and the environment occurs, exergy is destroyed. The destroyed exergy takes into account the immediate effect of the energy system on the environment. Consequently, the main objective of exergoenvironmental analysis is the identification and quantification of direct and/or indirect environmental impacts in correlation with exergy destroyed by the system.

Some simplified models can be used as an alternative, such as the ones mentioned by Lazzaretto and Toffolo [16], which relate the mass flow rate of CO and NOx emissions to fuel consumption. For carbon monoxide emissions, the correlation is as follows [16]:

$$\frac{\dot{m}_{\mathrm{CO}}}{\dot{m}_{\mathrm{f}}}=179{,}000\frac{\exp\left(\frac{7800}{T}\right)}{P^2\tau\left(\frac{\Delta P}{P}\right)^{0.5}} \tag{7.22}$$

For NOx emissions, the flowing equation is given as [16]

$$\frac{\dot{m}_{\mathrm{NOx}}}{\dot{m}_{\mathrm{f}}}=150{,}000\frac{\tau^{0.5}\exp\left(-\frac{71{,}100}{T}\right)}{P^{0.05}\tau\left(\frac{\Delta P}{P}\right)^{0.5}} \tag{7.23}$$

In these equations, the adiabatic flame temperature is denoted by T, combustion pressure is P and ΔP is the pressure drop through the combustion chamber, and τ is the residence time in the combustion zone. If the fuel contains sulfur, then an emission factor of sulfur dioxide can be determined. Usually, diesel fuel contains sulfur. The emission of sulfur dioxide and fuel combustion can be correlated as follows [16–18]:

$$\frac{\dot{m}_{\mathrm{SO_2}}}{\dot{m}_{\mathrm{f}}}=10^{-6}\left(\frac{M_{\mathrm{SO_2}}}{M_{\mathrm{S}}}\right)\cdot C_{\mathrm{S}}f_{\mathrm{c,SO_2}} \tag{7.24}$$

where $f_{\mathrm{c,SO_2}}$ is the conversion factor of sulfur in SO_2, C_S is the sulfur content of fuel in ppm, and M signifies molecular mass. Here, the factor $f_{\mathrm{c,\,SO_2}}$ is about 0.98 for ultra-low sulfur diesel (ULSD), and the sulfur content of the fuel is on the order of 15 ppm. The environmental impact can then be calculated based on a weighted average of the released impurities. For exergoenvironmental analysis, the chemical exergy of the pollutant can be used to form a weighted averaged environmental impact factor. When it is assumed that the polluting emissions are CO_2, NOx and SO_2, chemical exergy permits calculation of the total exergy of the pollutant stream as follows:

$$\dot{\mathrm{E}}\mathrm{x}_{\mathrm{pollutant}}=\dot{m}_{\mathrm{CO_2}}\mathrm{ex}^{\mathrm{ch}}_{\mathrm{NOx}}+\dot{m}_{\mathrm{NOx}}\mathrm{ex}^{\mathrm{ch}}_{\mathrm{NOx}}+\dot{m}_{\mathrm{SO_2}}\mathrm{ex}^{\mathrm{ch}}_{\mathrm{SO_2}} \tag{7.25}$$

where the chemical exergy of contaminants is given in Table 7.2.

The environmental impact factor is thus determined as the ratio between the exergy of a polluting stream divided by the mass flow rate of fuel combusted, $\dot{m}_{\mathrm{f}}$.

Furthermore, a dimensionless exergetic environmental impact factor can be derived for any system, attained by dividing the chemical exergy in a released pollutant by the exergy input into the system. Hence, the exergetic environmental impact factor is expressed as follows:

$$\mathrm{ExEI}=\frac{\dot{\mathrm{E}}\mathrm{x}_{\mathrm{pollutant}}}{\dot{\mathrm{E}}\mathrm{x}_{\mathrm{input}}} \tag{7.26}$$

The environmental impact can likewise be estimated based on exergy destruction. Actually, any output (beside the useful output) of an energy system such as emissions into the

TABLE 7.2 Chemical exergy of few atmospheric contaminants.

Pollutant	Molar weight (kg/kmol)	ex^{ch} (MJ/kmol)	ex^{ch} (MJ/kg)
CO_2	44	19.60	0.445
CH_4	16	831.66	52
NOx	38	72.4	2
SO_2	64	311	5
CO	28	275	10
VOCs	44	1233	28
PM	68	436	6

environment can be somewhat quantifiable by exergy destruction. In a dimensionless form, the exergetic sustainability impact can thus be determined as the ratio between exergy destruction and exergy input. Using this description, the exergetic environmental impact becomes equal to the relative exergy destruction, which is also known as the depletion factor, and is written as follows:

$$D_p = \frac{\dot{E}x_d}{\dot{E}x_{input}} \tag{7.27}$$

7.4 Life cycle assessment

The life cycle assessment (LCA) methodology is commonly used to investigate the life cycle emissions of a system, a process or a product from cradle to grave. This process is defined by the International Standards Organization norm ISO 14040 as the "compilation and evaluation of the inputs, outputs, and potential environmental impacts of a product system throughout its life cycle." The LCA is governed by international standards as listed in Table 7.3.

Once LCA is applied to an energy system, it determines an integrated efficiency of the complete process by considering final outputs over the lifespan and initial material inputs and the related energy flows. This type of cradle-to-grave analysis is principally essential for policy elaboration and decision making on means to achieve sustainable development. In the framework of sustainable energy, the LCA method can be applied for analyzing a particular technology for the entire life cycle, comparing various technologies, selecting a technology among several competitors, and designing more energy-efficient products and technologies [19].

An LCA study includes four main phases: (i) goal and scope definition, (ii) inventory analysis, (iii) impact assessment, and (iv) improvement analysis, as illustrated in Fig. 7.4. Each step can have improvement and interpretation phases to improve the accuracy of the results.

The goal and scope definition describes the targeted product, process or activity. It identifies the boundaries and environmental impacts to be studied for the assessment. Inventory

TABLE 7.3 ISO standards defining the life cycle assessment methodology.

Standard name	Title	Description
ISO 14040:2006	Environmental management-Life cycle assessment-Principles and framework	Describes the principles and framework for life cycle assessment (LCA) including: definition of the goal and scope of the LCA, the life cycle inventory analysis (LCI) phase, the life cycle impact assessment (LCIA) phase, the life cycle interpretation phase, reporting and critical review of the LCA, limitations of the LCA, the relationship between the LCA phases, and conditions for use of value choices and optional elements.
ISO 14044:2006	Environmental management-Life cycle assessment-Requirements and guidelines	Specifies requirements and provides guidelines for life cycle assessment (LCA) including: definition of the goal and scope of the LCA, the life cycle inventory analysis (LCI) phase, the life cycle impact assessment (LCIA) phase, the life cycle interpretation phase, reporting and critical review of the LCA, limitations of the LCA, relationship between the LCA phases, and conditions for use of value choices and optional elements.

Source: *Curran MA. Life cycle assessment handbook: a guide for environmentally sustainable products. Wiley, 2012; ISO 14044:2006, Environmental management—Life cycle assessment—Requirements and guidelines n.d. http://www.iso.org/iso/catalogue_detail?csnumber=38498 (Accessed 7 January 2017); ISO 14040:2006, Environmental management—life cycle assessment—principles and framework n.d. http://www.iso.org/iso/catalogue_detail?csnumber=37456 (Accessed 7 January 2017).*

section classifies and quantifies energy, water and materials usage, and environmental emissions. The impact assessment step assesses the human and environmental effects of energy, water, and material usage and the impact of environmental emissions identified in the inventory analysis. The improvement step considers the accuracy and completeness of the previous

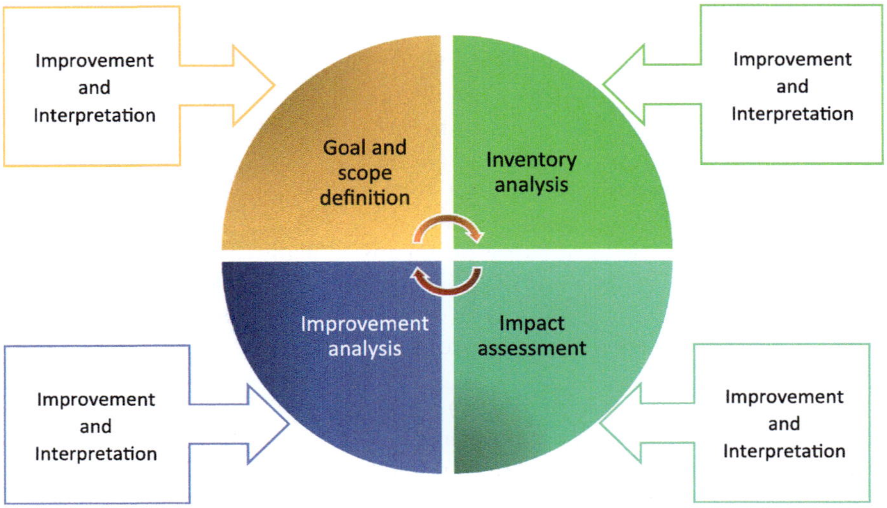

FIG. 7.4 Fundamental steps of performing a life cycle assessment study.

steps and enhances the results. The interpretation step evaluates the results of the inventory analysis and impact assessment to assess the chosen product, process or service.

The inventory analysis is the fundamental part of the LCA study, since it provides support for assessment and interpretation. In the first step of inventory analysis, the system boundary should be clearly described by classifying the flows of matter, pollution, energy, and costs (if life cycle costing is included) interacting with the system. A detailed flow digram can be quite beneficial in this step. It is also suggested to present data collection, compilation, evaluation, validation, and data quality estimation. The result of inventory analysis is the inventory table displaying the inputs and outputs of materials and nonmaterials, such as energy and cost flows, which cross the selected boundary of the system.

The LCA is a broadly recognized method to define the environmental impacts of the product, process or system throughout the entire life cycle [20]. LCA methodology can be useful for comprehensive information on the environmental impact of industrial products [21]. LCA can play a positive role in product designs and environmental management. Many analyses of the management of the entire product life cycle can only be accomplished at the expense of other aspects of simplification [20]. LCA analysis of integrated energy systems requires careful consideration for the definition of design, choice of functional units, reference system and system limitations.

7.5 Exergetic life cycle assessment

Exergetic life cycle assessment (ExLCA) has the potential to quantify the environmental impacts associated with the exergy losses and exergy destruction within a system, process or product, as depicted in Fig. 7.5. Increasing the exergy efficiencies can potentially reduce the environmental impacts. The framework of ExLCA is in fact quite similar to LCA. The main steps of ExLCA can be written as follows [22]:

- *Goal and scope definition*: The first step for ExLCA identifies the system boundary, target group and insights of interest. The scope of the system identifies the functional unit of the system. A functional unit is a measure of performance or end product such as 1 kWh of

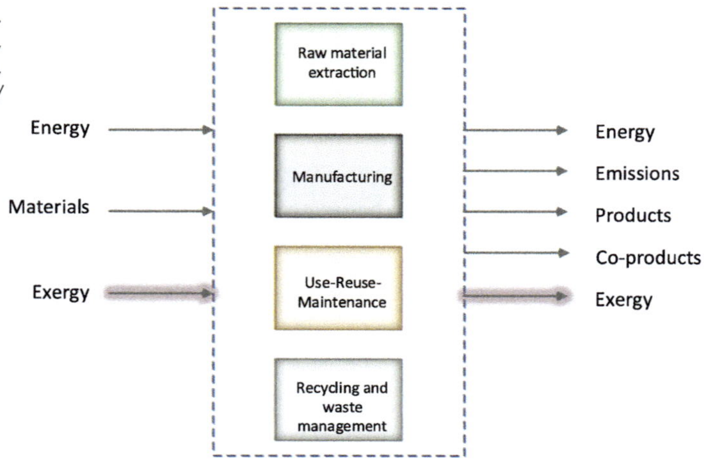

FIG. 7.5 Exergetic LCA framework. *(Adapted from Dincer I, Rosen MA. Exergy, Environment and Sustainable Development, Elsevier; 2013, p. 51–73. https://doi.org/10.1016/B978-0-08-097089-9.00004-8).*

electricity. The main purpose of the functional unit is to deliver a reference for relating the inputs and outputs.

- *Inventory analysis*: The inventory analysis of ExLCA is more detailed as compared to LCA. A complete identification of the mass, energy and exergy flows involved in each of the life cycle stages is required in ExLCA as represented in Fig. 7.5. The calculation of exergy values requires the identification of reference conditions and composition (e.g., reference temperature, reference chemical composition, etc.).
- *Impact assessment*: ExLCA aims to determine the environmental impacts of the exergy flows, the exergy destructions and exergy efficiencies of the overall process and its sub-processes. Calculation of exergy content of flows is often a combination of utilizing data reported in the literature and calculating data with exergy analysis.
- *Improvement analysis*: The improvement analysis in ExLCA is envisioned to decrease the life cycle irreversibilities. The sum of all exergy destructions in the life cycle, which are calculated using exergy analysis, identifies the life cycle irreversibility of the product or the process. The results of ExLCA are interpreted to attain conclusions and recommendations in alignment with the goal and scope of ExLCA.

In an ExLCA of a system, each step is evaluated by its material and energy flows (e.g., fossil-fuel consumption) and environmental impacts. The exergy consumption rate corresponding to fossil fuel use can be assessed with the following equation:

$$\dot{E}x^i_{LFC} = \dot{E}x^i_{direct} + \triangle\dot{E}x^i_{direct} + \triangle\dot{E}x^i_{indirect} \tag{7.28}$$

where $\dot{E}x^i_{LFC}$ is the life cycle fossil-fuel exergy consumption rate, $\dot{E}x^i_{direct}$ is the rate fuel exergy is directly transformed into final products, $\triangle\dot{E}x^i_{direct}$ is the rate at which fuel exergy is consumed to perform the transformation, and $\triangle\dot{E}x^i_{indirect}$ is the rate at which fuel exergy is consumed through being embodied in construction materials and equipment and during installation, operation, maintenance, decommissioning, and so forth.

The standard exergy values of most fuels are similar to their lower heating values (LHVs). The LHV is equal to the heat released by the complete burning of all fuel components to CO_2 and H_2O in the form of a vapor, as described in Chapter 2. The standard exergy of fuels Ex^0_f is equal to the maximum work available and can be evaluated as the negative of the standard Gibbs free energy change $-\triangle G^0$ (at $P_0 = 1$ atm, $T_0 = 298$ K) for the fuel combustion reaction:

$$Ex^0_f = -\triangle G^0 = -\left(\triangle H^0 - T_o \triangle S^0\right) \tag{7.29}$$

where $\triangle H^0$represents the changes of standard enthalpy and $\triangle S^0$ represent the changes in entropy in the reaction. When electricity, hydrogen, or other secondary energy carriers are the input exergy, the generally accepted efficiencies are usually applied to evaluate the direct input fossil-fuel exergy rates $\dot{E}x^i_{direct}$ and $\triangle\dot{E}x^i_{directr}$. For use of electricity, which is often produced from fossil fuels, $\triangle\dot{E}x^i_{direct}$ is expressible as [22]:

$$\triangle\dot{E}x^i_{direct} = \frac{\dot{W}_i}{\eta_{ex}} \tag{7.30}$$

where η_{ex} is the exergy efficiency for electricity generation from a fossil fuel source and $\dot{W}_i$ is the obtained work rate.

If the fossil-fuel exergy and LHV values are similar, the exergy and energy efficiencies for the processes of electricity, mechanical work, and hydrogen generation do not differ significantly, where the energy efficiency is:

$$\eta_{en} = \frac{\dot{W}_i}{LHV_i} \tag{7.31}$$

The indirect exergy $\Delta\dot{E}x^{i}_{indirect}$ is not equal to the exergy required to produce a given material or device or energy. The embodied exergy (useful energy) adequately reflects the environmental impact of the material extraction and material and device production stages, but it does not account for the economic cost of these products. In order to account for this, the exergy equivalent of construction materials and devices is considered by dividing the cost of materials or devices utilized in a given technological stage by the cost of a unit of the fossil-fuel exergy. Then, the indirect exergy consumption rate can be calculated as [23, 24]

$$\Delta\dot{E}x^{i}_{indirect} = \frac{\sum EEQ + EOP}{Lifetime} \tag{7.32}$$

where EEQ is the summation of the exergy equivalents of construction materials and devices associated with a given technological operation, EOP is the operation exergy, that is, the fossil-fuel exergy required for installation, construction, operation, maintenance, decommissioning, and so forth of equipment, and *Lifetime* is the lifetime of the unit performing a technological operation.

The accuracy of the obtained data, calculated/assumed efficiencies, costs, and many other factors can bring some level of uncertainties into LCA and ExLCA; however, they are useful tools for calculating the environmental impacts as well as accounting for exergy efficiencies of technological chains, including their construction and operating stages, and for determining the associated uncertainties.

7.6 Exergosustainability assessment

Actually, assessing the sustainability level of any process/product/system is subjective and a relatively challenging task. However, various well-established tools can be combined and employed for integrated sustainability assessment. Exergy plays an important role in sustainability assessment and defining the reference environment.

Regularly in policy planning, indicators are used to evaluate a country or geopolitical region from various points of view: economic, social, technical and so on. The aim of sustainability assessment indicators is to consider a base for decision-making and policy expansion toward sustainable development in view of the integrated nature-society structures for a certain temporal perspective. Overall, sustainability assessment indicators compute, in an integrative manner, the economic, social, environmental, and institutional development of a country or region. The sustainability indicators are mostly extensions of the environmental assessment indicators.

Some detailed features on the establishment of holistic sustainability assessment methods are listed as follows [13]:

- Sustainability assessment should integrate ecological condition assessment of humans and life system habitat with economic development, technical performance, social well-being, and with equity and disparity of the current population and future generations.
- Sustainability assessment should attain an agreement on the adoption of a time horizon satisfactorily long to be in harmony with ecosystem timescale and predicted society timescale.
- Sustainability indicators and indices need to be able to link sustainability classifications to society goals.
- Definition of a reference environment or state is needed for determination of the sustainability change.
- Sustainability assessment should also include a monetary assessment of environmental damage.
- Sustainability assessment needs to incorporate the following categories of models:
  - Stress-response models
  - Thermodynamic models
  - Multiple capital models
  - Economic models
  - Social-economic-environmental models
  - Human-ecosystem well-being models

One of the commonly used sustainability indicators based on mass and energy balances is the eco-efficiency indicator, which relates the ratio between the value in outputs and the environmental impact. For this reason, Linke et al. [25] presented sustainability assessment indices represented as the sustainability efficiency indicator (SEI). It is expressed as the ratio between a selected performance parameter and a quality parameter for a specified resource consumption as follows [25]:

$$\text{Sustainability efficiency indicator (SEI)} = \frac{\text{Performance Parameter}}{\text{Quality Parameter} \times \text{Resource Consumption}} \tag{7.33}$$

In addition to the quantity of any process/product, quality is a noteworthy aspect of sustainability. Hence, exergosustainability assessment is, in fact, a significant step to develop comprehensive sustainability assessment methods, since exergy is a measure of the usefulness of the system subjected to the analysis from the environment. The relationships among exergy, energy, and environmental impact of sustainable energy systems are detailed in many other studies of Dincer and Rosen [22, 26–28] as shown in Fig. 7.6.

Exergy, environmental impact and economic issues are better unified with sustainability. Fig. 7.6 depicts the relations of exergy and sustainability.

Fig. 7.7 shows the representation of a general energy system for sustainability analysis. The scope of the examination is the whole lifespan of the system. The energy from the source is converted to a valuable form such as power or highly exergetic synthetic fuel. The produced exergy corresponds to stream 5 in Fig. 7.7. A share of this stream is then used to maintain the system in a process for the lifetime span and to construct a newer system, ready for the process at the end of the lifespan of the old system.

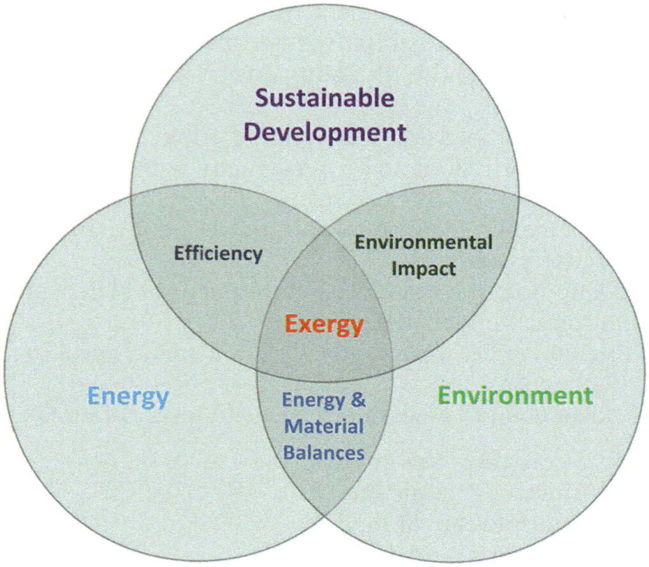

FIG. 7.6 Illustration of the exergy at the confluence of energy, environment and sustainable development. *(Modified after Dincer I, Rosen MA. Exergy: Energy, environment and sustainable development, Exergy. 2nd ed., New York, NY: Elsevier; 2013, p. 51–73. https://doi.org/10.1016/B978-0-08-097089-9.00004-8).*

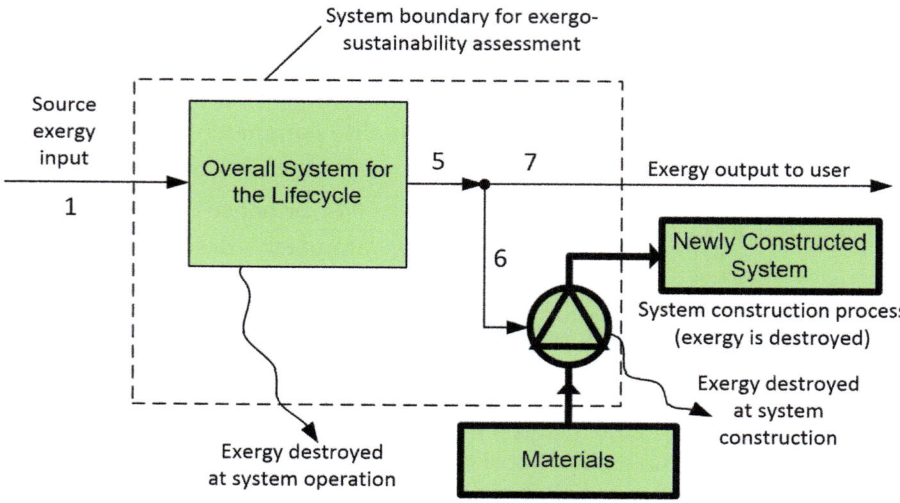

FIG. 7.7 Illustration of a sample energy system for exergosustainability assessment.

If the exergy output in 7 is positive, then an exergetic sustainability index can be defined as follows:

$$\text{ExSI} = \frac{\text{Ex}_7}{\text{Ex}_5} \tag{7.34}$$

7.7 Exergo-enviro-sustainability

A sustainable development process involves the energy security component, therefore exergy-based techniques is quite appropriate for assessing the sustainability index. By employing exergy analysis and exergosustainability assessment, improved efficiencies can be achieved with reduced exergy losses, environmental impact and cost as shown in Fig. 7.8. Increasing the exergy efficiency can yield more sustainable and more environmentally friendly systems. Exergo-enviro-sustainability assessment is based on mass, energy, entropy, exergy, cost and environment balance equations and leads to the determination of exergy destructions of the system and the components and their associated costs, environmental impacts as well as social impacts. In addition to pure thermodynamic exergy analysis, the exergoeconomic and exergoenvironmental analyses are integrated to find out the integrated sustainability index.

The exergoenvironmental analysis was developed to include the environmental impact of energy system and the environmental impacts that are also associated with inputs to the system [29]. Additional improvement of exergy analysis by considering energy, environmental and social aspects can assess the sustainability of energy systems, which is referred to exergo-enviro-sustainability analysis. The exergo-enviro-sustainability analysis tool is expected to be a comprehensive energy system assessment tool for sustainable energy systems because, it is based on the interdisciplinary character of exergy, energy, environment and sustainability as illustrated in Fig. 7.8.

Exergy efficiency can be interrelated with sustainability index (SI) and with the environmental impact index (EI). It can also be correlated with the life cycle sustainability index (LCSI). The difference between SI and LCSI contains the fact that SI discusses the system utilization phase only, whereas LCSI deliberates the complete system life cycle counting the construction and decommissioning phases. Further to achieving better sustainability, the exergo-enviro-sustainability analysis can provide the pathway toward designing systems with better cost-effectiveness under principal conditions and with regard to ecological, social, and ethical concerns. Actually, the economic component most likely restricts the design and governs the optimum exergy efficiency according to the law of diminishing return. Higher efficiency involves exponentially higher system cost with neither considerable

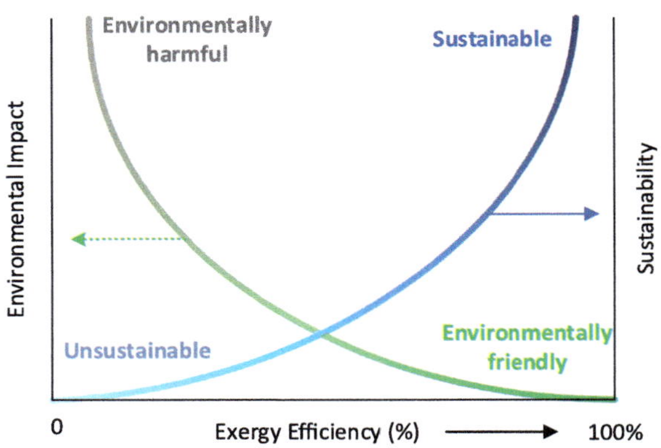

FIG. 7.8 The interrelationship among sustainability, exergy efficiency and environmental impact.

reduction of environmental impact nor significant enhancement of LCSI. Sustainable development necessitates a supply of energy resources that are readily and sustainably accessible at a reasonable cost and can be utilized for all essential tasks without producing negative societal impacts.

7.7.1 Illustrative example: Sustainability assessment of a PV-heat pump system

This illustrative example presents sustainability assessment of a sample energy system considering various aspects, namely energy, exergy, environmental impact, economic impact, technology, social aspect, education and the size factor associated with the energy system. The model of integrated sustainability index was previously developed by Abu-Rayash and Dincer [30]. It comprehensively assesses the sustainability index of energy systems.

In order to find out the final sustainability index score, multiple steps for data processing are required. Initially, the obtained data are normalized by the use of ratios and target values. Normalization is necessary as the various indexes alter in their units and nature, yet they are combined to reflect a single sustainability index score. The data are aggregated then using two main methods, the weighted geometric and arithmetic means. After the data are normalized and aggregated, various weighting schemes are applied for offsetting any possible subjectivity in the assessment (individualist, hierarchist, egalitarian, panel method, equal weighting). The integrated model utilizes the equations presented in Table 7.4.

Here, an integrated energy system is considered in which a residential heat pump driven by photovoltaic modules is taken into account, as shown in Fig. 7.9. The target total energy production is about 6.13 GWh. The case study considers 50% of south facing roofs, 25% east facing and 25% west facing roofs for PVs. The total size of the system is about 4.38 MW with approximately 13,681 solar panels to be installed. This comes to an average of 91 panels per household with a power rate of 320 W/panel. Further details regarding the considered system are available in Abu-Rayash and Dincer [30].

Because weighting and aggregation are subjective concepts, various weighting schemes and aggregation methods have been used in order to minimize any noise and in order to present results objectively as much as possible. Therefore, there are two aggregation schemes used, which are weighting arithmetic mean and weighting geometric mean. Fig. 7.10 shows the final sustainability index for the PV-heat pump system with respect to the characterization schemes applied and the aggregation methods used.

It is clear from the figure that that there is a minor difference between the uses of weighted geometric mean and the weighted arithmetic mean for calculating the sustainability index of the PV-heat pump system. Using weighted geometric mean, the sustainability index derived from the equal weighting and panel schemes is higher than the derived values using the weighted geometric mean.

All of the considered sustainability index factors affect the final score. Here, we present only the impact of exergy index on the final sustainability index, as illustrated in Fig. 7.11. It depicts the exergy index and its impact on the sustainability index. The equal weighting scheme suggests the highest sustainability index scores as the exergy index increases. The

TABLE 7.4 The fundamental equations utilized for integrated sustainability index model.

Sustainability index	Formulation	Details
Energy index	$Y_{ER} = (\eta_{en} W_{\eta}) + (Y_{Pr} W_{Pr})$	Y_{ER} denotes the total score of efficiency and productivity that is calculated by the addition of the scores of the two indicators. η_{en} denotes the score of the efficiency of the energy system. Y_{Pr} denotes the score of the productivity of the energy system.
Exergy index	$Y_{EX} = (\eta_{ex} W_{\psi}) + (Y_{ED} W_{ED})$	Y_{EX} denotes the total score for the exergy index. The score is a function of these two indicators. η_{ex} denotes the exergy efficiency of the system. Y_{ED} is the score of the exergy destruction indicator.
Environmental friendliness index	$Y_{ENV} = (Y_{GWP} W_{GWP})+(Y_{ODP} W_{ODP}) + (Y_{AP} W_{AP}) + (Y_{EP} W_{EP}) + (Y_{AT} W_{AT}) + (Y_{WE} W_{WE}) + (Y_{SA} W_{SA}) + (Y_{WC} W_{WC}) + (Y_{LU} W_{LU}) + (Y_{ADP} W_{ADP})$	GWP refers to the global warming potential, ODP to the stratospheric ozone depletion potential, AP to the acidification potential, EP to the eutrophication potential, AT to air toxicity, WE to water ecotoxicity, SA to smog air, WC to water consumption, LU to the land use and ADP to the abiotic depletion potential.
Economic index	$Y_{ECO} = (Y_{BCR} W_{BCR}) + (Y_{PBT} W_{PBT}) + (Y_{LCOE} W_{LCOE})$	Y_{BCR}, Y_{PBT}, and Y_{LCOE} denote the scores of benefit-cost ratio, payback time, and the levelized cost of energy/electricity, respectively.
Technology index	$Y_{TECH} = (Y_{COMM} W_{COMM}) + (Y_{TR} W_{TR}) + (Y_{IN} W_{IN})$	Y_{COMM}, Y_{TR} and Y_{IN} denote the scores of commercializability, technology readiness and innovation, respectively.
Social index	$Y_{SOC} = (Y_{JC} W_{JC}) + (Y_{PA} W_{PA}) + (Y_{SA} W_{SA}) + (Y_{SC} W_{SC}) + (Y_{HW} W_{HW}) + (Y_{HH} W_{HH})$	Y_{JC}, Y_{PA}, Y_{SA}, Y_{SC}, Y_{HW}, and Y_{HH} denote the scores of job creation, public awareness, social acceptance, social cost, human welfare, and human health, respectively.
Educational index	$Y_{EDU} = (Y_{TRAIN} W_{TRAIN}) + (Y_{EL} W_{EL}) + (Y_{EI} W_{EI})$	Y_{TRAIN}, Y_{EL}, and Y_{EI} denote the score for the number of trained people required by the industry, educational level, and educational innovation, respectively.
Sizing index	$Y_{MF} = (Y_M W_M) + (Y_{LU} W_{LU}) + (Y_V W_V)$	Y_M, Y_{LU}, and Y_V refer to the score for mass, land use, and volume, respectively.

Footnote: 'W' terms refer to the weighting factors associated with each indicator.
Source: *Abu-Rayash A, Dincer I. Sustainability assessment of energy systems: a novel integrated model. J Clean Prod 2019;212:1098–116. doi:10.1016/J.JCLEPRO.2018.12.090.*

sustainability index score increases from 0.48 to 0.63 when the exergy index approaches to 1 for the panel method. Similarly, the other methods range from 0.48 to 0.65. This means that there can be about 17% increase in the sustainability index if the exergetic performance of the system can be maximized.

The model developed transforms sustainability assessment of energy systems from a qualitative to a quantitative process. Furthermore, thermodynamic-based concepts such as energy

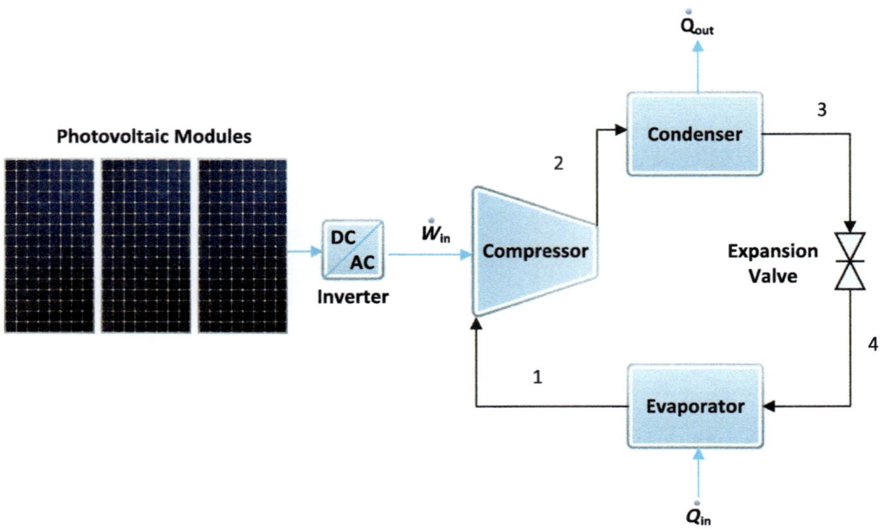

FIG. 7.9 Considered PV-heat pump system for sustainability index assessment.

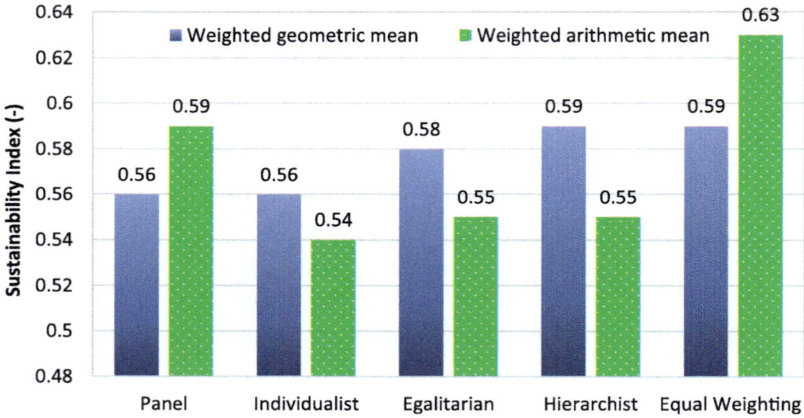

FIG. 7.10 Final integrated sustainability index number of a PV-heat pump system considering various aggregation and characterization methods. *(Data from Abu-Rayash A, Dincer I. Sustainability assessment of energy systems: a novel integrated model. J Clean Prod 2019;212:1098–116. https://doi.org/10.1016/J.JCLEPRO.2018.12.090).*

and exergy assessments are integrated for adding strength and objectivity to the model. The sustainability index for the PV-heat pump system ranged from 0.48 to 0.65 depending on the selected scheme. This integrated sustainability index score is convenient and beneficial for policy makers, researchers and the public. In addition, the weighting schemes have an effect on the final sustainability index with slight differences between weighted arithmetic mean and weighted geometric mean.

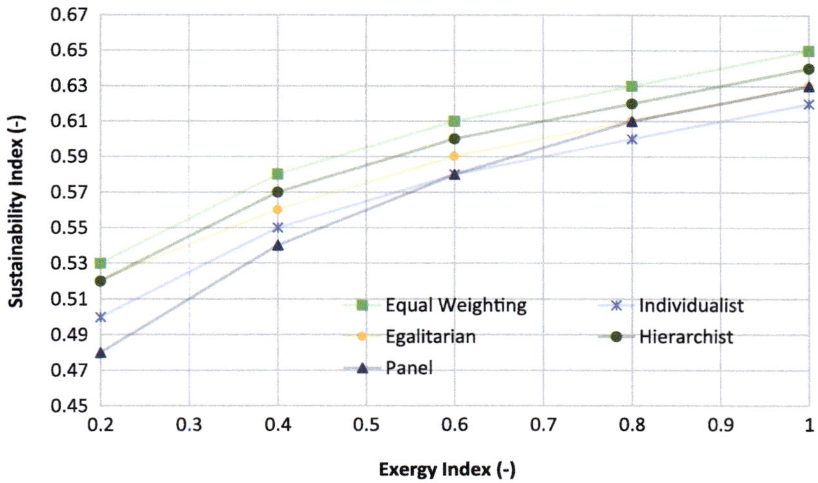

FIG. 7.11 The effects of exergy index on the final integrated sustainability index for a PV-heat pump system. *(Data from Abu-Rayash A, Dincer I. Sustainability assessment of energy systems: a novel integrated model. J Clean Prod 2019;212:1098–116. https://doi.org/10.1016/J.JCLEPRO.2018.12.090).*

7.8 Future directions of integrated energy systems

In this section, we elaborate on new directions for integrated energy systems by employing the 3S approach. Clean and renewable fuels are produced by converting renewable energy sources into chemical molecules for use in various applications with minimum greenhouse gas emissions or without adding net CO_2 to the atmosphere. The main advantage is the utilization of clean energy sources and employment of clean conversion technologies, as depicted in Fig. 7.12.

We make use of abundant renewable energy sources in a sustainable manner so that the natural resources are not depleted. This will contribute to transition to a low-carbon economy and help reduce the economic, environmental, and health risks associated with air pollution and climate change caused by conventional hydrocarbon fuels.

Renewable fuels have the potential to:

- reduce emissions by taking advantage of renewable and cleaner sources,
- lower energy input requirements by employing novel and advanced energy conversion technologies,
- increase system efficiencies by expanding useful outputs,
- reduce emissions and waste by recovering unused energy.

There are multiple pathways to produce renewable fuels, as illustrated in Fig. 7.13. Here, we mention some of them, ranging from photoelectrochemical to methanation. Under the renewable fuel category, we can list hydrogen (H_2), ammonia (NH_3), methanol (CH_3OH), ethanol (C_2H_5OH), dimethyl ether (DME) and synthetic natural gas (CH_4). Although there can be other potential fuels, these are currently the most promising and those receiving the greatest attention in the literature.

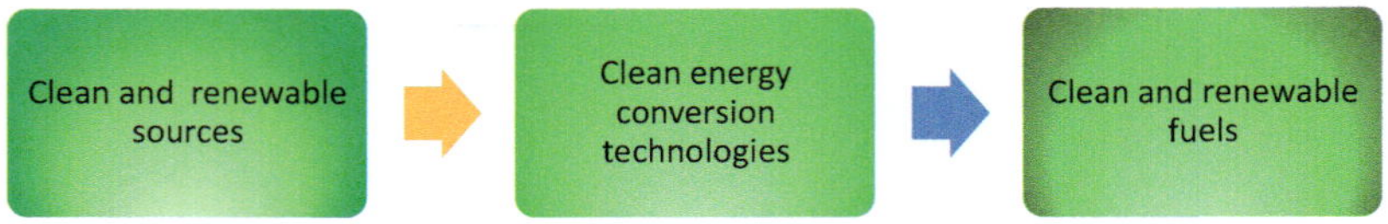

FIG. 7.12 Renewable fuel production from renewable energy resources and clean energy conversion technologies.

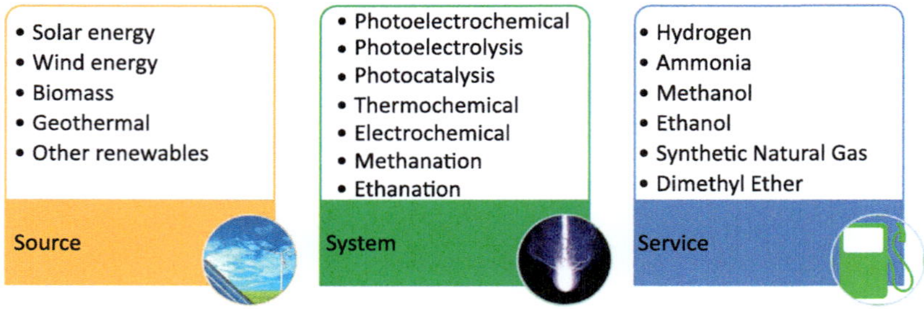

FIG. 7.13 3S approach for renewable fuel production

Based on the 3S approach for renewable fuels as shown in Fig. 7.13, the sources are selected from renewable energy options. When it comes to system steps, it is necessary to select the most efficient and clean energy conversion systems. The selected systems are desired to be effective, clean and novel. The service step includes clean fuels such as hydrogen, ammonia, methanol, ethanol, dimethyl ether, etc. In the process of renewable fuel production, there are many multigeneration opportunities such as heating, cooling and electricity, which can easily be integrated into the processes.

The initial step of ammonia, methanol, ethanol and synthetic natural gas production is generation of hydrogen, as illustrated in Fig. 7.14. Therefore, green hydrogen plays a significant role in the complete process. Later on, green hydrogen is used as feedstock for other fuels as shown in Fig. 7.14. Other required input gases such as CO, CO_2 and N_2 are also desired to be clean. CO_2 can be sourced from carbon capture technologies or chemical-looping. N_2 is currently obtained from air separation plants (e.g., cryogenic), but there are also novel options such as photo-fixation of nitrogen for direct ammonia synthesis from the air. In addition, novel electrochemical ammonia synthesis methods can accept direct air at atmospheric pressure.

The renewable fuels have diverse utilization options as depicted in Fig. 7.15. They can be used in heat engines, fueling systems, fuel cells, energy recovery systems, vehicles, power generators, pipeline transport, heating and cooling applications. Hydrogen, ammonia, methanol, synthetic natural gas, and ethanol can be directly used in fuel cells for electricity generation. By having synthetic methanol, ethanol and natural gas, the impurities within the feed gas can be eliminated, which enables direct feed to fuel cells without any pre-filtering process. In this way, the cost of operations can be lowered in addition to emission characteristics.

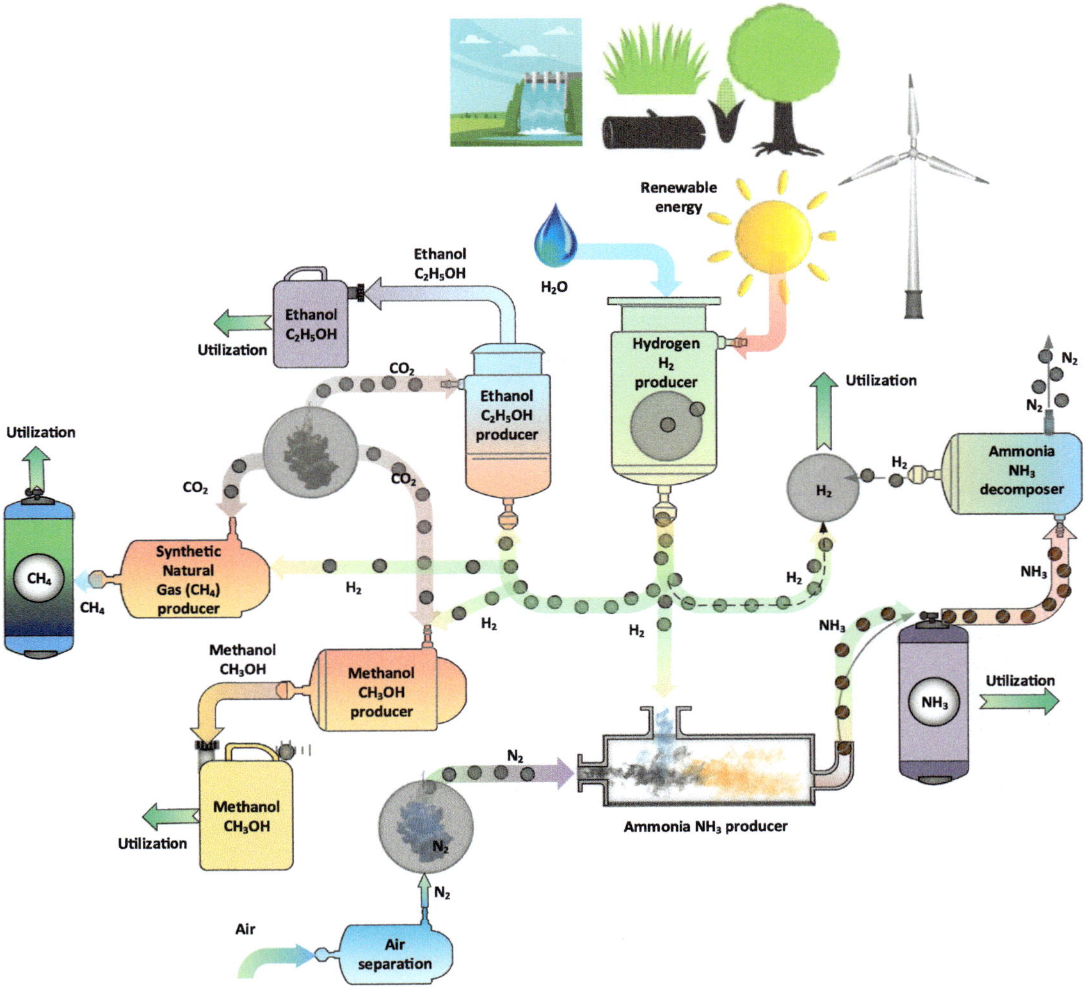

FIG. 7.14 Illustration of renewable fuel production from renewable energy sources.

Carbon-containing fuels such as natural gas (methane) or liquid fuels such as methanol or ethanol might be produced from sunlight, water, and carbon dioxide. Ammonia for use as fertilizer in agriculture can be made indirectly from solar hydrogen or directly as a solar fuel from sunlight, water, and nitrogen in the air. Fuels can be readily interconverted using well-known processes. Production of synthetic fuels from renewable resources using carbon capture and sequestration significantly reduces the carbon footprint of these fuels and decreases the dependency on hydrocarbon fuels. Although synthetic fuels include carbon atoms, they can be produced from renewable hydrogen and captured CO_2 from any industrial process, which eventually makes them environmentally friendly.

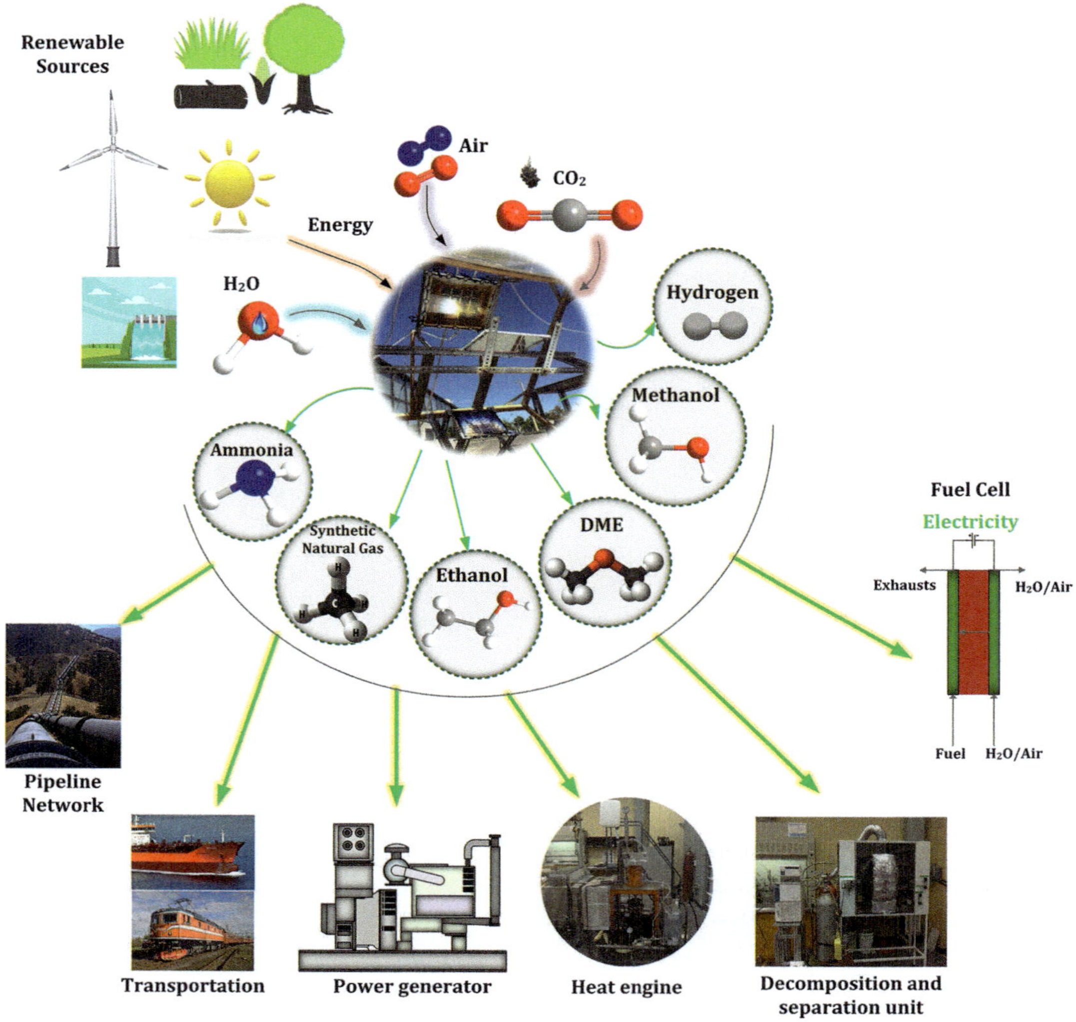

FIG. 7.15 Clean fuel production from renewables, carbon dioxide, air and water and various potential application areas.

7.8.1 Clean ammonia (NH_3)

Green and clean ammonia will have a significant role in sustainable energy systems because of the following main advantages:

- carbon-free fuel,
- sustainable energy carrier,
- excellent working fluid,
- easier storage and handling management,

- a fuel blending option for current gasoline and diesel engines,
- less volume and lower pressure levels compared to hydrogen on-board,
- potential candidate to replace hydrocarbon fuels in many sectors, ranging from power generation to transportation.
- promising potential for distributed power generation leading a non-centralized power generation and smart grid applications.

Ammonia is currently one of the most used chemicals throughout the world due to its many applications, such as fertilizers, cooling agents, chemical intermediates, etc. The Haber-Bosch process is the most dominant ammonia synthesis process, which requires very high temperatures and pressures to operate and consumes massive amounts of fossil fuels, mainly natural gas, leading a non-sustainable process in the long term. Ammonia is considered not only as a feedstock but also evaluated as an energy carrier. Due to its many advantages over hydrogen, it can be a medium to store and carry energy. Currently, over 90% of the world ammonia synthesis is realized by the Haber-Bosch synthesis process. This process is named for Fritz Haber and Carl Bosch who developed the method in 1913 [12, 13]. It is important to note, as previously mentioned, that natural gas is the main feedstock used for manufacturing ammonia worldwide [31]. It is an exothermic process combining hydrogen and nitrogen in a 3:1 ratio to yield ammonia. The reaction is assisted by the catalyst and the optimal temperature range is 450–600 °C [31–33]. The Haber–Bosch process was developed at the beginning of the 20th century to combine hydrogen and nitrogen thermo-catalytically according to the following reaction:

$$\frac{3}{2}H_2 + \frac{1}{2}N_2 \rightarrow NH_3 + 45.2\frac{kJ}{mol} \tag{7.35}$$

There are other newly developing ammonia synthesis methods categorized under electrochemical routes. For instance, solid state ammonia synthesis (SSAS) uses a solid state electrochemical process to produce ammonia from nitrogen, water, and electricity. A proton-conducting membrane is heated to about 550 °C. Under the same pressures, nitrogen and water vapor are supplied to each side of the membrane to initiate the reaction. The water vapor separates into protons and oxygen. By applying an external voltage, the protons are transferred through the membrane, and on the nitrogen side of the membrane, ammonia forms as a result of the nitrogen and proton reaction.

When hydrogen is produced from water electrolysis utilizing a renewable energy source such as wind or solar, environmentally pollutant emissions would noticeably diminish for ammonia production. Water can also be utilized as a source of hydrogen inside the electrolytic cell through its reaction in the electrochemical processes at various media such as (i) liquid electrolytes, (ii) molten salt electrolytes, (iii) composite electrolytes consisting of a traditional solid electrolyte mixed with a low melting salt, (iv) solid electrolytes.

For instance, a solar-based electrochemical ammonia synthesis in a molten salt medium is depicted in Fig. 7.16. Two reactants, nitrogen and hydrogen, mix in a eutectic mixture composed of KOH-NaOH. An electrochemical pathway is created to produce ammonia by the reaction of nitrogen and hydrogen in a molten hydroxide salt solution. The suspension of nano-Fe_3O_4 within the molten salt acts as a catalyst for the reaction to occur faster.

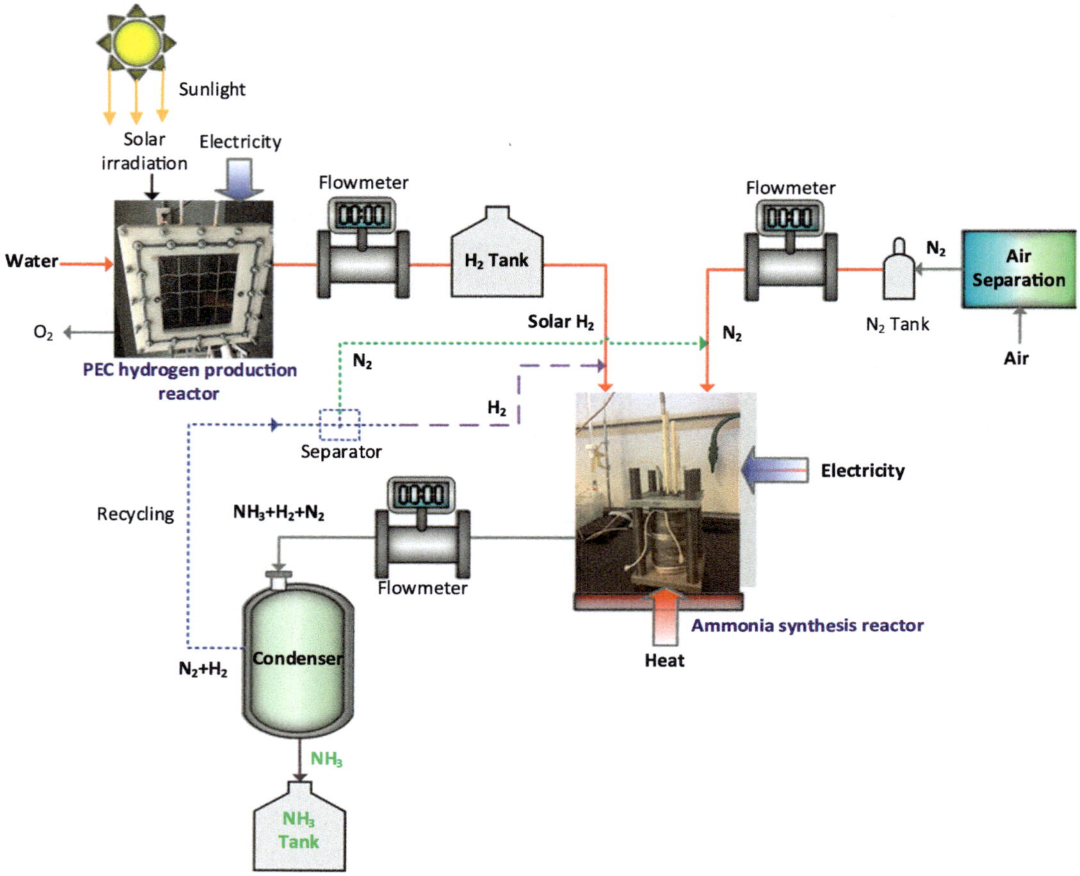

FIG. 7.16 Photoelectrochemical hydrogen-based ammonia synthesis from sunlight. *(Adapted from Bicer Y. Investigation of novel ammonia production options using photoelectrochemical hydrogen. PhD Thesis. University of Ontario Institute of Technology, 2017).*

Using green ammonia in various applications will enable a cleaner community. These systems are not necessarily centralized large-scale applications. In contrast, they can be standalone and small/medium scale applications, which can contribute decentralized smart energy systems. Ammonia can also be decomposed into hydrogen and nitrogen easily, using heat input for further usage.

7.8.2 Clean methanol (CH_3OH)

Methanol synthesized from green hydrogen has several advantages for sustainable energy systems as listed here:

- Methanol is already a commercially available and viable fuel,
- It has a very well-developed global distribution network,

- There is an easy handling experience,
- Methanol can be used in vehicles and power generation,

Technologies, which include possible reduction or conversion of CO_2, provide valuable advantages for protecting the environment. A promising technology is the power-to-gas methods, in which CO_2 is recycled and renewable methanol or synthetic natural gas can be generated. The power-to-gas process chain is based on water electrolysis and CO_2 methanation. With the advances of renewable energy as well as greater utilization of existing energy sources such as solar energy, methanol evolves not as an energy source but as an energy carrier with great potential. The process consists of combining hydrogen and carbon dioxide to produce methanol with the only by-product being oxygen. The idea is to produce methanol from carbon captured from the atmosphere, mainly from local emitters such as power plants and industrial facilities, but with improving technologies also from the atmosphere itself. To accommodate the need for hydrogen in the synthesis process, electrolysis of water is performed with electricity, or other renewable routes can be employed (e.g., photoelectrolysis, photocatalysis, etc.). The compressors within a methanol synthesis plant can also be powered by renewables for better sustainability.

7.8.3 Synthetic natural gas (SNG) (CH_4)

Natural gas can be produced synthetically by methanation of carbon dioxide, which is an exothermic catalytic reaction and is typically operated at temperatures between 200 °C and 550 °C depending on the catalyst:

$$CO_2 + 4H_2 \leftrightarrow CH_4 + 2H_2O, \ \Delta H = -165 \frac{kJ}{mol} \tag{7.36}$$

A schematic illustration of synthetic natural gas production from renewable sources and CO_2 is depicted in Fig. 7.17.

As shown in Fig. 7.17, power-to-methane can be divided into three main processes: conversion, storage, and utilization. The first one, the input conversion, produces a methane-rich gas called synthetic natural gas (SNG). The second process is the resulting energy carrier storage, which is followed by the utilization process that can comprise several usages. Electrical energy is converted into chemical energy in the form of H_2 using a few different electrolysis

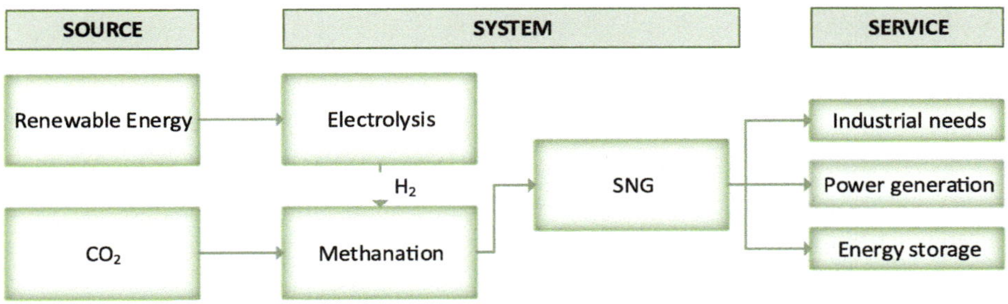

FIG. 7.17 Renewable energy based-SNG synthesis for various services.

technologies. Methanation is the conversion process of CO or CO_2 to methane using a source of hydrogen. SNG is a potential synthetic fuel and energy storage medium for countries having few or no natural gas reserves.

7.8.4 Clean ethanol (C_2H_5OH)

Most ethanol today is produced at high-temperature fermentation facilities that chemically convert corn, sugarcane and other plants into liquid fuel. However, growing crops for biofuel requires thousands of acres of land and vast quantities of fertilizer and water. Carbon dioxide from the atmosphere can be used to produce carbon monoxide, which, in turn, would be fed to a copper catalyst to make liquid ethanol fuel. The CO_2 that is released into the atmosphere during fuel combustion would be reused to make more carbon monoxide and more fuel in a closed-loop, emissions-free process. There are a couple of pathways for synthetic ethanol production, as illustrated in Fig. 7.18.

The first pathway consists of a direct synthesis of ethanol from syngas. The second pathway is syngas conversion to methanol, which is then converted to ethanol via a methanol homologation reaction. In the third pathway, syngas is converted to methanol over a methanol commercial synthesis catalyst followed by methanol carbonylation to acetic acid in the second step and, then, subsequent hydrogenation of acetic acid to ethanol.

Dissociation of CO_2 and H_2O separately in two different electrolysis cells or simultaneously in the same electrolysis cell is also possible. This can be done using renewable electricity instead of conventional reforming processes. The co-electrolysis of water vapor and carbon dioxide can be summarized by the following reaction:

$$H_2O + CO_2 \rightarrow H_2 + CO + O_2 \tag{7.37}$$

The first reaction to be considered is the reverse reaction of water gas shift:

$$CO_2 + H_2 \rightarrow CO + H_2O \tag{7.38}$$

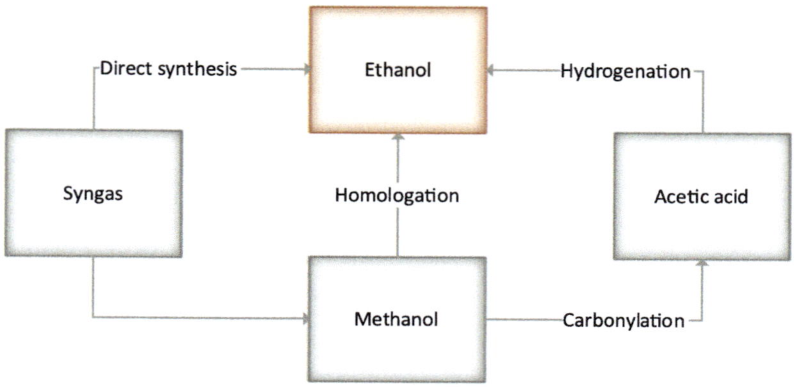

FIG. 7.18 Some ethanol synthesis pathways. *(Modified from Fouih Y El, Bouallou C. Recycling of carbon dioxide to produce ethanol. Energy Procedia 2013;37:6679–86. https://doi.org/10.1016/j.egypro.2013.06.600).*

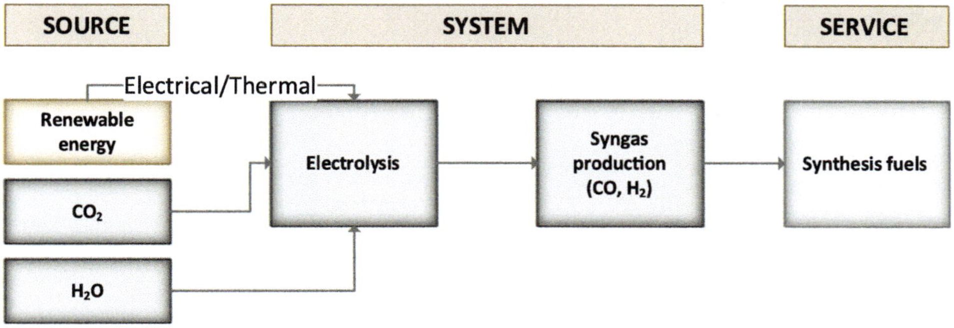

FIG. 7.19 CO_2 and H_2O co-electrolysis for synthetic fuels production.

Another side reaction that can also affect the electrolysis operation is the formation of coke as follows:

$$CO \rightarrow C + \frac{1}{2}O_2 \tag{7.39}$$

Direct synthesis of ethanol is governed by the following reaction:

$$2CO + 4H_2 \rightarrow C_2H_5OH + H_2O \tag{7.40}$$

A sample illustration of synthetic fuel production via co-electrolysis of CO_2 and H_2O is shown in Fig. 7.19.

In addition, a sample synthetic ethanol synthesis option from solar hydrogen is depicted in Fig. 7.20.

The reduced cost of renewable fuels allows them to compete with current hydrocarbon fuels, and they will eventually replace hydrocarbons with carbon-free fuels and synthetic fuels in an cost-effective and efficient manner. Methanol and ethanol can be produced from clean hydrogen, making the entire process more environmentally friendly.

7.8.5 Clean dimethyl ether (DME) (CH_3OCH_3)

Dimethyl ether (DME) is considered to be one of the promising renewable alternative fuels. Dimethyl ether offers substantial benefits in terms of hydrogen storage and utilization. It can reduce the dependence on fossil fuels, minimize the environmental impacts of fuel combustion and implement more environmentally friendly fuel alternatives.

DME has been produced traditionally in two steps in which methanol is synthesized in the first step from syngas; after the methanol is purified, it is converted to DME in another reactor through a process called methanol dehydration [34, 35]. The methanol dehydration process is typically preferred to be at a lower temperature to evade the creation of other byproducts.

The process reactions for DME indirect synthesis can be written as follows:

Methanol synthesis from CO_2:

$$CO_2 + 3H_2 \rightarrow CH_3OH + H_2O \tag{7.41}$$

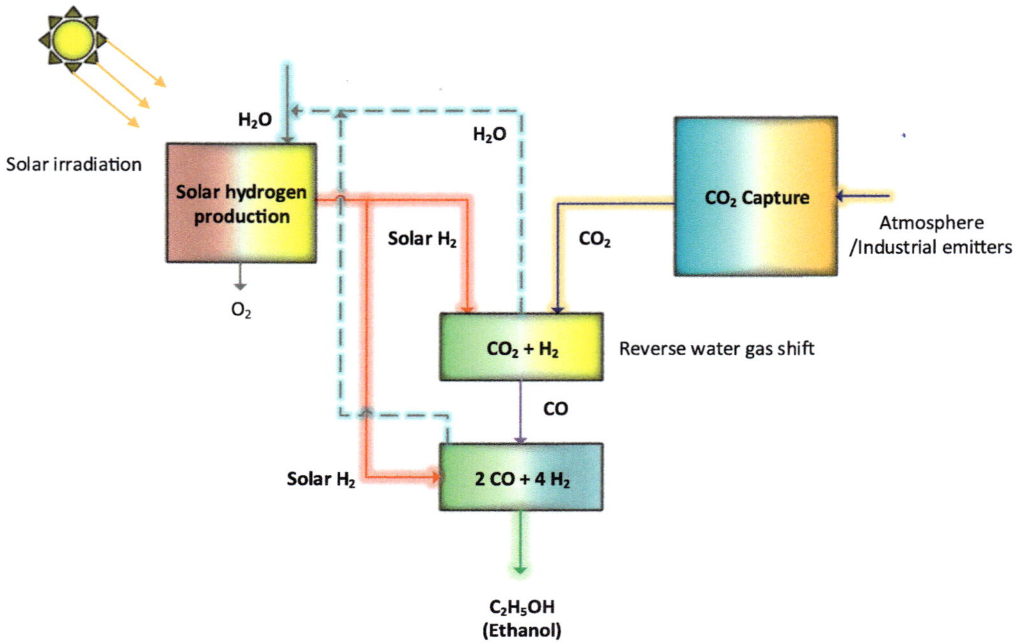

FIG. 7.20 Solar hydrogen-based synthetic ethanol production.

Methanol dehydration:

$$2CH_3OH \rightarrow CH_3OCH_3 + H_2O \tag{7.42}$$

Alternatively, both steps (methanol synthesis and methanol dehydration) can be combined in a single process [36]. In this case, it is called direct DME synthesis. The process reactions for DME direct synthesis can be written as follows:

$$3CO + 3H_2 \rightarrow CH_3OCH_3 + CO_2 \tag{7.43}$$

$$2CO + 4H_2 \rightarrow CH_3OCH_3 + H_2O \tag{7.44}$$

In a DME synthesis process, methanol is transformed into DME by a methanol dehydration process. Methanol dehydration is essentially removal of water from methanol to produce DME. In a typical plant, the methanol-DME mixture is heated to about 260 °C by a heat exchanger before entering the DME reactor which is at a pressure of about 10–15 bar [37]. The final products of the reactor are DME, unreacted methanol and water. The product stream exits at a temperature in the range of 275–300 °C since the reaction is exothermic. The product stream passes through a heat exchanger for cooling and heating the feed stream to DME reactor. The cooled product stream is then fed to the second distillation column to detach the formed DME from the mixture at a pressure of about 10 bar [37]. The remaining mixture is led to the third distillation column in which methanol is separated from water at a pressure of

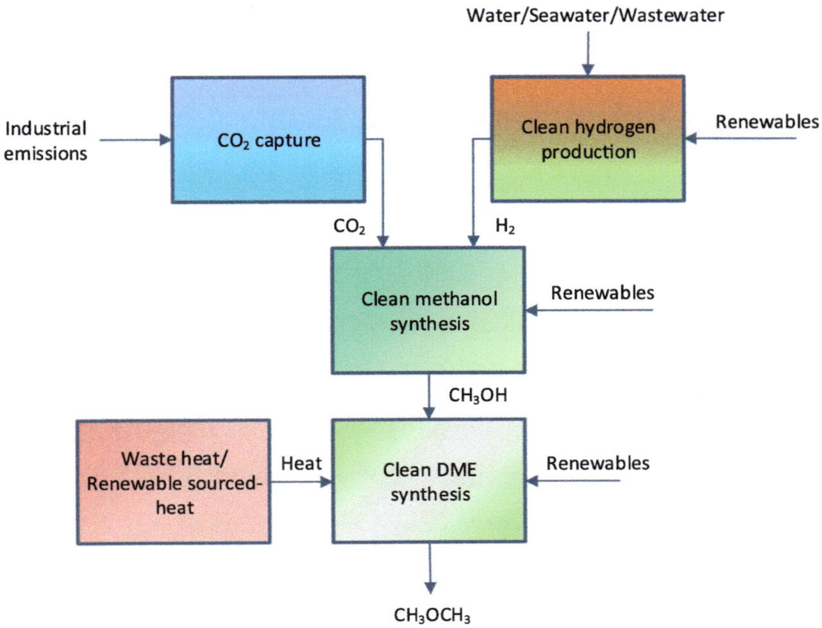

FIG. 7.21 Clean DME synthesis from waste heat and renewable sources.

about 2.5–3 bar [37]. The extracted pure methanol from the top of the distillation column is recycled back to the DME reactor.

The clean DME production starts with clean hydrogen generation, continues with clean methanol synthesis and ends with clean DME synthesis using renewable energy sources. If the thermal and electrical energy requirements of the entire process can be originated from renewables or waste heat, clean DME can be achieved, as illustrated in Fig. 7.21.

DME is considered to be a clean fuel since it has low exhaust emissions, does not produce particulate matter, has lower CO_2 emissions and has high thermal efficiency. Additionally, it can be used for different applications such as transportation, residential heating and power generation. DME is likely to be utilized as a substitute for diesel fuel in diesel engines and gas turbines. Besides, it can act as a dual fuel with LPG and gasoline to reduce the environmental impacts [38, 39]. DME is portable, implying that it can be stored, transported and used easily as long as the handling is properly realized. It can be synthesized from a variety of feed stocks like biomass, coal, natural gas, and hydrogen.

7.8.6 Novel options for system integration

In this section, we list many novel approaches for system integration, as shown in Fig. 7.22. These systems have promising future potential.

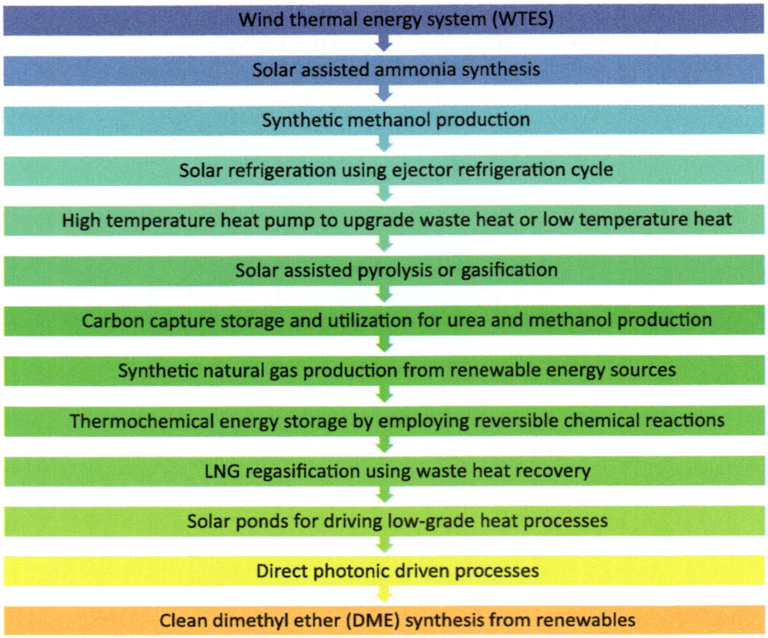

FIG. 7.22 Several novel system integration options for future sustainable energy systems.

7.9 Conclusions

In this chapter, sustainability aspects of energy are reviewed with an emphasis on the development of assessment tools and exergy methods for sustainability. The role of sustainability indicators on sustainability assessment and policy-making is described. There is a great interdependence of issues affecting the sustainability of energy systems and technologies, such as environmental, economic, and social factors. The linkage of sustainability with energy conservation is also presented.

The objective of achieving a society that can sustain itself over the long term is usually regarded as a positive evolutionary course. One of the issues with this target is finding a quantitative assessment of the sustainability of a system/product/process. Henceforth, an integrated sustainability index assessment is needed that incorporates fundamental thermodynamic, economic, environmental, social, technological, and size constraints to form a single-score evaluation of sustainability from multiple dimensions.

Ecological footprint analysis is an accounting tool for the estimation of resource consumption and waste assimilation requirements of a defined human population or economy in terms of corresponding productive land use. Risk assessment can estimate the likelihood of potential impacts and the degree of uncertainty in both the impact and the likelihood of the risk for sustainable energy systems. Material flux analysis is a materials accounting tool that can be used to track the movement of elements of concern throughout a specified system boundary.

Exergoeconomic analysis can reveal the cost of exergy losses and exergy destructions within a system or process. This can become a basis for optimizing the energy systems from both performance and cost point of view.

Exergoenvironmental analysis can determine and calculate the direct and/or indirect environmental impacts in correlation with exergy destroyed within the process or system. After determination of the impacts, they can be minimized by taking suitable precautions and improvements.

The execution of a life cycle cost analysis (LCCA) allows the potential cost drivers and cost savings of a product or service to be identified over its entire life cycle. An integrated life cycle assessment and life cycle costing (LCA-LCCA) model can also be employed to deliver a comprehensive assessment.

Exergetic life cycle assessment (ExLCA) can calculate the environmental impacts associated with the exergy losses and exergy destructions within a system, process or product, which is a basis for observing the weak points of the system as well as identifying the exergetic life cycle performances.

Exergosustainability assessment of energy systems considers the entire lifespan of the system and investigates the flows of exergy. It is then possible to define an exergetic sustainability index, which accounts for system construction and for mitigation of any environmental pollution. Scaling factors and pollution cost become important in the analysis.

Increasing the exergy efficiency can yield more sustainable and more environmentally friendly systems. Exergo-enviro-sustainability assessment is based on mass, energy, entropy, exergy, cost and environment balance equations and leads to the determination of exergy destructions of the system and the components and their associated costs, environmental impacts and social impacts. Exergo-enviro-sustainability assessment can be used for calculating the integrated sustainability index.

An integrated sustainability assessment model can involve several important analyses, mainly (i) exergy analysis, (ii) exergoeconomic analysis, (iii) exergoenvironmental analysis, (iv) exergosustainability analysis, (v) life cycle assessment, (vi) exergetic life cycle assessment, (vii) risk assessment, (viii) ecological footprints, (ix) life cycle costing, (x) material flux analysis and so forth. Overall, an integrated sustainable assessment framework is necessary to develop a novel, multidimensional assessment of sustainable energy systems.

Clean and renewable fuels are produced by converting renewable energy sources into chemical molecules for use in various applications with minimum greenhouse gas emissions or without adding net CO_2 to the atmosphere. Hydrogen, ammonia, methanol, ethanol, DME and synthetic natural gas emerge as alternative and clean fuels for various sectors. Renewable fuels can be used in heat engines, fueling systems, energy recovery systems, fuel cells, vehicles, power generators, pipeline transport, heating and cooling applications, and they can significantly help achieve more sustainable and livable world.

Nomenclature

C	cost ($)
CRF	capital recovery factor
E	energy (kJ)
EI	environmental impact
Ex	exergy (kJ)
ExSI	exergetic sustainability index
f_c	conversion factor
f	exergoeconomic factor
H	total enthalpy (kJ)

i	interest rate (%)
n	system lifetime (year)
$n_{\mathbf{hours}}$	operating hours in a year (hours)
P	pressure (kPa)
$\dot{Q}$	heat rate (kW)
R	loss ratio
r	real discount rate
S	entropy (kJ/kg K)
T	temperature (°C or K)
t	time (s)
$\dot{W}$	work rate (kW)

Acronyms

CF	cost factor
D_p	depletion factor
EEQ	exergy equivalents
EOP	operation exergy
LCA	life cycle assessment
LCSI	life cycle sustainability index
LHV	lower heating values
O&M	operation and maintenance
PM	particulate matter
SEI	sustainability efficiency indicator
SNG	synthetic natural gas
SSAS	solid state ammonia synthesis
ULSD	ultra-low sulfur diesel
VOC	volatile organic compound
WTES	wind thermal energy system

Greek letters

η	efficiency
$\triangle$	change
$\triangle G^0$	Gibbs free energy

Subscripts and superscripts

dest	destruction
en	energy
ex	exergy
in	input
LFC	life cycle fossil-fuel exergy consumption
out	output
sys	system

References

[1] Bejan A. Entropy generation through heat and fluid flow. New York, NY: Wiley; 1982 https://doi.org/10.1115/1.3167072.

[2] Bejan A. Entropy generation minimization: The method of thermodynamic optimization on finite-size systems and finite-time processes. CRC Press; 1995.

[3] Bejan A. Advanced engineering thermodynamics. 3rd ed New York, NY: Wiley; 2006 https://doi.org/10.1109/TNSRE.2003.814453.

[4] Tsatsaronis G. A review of exergoeconomic methodologies. Second Law Anal Therm Syst 1987;81–7.

[5] Tsatsaronis G, Moran MJ. Exergy-aided cost minimization. Energ Conver Manage 1997;38:1535–42. https://doi.org/10.1016/S0196-8904(96)00215-4.

[6] Szargut J, Morris DR, Steward FR. Exergy analysis of thermal, chemical, and metallurgical processes. 1st ed. Hemisphere; 1988.

[7] Dincer I, Rosen MA. Exergy: Energy, environment and sustainable development. In: Exergy. 2nd ed. New York, NY: Elsevier; 2013. p. 51–73. https://doi.org/10.1016/B978-0-08-097089-9.00004-8.

[8] Lozano MA, Valero A. Theory of the exergetic cost. Energy 1993;18:939–60. https://doi.org/10.1016/0360-5442(93)90006-Y.

[9] Kotas TJ. The exergy method of thermal plant analysis. 1st ed. Essex, Great Britain: Elsevier; 1985 https://doi.org/10.1016/C2013-0-00894-8.

[10] Rosen MA. The development and application of a process analysis methodology and code based on exergy, cost, energy and mass. Toronto, Canada: University of Toronto; 1986.

[11] Rosen MA, Dincer I. Exergy–cost–energy–mass analysis of thermal systems and processes. Energ Conver Manage 2003;44:1633–51.

[12] Rosen M, Dincer I. Exergoeconomic analysis of power plants operating on various fuels. Appl Therm Eng 2003;23:643–58. https://doi.org/10.1016/S1359-4311(02)00244-2.

[13] Dincer I, Zamfirescu C. Drying phenomena: Theory and applications. 1st ed Wiley; 2016 https://doi.org/10.1017/CBO9781107415324.004.

[14] Lazzaretto A, Tsatsaronis G. SPECO: a systematic and general methodology for calculating efficiencies and costs in thermal systems. Energy 2006;31:1257–89. https://doi.org/10.1016/j.energy.2005.03.011.

[15] Meyer L, Castillo R, Buchgeister J, Tsatsaronis G. Application of exergoeconomic and exergoenvironmental analysis to an SOFC system with an Allothermal biomass Gasifier. Int J Thermodyn 2009;12:177–86.

[16] Lazzaretto A, Toffolo A. Prediction of performance and emissions of a two-shaft gas turbine from experimental data. Appl Therm Eng 2008;28:2405–15. https://doi.org/10.1016/J.APPLTHERMALENG.2008.01.021.

[17] Gülder OL. Flame temperature estimation of conventional and future jet fuels. J Eng Gas Turbines Power 1986;108:376. https://doi.org/10.1115/1.3239914.

[18] Ahmadi P, Dincer I. Exergoenvironmental analysis and optimization of a cogeneration plant system using multimodal genetic algorithm (MGA). Energy 2010;35:5161–72. https://doi.org/10.1016/J.ENERGY.2010.07.050.

[19] Dincer I, Bicer Y. 1.27 Life Cycle Assessment of Energy. In: Compr. Energy Syst. Elsevier; 2018. p. 1042–84. https://doi.org/10.1016/B978-0-12-809597-3.00134-6.

[20] Guinée JB. Handbook on life cycle assessment: Operational guide to the ISO standards. Kluwer Academic Publishers 2002;.

[21] de Eicker MO, Hischier R, Kulay LA, Lehmann M, Zah R, Hurni H. The applicability of non-local LCI data for LCA. Environ Impact Assess Rev 2010;30:192–9.

[22] Dincer I, Rosen MA. Exergy, Environment and Sustainable Development, Elsevier; 2013, p. 51–73. https://doi.org/10.1016/B978-0-08-097089-9.00004-8.

[23] Granovskii M, Dincer I, Rosen MA. Life cycle assessment of hydrogen fuel cell and gasoline vehicles. Int J Hydrogen Energy 2006;31:337–52. https://doi.org/10.1016/j.ijhydene.2005.10.004.

[24] Granovskii M, Dincer I, Rosen MA. Environmental and economic aspects of hydrogen production and utilization in fuel cell vehicles. J Power Sources 2006;157:411–21. https://doi.org/10.1016/J.JPOWSOUR.2005.07.044.

[25] Linke B, Das J, Lam M, Ly C. Sustainability indicators for finishing operations based on process performance and part quality. Procedia CIRP 2014;14:564–9. https://doi.org/10.1016/J.PROCIR.2014.03.017.

[26] Rosen MA, Dincer I. On exergy and environmental impact. Int J Energy Res 1997;21:643–54. https://doi.org/10.1002/(SICI)1099-114X(19970610)21:7<643::AID-ER284>3.0.CO;2-I.

[27] Rosen MA, Dincer I. Exergy as the confluence of energy, environment and sustainable development. Int J Exergy 2001;1:3–13.

[28] Dincer I, Rosen MA. Energy, environment and sustainable development. Appl Energy 1999;64:427–40. https://doi.org/10.1016/S0306-2619(99)00111-7.

[29] Dincer I, Rosen MA, Al-Zareer M. 1.9 Exergoenvironmental analysis. In: Dincer I, editor. Compr. Energy Syst. 1st ed. vol. 1. New York, NY: Elsevier; 2018. p. 377.

[30] Abu-Rayash A, Dincer I. Sustainability assessment of energy systems: a novel integrated model. J Clean Prod 2019;212:1098–116. https://doi.org/10.1016/J.JCLEPRO.2018.12.090.
[31] Bartels JR. A feasibility study of implementing an ammonia economy. Iowa State University; 2008.
[32] Appl M. Ammonia: Principles and industrial practice. VCH Verlagsgesellschaft MBH; 1999.
[33] Appl M. Chemical reactions and uses of Ammonia. Ammonia. Wiley-VCH Verlag GmbH; 2007231–4. https://doi.org/10.1002/9783527613885.ch12.
[34] Hosseininejad S, Afacan A, Hayes RE. Catalytic and kinetic study of methanol dehydration to dimethyl ether. Chem Eng Res Des 2012;90:825–33. https://doi.org/10.1016/J.CHERD.2011.10.007.
[35] Bai Z, Ma H, Zhang H, Ying W, Fang D. Process simulation of dimethyl ether synthesis via methanol vapor phase dehydration. Polish J Chem Technol 2013;15:122–7. https://doi.org/10.2478/pjct-2013-0034.
[36] Ng KL, Chadwick D, Toseland BA. Kinetics and modelling of dimethyl ether synthesis from synthesis gas. Chem Eng Sci 1999;54:3587–92. https://doi.org/10.1016/S0009-2509(98)00514-4.
[37] DinAli MN. Development and assessment of renewable energy based integrated systems for dimethyl ether production. Master's Thesis. University of Ontario Institute of Technology, 2018.
[38] Chen W-H, Lin B-J, Lee H-M, Huang M-H. One-step synthesis of dimethyl ether from the gas mixture containing CO2 with high space velocity. Appl Energy 2012;98:92–101. https://doi.org/10.1016/J.APENERGY.2012.02.082.
[39] Tokay KC, Dogu T, Dogu G. Dimethyl ether synthesis over alumina based catalysts. Chem Eng J 2012;184:278–85. https://doi.org/10.1016/J.CEJ.2011.12.034.

Further reading

[40] Curran MA. Life cycle assessment handbook: a guide for environmentally sustainable products. Wiley; 2012.
[41] ISO 14044:2006. Environmental management—Life cycle assessment—Requirements and guidelines n.d, http://www.iso.org/iso/catalogue_detail?csnumber=38498. Accessed 7 January 2017.
[42] ISO 14040:2006. Environmental management—life cycle assessment—principles and framework n.d, http://www.iso.org/iso/catalogue_detail?csnumber=37456. Accessed 7 January 2017.
[43] Gálvez ME, Halmann M, Steinfeld A. Ammonia production via a two-step Al2O3/AlN thermochemical cycle. 1. Thermodynamic, environmental, and economic analyses. Ind Eng Chem Res 2007;46:2042–6. https://doi.org/10.1021/ie061550u.
[44] Kool A, Marinussen M, Blonk H. LCI data for the calculation tool Feedprint for greenhouse gas emissions of feed production and utilization. In: GHG emissions of N, P, and K Fertilizer production. Vol. 34. Gravin Beatrixstraat; 2012. p. 2805.
[45] Bicer Y. Investigation of novel ammonia production options using photoelectrochemical hydrogen. PhD Thesis. University of Ontario Institute of Technology; 2017.
[46] El FY, Bouallou C. Recycling of carbon dioxide to produce ethanol. Energy Procedia 2013;37:6679–86. https://doi.org/10.1016/j.egypro.2013.06.600.

Index

Note: Page numbers followed by *f* indicate figures and *t* indicate tables.

D

E

I

K

L

M

N

U

V

W

X

Z

CPI Antony Rowe
Eastbourne, UK
September 19, 2019